Monte Carlo Techniques in Radiation Therapy

Imaging in Medical Diagnosis and Therapy

Series Editors:
Bruce R. Thomadsen and David W. Jordan

Beam's Eye View Imaging in Radiation Oncology
Ross I. Berbeco, Ph.D.

Principles and Practice of Image-Guided Radiation Therapy of Lung Cancer
Jing Cai, Joe Y. Chang, and Fang-Fang Yin

Radiochromic Film
Role and Applications in Radiation Dosimetry
Indra J. Das

Clinical 3D Dosimetry in Modern Radiation Therapy
Ben Mijnheer

Hybrid Imaging in Cardiovascular Medicine
Yi-Hwa Liu and Albert J. Sinusas

Observer Performance Methods for Diagnostic Imaging
Foundations, Modeling, and Applications with R-Based Examples
Dev P. Chakraborty

Ultrasound Imaging and Therapy
Aaron Fenster and James C. Lacefield

Dose, Benefit, and Risk in Medical Imaging
Lawrence T. Dauer, Bae P. Chu, and Pat B. Zanzonico

Big Data in Radiation Oncology
Jun Deng and Lei Xing

Monte Carlo Techniques in Radiation Therapy
Introduction, Source Modelling, and Patient Dose Calculations, Second Edition
Frank Verhaegen and Joao Seco

Monte Carlo Techniques in Radiation Therapy
Applications to Dosimetry, Imaging, and Preclinical Radiotherapy, Second Edition
Joao Seco and Frank Verhaegen

For more information about this series, please visit:
https://www.crcpress.com/Series-in-Optics-and-Optoelectronics/book-series/TFOPTICSOPT

Monte Carlo Techniques in Radiation Therapy

Introduction, Source Modelling, and Patient Dose Calculations

Second Edition

Edited by
Frank Verhaegen and Joao Seco

CRC Press
Taylor & Francis Group
Boca Raton London New York

CRC Press is an imprint of the
Taylor & Francis Group, an **informa** business

Second edition published 2022
by CRC Press
6000 Broken Sound Parkway NW, Suite 300, Boca Raton, FL 33487-2742

and by CRC Press
2 Park Square, Milton Park, Abingdon, Oxon, OX14 4RN

First edition published by CRC Press 2013

CRC Press is an imprint of Taylor & Francis Group, LLC

ISBN: 9781032078526 (hbk)
ISBN: 9781032078564 (pbk)
ISBN: 9781003211846 (ebk)

DOI: 10.1201/9781003211846

Typeset in Minion Pro
by codeMantra

To all who made this book possible

Contents

PART I Introduction

PART II   Source Modelling

Series Preface

Since their inception over a century ago, advances in the science and technology of medical imaging and radiation therapy are more profound and rapid than ever before. Further, the disciplines are increasingly cross-linked as imaging methods become more widely used to plan, guide, monitor, and assess treatments in radiation therapy. Today, the technologies of medical imaging and radiation therapy are so complex and computer-driven that it is difficult for the people (physicians and technologists) responsible for their clinical use to know exactly what is happening at the point of care, when a patient is being examined or treated. The people best equipped to understand the technologies and their applications are medical physicists, and these individuals are assuming greater responsibilities in the clinical arena to ensure that what is intended for the patient is actually delivered in a safe and effective manner.

The growing responsibilities of medical physicists in the clinical arenas of medical imaging and radiation therapy are not without their challenges, however. Most medical physicists are knowledgeable in either radiation therapy or medical imaging, and expert in one or a small number of areas within their disciplines. They sustain their expertise in these areas by reading scientific articles and attending scientific talks at meetings. In contrast, their responsibilities increasingly extend beyond their specific areas of expertise. To meet these responsibilities, medical physicists periodically must refresh their knowledge of advances in medical imaging or radiation therapy, and they must be prepared to function at the intersection of these two fields. How to accomplish these objectives is a challenge.

At the 2007 annual meeting of the American Association of Physicists in Medicine in Minneapolis, this challenge was the topic of conversation during a lunch hosted by Taylor & Francis Publishers and involving a group of senior medical physicists (Arthur L. Boyer, Joseph O. Deasy, C.-M. Charlie Ma, Todd A. Pawlicki, Ervin B. Podgorsak, Elke Reitzel, Anthony B. Wolbarst, and Ellen D. Yorke). The conclusion of this discussion was that a book series should be launched under the Taylor & Francis banner, with each volume in the series addressing a rapidly advancing area of medical imaging or radiation therapy of importance to medical physicists. The aim would be for each volume to provide medical physicists with the information needed to understand technologies driving a rapid advance and their applications to safe and effective delivery of patient care.

Each volume in the series is edited by one or more individuals with recognized expertise in the technological area encompassed by the book. The editors are responsible for selecting the authors of individual chapters and ensuring that the chapters are comprehensive and intelligible to someone without such expertise. The enthusiasm of volume editors and chapter authors has been gratifying and reinforces the conclusion of the Minneapolis luncheon that this series of books addresses a major need of medical physicists.

William R. Hendee
Founding Series Editor

Preface to the First Edition

Monte Carlo simulation techniques made a slow entry in the field of radiotherapy in the late 1970s. Since then, they have gained enormous popularity, judging by the number of papers published and PhDs obtained on the topic. Calculation power has always been an issue, so initially only simple problems could be addressed. They led to insights, though, that could not have been obtained by any other method. Recently, fast-forwarding some 30 years, Monte Carlo-based treatment planning tools have now begun to be available from some commercial treatment planning vendors, and it can be anticipated that a complete transition to Monte Carlo-based dose calculation methods may take place over the next decade. The progress of image-guided radiotherapy further advances the need for Monte Carlo simulations, in order to better understand and compute radiation dose from imaging devices and make full use of the four-dimensional information now available. Exciting new developments in in-beam imaging in light ion beams are now also being vigorously investigated. Many new discoveries await the use of the Monte Carlo technique in radiotherapy in the coming decades.

The book addresses the application of the Monte Carlo particle transport simulation technique in radiation therapy, mostly focusing on external beam radiotherapy and brachytherapy. It includes a presentation of the mathematical and technical aspects of the technique in particle transport simulations. It gives practical guidance relevant for clinical use in radiation therapy, discussing modelling of medical linacs and other irradiation devices, issues specific to electron, photon, proton/particle beams, and brachytherapy, utilization in the optimization of treatment planning, radiation dosimetry, and quality assurance (QA).

We have assembled this book—a first of its kind—to be useful to clinical physicists, graduate students, and researchers who want to learn about the Monte Carlo method; we hope you will benefit from the collective knowledge presented herein. In addition, the editors wish to sincerely thank all the outstanding contributing authors, without whom this work would not have been possible. They would also like to thank the series editor, Dr. Bill Hendee, for the opportunity. We sincerely thank the editorial and production staff at Taylor & Francis for the smooth collaboration and for the pleasant interactions.

Preface to the Second Edition

Fast-forwarding again another ten years since the first edition of this book appeared, it is clear that the field of Monte Carlo simulations, applied to radiotherapy, has flourished tremendously. The amount of studies published, the number of Monte Carlo codes and user interfaces that kept growing, and the novel fields where Monte Carlo found applications, demonstrate the field is very healthy. Never before have so many young scientists enthusiastically joined the long list of users and developers. We can therefore conclude that the Monte Carlo method still has a very bright future ahead.

Among the many new chapters and completely updated chapters in this book, we find novel applications that were absent or barely mentioned in the first edition. These include applications in magnetic fields such as in MR-linacs, microbeams, preclinical precision dosimetry, microdosimetry, DNA modelling, total skin irradiation, and the completely novel combination of artificial intelligence with Monte Carlo. Chapters in the first edition that combined topics have now been split in two since so much more material had to be covered (brachytherapy, MV imaging, kV imaging, proton beam modelling, prompt gamma verification, proton CT, etc.). Indeed, the book now consists of two volumes, testifying to the rapid growth of applications.

What is completely clear to us is that Monte Carlo simulations are still slow, simply because we move on to always bigger and more demanding problems. Simulation of motion of patients, beam delivery systems, or both simultaneously have become much more common. So has simulating the tracking of many secondary particles in ion beams, or the behaviour of electrons in magnetic fields. Nevertheless, finally, computer technology seems to have advanced to the point that Monte Carlo treatment planning has become possible in reasonable times. And perhaps the combination of artificial intelligence and Monte Carlo methods may cause a real quantum leap here. We have not yet seen the end of potential applications, guaranteeing endless fun for generations to come.

The new edition was of course in the first place made for Monte Carlo specialists and novices, but also for those who simply are curious about this field, still considered exotic by many. Also this time we deeply thank all contributing authors, of which there were many more than in the first edition. Without those authors, this book would not exist. We would also like to thank the series editors and the publisher for the opportunity to expand the first edition with so much new material.

MATLAB® is a registered trademark of The MathWorks, Inc. For product information,
please contact:

The MathWorks, Inc.
3 Apple Hill Drive
Natick, MA 01760-2098 USA
Tel: 508-647-7000
Fax: 508-647-7001
E-mail: info@mathworks.com
Web: www.mathworks.com

Editors

Frank Verhaegen is Head of Clinical Physics Research at the Maastro Clinic in Maastricht, the Netherlands. He holds a professorship at the University of Maastricht. Formerly, he held an Associate Professorship at McGill University in Montréal, Canada. He earned his PhD from the University of Ghent in Belgium in 1996. He held research positions at the Royal Marsden Hospital and the National Physical Laboratory (UK) for several years. Dr Verhaegen is a Fellow of the Institute of Physics and Engineering in Medicine and the Institute of Physics. His group has published about 250 research papers, a significant fraction of them about Monte Carlo modelling, and was the recipient of the Sylvia Fedoruk Prize for best Canadian Medical Physics paper in 2007. His interests range broadly in imaging and dosimetry for photon, proton and electron therapy, brachytherapy, particle therapy, and small animal radiotherapy. He also founded a company that offers Monte Carlo-based treatment planning for preclinical precision radiation research. Dr Verhaegen has been passionate about Monte Carlo simulations since the days of his master's thesis in the late 1980s.

Joao Seco is a Professor in the Department of Physics and Astronomy, Heidelberg University and the Head of the division BioMedical Physics in Radiation Oncology at the DKFZ – German Cancer Research Center, Heidelberg, Germany. He earned his PhD from the Institute of Cancer Research, University of London, UK, in 2002. He held research positions at the Royal Marsden Hospital, UK, and the Harvard Medical School, Boston, Massachusetts, for several years, where he was an Assistant Professor in radiation oncology. While at Harvard Medical School in Boston, he was the recipient of the Harvard Club of Australia Foundation Award in 2012 on the development of Monte Carlo and optimization techniques for use in radiation therapy of lung cancer. In 2016 he was appointed Professor and Department Head at the DKFZ and University of Heidelberg, Germany. His interests range from proton imaging and therapy to photon beam modelling with Monte Carlo, electronic portal imaging, and 4D Monte Carlo proton and photon dosimetry. He started working on Monte Carlo simulations while a master's student at the Laboratory of Instrumentation and Experimental Particle Physics (LIP), a Portuguese research institute part of the CERN (the European Organization for Nuclear Research) worldwide network for particle physics research.

Contributors

Alex F. Bielajew
Department of Nuclear Engineering
and Radiological Sciences
University of Michigan
Ann Arbor, Michigan

Åsa Carlsson Tedgren
Department of Medicine, Health and
Caring Sciences
Linköping University
Linköping, Sweden
and
Department of Medical Radiation
Physics and Nuclear Medicine
Karolinska University Hospital
Stockholm, Sweden

Joanna E. Cygler
Department of Medical Physics
The Ottawa Hospital Cancer Centre
Ottawa, Ontario, Canada

George X. Ding
Department of Radiation Oncology
Vanderbilt University School of
Medicine
Nashville, Tennessee

Bruce A. Faddegon
Radiation Oncology Department
University of California, San Francisco
San Francisco, California

Matthias Fippel
R&D Radiosurgery
Brainlab AG
Munich, Germany

Michael K. Fix
Division of Medical Radiation Physics
Inselspital, University Hospital Bern,
University of Bern
Bern, Switzerland

Gabriel Fonseca
Department of Radiation Oncology
(MAASTRO), GROW, School for
Oncology and Developmental
Biology, Maastricht University
Medical Center, Maastricht

Maggy Fragoso
Department of Communication and
Strategy
Alfa-Comunicações
Praia, Santiago Island, Cape Verde

Loïc Grevillot
Department of Medical Physics
MedAustron Ion Therapy Center
Wiener Neustadt, Austria

Emily Heath
Carleton University
Ottawa, Ontario, Canada

N. Krah
Université de Lyon
CREATIS, CNRS UMR5220, Inserm
U1294, INSA-Lyon, Université Lyon 1
Lyon, France

Guillaume Landry
Ludwig Maximilian's University
Munich, Germany
and
Maastro Clinic
Maastricht, the Netherlands

JinSheng Li
Department of Radiation Oncology
Fox Chase Cancer Center
Philadelphia, Pennsylvania

C.-M. Charlie Ma
Department of Radiation Oncology
Fox Chase Cancer Center
Philadelphia, Pennsylvania

Harald Paganetti
Department of Radiation Oncology
Massachusetts General Hospital,
Harvard Medical School
Boston, Massachusetts

Tony Popescu
University of British Columbia
Vancouver, British Columbia, Canada

Brigitte Reniers
University Hasselt
Diepenbeek, Belgium

Mark J. Rivard
Rhode Island Hospital
Providence, Rhode Island
and
Department of Radiation Oncology
Tufts University School of Medicine
Boston, Massachusetts

D. Sarrut
Université de Lyon
CREATIS, CNRS UMR5220, Inserm
U1294, INSA-Lyon, Université Lyon 1
Lyon, France

Joao Seco
DKFZ - German Cancer Research
    Center
Heidelberg, Germany
and
Department of Physics and Astronomy,
    University of Heidelberg
Heidelberg, Germany

Rowan M. Thomson
Department of Physics
Carleton University
Ottawa, Ontario, Canada

Frank Verhaegen
Radiotherapy Physics Division
Maastro Clinic
Maastricht, the Netherlands

Jeffrey F. Williamson
Washington University
St Louis, Missouri
and
VCU Massey Cancer Center
Richmond, Virginia

Introduction

I

1

History of Monte Carlo

Alex F. Bielajew
University of Michigan

It is still an unending source of surprise for me to see how a few scribbles on a blackboard or on a sheet of paper could change the course of human affairs.

Stan Ulam
*Founder of the modern Monte Carlo method,
in his 1991 autobiography (1991)*

1.1 Motivating Monte Carlo

Generally speaking, the Monte Carlo method provides a numerical solution to a problem that can be described as a temporal evolution ("translation/reflection/mutation") of objects ("quantum particles" [photons, electrons, neutrons, protons, charged nuclei, atoms, and molecules], in the case of medical physics) interacting with other objects based upon object–object interaction relationships ("cross sections"). Mimicking nature, the rules of interaction are processed randomly and repeatedly, until numerical results converge usefully to estimated means, moments, and their variances. Monte Carlo represents an attempt to model nature through a direct simulation of the essential dynamics of the system in question. In this sense, the Monte Carlo method is, in principle, simple in its approach—a solution to a macroscopic system through simulation of its microscopic interactions and therein is the advantage of this method. All interactions are microscopic in nature. The geometry of the environment, so critical in the development of macroscopic solutions, plays little role except to define the local environment of objects interacting at a given place at a given time.

The scientific method is dependent on the observation (measurement) and hypothesis (theory) to explain nature. The conduit between these two is facilitated by a myriad of mathematical, computational, and simulation techniques. The Monte Carlo method exploits all of them. Monte Carlo is often seen as a "competitor" to other methods of macroscopic calculation, which we will call the deterministic and/or analytic methods. Although the proponents of either method sometimes approach a level of fanaticism in their debates, a practitioner of science should first ask, "What do I want to accomplish?" followed by "What is the most efficient way to do it?," and then, "What serves science the best?" Sometimes the correct answer will be "Deterministic," and other times it will be "Monte Carlo." The most successful scientist will avail himself or herself of more than one method of approach.

There are, however, two inescapable realities. The first is that macroscopic theory, particularly transport theory, provides deep insight and allows one to develop sophisticated intuition as to how macroscopic particle fields can be expected to behave. Monte Carlo cannot compete very well with this. In discovering the properties of macroscopic field behavior, Monte Carlo practitioners operate very much like experimentalists. Without theory to provide guidance, discovery is made via trial and error, guided perhaps, by some brilliant intuition.

However, complexity is measured, and when it comes to developing an understanding of a physical problem, Monte

DOI: 10.1201/9781003211846-2

Carlo techniques become, at some point, the most advantageous. A proof is given, in the appendix of this chapter, that the Monte Carlo method is more advantageous in the evolution of five and higher dimensional systems. The dimensionality is just one measure of a problem's "complexity." The problems in radiotherapy target practice (RTP) and dosimetry are typically of dimension $6.\varepsilon$ or $7.\varepsilon$. That is, particles move in Cartesian space, with position $\vec{x}$, that varies continuously, except at particle inception or expiration. They move with momentum, $\vec{P}$, that varies both discretely and continuously. The dimension of time is usually ignored for static problems, though it cannot be for nonlinear problems, where a particle's evolution can be affected by the presence of other particles in the simulation. (The "space-charge" effect is a good example of this.) Finally, the ε is a discrete dimension that can encompass different particle species, as well as intrinsic spin.

This trade-off, between complexity and time to solution, is expressed in Figure 1.1.

Although the name "Monte Carlo method" was coined in 1947, at the start of the computer age, stochastic sampling methods were known long before the advent of computers. The first reference known to this author is that of Comte de Buffon (1777) who proposed a Monte Carlo-like method to determine the outcome of an "experiment" consisting of repeatedly tossing a needle onto a ruled sheet of paper, to determine the probability of the needle crossing one of the lines. This reference goes back to 1777, well before the contemplation of automatic calculating machines. Buffon further calculated that a needle of length L tossed randomly on a plane ruled with parallel lines, distance d apart, where $d > L$, would have a probability of crossing one of the rules lines of

$$p = \frac{2L}{\pi d}. \tag{1.1}$$

Monte Carlo versus deterministic/analytic methods

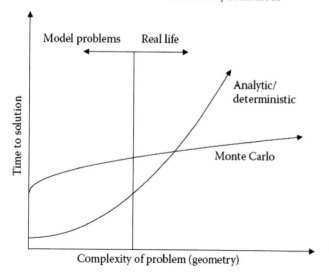

FIGURE 1.1 Time to solution using Monte Carlo versus deterministic/analytic approaches.

Much later, Laplace (1886) suggested that this procedure could be employed to determine the value of π, albeit slowly. Several other historical uses of Monte Carlo predating computers are cited by Kalos and Whitlock (2008).

The idea of using stochastic sampling methods first occurred to Ulam,[*] who, while convalescing from an illness, played solitaire repeatedly, and then wondered if he could calculate the probability of success by combinatorial analysis. It occurred to him, that it would be possible to do so by playing a large number of games, tallying the number of successful plays (Metropolis, 1987; Eckhart, 1987), and then estimating the probability of success. Ulam communicated this idea to von Neumann who, along with Ulam and Metropolis, was working on theoretical calculations related to the development of thermonuclear weapons. Precise calculations of neutron transport are essential in the design of thermonuclear weapons. The atomic bomb was designed by experiments, mostly with modest theoretical support. The trigger for a thermonuclear weapon is an atomic bomb, and the instrumentation is destroyed before useful signals can be extracted.[†] Von Neumann was especially intrigued with the idea. The modern Monte Carlo age was ushered in later, when the first documented suggestion of using stochastic sampling methods applied to radiation transport calculations appeared in correspondence between von Neumann and Richtmyer (Metropolis, 1987; Eckhart, 1987), on March 11, 1947. (Richtmyer was the leader of the Theoretical Division at Los Alamos National Laboratories [LANL].) This letter suggested the use of LANL's ENIAC computer to do the repetitive sampling. Shortly afterward, a more complete proposal was written (von Neumann and Richtmyer, 1947). Although this report was declassified as late as 1988, inklings of the method, referred to as a "mix of deterministic and random/stochastic processes," started to appear in the literature, as published abstracts (Ulam and von Neumann, 1945, 1947). Then in 1949, Metropolis and Ulam published their seminal, founding paper, "The Monte Carlo Method" (Metropolis and Ulam, 1949), which was the first unclassified paper on the Monte Carlo methods, and the first to have the name, "Monte Carlo" associated with stochastic sampling.

Already by the 1949, symposia on the Monte Carlo methods were being organized, focusing primarily on mathematical techniques, nuclear physics, quantum mechanics, and general statistical analysis. A later conference, the *Symposium on Monte Carlo Methods*, held at the University of Florida in 1954 (Meyer, 1981) was especially important. There were 70 attendees, many of whom would be recognized as "founding fathers" by Monte Carlo practitioners in the radiological sciences. Twenty papers were presented, including two involving gamma rays, spanning 282 pages in the proceedings. This

---

[*] The direct quote from Ulam's autobiography (Ulam, 1991). (p. 196, 1991 edition): "The idea for what was later called the Monte Carlo method occurred to me when I was playing solitaire during my illness."

[†] The book *Dark Sun*, by Richard Rhodes, is an excellent starting point for the history of that topic (Rhodes, 1988).

proceedings also includes a 95-page bibliography, a grand summary of the work-to-date, with many references having their abstracts and descriptions published in the proceedings.

The rest, to quote an overused expression, is history. It is interesting to note the wonderful irony: This mathematical method was created for destruction by means of the most terrible weapon in history, the thermonuclear bomb. Fortunately, this weapon has never been used in conflict. Rather, millions have benefited from the development of Monte Carlo methods for medicine. That topic, at least a small subset of it, will occupy the rest of this chapter.

As of this writing, with the data from 2020 still incomplete, we have found that about 900,000 papers have been published on the Monte Carlo method. If we restrict this search to only those papers related to medicine, the number of publications is almost 55,000. The 10%–20% contribution to the Monte Carlo method seems to be consistent over time, at least since 1970. That represents an enormous investment in human capital to develop this most useful tool. The temporal evolution of this human effort is shown in Figure 1.2. Before 2005, the growth in both areas appears exponential in nature. The total effort shows three distinct areas of slope, with sudden changes, currently unexplained, though it may be due to the sudden emergence of "vector" and "massively parallel" machines, and the increase in research associated with this fundamentally new computer architecture. The growth in the "medicine" area has been constant.

Since 2005, both areas are still growing but the rate is slowing down, with the "medicine" curve leveling out, at greater than 3,000 publications/year. It appears that this communication is being written at the pinnacle of this scientific endeavor!

1.2 Monte Carlo in Medical Physics

Every historical review has its biases, and the one employed here will restrict the discussion to the applications of radiotherapy and radiation dosimetry. Moreover, the focus will

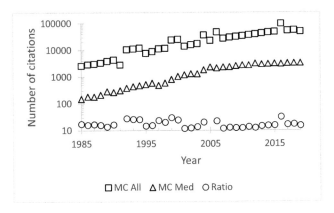

FIGURE 1.2 The number of papers published per year on Monte Carlo garnered from CU Boulder OneSearch ("MC All": from physics, medicine, chemistry, biology, engineering) and MedLine ("MC Med": mostly from applications of physics in medicine).

be on the development of electron Monte Carlo, for reasons explained in the following paragraph. There is an abundance of reviews on the use of Monte Carlo in medical physics. A few of the more recent ones that discuss radiotherapy physics and dosimetry are Andreo (1985, 1991), Mackie (1990), Rogers and Bielajew (1990a), Ma and Jiang (1999), Verhaegen and Seuntjens (2003), and Rogers (2006).

The rise of the use of linear electron accelerators (LINACs) for radiotherapy also ushered in the need to develop Monte Carlo methods for the purpose of dose prediction and dosimetry. The LINACs employed in radiotherapy provide an energetic and penetrating source of photons that enter deep into tissue, sparing the surface and attenuating considerably less rapidly than ^{60}Co or ^{137}Cs beams. Relativistic electrons have a range of about 1 cm for each 2 MeV of kinetic energy in water. At its maximum, starting with a pencil beam of electrons, the diameter of the electron energy deposition, pear-shaped "plume" is also about 1 cm per 2 MeV of initial kinetic energy. These dimensions are commensurate with the organs being treated, as well as the organs at risk. The treatment areas are heterogeneous, with differences in composition and density. Moreover, the instruments used to meter dose are even more diverse. It is true now, as it was then, that the Monte Carlo method provides the only prediction of radiometric quantities that satisfies the accuracy demand of radiotherapy.

Thus, the history of the utility of the Monte Carlo method in medical physics is inextricably tied to the development of Monte Carlo methods of electron transport in complex geometries and in the description of electromagnetic cascades.[*]

The first papers employing the Monte Carlo method using electron transport were authored by Robert R. Wilson (1950, 1951, 1952, p. 261), who performed his calculations using a "spinning wheel of chance."[†] Although apparently quite tedious, Wilson's method was still an improvement over the analytic methods of the time—particularly in studying the average behavior and fluctuations about the average (Rossi, 1952). Hebbard and P. R. Wilson (1955) used computers to investigate electron straggling and energy loss in thick foils. The first use of an electronic digital computer in simulating high-energy cascades by Monte Carlo methods was reported by Butcher and Messel (1958, 1960), and independently by Varfolomeev and Svetlolobov (1959). These two groups collaborated in a much publicized work (Messel et al., 1962) that eventually led to an extensive set of tables describing the shower distribution functions (Messel and Crawford, 1970)—the so-called "shower book."

For various reasons, two completely different codes were written in the early-to-mid 1960s to simulate electromagnetic cascades. The first was written by Zerby and Moran (1962a,

---

[*] Certainly there are important applications in brachytherapy and imaging that ignore electron transport. However, we shall leave that description to other authors.

[†] R. R. Wilson is also acknowledged as the founder of proton radiotherapy (Wilson, 1946).

b, 1963) of the Oak Ridge National Laboratory, motivated by the construction of the Stanford Linear Accelerator Center (SLAC). Many physics and engineering problems were anticipated as a result of high-energy electron beams showering in various devices and structures at that facility. This code had been used by Alsmiller and others (Alsmiller and Moran, 1966, 1967, 1968, 1969; Alsmiller and Barish, 1969, 1974; Alsmiller et al., 1974) for a number of studies since its development.[*]

The second code was developed by Nagel (Nagel and Schlier, 1963; Nagel, 1964, 1965; Völkel, 1966) and several adaptations have been reported (Völkel, 1966; Nicoli, 1966; Burfeindt, 1967; Ford and Nelson, 1978). The original Nagel version, which Ford and Nelson called SHOWER1, was a FORTRAN code written for high-energy electrons (≤1,000 MeV) incident upon lead in cylindrical geometry. Six significant electron and photon interactions (Bremsstrahlung, electron–electron scattering, ionization loss, pair production, Compton scattering, and the photoelectric effect) plus multiple Coulomb scattering were accounted for. Except for annihilation, positrons and electrons were treated alike and were followed until they reached a cutoff energy of 1.5 MeV (total energy). Photons were followed down to 0.25 MeV. The cutoff energies were as low as or lower than those used by either Messel and Crawford or by Zerby and Moran. The availability of Nagel's dissertation (1964) and a copy of his original shower program provided the incentive for Nicoli (Nicoli, 1966) to extend the dynamic energy range and flexibility of the code in order for it to be made available as a practical tool for the experimental physicist. It was this version of the code that eventually became the progenitor of the electron gamma shower (EGS) code systems (Ford and Nelson, 1978; Nelson et al., 1985; Bielajew and Rogers, 1987; Kawrakow and Rogers, 2000; Hirayama et al., 2005).

On a completely independent track, and apparently independent from the electromagnetic cascade community, was Berger's e–γ code. It was eventually released to the public as ETRAN in 1968 (Berger and Seltzer, 1968), though it is clear that internal versions were being worked on at NBS (now NIST) (Seltzer, 1989) since the early 1960s, on the foundations laid by Berger's landmark paper (Berger, 1963). The ETRAN code then found its way, being modified somewhat, into the Sandia codes, EZTRAN (Halbleib and Vandevender, 1971), EZTRAN2 (Halbleib and Vandevender, 1973), SANDYL (Colbert, 1973), TIGER (Halbleib and Vandevender, 1975), CYLTRAN (Halbleib and Vandevender, 1976), CYLTRANNM (Halbleib and Vandevender, 1977), CYLTRANP (unpublished), SPHERE (Halbleib, 1978), TIGERP (Halbleib and Morel, 1979), ACCEPT (Halbleib, 1980), ACCEPTTM (Halbleib et al., 1981), SPHEM (Miller et al., 1981), and finally the all-encompassing ITS (Halbleib and Mehlhorn, 1986; Halbleib et al., 1992) codes. The ITS electron transport code was incorporated into the

Monte Carlo N-particle (MCNP) code at Version 4, in 1990 (Hendricks and Briesmeister, 1991). The MCNP code lays claim to being a direct descendant of the codes written by the originators of the Monte Carlo method, Fermi, von Neumann, Ulam, as well as Metropolis and Richtmyer (Briesmeister, 1986).

Much of the early work is summarized in the first book to appear on Monte Carlo by Cashwell and Everett in 1957.[†] Shortly thereafter the first Monte Carlo neutron transport code MCS was written, followed in 1967 by MCN. The photon codes MCC and MCP were then added and in 1973 MCN and MCC were merged to form MCNG. The above work culminated in Version 1 of MCNP in 1977. The first two large user manuals were published by W. L. Thompson in 1979 and 1981. This manual draws heavily from its predecessors.

The first appearance of electron transport in MCNP occurred with Version 4, in 1990 (Hendricks and Briesmeister, 1991). After that time, MCNP became an important player in medical-related research, to be discussed later.

Berger's contribution (1963) is considered to be the *de facto* founding paper (and Berger the founding father) of the field of Monte Carlo electron and photon transport. That article, 81 pages long, established a framework for the next generation of Monte Carlo computational physicists. It also summarized all the essential theoretical physics for Monte Carlo algorithm development. Moreover, Berger introduced a specialized method for electron transport. Electron transport and scattering, for medical physics, dosimetry, and many other applications, are subject to special treatment. Rather than modelling every discrete electron interaction (of the order of 10^6 for relativistic electrons), cumulative scattering theories, whereby 10^3–10^5 individual elastic and inelastic events are "condensed" into single "virtual" single-scattering events, enable a speedup by factors of hundreds, typically. Nelson, the originator of the EGS code system, is quoted as saying (W. R. Nelson, personal communication, 2011), "Had I known about Berger's work, I may not have undertaken the work on EGS!"

As for general-purpose uses in medical-related fields, with multi-material, combinatorial geometries, the two historically dominant players in RTP/dosimetry,[‡] are the EGS and MCNP codes, introduced above. In the last decade, GEANT (Brun et al., 1982; Allison et al., 2006) has also made significant contributions as well, presently equal in use to MCNP. A plot of the number of papers published using these methods is charted in Figure 1.3. Once MCNP introduced electron transport, we see, from Figure 1.3, that usage of MCNP experienced exponential growth in its use in medical-related areas. That exponential growth ended in about 2000. Since then, both the

---

[*] According to Alsmiller (R. G. Alsmiller Jr. private communication. [conversation with W. R. Nelson], 1970), the Zerby and Moran source code vanished from ORNL and they were forced to work with an octal version.

[†] 1959, to be exact (Cashwell and Everett, 1959).

[‡] There are some very relevant, alternative approaches that the reader should be aware of, namely FLUKA (Aarnio et al., 1984; Fasso et al., 2005; Battistoni et al., 2007) (that traces its roots to 1964 [Ranft, 1964]), and the Penelope code (Salvat and Parrellada, 1984; Baró et al., 1995). As of this writing, the number of papers produced using these codes in medical areas is about 240, about half that of MCNP.

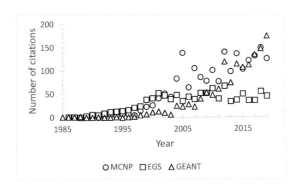

FIGURE 1.3 Papers using Monte Carlo codes EGSx, MCNPx, and GEANTx, as captured on PubMed.

EGS and MCNP code systems seem to be experiencing steady use, with GEANT still, arguably, on the increase.

If one considers all of the non-medical literature related to Monte Carlo, MCNP and GEANT are the most heavily cited codes, with an almost equal number of citations, nearly a factor of 2 over EGS.

t should be emphasized that these two code systems are very different in nature. EGS has specifically targeted the medical area, since 1984, though it has enjoyed some use in other areas of physics as well. Some features are genuinely unique, such as the tracking of separate electron spins, a feature introduced (Namito et al., 1993) in EGS5 (Hirayama et al., 2005), as well as Doppler broadening (Namito et al., 1994), both inclusions of great interest to those doing research in synchrotron radiation light sources. Overall, however, considering Monte Carlo uses over all areas, MNCP's state-of-the-art neutron transport makes it the world leader in the nuclear and radiological sciences conglomerate. The EGS code systems are supported by practitioners representing a panoply of scientific disciplines: medical physicists, radiological scientists, pure and applied physicists. MCNP serves these communities as well, but also enjoys the support of a vibrant nuclear engineering profession, where transport theory of neutrons is a rich and active research area.

1.3 EGSx Code Systems

The history of how EGS, a code developed primarily for high-energy physics shielding and detector simulations, came to be used in medical physics has never appeared in text, in its entirety. Rogers' (2006) humility probably interfered with its exposition in his article. However, I was an observer at those early events and think I may offer some insights. In 1978, SLAC published EGS3 (Ford and Nelson, 1978), and Rogers employed it in several important publications (Rogers, 1982, 1983a, b, 1984a, b). Of particular importance was the publication that offered a patch to the EGS3 algorithms to mimic a technique employed in ETRAN, making electron-dependent calculations reliable, by shortening the steps to virtual interactions. At the time, electron transport step-size artifacts

were completely unexplained. Shortening the steps is understood to solve the problem, as explained most eloquently by Larsen (1992), but at the cost of added computational time. These "step-size artifacts" attracted the attention of Nelson, who invited Rogers to participate in authoring the next version of EGS, EGS4 (Nelson et al., 1985), along with Hirayama, a research scientist at the KEK, the High Energy Accelerator Research Organization, Tsukuba, Japan.

Following its release in December 1985, Rogers' institution (a Radiation Standards Laboratory, and a sister laboratory to Berger's NIST) became a nucleus of medical physics and dosimetry Monte Carlo dissemination. It took over support and distribution of the EGS4 code and began offering training courses all over the world. Hirayama was engaged in similar efforts in the Asian regions.

Yet, the step-size artifacts in EGS4 remained unexplained. Nahum, who was interested in modelling ionization chamber response, visited Rogers' laboratory in the spring of 1984, to collaborate on this topic. Nahum already had a scholarly past in electron Monte Carlo (Nahum, 1976), producing what would eventually be realized, through the Lewis (1950) moments analysis (Kawrakow and Bielajew, 1998b), to be a far superior electron transport algorithm. While using EGS4 to predict the ionization chamber response, EGS4 would predict responses that could be 60% too low. Quoting Nahum, "How could a calculation that one could sketch on the back of an envelope, and get correct to within 5%, be 60% wrong using Monte Carlo?" Step-size reduction solved the problem (Bielajew et al., 1985; Rogers et al., 1985a), but the search for a resolution to step-size anomalies was commenced, resulting in the PRESTA algorithm (Bielajew and Rogers, 1987, 1989). The release of PRESTA was followed by the demonstration of various small, but important, shortcomings (Rogers, 1993; Foote and Smyth, 1995). There were improvements over the years (Bielajew and Kawrakow, 1997a, b, c, Kawrakow and Bielajew, 1998b), eventually resulting in a revision of the EGS code, known as EGSnrc (Kawrakow and Rogers, 2000) and EGS5 (Hirayama et al., 2005). A PRESTA-like improvement of ETRAN (Seltzer, 1991; Kawrakow, 1996) was even developed.

1.4 Application: Ion Chamber Dosimetry

The founding paper for applying Monte Carlo methods to ionization chamber response is attributed to Bond et al. (1978), who, using an in-house Monte Carlo code, calculated ionization chamber response as a function of wall thickness, to ^{60}Co γ-irradiation. While validating the EGS code for this application, it was found that the EGS code had fundamental algorithmic difficulties with this low-energy regime, as well as this application. The resolution of these difficulties, not patent in other general-purpose Monte Carlo codes, became of great interest to this author. While general improvements to electron transport ensued, the fundamental problem was quite subtle and was eventually described elegantly by Foote and Smythe (1995). In a nutshell, the underlying algorithmic reason that was identified arose from electron tracks being stopped at material boundaries, where cross sections change. EGS used the partial electron path to model a deflection of the electron, from the accumulated scattering power. The result was a rare, but important effect, the spurious generation of fluence singularities.

The literature on ionization chamber dosimetry is extensive. A partial compilation of very early contributions is Bielajew et al. (1985), Rogers et al. (1985a), Andreo et al. (1986), Bielajew (1990), Rogers and Bielajew (1990b), and Ma and Nahum (1991).

Presently, the calculation of ionization chamber corrections is a very refined enterprise, with results being calculated to better than 0.1%. The literature on this topic is summarized by Bouchard and Seuntjens in their chapter in this book, "Applications of Monte Carlo to radiation dosimetry." The chapter also summarizes the contribution of Monte Carlo to dosimetry protocol and basic dosimetry data, some of the earliest applications of Monte Carlo to medicine.

1.5 Early Radiotherapy Applications

For brevity, only the earliest papers are cited in this section, and the reader is encouraged to employ the comprehensive reviews already cited earlier in this article. Some of the very early history of radiotherapy applications is rarely mentioned, and I have attempted to gather them here:

The Monte Carlo modelling of Cobalt-60 therapy units was first mentioned in the ICRU Report # 18 (1971). However, a more complete descriptive work followed somewhat later (Rogers et al., 1985b; Han et al., 1987).

The modelling of LINAC therapy units was first accomplished by Petti et al. (1983a, b) and then, soon after by Mohan et al. (1985).

Photoneutron contamination from a therapy unit was first described by Ing et al. (1982), although the simulation geometry was simplified.

Mackie et al. pioneered the convolution method (Mackie et al., 1985) and then with other collaborators, generated the first database of "kernels" or "dose-spread arrays" for use in radiotherapy (Mackie et al., 1988). Independently, these efforts were being developed by Ahnesjö et al. (1987). These are still in use today.

Modelling of electron beams from medical LINACs was first accomplished by Teng et al. (1986), Hogstrom et al. (1986), and then Manfredotti et al. (1990).

An original plan to use Monte Carlo calculation for "target button to patient dose" was proposed by Mackie et al. (1990). That effort became known as "the OMEGA (Ottawa Madison Electron Gamma Algorithm) Project." However, early on in that project, a "divide-and-conquer" approach was adopted, whereby the fixed machine outputs ("phase-space files") were used as inputs to a patient-specific target (applicators and patient), to generate a full treatment plan. This bifurcation spawned two industries, treatment head modelling, of which the BEAM/EGSx code is the most refined (Rogers et al., 1995) and is the most cited paper in the "Web of knowledge" with "Monte Carlo" in the title, and "radiotherapy" as a topic. The second industry spawned by the OMEGA project was the development of fast patient-specific Monte Carlo–based dose-calculation algorithms (Kawrakow et al., 1995, 1996; Sempau et al., 2000; Gardner et al., 2007). For more discussion on the current fast Monte Carlo methods in use, the reader is encouraged to see the excellent review by Spezi and Lewis (2008).

1.6 The Future of Monte Carlo

The first step in predicting the future is to look where one has been, extrapolate the process, thereby predicting the future. The second step in predicting the future is to realize that the first step involves some very specious and problematic reasoning!

The progress of time, with the events that it contains, is intrinsically "catastrophic" in nature. A scientific discovery can be so earth-shattering, that new directions of research are spawned, while others dissolve into irrelevance. Yet, we persist in the practice of prediction. Therefore, allow me to be very modest in this effort.

Amdahl's law (1967): Multiprocessor intercommunication bottlenecks will continue to be a limiting factor, for massively parallel machines. However, gains in traditional, monolayer, single-chip speeds do *not* appear to be slowing, following Moore's law.[*]

Harder to predict is algorithm development, specific to Monte Carlo applications in RTP. There is a historical precedent for this in the citation data. In 1991, there was a 2.8 factor increase in productivity in only one year, followed by another in 1998, by a factor of 1.6. These increases are large and unexplained. Yet, they illustrate the chaotic nature of the field.

Around 2005, there was a suggestion that research in Monte Carlo was saturating. Since then new codes have appeared, mostly user interfaces to existing codes such as TOPAS and GATE. These will be discussed extensively in other parts of the books. The number of published papers on Monte Carlo is still increasing rapidly and codes like GEANT4 are seeing a strong growth in their use. The fact that a Monte Carlo method is the "engine" beneath a computational algorithm should be transparent, but not invisible, to the researcher using the tool. It may be that the development of the Monte Carlo method is somewhat unpredictable as it matures, but it will remain an essential component of our scientific infrastructure, forever.

---

[*] In 1965, Gordon E. Moore, co-founder of Intel Corporation, predicted that areal transistor density would double every 18 months. One can infer a commensurate increase in computer speed. Moore changed his prediction to "doubling every two years" in 1975. Current chips designs follow Moore's law until 2015 at least.

Appendix: Monte Carlo and Numerical Quadrature

In this appendix, we present a mathematical proof that the Monte Carlo method is the most efficient way of estimating tallies in three spatial dimensions when compared to first-order deterministic (analytic, phase-space evolution) methods. Notwithstanding the opinion that the Monte Carlo method is thought of as providing the most accurate calculation, the argument may be made in such a way, that it is independent of the physics content of the underlying algorithm or the quality of the incident radiation field.

A.1 Dimensionality of Deterministic Methods

For the purposes of estimating tallies from initiating electrons, photons, or neutrons, the transport process that describes the trajectories of particles is adequately described by the linear Boltzmann transport equation (Duderstadt and Martin, 1979):

$$\left[\frac{\partial}{\partial s} + \frac{p}{|p|}\cdot\frac{\partial}{\partial x} + \mu(x,p)\right]\psi(x,p,s)$$

$$= \int dx'\int dp'\,\mu(x,p,p')\psi(x',p',s), \tag{A.1}$$

where x is the position, p is the momentum of the particle, $(p\wedge p\backslash)\cdot\partial/\partial x$ is a directional derivative (in three dimensions $\vec{\Omega}\cdot\vec{\nabla}$, for example), and s is a measure of the particle pathlength. We use the notation that x and p are multi-dimensional variables of dimensionality N_x and N_p. Conventional applications span the range $1\leq N_{px}\leq 3$. The macroscopic differential scattering cross section (probability per unit length) $\mu(x,p,p')$ describes scattering from momentum p' to p at location x, and the total macroscopic cross section is defined by

$$\mu(x,p) = \int dp'\,\mu(x,p,p'). \tag{A.2}$$

$\psi(x, p, s)\,dx\,dp$ is the probability of there being a particle in dx about x, in dp about p, and at pathlength s. The boundary condition to be applied is

$$\psi(x,p,0) = \delta(x)\delta(p_0 - p)\delta(s), \tag{A.3}$$

where p_0 represents the starting momentum of a particle at $s=0$. The essential feature of Equation A.1, insofar as this proof is concerned, is that the solution involves the computation of a (N_x+N_p)-dimensional integral.

A general solution may be stated formally:

$$\psi(x,p,s) = \int dx'\int dp'\,G(x,p,x',p',s)Q(x',p'), \tag{A.4}$$

where $G(x, p, x', p', s)$ is the Green's function and $Q(x', p')$ is a source. The Green's function encompasses the operations of transport (drift between points of scatter, $x' \to x$), scattering

(i.e., change in momentum), and energy loss, $p' \to p$. The interpretation of $G(x, p, x', p', s)$ is that it is an operator that moves particles from one point in (N_x+N_p)-dimensional phase space (x', p') to another (x, p) and can be computed from the kinematical and scattering laws of physics.

Two forms of Equation A.4 have been employed extensively for general calculation purposes. Convolution methods integrate Equation A.4 with respect to pathlength s and further assume (at least for the calculation of the Green's function) that the medium is effectively infinite. Thus,

$$\psi(x,p) = \int dx'\int dp'G\left(|x-x'|,\left[\frac{p}{|p|}\cdot\frac{p'}{|p'|}\right],|p'|\right)Q(x',p'), \tag{A.5}$$

where the Green's function is a function of the distance between the source point x' and x, the angle between the vector defined by the source p' and p, and the magnitude of the momentum of the course, $|p'|$, or equivalently, the energy.

To estimate a tally using Equation A.5, we integrate $\psi(x, p)$ over p, with a response function, $\mathcal{R}(x, p)$ (Shultis and Faw, 1996):

$$T(x) = \int dx'\int dp'F(|x-x'|,p')Q(x',p'), \tag{A.6}$$

where the "kernel," $F(|x-x'|, p')$, is defined by

$$F(|x-x'|,p') = \int dp\mathcal{R}(x,p)G\left(|x-x'|,\left[\frac{p}{|p|}\cdot\frac{p'}{|p'|}\right],|p'|\right). \tag{A.7}$$

$F(|x-x'|, p')$ has the interpretation of a functional relationship that connects particle fluence at phase-space location x', p' to a tally calculated at x. This method has a known difficulty—its treatment of heterogeneities and interfaces. Heterogeneities and interfaces can be treated approximately by scaling $|x-x'|$ by the collision density. This is an exact for the part of the kernel that describes the first scatter contribution but approximate for higher-order scatter contributions. It can also be approximate, to varying degrees, if the scatter produces other particles with different scaling laws, such as the electron set in motion by a first Compton collision of a photon.

For calculation methods that are concerned with primary charged particles, the heterogeneity problem is more severe. The true solution in this case is reached when the pathlength steps, s, in Equation A.4 are made small (Larsen, 1992) and so, an iterative scheme is set up:

$$\psi_1(x,p) = \int dx'\int dp'G(x,p,x',p',\Delta s)Q(x',p')$$

$$\psi_2(x,p) = \int dx'\int dp'G(x,p,x',p',\Delta s)\psi_1(x',p')$$

$$\psi_3(x,p) = \int dx'\int dp'G(x,p,x',p',\Delta s)\psi_2(x',p') \tag{A.8}$$

$$\vdots$$

$$\psi_N(x,p) = \int dx'\int dp'G(x,p,x',p',\Delta s)\psi_{N-1}(x',p')$$

which terminates when the largest energy in $\psi_N(x, p)$ has fallen below an energy threshold or there is no x remaining within the target. The picture is of the phase space represented by $\psi(x, p)$ "evolving" as s accumulates. This technique has come to be known as the "phase-space evolution" model. Heterogeneities are accounted for by forcing Δs to be "small" or of the order of the dimensions of the heterogeneities and using a $G()$ that pertains to the atomic composition of the local environment. The calculation is performed in a manner similar to the one described for convolution. That is,

$$T(x) = \sum_{i=1}^{N} \int dx' \int dp' F(x, x', p, p', \Delta s)\psi_i(x', p') \quad (A.9)$$

where the "kernel," $F(x, x', p, p', \Delta s)$, is defined by

$$F = (x, x', p, p', \Delta s) = \int dp \mathcal{R}(x, p)G(x, p, x', p', \Delta s). \quad (A.10)$$

In the following analysis, we will not consider further any systematic errors associated with the treatment of heterogeneities in the case of the convolution method, nor with the "stepping errors" associated with incrementing s using Δs in the phase-space evolution model. Furthermore, we assume that the Green's functions or response kernels can be computed "exactly"—that there is no systematic error associated with them. The important result of this discussion is to demonstrate that the dimensionality of the analytic approach is $N_x + N_p$.

A.2 Convergence of Deterministic Solutions

The discussion of the previous section indicates that deterministic solutions are tantamount to solving a D-dimensional integral of the form:

$$I = \int_D du \, H(u). \quad (A.11)$$

In D dimensions, the calculation is no more difficult than in two dimensions, only the notation is more cumbersome. One notes that the integral takes the form:

$$I = \int_{u_1,\min}^{u_1,\max} du_1 \int_{u_2,\min}^{u_2,\max} du_2 \cdots \int_{uD,\min}^{uD,\max} du_D \, H(u_1, u_2 \ldots u_D)$$

$$= \sum_{i_1=1}^{N_{\text{cell}}^{1/D}} \int_{u_{i1}-\Delta u_1/2}^{u_{i1}+\Delta u_1/2} du_1 \sum_{i_2=1}^{N_{\text{cell}}^{1/D}} \int_{u_{i2}-\Delta u_2/2}^{u_{i2}+\Delta u_2/2} du_2 \cdots \quad (A.12)$$

$$\int_{u_{iD}-\Delta u_D/2}^{u_{iD}+\Delta u_D/2} du_D \times \sum_{i_D=1}^{N_{\text{cell}}^{1/D}} H(u_1, u_2 \ldots u_D)$$

The Taylor expansion takes the form

$$H(u_1, u_2 \ldots u_D) = H(u_{i_1}, u_{i_2} \ldots u_{i_D}) + \sum_{j=1}^{D}(u_i - u_{ij}) \times \partial H(u_{i_1}, u_{i_2} \ldots u_{i_D})/\partial u_j$$

$$+ \sum_{j=1}^{D}\frac{(u_i - u_{ij})^2}{2} \times \partial^2 H(u_{i_1}, u_{i_2} \ldots u_{i_D})/\partial u_j^2 \quad (A.13)$$

$$+ \sum_{j=1}^{D}\sum_{k\neq j=1}^{D}(u_i - u_{ij})(u_i - u_{ik}) \times \partial^2 H(u_{i_1}, u_{i_2} \ldots u_{i_D})/\partial u_i \partial u_j \ldots$$

The linear terms of the form $(u_i - u_{ji})$ and the bilinear terms of the form $(u_i - u_{ij})(u_i - u_{ik})$ for $k \neq j$ all vanish by symmetry and a relative $N^{-2/D}$ is extracted from the quadratic terms after integration. The result is that

$$\frac{\Delta I}{I} = \frac{1}{24 N_{\text{cell}}^{2/D}} \frac{\displaystyle\sum_{i_1=1}^{N_{\text{cell}}^{1/D}}\sum_{i_2=1}^{N_{\text{cell}}^{1/D}}\cdots\sum_{i_D=1}^{N_{\text{cell}}^{1/D}}\sum_{d=1}^{D}(u_{d,\max} - u_{d,\max})^2 \times \partial^2 H(u_{i_1}, u_{i_2} \ldots u_{i_D})/\partial u_d^2}{\displaystyle\sum_{i_1=1}^{N_{\text{cell}}^{1/D}}\sum_{i_2=1}^{N_{\text{cell}}^{1/D}}\cdots\sum_{i_D=1}^{N_{\text{cell}}^{1/D}} H(u_{i_1}, u_{i_2} \ldots u_{iD})} \quad (A.14)$$

Note that the one- and two-dimensional results can be obtained from the above equation. The critical feature to note is the overall $N_{\text{cell}}^{2/D}$ convergence rate. The more dimensions in the problem, the slower the convergence for numerical quadrature.

A.3 Convergence of Monte Carlo Solutions

An alternative approach to solving Equation A.1 is the Monte Carlo method whereby N_{hist} particle histories are simulated. In this case, the Monte Carlo method converges to the true answer according to the central limit theorem (Feller, 1967) which is expressed as

$$\frac{\Delta T_{\text{MC}}(x)}{T_{\text{MC}}(x)} = \frac{1}{\sqrt{N_{\text{hist}}}}\frac{\sigma_{\text{MC}}(x)}{T_{\text{MC}}(x)}, \quad (A.15)$$

where $T_{\text{MC}}(x)$ is the tally calculated in a voxel located at x as calculated by the Monte Carlo method and $\sigma_{\text{MC}}(x)$ is the variance associated with the distribution of $T_{\text{MC}}(x)$. Note that this variance $\sigma_{\text{MC}}(x)$ is an intrinsic feature of how the particle trajectories deposit energy in the spatial voxel. It is a "constant" for a given set of initial conditions and is conventionally estimated from the sample variance. It is also assumed, for the purpose of this discussion, that the sample variance exists and is finite.

A.4 Comparison between Monte Carlo and Numerical Quadrature

The deterministic models considered in this discussion pre-calculate $F(|x - x'|, p')$ of Equation A.7 or $F(x, x', p, p', \Delta s)$ of Equation A.10 storing them in arrays for iterative use. Then, during the iterative calculation phase, a granulated matrix operation is performed. The associated matrix product is mathematically similar to the "mid-point" $N_x + N_p$-multidimensional integration discussed previously:

$$T(x) = \int_D du H(u, x), \quad (A.16)$$

where $D = N_x + N_p$ and $u = (x_1, x_2 \ldots x_{N_x}, p_1, p_2 \ldots p_{N_p})$. That is, u is a multi-dimensional variable that encompasses both space and momentum. In the case of photon convolution, $H(u, x)$ can be inferred from Equation A.6 and takes the explicit form:

$$H(u,x) = \int dp F(|x - x'|, p') Q(x', p'). \quad \text{(A.17)}$$

There is a similar expression for the phase-space evolution model.

The "mid-point" integration represents a "first-order" deterministic technique and is applied more generally than the convolution or phase-space evolution applications. As shown previously, the convergence of this technique obeys the relationship:

$$\frac{\Delta T_{\text{NMC}}(x)}{T_{\text{NMC}}(x)} = \frac{1}{N_{\text{cell}}^{2/D}} \frac{\sigma_{\text{NMC}}(x)}{T_{\text{NMC}}(x)}, \quad \text{(A.18)}$$

where $T_{\text{NMC}}(x)$ is the tally in a spatial voxel in an arbitrary N_x-dimensional geometry calculated by a non-Monte Carlo method where N_p momentum components are considered. The D-dimensional phase space has been divided into N_{cell} "cells" equally divided among all the dimensions so that the "mesh-size" of each phase-space dimension is $N_{\text{cell}}^{1/D}$. The constant of proportionality as derived previously is

$$\sigma_{\text{NMC}}(x) = \frac{1}{24} \sum_{i_1=1}^{N_{\text{cell}}^{1/D}} \sum_{i_2=1}^{N_{\text{cell}}^{1/D}} \cdots \sum_{i_D=1}^{N_{\text{cell}}^{1/D}}$$
$$\times \sum_{d=1}^{D} (u_{d,\max} - u_{d,\min})^2 \, \partial^2 H(u_{i_1}, u_{i_2} \ldots u_{i_D}) / \partial u_d^2, \quad \text{(A.19)}$$

where the u-space of $H(u)$ has been partitioned in the same manner as the phase space described above. $u_{d,\min}$ is the minimum value of u_d while $u_{d,\max}$ is its maximum value. u_{ij} is the mid-point of the cell in the jth dimension at the i_jth mesh index.

The equation for the proportionality factor is quite complicated. However, the important point to note is that it depends only on the second derivatives of $H(u)$ with respect to the phase-space variables, u. Moreover, the non-Monte Carlo proportionality factor is quite different from the Monte Carlo proportionality factor. It would be difficult to predict which would be smaller and, almost certainly, would be application dependent.

We now assume that the computation time in either case is proportional to N_{hist} or N_{cell}. That is, $T_{\text{MC}} = \alpha_{\text{MC}} N_{\text{hist}}$ and $T_{\text{NMC}} = \alpha_{\text{NMC}} N_{\text{cell}}$. In the Monte Carlo case, the computation time is simply N_{hist} times the average computation time/history. In the non-Monte Carlo case, the matrix operation can potentially attempt to connect every cell in the D-dimensional phase space to the tally at point x. Thus, a certain number of floating-point and integer operations are required for each cell in the problem.

Consider the convergence of the Monte Carlo and non-Monte Carlo method. Using the above relationships, one can show that:

$$\frac{\Delta T_{\text{MC}}(x)/T_{\text{MC}}(x)}{\Delta T_{\text{NMC}}(x)/T_{\text{NMC}}(x)} = \left(\frac{\sigma_{\text{NMC}}(x)}{\sigma_{\text{MC}}(x)} \right) \left(\frac{\alpha_{\text{NMC}}^D}{\alpha_{\text{MC}}} \right)^{1/2} t^{(4-D)/2D}, \quad \text{(A.20)}$$

where t is the time measuring computational effort for either method. We have assumed that the two calculational techniques are the same. Therefore, given enough time, $\sigma_{\text{MC}}(x) \approx \sigma_{\text{NMC}}(x)$. One sees that, given long enough, the Monte Carlo method is always more advantageous for $D > 4$. We also note that inefficient programming in the non-Monte Carlo method is severely penalized in this comparison of the two methods.

Assume that one desires to do a calculation to a prescribed $\varepsilon = \Delta T(x)/T(x)$. Using the relations derived so far, we calculate the relative amount of time to execute the task to be:

$$\frac{t_{\text{NMC}}}{t_{\text{MC}}} = \left(\frac{\alpha_{\text{MC}}}{\alpha_{\text{NMC}}} \right) \left(\frac{[\sigma_{\text{NMC}}(x)/T_{\text{NMC}}(x)]^{D/2}}{\sigma_{\text{MC}}(x)/T_{\text{MC}}(x)} \right) \varepsilon^{(4-D)/2}, \quad \text{(A.21)}$$

which again shows an advantage for the Monte Carlo method for $D > 4$. Of course, this conclusion depends somewhat upon assumptions of the efficiency ratio $\alpha_{\text{MC}}/\alpha_{\text{NMC}}$ which would be dependent on the details of the calculational technique. Our conclusion is also dependent on the ratio $(\sigma_{\text{NMC}}(x)/T_{\text{NMC}}(x)^{D/2})/(\sigma_{\text{MC}}(x)/T_{\text{MC}}(x))$ which relates to the detailed shape of the response functions. For distributions that can vary rapidly, the Monte Carlo method is bound to be favored. When the distributions are flat, non-Monte Carlo techniques may be favored.

Nonetheless, at some level of complexity (large number of N_{cell}'s required), Monte Carlo becomes more advantageous. Whether or not one's application crosses this complexity "threshold" has to be determined on a case-by-case basis.

Smaller dimensional problems will favor the use of non-Monte Carlo techniques. The degree of the advantage will depend on the details of the application.

Acknowledgment

The author would like to thank Dr. Frank Verhaegen (Maastro Clinic, Netherlands), editor of *Monte Carlo Techniques in Radiation Therapy: Introduction, Source Modelling and Patient Dose*, Second Edition, for bringing the manuscript up-to-date from the previous version. This new data provided the incentive to arrive at a new and exciting conclusion regarding the future development and use of the Monte Carlo method for medical physics.

References

Aarnio P. A., Ranft J., and Stevenson G. R. A long write-up of the FLUKA82 program. CERN Divisional Report, TIS-RP/106-Rev., 1984.

Ahnesjö A., Andreo P., and Brahme A. Calculation and application of point spread functions for treatment planning with high energy photon beams. *Acta Oncol.*, 26:49–57, 1987.

Allison J. et al. GEANT4 development and applications. *IEEE T. Nucl. Sci.*, 53(1):250–303, 2006.

Alsmiller R. G., Jr and Barish J. High-energy (<18 GeV) muon transport calculations and comparison with experiment. *Nucl. Instrum. Methods*, 71:121–124, 1969.

Alsmiller R. G., Jr and Barish J. Energy deposition by 45 GeV photons in H, Be, Al, Cu, and Ta. Report ORNL-4933, Oak Ridge National Laboratory, Oak Ridge, Tennessee, 1974.

Alsmiller R. G., Jr, Barish J., and Dodge S. R. Energy deposition by high-energy electrons (50 to 200 MeV) in water. *Nucl. Instrum. Methods*, 121:161–167, 1974.

Alsmiller R. G., Jr and Moran H. S. Electron-photon cascade calculations and neutron yields from electrons in thick targets. Report ORNL-TM-1502, Oak Ridge National Laboratory, Oak Ridge, Tennessee, 1966.

Alsmiller R. G., Jr and Moran H. S. Electron-photon cascade calculations and neutron yields from electrons in thick targets. *Nucl. Instrum. Methods*, 48:109–116, 1967.

Alsmiller R. G., Jr and Moran H. S. The electron-photon cascade induced in lead by photons in the energy range 15 to 100 MeV. Report ORNL-4192, Oak Ridge National Laboratory, Oak Ridge, Tennessee, 1968.

Alsmiller R. G., Jr and Moran H. S. Calculation of the energy deposited in thick targets by high-energy (1 GeV) electron– photon cascades and comparison with experiment. *Nucl. Sci. Eng.*, 38:131–134, 1969.

Amdahl G. M. Validity of the single processor approach to acheiving large scale computing facilities. In *Proceedings of the April 18–20, 1967, Spring Joint Computer Conference*, 1967.

Andreo P. Monte Carlo simulation of electron transport. In *The Computation of Dose Distributions in Electron Beam Radiotherapy*, ed. Nahum A. E., Umea University, Umea, Sweden, pp. 80–97, 1985.

Andreo P. Monte Carlo techniques in medical radiation physics. *Phys. Med. Biol.*, 36:861–920, 1991.

Andreo P., Nahum A. E., and Brahme A. Chamber-dependent wall correction factors in dosimetry. *Phys. Med. Biol.*, 31:1189–1199, 1986.

Baró J., Sempau J., Fernandez-Varea J. M., and Salvat F. PENELOPE: An algorithm for Monte Carlo simulation of the penetration and energy loss of electrons and positrons in matter. *Nucl. Instrum. Methods*, B100:31–46, 1995.

Battistoni G., Muraro S., Sala P. R., Cerutti F., Ferrari A., Roestler S., Fasso A., and Ranft J. The FLUKA code: Description and benchmarking. *Proceedings of Hadronic Simulation Workshop*, Fermilab 6–8 September 2006, eds. Albrow M. and Baja R., Melville, New York, pp. 31–49, 2007.

Berger M. J. Monte Carlo calculation of the penetration and diffusion of fast charged particles. *Methods Comput. Phys.*, 1:135–215, 1963.

Berger M. J. and Seltzer S. M. ETRAN Monte Carlo code system for electron and photon transport through extended media. *Radiation Shielding Information Center, Computer Code Collection, CCC-107*, 1968.

Bielajew A. F. Correction factors for thick-walled ionisation chambers in point-source photon beams. *Phys. Med. Biol.*, 35:501–516, 1990.

Bielajew A. F. and Kawrakow I. The EGS4/PRESTA-II electron transport algorithm: Tests of electron step-size stability. In *Proceedings of the XII'th Conference on the Use of Computers in Radiotherapy*, Medical Physics Publishing, Madison, Wisconsin, pp. 153–154, 1997a.

Bielajew A. F. and Kawrakow I. From "black art" to "black box": Towards a step-size independent electron transport condensed history algorithm using the physics of EGS4/PRE STA-II. In *Proceedings of the Joint International Conference on Mathematical Methods and Supercomputing for Nuclear Applications*, American Nuclear Society Press, La Grange Park, Illinois, USA, pp. 1289–1298, 1997b.

Bielajew A. F. and Kawrakow I. PRESTA-I 25⇒ PRESTA-II: The new physics. In *Proceedings of the First International Workshop on EGS4*, Technical Information and Library, Laboratory for High Energy Physics, Japan, pp. 51–65, 1997c.

Bielajew A. F. and Rogers D. W O. PRESTA: The parameter reduced electron-step transport algorithm for electron Monte Carlo transport. *Nucl. Instrum. Methods*, B18:165–181, 1987.

Bielajew A. F. and Rogers D. W O. Electron step-size artefacts and PRESTA. In *Monte Carlo Transport of Electrons and Photons*, eds. T. M. Jenkins, W. R. Nelson, A. Rindi, A. E. Nahum, and D.W O. Rogers. Plenum Press, New York, pp. 115–137, 1989.

Bielajew A. F., Rogers D. W O., and Nahum A. E. Monte Carlo simulation of ion chamber response to ^{60}Co—Resolution of anomalies associated with interfaces. *Phys. Med. Biol.*, 30:419–428, 1985.

Bond J. E., Nath R., and Schulz R. J. Monte Carlo calculation of the wall correction factors for ionization chambers and A_{eq} for ^{60}Co γ rays. *Med. Phys.*, 5:422–425, 1978.

Briesmeister J. MCNP—A general purpose Monte Carlo code for neutron and photon transport, Version 3A. Los Alamos National Laboratory Report LA-7396-M, Los Alamos, NM, 1986.

Brun R., Hansroul M., and Lassalle J. C. GEANT Users Guide. CERN Report DD/EE/82, 1982.

Burfeindt H. Monte-Carlo-Rechnung für 3 GeV-Schauer in Blei. Deutsches Elektronen-Synchrotron Report Number DESY-67/24, 1967.

Butcher J. C. and Messel H. Electron number distribution in electron–photon showers. *Phys. Rev.*, 112:2096–2106, 1958.

Butcher J. C. and Messel H. Electron number distribution in electron–photon showers in air and aluminum absorbers. *Nucl. Phys.*, 20:15–128, 1960.

Cashwell E. D. and Everett C. J. *Monte Carlo Method for Random Walk Problems*. Pergamon Press, New York, 1959.

Colbert H. M. SANDYL: A computer program for calculating combined photon-electron transport in complex systems. Sandia Laboratories, Livermore, Report Number SCL-DR-720109, 1973.

Comte de Buffon G. *Essai d'arithmétique morale*, Vol. 4. Supplément à l'Histoire Naturelle, 1777.

Duderstadt J. J. and Martin W M. *Transport Theory*. Wiley, New York, 1979.

Eckhart R. Stan Ulam, John von Neumann, and the Monte Carlo method. *Los Alamos Science (Special Issue)*, 131–141, 1987.

Fasso A., Ferrari A., Ranft J., and Sala P. R. FLUKA: A multiparticle transport code. CERN 2005–10, INFN/TC_05/11, SLAC-R-773, 2005.

Feller W. *An Introduction to Probability Theory and Its Applications*, Vol. I, 3rd Edition. Wiley, New York, 1967.

Foote B. J. and Smyth V. G. The modelling of electron multiple-scattering in EGS4/PRESTA and its effect on ionization-chamber response. *Nucl. Instrum. Methods*, B100:22–30, 1995.

Ford R. L. and Nelson W. R. The EGS code system—Version 3. Stanford Linear Accelerator Center Report SLAC-210, 1978.

Gardner J., Siebers J., and Kawrakow I. Dose calculation validation of VMC++ for photon beams. *Med. Phys.*, 34:18091818, 2007.

Halbleib J. A. SPHERE: A spherical geometry multimaterial electron/photon Monte Carlo transport code. *Nucl. Sci. Eng.*, 66:269, 1978.

Halbleib J. A. ACCEPT: A Three-dimensional multilayer electron/photon Monte Carlo transport code using combinatorial geometry. *Nucl. Sci. Eng.*, 75:200, 1980.

Halbleib J. A., Hamil R., and Patterson E. L. Energy deposition model for the design of REB-driven large-volume gas lasers. *IEEE International Conference on Plasma Science (abstract)*, IEEE Conference Catalogue No. 81CH1640–2 NPS:117, 1981.

Halbleib J. A., Kensek R. P., Mehlhorn T. A., Valdez G. D., Seltzer S. M., and Berger M. J. ITS Version 3.0: The integrated TIGER Series of coupled electron/photon Monte Carlo transport codes. Sandia report SAND91-1634, 1992.

Halbleib J. A. and Mehlhorn T. A. ITS: The integrated TIGER series of coupled electron/photon Monte Carlo transport codes. *Nucl. Sci. Eng.*, 92(2):338, 1986.

Halbleib J. A., Sr. and Morel J. E. TIGERP, A one-dimensional multilayer electron/photon Monte Carlo transport code with detailed modelling of atomic shell ionization and relaxation. *Nucl. Sci. Eng.*, 70:219, 1979.

Halbleib J. A. and Vandevender W. H. EZTRAN—A user-oriented version of the ETRAN-15 electron-photon Monte Carlo technique. Sandia National Laboratories Report, SC-RR-71-0598, 1971.

Halbleib J. A. and Vandevender W. H. EZTRAN 2: A User-oriented version of the ETRAN-15 electron–photon Monte Carlo technique. Sandia National Laboratories Report, SLA-73-0834, 1973.

Halbleib J. A., Sr. and Vandevender W. H. TIGER, A onedimensional multilayer electron/photon Monte Carlo transport code. *Nucl. Sci. Eng.*, 57:94, 1975.

Halbleib J. A. and Vandevender W H. CYLTRAN: A cylindrical-geometry multimaterial electron/photon Monte Carlo transport Code. *Nucl. Sci. Eng.*, 61:288–289, 1976.

Halbleib J. A., Sr. and Vandevender W. H. Coupled electron photon collisional transport in externally applied electromagnetic fields. *J. Appl. Phys.*, 48:2312–2319, 1977.

Han K., Ballon D., Chui C., and Mohan R. Monte Carlo simulation of a cobalt-60 beam. *Med. Phys.*, 14:414–419, 1987.

Hebbard D. F. and Wilson P. R. The effect of multiple scattering on electron energy loss distributions. *Australian J. Phys.*, 1:90–97, 1955.

Hendricks J. S. and Briesmeister J. F. Recent MCNP Developments. Los Alamos National Laboratory Report LA-UR-91-3456 (Los Alamos, NM), 1991.

Hirayama H., Namito Y., Bielajew A. F., Wilderman S. J., and Nelson W. R. The EGS5 Code System. Report KEK 2005–8/SLAC-R-730, High Energy Accelerator Research Organization/Stanford Linear Accelerator Center, Tskuba, Japan/Stanford, USA, 2005.

Hogstrom K. R. Evaluation of electron pencil beam dose calculation. *Medical Physics Monograph (AAPM)* No. 15, 532–561, 1986.

ICRU. Specification of high-activity gamma-ray sources. ICRU Report 18, ICRU, Washington, DC, 1971.

Ing H., Nelson W. R., and Shore R. A. Unwanted photon and neutron radiation resulting from collimated photon beams interacting with the body of radiotherapy patients. *Med. Phys.*, 9:27–33, 1982.

Kalos M. H. and Whitlock P. A. *Monte Carlo Methods,* 2nd Edition. John Wiley and Sons-VCH, Weinnheim, Germany, 2008.

Kawrakow I. Electron transport: Longitudinal and lateral correlation algorithm. *Nucl. Instrum. Methods*, B114:307–326, 1996.

Kawrakow I. and Bielajew A. F. On the representation of electron multiple elastic-scattering distributions for Monte Carlo calculations. *Nucl. Instrum. Methods*, B134:325–336, 1998a.

Kawrakow I. and Bielajew A. F. On the condensed history technique for electron transport. *Nucl. Instrum. Methods*, B142:253–280, 1998b.

Kawrakow I., Fippel M., and Friedrich K. The high performance Monte Carlo algorithm VMC. *Medizinische Physik, Proceedings*, 256–257, 1995.

Kawrakow I., Fippel M., and Friedrich K. 3D electron dose calculation using a voxel based Monte Carlo algorithm. *Med. Phys.*, 23:445–457, 1996.

Kawrakow I. and Rogers D. W. O. The EGSnrc Code System: Monte Carlo simulation of electron and photon transport. Technical Report PIRS-701, National Research Council of Canada, Ottawa, Canada, 2000.

Laplace P. S. Theorie analytique des probabilités, Livre 2. In *Oeuvres complètes de Laplace*, Vol. 7, Part 2, pp. 365–366. L'académie des Sciences, Paris, 1886.

Larsen E. W. A theoretical derivation of the condensed history algorithm. *Ann. Nucl. Energy*, 19:701–714, 1992.

Lewis H. W. Multiple scattering in an infinite medium. *Phys. Rev.*, 78:526–529, 1950.

Ma C. -M. and Jiang S. B. Monte Carlo modelling of electron beams for Monte Carlo treatment planning. *Phys. Med. Biol.*, 44:R157–R189, 1999.

Ma C. M. and Nahum A. E. Bragg–Gray theory and ion chamber dosimetry in photon beams. *Phys. Med. Biol.*, 36:413–428, 1991.

Mackie T. R. Applications of the Monte Carlo method in radiotherapy. In *Dosimetry of Ionizing Radiation*, Vol. III, eds. K. Kase, B. Bjärngard and F. H. Attix. Academic Press, New York, pp. 541–620, 1990.

Mackie T. R., Bielajew A. F., Rogers D. W. O., and Battista J. J. Generation of energy deposition kernels using the EGS Monte Carlo code. *Phys. Med. Biol.*, 33:1–20, 1988.

Mackie T. R., Kubsad S. S., Rogers D. W. O., and Bielajew A. F. The OMEGA project: Electron dose planning using Monte Carlo simulation. *Med. Phys. (abs)*, 17:730, 1990.

Mackie T. R., Scrimger J. W., and Battista J. J. A convolution method of calculating dose for 15 MV x-rays. *Med. Phys.*, 12:188–196, 1985.

Manfredotti C., Nastasi U., Marchisio R., Ongaro C., Gervino G., Ragona R., Anglesio S., and G. Sannazzari. Monte Carlo simulation of dose distribution in electron beam radiotherapy treatment planning. *Nucl. Instrum. Methods*, A291:646–654, 1990.

Messel H. and Crawford D. F. *Electron–Photon Shower Distribution Function*. Pergamon Press, Oxford, 1970.

Messel H., Smirnov A. D., Varfolomeev A. A., Crawford D. F., and Butcher J. C. Radial and angular distributions of electrons in electron-photon showers in lead and in emulsion absorbers. *Nucl. Phys.*, 39:1–88, 1962.

Metropolis N. The beginning of the Monte Carlo method. *Los Alamos Science (Special Issue)*, 125–130, 1987.

Metropolis N. and Ulam S. The Monte Carlo method. *Amer. Stat. Assoc.*, 44:335–341, 1949.

Meyer H. A., ed. *Symposium on Monte Carlo Methods*. John Wiley and Sons, New York, 1981.

Miller P. A., Halbleib J. A., and Poukey J. W. SPHEM: A three-dimensional multilayer electron/photon Monte Carlo transport code using combinatorial geometry. *J. Appl. Phys.*, 52:593–598, 1981.

Mohan R., Chui C., and L. Lidofsky. Energy and angular distributions of photons from medical linear accelerators. *Med. Phys.*, 12:592–597, 1985.

Nagel H. H. Die Berechnung von Elektron-Photon-Kaskaden in Blei mit Hilfe der Monte-Carlo Methode. *Inaugural-Dissertation zur Erlangung des Doktorgrades der Hohen Mathematich-Naturwissenschaftlichen Fakultät der Rheinischen Friedrich-Wilhelms-Universtät zu Bonn*, 1964.

Nagel H. H. Elektron-Photon-Kaskaden in Blei: Monte-Carlo-Rechnungen für Primärelektronenergien zwischen 100 und 1000 Me. *Physik V. Z.*, 186:319–346, 1965 (English translation Stanford Linear Accelerator Center Report Number SLAC-TRANS-28, 1965.

Nagel H. H. and Schlier C. Berechnung von Elektron-PhotonKaskaden in Blei für eine Primärenergie von 200 MeV. *Z. Phys.*, 174:464–471, 1963.

Nahum A. E. Calculations of electron flux spectra in water irradiated with megavoltage electron and photon beams with applications to dosimetry. PhD thesis, University of Edinburgh, UK, 1976.

Namito Y., Ban S., and Hirayama H. Implementation of linearly-polarized photon scattering into the EGS4 code. *Nucl. Instrum. Methods*, A322:277–283, 1993.

Namito Y., Ban S., and Hirayama H. Implementation of Doppler broadening of Compton-scattered photons into the EGS4 code. *Nucl. Instrum. Methods*, A349:489–494, 1994.

Nelson W. R., Hirayama H., and Rogers D. W. O. The EGS4 Code System. Report SLAC-265, Stanford Linear Accelerator Center, Stanford, CA, 1985.

Nicoli D. F. The application of Monte Carlo cascade shower generation in lead. Submitted in partial fulfillment of the requirement for the degree of Bachelor of Science at the Massachusetts Institute of Technology, 1966.

Petti P. L., Goodman M. S., Gabriel T. A., and Mohan R. Investigation of buildup dose from electron contamination of clinical photon beams. *Med. Phys.*, 10:18–24, 1983a.

Petti P. L., Goodman M. S., Sisterson J. M., Biggs P. J., Gabriel T. A., and Mohan R. Sources of electron contamination for the Clinac–35 25–MV photon beam. *Med. Phys.*, 10:856–861, 1983b.

Ranft J. Monte Carlo calculation of the nucleon–meson cascade in shielding materials by incoming proton beams with energies between 10 and 1000 GeV. CERN Yellow Report 64–67, 1964.

Rhodes R. *The Making of the Hydrogen Bomb*. Touchstone (Simon & Schuster Inc.), New York, 1988.

Rogers D. W. O. More realistic Monte Carlo calculations of photon detector response functions. *Nucl. Instrum. Methods*, 199:531–548, 1982.

Rogers D. W. O. The use of Monte Carlo techniques in radiation therapy. *Proceedings of CCPM Course on Computation in Radiation Therapy*, Canadian College of Physicists in Medicine, London, Ontario, 1983a.

Rogers D. W. O. A nearly mono-energetic 6 to 7 MeV photon calibration source. *Health Phys.*, 45:127–137, 1983b.

Rogers D. W. O. Fluence to dose equivalent conversion factors calculated with EGS3 for electrons from 100 keV to 20 GeV and photons from 20 keV to 20 GeV. *Health Phys.*, 46:891–914, 1984a.

Rogers D. W. O. Low energy electron transport with EGS. *Nucl. Instrum. Methods*, 227:535–548, 1984b.

Rogers D. W. O. How accurately can EGS4/PRESTA calculate ion chamber response? *Med. Phys.*, 20:319–323, 1993.

Rogers D. W. O. Fifty years of Monte Carlo simulations for medical physics. *Phys. Med. Biol.*, 51:R287–R301, 2006.

Rogers D. W. O. and Bielajew A. F. Monte Carlo techniques of electron and photon transport for radiation dosimetry. In *The Dosimetry of Ionizing Radiation*, Vol III, eds. K. R. Kase, B. E. Bjärngard, and F. H. Attix. Academic Press, New York and London, pp. 427–539. 1990a.

Rogers D. W. O. and Bielajew A. F. Wall attenuation and scatter corrections for ion chambers: Measurements versus calculations. *Phys. Med. Biol.*, 35:1065–1078, 1990b.

Rogers D. W. O., Bielajew A. F., and Nahum A. E. Ion chamber response and A_{wall} correction factors in a ^{60}Co beam by Monte Carlo simulation. *Phys. Med. Biol.*, 30:429–443, 1985a.

Rogers D. W. O., Ewart G. M., Bielajew A. F., and G. van Dyk. Calculation of contamination of the ^{60}Co beam from an AECL therapy source. NRC Report PXNR-2710, 1985b.

Rogers D. W. O., Faddegon B. A., Ding G. X., Ma C. M., Wei J., and Mackie T. R. BEAM: A Monte Carlo code to simulate radiotherapy treatment units. *Med. Phys.*, 22:503–524, 1995.

Rossi B. B. *High Energy Particles*. Prentice-Hall, New York, 1952.

Salvat F. and Parrellada J. Penetration and energy loss of fast electrons through matter. *J. Appl. Phys. D: Appl. Phys.*, 17(7): 1545–1561, 1984.

Seltzer S. M. An overview of ETRAN Monte Carlo methods. In: *Monte Carlo Transport of Electrons and Photons*, eds. T. M. Jenkins, W. R. Nelson, A. Rindi, A. E. Nahum, and D. W. O. Rogers. Plenum Press, New York, pp. 153–182, 1989.

Seltzer S. M. Electron–photon Monte Carlo calculations: The ETRAN code. *Int. J. Appl. Radiat. Isotopes*, 42:917–941, 1991.

Sempau J., Wilderman S. J., and Bielajew A. F. DPM, a fast, accurate Monte Carlo code optimized for photon and electron radiotherapy treatment planning dose calculations. *Phys. Med. Biol.*, 45:2263–2291, 2000.

Shultis J. K. and Faw R. E. *Radiation Shielding*. Prentice-Hall, Upper Saddle River, NJ, 1996.

Spezi E. and Lewis G. An overview of Monte Carlo treatment planning for radiotherapy. *Radiat. Protect. Dosim.*, 131:123129, 2008.

Teng S. P., Anderson D. W., and Lindstrom D. G. Monte Carlo electron-transport calculations for clinical beams using energy grouping. *Appl. Radiat. Isotopes*, 37:1189–1194, 1986.

Ulam S. M. *Adventures of a Mathematician*, 2nd Edition. University of California Press, Berkeley, 1991.

Ulam S. M. and von Neumann J. Random ergodic theorems. *Bull. Amer. Math. Soc. (abstract)*, 51:660, 1945.

Ulam S. M. and von Neumann J. On combination of stochastic and deterministic processes. *Bull. Amer. Math. Soc. (abstract)*, 53:1120, 1947.

Varfolomeev A. A. and Svetlolobov I. A. Monte Carlo calculations of electromagnetic cascades with account of the influence of the medium on bremsstrahlung. *Sov. Phys. JETP*, 36:1263–1270, 1959.

Verhaegen F. and Seuntjens J. Monte Carlo modelling of external radiotherapy photon beams. *Phys. Med. Biol.*, 48:R107–R164, 2003.

Völkel U. Elektron-Photon-Kaskaden in Blei für Primärteilchen der Energie 6 GeV. Deutsches Elektronen-Synchrotron Report Number DESY-65/6, 1965 (English translation Stanford Linear Accelerator Center Report Number SLAC-TRANS-41, 1966.

von Neumann J. and Richtmyer R. Statistical methods in neutron diffusion. Technical Report LAMS-551, Los Alamos National Laboratory, 1947.

Wilson R. R. Radiological use of fast protons. *Radiology*, 47:487–491, 1946.

Wilson R. R. Monte Carlo calculations of showers in lead. *Phys. Rev. (abstract)*, 79:204, 1950.

Wilson R. R. The range and straggling of high energy electrons. *Phys. Rev.*, 84:100–103, 1951.

Wilson R. R. Monte Carlo study of shower production. *Phys. Rev.*, 86:261–269, 1952.

Zerby C. D. and Moran H. S. A Monte Carlo calculation of the three-dimensional development of high-energy electron-photon cascade showers. Report ORNL-TM-422, Oak Ridge National Laboratory, Oak Ridge, Tennessee, 1962a.

Zerby C. D. and Moran H. S. Studies of the longitudinal development of high-energy electron–photon cascade showers in copper. Report ORNL-3329, Oak Ridge National Laboratory, Oak Ridge, Tennessee, 1962b.

Zerby C. D. and Moran H. S. Studies of the longitudinal development of electron–photon cascade showers. *J. Appl. Phys.*, 34:2445–2457, 1963.

2

Basics of Monte Carlo Simulations

Matthias Fippel

Brainlab AG

2.1 Monte Carlo Method

Monte Carlo (MC) techniques are widely used in natural and social sciences. There are many different "flavors" of how to work with these techniques. As we have seen in the previous chapter, there is a long tradition of using MC methods in different areas, including medical physics. Therefore, it is difficult to provide a general definition of the MC method. Consequently, a literature search results in many definitions. Some examples of introducing literature, useful to read before working with MC methods, are the corresponding chapters in the textbooks *Numerical Recipes in C* (Press et al., 1992) or "The Review of Particle Physics" (Nakamura et al., 2010) of the Particle Data Group. A nice introduction is also provided by James in "Monte Carlo Theory and Practice" (James, 1980). More references on MC techniques and random number sampling can be found in these reviews.

For our purposes, we define the MC method in the following short way:

Monte Carlo is a numerical method to solve equations or to calculate integrals based on random number sampling.

The two aspects of this definition, random number sampling and numerical integration, are outlined in detail in the next two sections.

2.1.1 Random Number Sampling

Monte Carlo algorithms use a computer program, a procedure, or a subroutine, called "random number generator" (RNG). However, computers cannot really generate "random" numbers because the output of any program is—by definition—predictable; hence, it is not truly "random." Therefore, the result of these generators shall correctly be termed "pseudorandom numbers."

A huge sequence of these pseudorandom numbers is required to solve a complex problem. The numbers within a sequence of random numbers shall be uncorrelated, that is, they must not depend on each other. Because this is impossible with a computer program, they should at least appear independent. In other words, any statistical test program shall show that the numbers within the sequence are uncorrelated and any computer code that requires independent random numbers shall produce the same result with different sequences. If this is the case within the uncertainty of the simulation, then these sequences can be called pseudorandom. To keep the notation short throughout this book, we will nevertheless call them just "random numbers." However, we have to keep in mind the real character of these numbers.

A pseudo-RNG must be examined carefully before it can be used for a specific purpose. A useful generator for simulations in radiation therapy must provide two important features:

- The period of the sequence shall be large enough. Otherwise, if the sequence is reused several times, the results of the MC simulation are correlated.
- They must be uniformly distributed in multiple dimensions. That means random vectors created from an n-tuple of random numbers must be uniformly distributed in the n-dimensional space. Typically, it is not obvious how to detect correlations in higher dimensions.

Most of the generators produce uniformly distributed random numbers in some intervals, typically in [0,1]. It is useful to have a look at simple RNGs to understand their operating

DOI: 10.1201/9781003211846-3

principle. One class of simple RNGs is called *linear congruential generators*. They generate a sequence of integers I_1, I_2, I_3, ..., each between 0 and $m-1$ by the recurrence relation:

$$I_{j+1} = aI_j + c \pmod{m} \tag{2.1}$$

with the parameters

   a: multiplier
   c: increment
   m: modulus

An example is the quick and dirty generator used by old EGS4 (Nelson et al., 1985) implementations:

$$I_{j+1} = aI_j \tag{2.2}$$

with

   $a = 663608941$
   $c = 0$
   $m = 2^{32}$

On a machine with 32-bit integer representation, the product of two unsigned integers is the low-order 32 bits of the true 64-bit result. Because of $m = 2^{32}$, the modulus is taken into account automatically. Therefore, this RNG is very fast. However, its sequence length of 2^{32} is not enough for MC applications in radiotherapy.

Computer operating system-supplied RNGs are typically of the same *linear congruential* type. Hence, it is not recommended to trust them. For MC applications in radiotherapy, long-sequence RNGs are required. They also should be portable, that is, the same sequence should be produced on different machines. A good source of high-quality RNGs including test programs to examine the created sequences in multiple dimensions is the CERN program library (http://www.cern.ch).

A class of long-sequence RNGs called "subtract-with-borrow" algorithms has been developed by Marsaglia and Zaman (1991). The CERN library function RANMAR for instance has a sequence length of 2^{144}. It is used, for example, by EGS4 (Nelson et al. 1985) and XVMC (Fippel, 1999).

In EGSnrc (Kawrakow, 2000a), the use of RANLUX (Lüscher, 1994) is recommended. It allows different luxury levels between 0 and 4. The quality and also the simulation time increase with increasing luxury levels. RANMAR can also be selected in EGSnrc. According to the CERN library documentation, its quality corresponds to RANLUX with luxury level between 1 and 2. According to the EGSnrc manual (Kawrakow et al., 2011), with luxury level 1 or higher, no problems have been discovered in practical EGSnrc calculations.

2.1.2 Numerical Integration

Function $y = f(x)$ shall be integrated in interval $[a, b]$, that is, the area A enclosed by function $f(x)$, the x-axis as well as the interval limits a and b shall be calculated (see Figure 2.1):

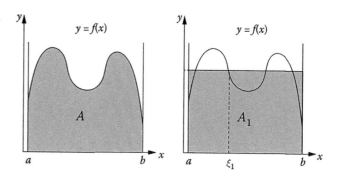

FIGURE 2.1 The left plot shows area A as calculated by integrating function $y = f(x)$ in interval $[a, b]$. The right plot shows a very rough estimate of area A as given by the area of the rectangle $A_1 = (b-a)f(\xi_1)$

$$A = \int_a^b f(x)\,dx. \tag{2.3}$$

If this is impossible analytically, some numerical method must be applied. One of the many numerical options is called MC integration because it is based on a sequence of uniformly distributed random numbers. Generally, a computer-generated random number η_i is uniformly distributed in interval $[0,1]$. It can be scaled to interval $[a, b]$ by

$$\xi_i = (b-a)\eta_i + a, \tag{2.4}$$

that is, ξ_i is uniformly distributed in $[a, b]$. Some first rough estimate of the real area A is now given by (see Figure 2.1)

$$A_1 = (b-a)f(\xi_1), \tag{2.5}$$

that is, the rectangle given by the function value at random point ξ_1 and the length of the interval. This estimate is really rough; therefore, we should do this a second time:

$$A_2 = \frac{1}{2}\left\{(b-a)f(\xi_2) + A_1\right\} = \frac{b-a}{2}\left\{f(\xi_1) + f(\xi_2)\right\}. \tag{2.6}$$

Note that we have averaged the areas from both runs providing in this way a better estimate of the real integral. The generalization is now obvious; after N runs (with N random numbers), we obtain

$$A_N = \frac{b-a}{N}\sum_{i=1}^{N} f(\xi_i) = (b-a)\langle f(x)\rangle \tag{2.7}$$

with the average function value for N samples:

$$\langle f(x)\rangle \equiv \frac{1}{N}\sum_{i=1}^{N} f(\xi_i). \tag{2.8}$$

The basic theorem of MC integration (Press et al., 1992) also provides information on the uncertainty of the estimate

$$A = A_N \pm (b-a)\sqrt{\frac{\langle f^2(x)\rangle - \langle f(x)\rangle^2}{N}}. \tag{2.9}$$

The average of the function value squared is defined by

$$\langle f^2(x)\rangle \equiv \frac{1}{N}\sum_{i=1}^{N} f^2(\xi_i). \tag{2.10}$$

The estimated area A_N converges to the real integral A in the limit $N \to \infty$. The convergence is slow because of the $1/\sqrt{N}$ behavior, that is, the statistical uncertainty is reduced by a factor of 2 if the number of random points N (and the calculation time) is increased by a factor of 4. Therefore, the MC method should not be used for this simple type of numerical integration. There are better options (see, e.g., Press et al. (1992)). However, the MC method comes into play if all other methods fail, for example, if the dimensionality of the problem becomes very large, that is, the integral shall be calculated in a space with 10, 100, or even an infinite number of dimensions.

Let us assume that function $f(\vec{x})$ shall be integrated in volume V of a space with D dimensions. Instead of random numbers for MC integration, random points (or vectors) uniformly distributed in the multidimensional volume V are required. Because volume V has D dimensions, we need D random numbers to form one random point. To sample N random points in this volume, we need $D \times N$ random numbers. Therefore, the quality of the RNG should be checked for higher dimensions, not just for the 1D, 2D, or 3D case. Often, correlations can be observed in higher dimensions even if the RNG looks uniform in lower dimensions.

Now, let us assume that we have sampled N random points $\vec{\xi}_1,\ldots,\vec{\xi}_N$ uniformly distributed in multidimensional volume V. Then, the basic theorem of MC integration is given by (Press et al., 1992)

$$\int dV f(\vec{x}) \approx V\langle f(\vec{x})\rangle \pm \sqrt{\frac{\langle f^2(\vec{x})\rangle - \langle f(\vec{x})\rangle^2}{N}} \tag{2.11}$$

with

$$\langle f(\vec{x})\rangle \equiv \frac{1}{N}\sum_{i=1}^{N} f(\vec{\xi}_i) \quad \text{and} \quad \langle f^2(\vec{x})\rangle \equiv \frac{1}{N}\sum_{i=1}^{N} f^2(\vec{\xi}_i). \tag{2.12}$$

Multidimensional numerical integration is necessary to solve the system of coupled transport equations for problems in radiation therapy, for example, for dose calculation. It is a system of equations because the transport problem for photons and electrons (positrons) must be solved. The system is coupled because electrons influence the photon transport (bremsstrahlung) and vice versa (Compton scatter, photoelectric absorption, and pair production). Theoretically, the problem has an infinite parameter space because the number of secondary photons and electrons is physically unlimited when we start with a primary particle of definite energy. Therefore, the numerical integration has to be performed in a space with infinite dimensions. Practically, the dimensionality is limited because the region of interest is limited and we usually stop the simulation if the photon or electron energy falls below some minimum energy.

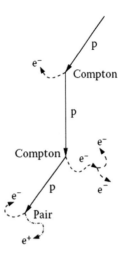

FIGURE 2.2 Example of a particle history starting with a primary photon p (straight line) via Compton interactions and pair production events leading to secondary photons p (straight lines) and secondary electrons e⁻ (dashed lines) and positrons e⁺ (dashed dotted line).

Nevertheless, for numerical integration using MC, a random point in a high-dimensional parameter space must be sampled. This point can be demonstrated by a so-called particle history, a shower of secondary particles generated by a primary particle including all daughter particles. An example of a particle history is shown in Figure 2.2. This shows schematically what happens during an MC radiation transport simulation. It starts with a primary particle emitted from a particle source. The simulation takes into account the geometry of the problem, for example, the linac head geometry given by the technical specifications and/or the patient anatomy given by CT images. It also takes into account the material transport properties given by cross-sectional data. For the primary photon in the example of Figure 2.2, the distance to the first interaction site must be sampled based on the total cross section of the corresponding medium. At the interaction site, the type of interaction (Compton scatter in this case) must be sampled. All secondary particle parameters (secondary particle energies and scattering angles) are determined using differential cross sections and the corresponding probability distribution functions. These steps are repeated until the primary and secondary particles have left the simulation geometry or the particle energy falls below some minimum energy. Dose, for example, is calculated by accumulating the absorbed energy per region. It is obvious that a huge number of particle histories have to be simulated to obtain a result of low noise. Because of Equation 2.11, the noise level and the statistical uncertainty can be reduced by a factor of 2 if the number of histories is increased by a factor of 4. This causes very long calculation times for transport simulation problems in radiation therapy.

2.1.3 Nonuniform Sampling Methods

As we have learned in Section 2.1.1, RNGs usually generate uniformly distributed random numbers. However, for MC transport simulations in radiation therapy, random numbers distributed according to specific probability weight distribution functions $p(x)$ are required many times within the algorithm.

Let us assume that we want to generate a random number ξ in interval $[a, b]$ and distributed according to the nonuniform probability weight function $p(x)$. Available is a RNG generating only uniformly distributed random numbers η in interval $[0,1]$. How can ξ be sampled from $p(x)$? This is possible using the cumulative distribution function $P(x)$ defined by the integral

$$P(x) = \int_a^x dx' p(x'), \quad a \le x \le b, \quad P(a) = 0, \quad P(b) = 1. \quad (2.13)$$

Function $P(x)$ is monotonically increasing in interval $[a, b]$. The function values $y = P(x)$ are limited to interval $[0,1]$. If formula $y = P(x)$ can be transposed with respect to x, that is, the inverse can be determined as

$$x = P^{-1}(y), \quad (2.14)$$

then it can be shown that

$$\xi = P^{-1}(\eta), \quad (2.15)$$

is distributed according to probability weight function $p(x)$. For a proof, see, for example, Nelson et al. (1985). This method of nonuniform random number sampling is called the *direct* or *transformation* method. It depends on the fact that the inverse of $P(x)$ can be easily and efficiently calculated.

If this is impossible, the *indirect* or *rejection* method (Nelson et al. 1985) has to be applied instead. It takes into account an adequately chosen comparison function $g(x)$ with $g(x) > p(x)$ in interval $[a, b]$. A further condition of $g(x)$ is that a random number ξ_0 can be sampled easily from $g(x)$, for example, using the transformation method. Now, we sample a new uniform random number η_0 from interval $[0, g(\xi_0)]$. We accept ξ_0 as a valid random number if $\eta_0 < p(\xi_0)$. We reject ξ_0 if $p(\xi_0) \le \eta_0 \le g(\xi_0)$. In case of rejection, we start with sampling a new random number ξ_0 from $g(x)$ and so on. Again, we do not prove here that ξ_0 sampled in this way is distributed according to $p(x)$. But the rejection method can be demonstrated using the simple example of a constant rejection function $g(x) = p(x_{max}) = $ const, with x_{max} as the position of the maximum of $p(x)$ in interval $[a, b]$. In this case, ξ_0 is just sampled from a uniform distribution in $[a, b]$. With the rejection random number $\eta_0 < p(\xi_0)$, it is obvious that ξ_0 is distributed according to $p(x)$.

Of course, the rejection method should be used with care. That is, comparison function $g(x)$ should not differ too much from $p(x)$. If the deviation is too large, the number of rejections can become too large and the method becomes inefficient.

2.2 Monte Carlo Transport in Radiation Therapy

The basic concepts to simulate the transport of photons, electrons, positrons, neutrons, protons, or heavy ions in the energy range

of radiation therapy are complex. Therefore, this chapter can only provide the fundamentals of the methodology. To understand all techniques in full detail, the reader is referred to corresponding literature, like the textbook *Monte Carlo Transport of Electrons and Photons* (Jenkins et al., 1988). A lot of information is available also in the user manuals of frequently used MC packages, like EGS4 (Nelson et al., 1985), EGSnrc (Kawrakow, 2000a; Kawrakow et al., 2011), Penelope (Salvat et al., 2011), MCNP (Briesmeister, 1997), and GEANT4 (Agostinelli et al., 2003). More basic literature is listed in review papers about MC treatment planning (Chetty et al., 2007; Reynaert et al., 2007).

In the next sections, we show how particle transport collision by collision is simulated in detail. This so-called analog particle transport simulation scheme is demonstrated using the photon as an example. A further section introduces the condensed history (CH) technique that allows efficient simulations of charged particle transport in radiotherapy.

2.2.1 Analog Particle Transport

Let us assume that a photon of energy E hits the surface of a homogeneous medium. Then, the probability $p(s)$ that this photon interacts after path length s with the medium is given by the attenuation law:

$$p(s) ds = \mu(E) e^{-\mu(E)s} ds. \quad (2.16)$$

The parameter $\mu(E)$ is called the linear attenuation coefficient of the medium for photons of energy E. The mean free path length s until interaction can be calculated from this distribution function if the medium is extended infinitely below the surface:

$$\langle s \rangle = \int_0^\infty ds\, s\, p(s) = \mu(E) \int_0^\infty ds\, s\, e^{-\mu(E)s} = \frac{1}{\mu(E)}. \quad (2.17)$$

This allows us to express the attenuation law (2.16) in terms of the number of mean free path lengths:

$$\lambda = \frac{s}{\langle s \rangle} = \mu(E) s, \quad (2.18)$$

that is

$$p(\lambda) d\lambda = e^{-\lambda} d\lambda. \quad (2.19)$$

The advantage of this notation is that it also works for heterogeneous geometries if the number of mean free path lengths is defined by

$$\lambda = \sum_{start}^{P} \mu_i(E) s_i. \quad (2.20)$$

To calculate λ, the photon must be traced on a straight line from the *Start* position on the surface through different regions i containing different materials until the interaction point P. In each region i with linear attenuation coefficient $\mu_i(E)$, the corresponding line segment s_i must be determined. This tracing algorithm to calculate λ is an essential part of MC simulations in radiation therapy. The calculation time can be unnecessarily long if it is implemented inefficiently.

The attenuation law (2.19) provides the probability weight distribution function $p(\lambda)$. The cumulative distribution function is given by

$$P(\lambda) = \int_0^\lambda d\lambda' p(\lambda') = \int_0^\lambda d\lambda' e^{-\lambda'} = 1 - e^{-\lambda}, \quad P(0) = 0, \quad P(\infty) = 1. \quad (2.21)$$

This function is monotonically increasing in $[0,\infty]$. We can now sample λ_1, the distance to the first interaction site, using the transformation method and a uniform random number ξ_1 from the half-open interval $[0,1)$:

$$\xi_1 = 1 - e^{-\lambda_1}, \Rightarrow \lambda_1 = -\ln(1 - \xi_1). \quad (2.22)$$

Note the special notation of the interval limits above. It means that number 1 must not be included in the sequence of random numbers. If it is included, the logarithm in Equation 2.22 is undefined.

Taking into account the geometric setup of the simulation, the photon is tracked λ_1 mean free path lengths to the first interaction point. Then, the type of the interaction has to be sampled. In the energy range of radiation therapy, four processes are most common, *photoelectric absorption, Raleigh scatter, Compton scatter,* and *pair production*. They are represented by the corresponding interaction coefficients as material parameters at the interaction site:

$$\mu(E) \equiv \mu_{tot}(E) = \mu_A(E) + \mu_R(E) + \mu_C(E) + \mu_P(E) \quad (2.23)$$

and they are used to divide the interval $[0,1]$ into four parts:

$$
\begin{array}{lll}
[P_0, P_1] & : & \text{photoelectric absorption} \\
[P_1, P_2] & : & \text{Raleigh scatter} \\
[P_2, P_3] & : & \text{Compton scatter} \\
[P_3, P_4] & : & \text{pair production}
\end{array} \quad (2.24)
$$

with

$$P_0 = 0, \quad P_1 = P_0 + \frac{\mu_A}{\mu_{tot}}, \quad P_2 = P_1 + \frac{\mu_R}{\mu_{tot}}, \quad P_3 = P_2 + \frac{\mu_C}{\mu_{tot}}, \quad P_4 = 1. \quad (2.25)$$

The interaction type is sampled by using a second uniform random number ξ_2 from interval $[0,1]$ and by checking in which subinterval ξ_2 is located.

With known interaction type, the parameters of all secondary particles can be determined. These parameters, that is, energy and scattering angles, are sampled using the probability distributions given by the corresponding differential cross sections. Furthermore, kinematic conservation laws must be taken into account. In general, for this purpose, the

transformation method does not work, that is, the rejection technique is the method of choice.

After that, everything is known to repeat the three steps with the secondary particles. Even electrons and positrons could be simulated in this analog manner. The whole particle history (see an example in Figure 2.2) is simulated including all secondary particles and their daughter particles. The transport simulation of a particle stops if it leaves the geometry of interest or its energy falls below some predefined minimum energy. These cutoff parameters are usually denoted as P_{cut} for photons or E_{cut} for charged particles. During each step of the history, the values of interest are calculated for accumulation. For example, the absorbed energy per region (voxel) is accumulated if the dose is calculated. According to Equation 2.11, the number of simulated particle histories determines the statistical accuracy and thus the calculation time.

2.2.2 Charged Particle Transport

2.2.2.1 Condensed History (CH) Technique

Section 2.2.1 outlined the fundamental procedure to simulate the transport of any particle type through matter. In general, this is the standard simulation method for neutral particles because the free path length between two interactions is in the order of the size of the simulation geometry. For example, the mean free path lengths of photons in the therapeutic energy range are in the order of 10 cm in water and human tissue. The region of interest for dose calculation in radiation therapy has a size of about 30 cm. Therefore, on average, very few photon interactions are simulated.

This is completely different for charged particles like electrons or protons. For radiation therapy energies, they undergo a very large number of single interactions. Consequently, the simulation of one electron (positron or proton) history would require a much longer calculation time than the simulation of one photon history. Hence, this approach is impractical for most of the transport problems in radiation therapy.

Fortunately, almost all of these interactions are elastic or semielastic. That means no energy or a small amount of energy is transferred from the charged particle to the surrounding matter. Furthermore, the particle direction changes in general only by small scattering angles. This allows us to group many of these elastic and semielastic events into one CH step. The method has been introduced in 1963 by Berger (1963) and is called CH technique. The majority of MC algorithms, applied in radiotherapy, perform electron, positron, proton, or heavier charged particle transport using the CH technique.

Present-day CH implementations divide all interactions of one charged particle history into hard and soft collisions as well as hard and soft bremsstrahlung production events. The two collision types are distinguished by an arbitrary kinetic energy loss threshold E_c. Hard and soft bremsstrahlung production is distinguished using the parameter k_c. Collision events with an energy transfer lower than E_c to secondary electrons are called soft collisions. These soft collisions are simulated implicitly by continuous energy transfer from the charged particle onto the matter surrounding the particle track. The direction change of the particle due to many small angles is simulated by one large multiple scattering angle. All hard collisions are simulated explicitly as in the case of photons. The minimum energy of secondary particles created during hard collisions is equivalent to E_c. It also provides the maximum energy and consequently the maximum range of charged secondary particles produced during soft collisions. This range has to be smaller than the spatial resolution of the simulation geometry. The meaning of the bremsstrahlung production threshold k_c is similar. Therefore, the arbitrary parameters E_c and k_c must be chosen with care. The simulation result can be influenced negatively if they are too large. On the contrary, the simulation can last too long if these parameters are chosen too small. Please note that the parameter E_c must not be confused with the particle track end energy E_{cut}. However, in many MC simulations, both parameters are chosen to be equal.

Because of the approximate nature of the CH transport, it is furthermore useful to limit the maximum distance traveled in one CH step. This could be realized by another arbitrary user parameter, the global maximum step size s_{max}, or by material- and mass density-dependent parameters s_{max}^i. In many MC algorithms, the maximum step size is determined based on the percentage maximum energy loss E_{step}. This way, the maximum spatial step size automatically depends on the stopping power and the mass density of the present material.

With all of these possibilities, the end of one CH step is determined either by the maximum step size or by the next hard interaction. As a result, a complete electron history may look like the example shown in Figure 2.3. Electrons move in general on straight lines during the CH step. They change the direction due to multiple scattering at the end of the step (see Figure 2.4) or (as in the case of Figures 2.3 and 2.5) between

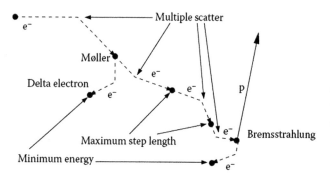

FIGURE 2.3 Example of a particle history starting with a primary electron e⁻ (dashed line) via multiple scatter, Møller interactions, and bremsstrahlung production events leading to secondary (delta) electrons e⁻ (dashed lines) and secondary photons p (straight lines).

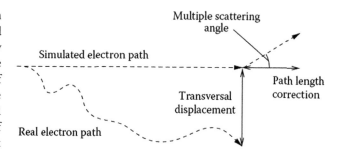

FIGURE 2.4 Example of a simulated electron path using the CH technique and multiple scatter compared to a possible real electron path (or simulated using the analog technique). As a result of the CH transport, the simulated path length must be corrected and a transverse displacement has to be taken into account.

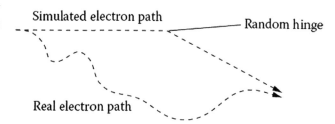

FIGURE 2.5 Same as Figure 2.4, but this time the electron is simulated using the random hinge method. This way, PLCs and transverse displacements are approximately taken into account.

step limits. In the second case, the position of the multiple scattering can be determined, for example, randomly using the so-called random hinge method. Figure 2.3 shows that the first CH step is limited by a hard Møller interaction. Møller interactions result in secondary electrons, also called delta electrons. The delta electrons are simulated in the same way using the CH technique until their energy falls below E_{cut} or if they leave the region of interest. Figure 2.3 also shows that the CH steps can be limited by the maximum step length or by hard bremsstrahlung events. Secondary bremsstrahlung photons are simulated like all other photons using the analog technique (see Section 2.2.1).

The various components of the CH technique are outlined in the following subsections using the electron as an example.

2.2.2.2 Continuous Energy Loss

During a CH step, the charged particle continuously loses energy due to soft interactions. The average energy loss dE per CH step length ds at point $\vec{r}$ is given by the restricted linear stopping power:

$$L(\vec{r}, E, E_c, k_c) \equiv -\left(\frac{dE}{ds}\right)_{res} = L_{col}(\vec{r}, E, E_c) + L_{rad}(\vec{r}, E, k_c) \quad (2.26)$$

with

$$L_{col}(\vec{r}, E, E_c) \equiv -\left(\frac{dE}{ds}\right)_{res,col} \quad \text{and} \quad L_{rad}(\vec{r}, E, k_c) \equiv -\left(\frac{dE}{ds}\right)_{res,rad} \quad (2.27)$$

defined as restricted linear collision and radiation stopping powers. They can be calculated using the collision cross section $\sigma_{col}(\vec{r},E,E')$ and the bremsstrahlung production cross section $\sigma_{rad}(\vec{r},E,k')$ by

$$L_{col}(\vec{r},E,E_c) = N(\vec{r})\int_0^{E_c} dE'\, E'\sigma_{col}(\vec{r},E,E') \qquad (2.28)$$

$$L_{rad}(\vec{r},E,k_c) = N(\vec{r})\int_0^{k_c} dk'\, k'\sigma_{rad}(\vec{r},E,k')$$

$N(\vec{r})$ is the number of scattering targets per unit volume at point $\vec{r}$. Note the stopping power integration is restricted to energies below E_c and k_c. That means the energy transfer to secondary charged particles is restricted to be below E_c and the energy transfer to secondary photons is restricted to be below k_c. Interactions with higher energy transfer are simulated explicitly within this CH scheme.

The step length s for an electron with initial energy E_0 that loses energy ΔE due to CH transport can be calculated by integrating Equation 2.26:

$$s = -\int_{E_0}^{E_1} \frac{dE}{L(\vec{r},E,E_c,k_c)} = \int_{E_1}^{E_0} \frac{dE}{L(\vec{r},E,E_c,k_c)} \qquad (2.29)$$

with $E_1 = E_0 - \Delta E$ being the electron energy at the end of the step. The minus sign in the definition (2.26) is used here to exchange the upper and lower integration limits. The integration (2.29) is performed along the simulated electron path and takes into account the spatial dependency of the stopping power on $\vec{r}$. That means function $L(\vec{r},E,E_c,k_c)$ changes accordingly if a boundary to a region with different material is crossed.

Equation 2.29 provides a unique functional dependency of the CH step length s and the step end energy E_1. Therefore, all electrons with initial energy E_0 and transported along the same step will reach the step end with the same energy E_1. Here, it becomes clear that this approach is an approximation because in reality, only the mean step end energy can be calculated by Equation 2.29. Analog MC simulations show that the real step end energies are random according to a distribution with a mean value of E_1. This effect is known as energy straggling. In the CH simulation scheme defined here, we have to distinguish between soft energy straggling and hard energy straggling. Hard energy straggling is simulated explicitly and the corresponding effects are correctly taken into account. Soft or subthreshold energy straggling is either neglected or it is simulated by an adequate straggling distribution function.

Indeed, subthreshold straggling can be neglected during a CH MC simulation of specific problems if we choose the parameter E_c and the step size small enough. In this case, energy straggling is dominated by the explicitly modeled hard ionization events with energy transfer larger than E_c and the low energy fluctuations have a negligible influence on the final result. Furthermore, in many cases, subthreshold energy straggling is additionally insignificant compared to the influence of the range straggling (see Section 2.2.2.4).

A potential disadvantage of neglecting soft energy fluctuations is that the calculation time can increase unnecessarily. Therefore, the implementation of some subthreshold straggling theory can help to speed up the simulation or to avoid artifacts. For example, a Gaussian distribution function according to an approach by Bohr (1948) can be used to model energy straggling. More accurate are the theories of Landau (1944) and Vavilov (1957).

2.2.2.3 Multiple Scattering

In contrast to reality and as demonstrated in Figure 2.4, in a CH approach, charged particles move on straight lines during one step. The combined effect of many small-angle elastic and semielastic collisions during one step is simulated by sampling the angular deflection based on a dedicated multiple scattering theory. An example is the theory developed by Fermi and Eyges (Eyges, 1948). This theory models the probability $p(\theta,\varphi)\,d\theta\,d\varphi$ that the electron is scattered within the solid multiple scattering angular section ($[\theta,\theta+d\theta]$, $[\varphi,\varphi+d\varphi]$) as 2D Gaussian distribution:

$$p(\theta,\varphi)d\theta\,d\varphi = \frac{\theta}{\pi\overline{\theta^2}(s)}\exp\left(-\frac{\theta^2}{\overline{\theta^2}(s)}\right)d\theta\,d\varphi, \qquad (2.30)$$

where θ is the azimuthal multiple scattering angle, φ is the polar multiple scattering angle, and $\overline{\theta^2}(s)$ is the mean square deflection angle after step length s. The distribution (2.30) is not normalized because θ is limited to the interval $[0,\pi]$ instead of $[0,\infty]$. Equation 2.30 results in two separate cumulative distribution functions:

$$P_\theta(\theta) = 1 - \exp\left(-\frac{\theta^2}{\overline{\theta^2}(s)}\right), \quad P_\varphi(\varphi) = \frac{\varphi}{2\pi}. \qquad (2.31)$$

The transformation method from Section 2.1.3 and a uniform random number ξ_θ can be applied to sample θ:

$$\theta = \sqrt{-\overline{\theta^2}(s)\ln(1-\xi_\theta)}. \qquad (2.32)$$

It has to be noted that ξ_θ is uniformly distributed in an interval $\left[0,\xi_\theta^{max}\right]$ with $\xi_\theta^{max} < 1$ to ensure $\theta \leq \pi$. Another option is to sample ξ_θ from $[0,1]$ and to reject θ if $\theta > \pi$. Because of the rotational symmetry, the polar scattering angle φ is determined from a uniform distribution in $[0,2\pi]$.

The quantity $\overline{\theta^2}(s)$ is calculated using the linear scattering power $T_s(\vec{r},E)$ at point $\vec{r}$:

$$\overline{\theta^2}(s) = \int_0^s ds'\, T_s(s',E). \qquad (2.33)$$

The material parameter $T_s(\vec{r},E)$ depends on the atomic composition at point $\vec{r}$ but also on the electron energy E. This must be taken into account in Equation 2.33 because the electron loses energy between the beginning and the end of the step.

The Gaussian (2.30) is a good approximation of the reality for small cumulative scattering angles θ. However, large scattering angles are underestimated using this distribution. Therefore, very often in the past (e.g., in EGS4 (Nelson et al., 1985)) the multiple scattering distribution of Molière (1948) was used. Nevertheless, this improvement is also based on the small-angle approximation, and especially for large angles, the Molière theory still has limitations.

Most suitable for MC simulations are algorithms based on the exact theory of Goudsmit and Saunderson (1940a, b). Examples are the multiple scattering algorithms as implemented in MCNP (Briesmeister, 1997), Penelope (Salvat et al., 2011), and EGSnrc (Kawrakow et al., 2011). These algorithms sample multiple scattering angles close to reality even for large angles.

2.2.2.4 Transport Mechanics

Figure 2.4 compares a CH electron step to some possible real electron path or to an electron path simulated using analog MC transport. The figure shows that several problems can be expected from a CH simulation. For better comprehension, in this section, we neglect energy loss during the CH step. That means here the electron energy has not changed between the beginning and the end of the step and all interactions during the step are assumed to be elastic. As shown in Figure 2.4, many MC codes sample the multiple scattering angle at the end of the step, that is, the electron moves on a straight line until the final position, and then, it changes the direction due to multiple scatter. It is obvious that in this case, the electron distance is overestimated. Also shown in Figure 2.4, the real electron path is curved. Therefore, the real electron range is shorter if we assume that both electrons (the real electron and the CH electron) move with the same path length. Furthermore, the real electron range fluctuates around some mean value. This effect is called range straggling and should not be confused with energy straggling as discussed in Section 2.2.2.2. Range straggling is independent of the energy loss of the electron.

Many CH history algorithms employ a path length correction (PLC) algorithm to take into account range overestimation and range straggling. Besides this, a transverse displacement (TD) algorithm is useful to take into account transverse fluctuations of the real electron end position relative to the lateral position as simulated during the CH step. A very simple PLC and TD approach, called random hinge method, is demonstrated in Figure 2.5. It is implemented, for example, in Penelope (Salvat et al., 2011) or XVMC (Kawrakow and Fippel, 2000). In this method, the whole step length s is subdivided using a uniformly distributed random number ξ from interval [0,1] into two substeps $(1-\xi)s$ and $(1-\xi)s$. The multiple scattering angle is sampled between the two substeps instead at the end of the step. The average longitudinal (PLC) and lateral (TD) displacements simulated in this manner are very close to the exact values. More accurate is the parameter-reduced electron-step transport algorithm developed by Kawrakow and Bielajew (1998) (sometimes called PRESTA-II) as implemented in EGSnrc (Kawrakow, 2000a).

This algorithm reproduces first- and second-order spatial moments to within 0.1%.

The problems and algorithms discussed in this section could be neglected completely if the charged particle transport steps are small enough. In the limit of infinitesimal small step sizes, any CH algorithm will converge to the correct answer. However, in this way, the simulation becomes extremely inefficient. Therefore, with the implementation of a sophisticated algorithm for the transport mechanics, charged particle transport can be used efficiently for MC simulations in radiation therapy.

2.2.2.5 Boundary Crossing

Theoretically, charged particle MC transport using the CH technique with multiple scattering works only in homogeneous regions containing one definite medium. The problem of conditions with more than one material is shown schematically in Figure 2.6. It shows two regions consisting of different materials as well as two electron tracks, a real electron track calculated using the single scatter scheme and a CH-simulated electron path. By definition, the CH technique shall simulate the condensed effect of a huge number of real electron paths, including the real path shown in Figure 2.6. This path is partly located in material II and has to take into account (for this part of the path) the interaction properties of material II. For the rest of the path, it has to consider the properties of material I. Other real electrons may be tracked completely in material I.

Figure 2.6 clearly demonstrates that, for arbitrarily shaped material interfaces, an exact theory of multiple scattering does not exist. Consequently, the CH technique can only be used in regions where the distance to the next material boundary is much larger than the size of the present electron step. Conversely, this means that the maximum step size s_{max} must be decreased if an electron comes close to a material interface and it can be increased again if the electron moves away from the boundary. This kind of step-size variation has been introduced, for example, in the original version of PRESTA (Bielajew and Rogers 1987). For geometries with many small regions of different materials, this leads to small step sizes and long calculation times. A further problem is that before an

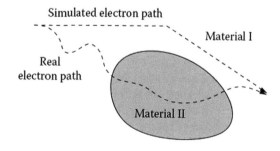

FIGURE 2.6 Example of an electron path simulated using the random hinge method. In contrast to Figures 2.4 and 2.5, two different materials are involved. This can cause simulation artifacts because the electron path simulated using the CH technique is not influenced by material II; however, the real electron path is influenced by material II.

electron crosses the boundary, its CH step size becomes infinitesimally small. To permit boundary crossing anyway, the step-size reduction must be switched off if a minimum step size is achieved. However, in this case, the step size becomes similar to the boundary distance and step-size artifacts may occur as shown in Figure 2.6 and the discussion above.

To avoid step-size artifacts, a more accurate algorithm has been implemented in EGSnrc (Kawrakow, 2000a). Here, a minimum distance d_{min} to the next material interface can be defined by the user and CH transport is performed only if the electron is farther away from the boundary than this minimum distance. If the electron comes closer to the boundary than d_{min}, the electron is transported in single scatter mode using the analog technique. This algorithm is free of step-size artifacts. However, it also means that electrons are simulated in pure single scatter mode within geometries with very small regions, that is, if the average diameter of one region is smaller than d_{min}. Consequently, calculation times can increase dramatically. Therefore, this technique including an appropriate choice of parameter d_{min} makes sense only for simulations with high accuracy requirements. An example is the simulation of ion chamber responses (Kawrakow, 2000b).

For the majority of MC simulations in radiation therapy, artifacts due to boundary crossing can be ignored because of their negligible influence on the result. A useful approach is, for example, to simulate electron boundary crossing without stopping at the interface using the random hinge technique. Here, the multiple scattering properties of material I are used if the "hinge" is sampled in material I; otherwise, the properties of material II are used. The total electron-step length depends on the stopping powers and the corresponding step segments in both materials. According to the Penelope user guide (Salvat et al., 2011), this approach provides a "fairly accurate description of interface crossing." It is, for example, an efficient and accurate method for linear accelerator head modelling or an MC-based dose calculation algorithm in radiation therapy treatment planning.

2.2.3 Cross Sections

2.2.3.1 Photon Interaction Coefficients

In Section 2.2.1, the linear attenuation coefficient $\mu(E)$ has been introduced to sample the distance between two photon interactions. For elements, compounds, and mixtures, this material parameter can be calculated from the total cross sections $\sigma_i(E)$ of the *i*th element in the material as

$$\mu(E) = \sum_i N_i(\vec{r})\sigma_i(E). \tag{2.34}$$

$N_i(\vec{r})$ is the number of atoms of element *i* per unit volume at point $\vec{r}$ and is calculated by

$$N_i(\vec{r}) = \frac{\rho(\vec{r})w_i(\vec{r})}{m_u A_i(\vec{r})} = \frac{\rho(\vec{r})N_A w_i(\vec{r})}{M_i(\vec{r})}. \tag{2.35}$$

where

$\rho(\vec{r})$ is the mass density at point $\vec{r}$, $w_i(\vec{r})$ is the weight fraction of the *i*th element at point $\vec{r}$, $A_i(\vec{r})$ is the relative atomic mass of the *i*th element at point $\vec{r}$, $M_i(\vec{r})$ is molar mass of the *i*th element at point $\vec{r}$, $m_u = 1.6605388 \cdot 10^{-27}$ kg is the atomic mass unit, and $N_A = 6.0221418 \cdot 10^{23}\,\text{mol}^{-1}$ is the Avogadro constant (Nakamura et al. 2010).

Corresponding to Equation 2.23, the total atomic cross sections $\sigma(E)$ are given by the sum of the cross sections for the different processes:

$$\sigma_i(E) = \sigma_{i,A}(E) + \sigma_{i,R}(E) + \sigma_{i,C}(E) + \sigma_{i,P}(E), \tag{2.36}$$

that is, photoelectric absorption (*A*), Raleigh scatter (*R*), Compton scatter (*C*), and pair production (*P*).

If the material composition is exactly known, then the cross sections for the different photon interactions can be calculated using some database, such as XCOM (Berger and Hubbell, 1987) or EPDL97 (Cullen et al., 1997). These databases are available online, and they are updated from time to time to include the most recent measurement and calculation results. Most of the general-purpose and specific MC codes are based on one or more of these databases.

Especially in radiation therapy, the material compositions are not always available. Very often, only a CT number (or Hounsfield unit, HU) is known. Therefore, the International Commission on Radiation Units and Measurements (ICRU) has compiled a list of about 100 human tissue types, including material compositions and interaction data (e.g., photon cross sections). These data are published in ICRU Report 46 (*Photon, Electron, Proton and Neutron Interaction Data for Body Tissues*, 1992). With an adequate CT calibration, an HU number can be mapped to a specific body tissue type. The use of a very limited number of materials, for example, six materials such as air, lung, water, soft tissue, soft bone, and hard bone, should be avoided. Linear interpolation between these supporting points can lead to large cross-sectional inaccuracies.

The interaction probabilities $\mu_k(E, \rho)$ ($k = A, R, C, P$) of Equation 2.23 for human tissue can also be determined more directly, that is, without knowing the material compositions, using an approach presented for MC dose calculation (Kawrakow et al., 1996; Fippel, 1999):

$$\mu_k(E, \rho) = \frac{\rho}{\rho^w} f_k(\rho)\mu_k^w(E). \tag{2.37}$$

where

$\mu_k^w(E)$ are the interaction coefficients of water and ρ^w is the mass density of water. As an example, in Figure 2.7, the Compton cross-sectional ratios $f_C(\rho)$ for all materials of ICRU Report 46 are plotted. It shows that most of the data points (crosses) are located close to the solid line representing the fit function:

$$f_C(\rho) \approx \begin{cases} 0.99 + 0.01\rho/\rho^w, & \rho \leq \rho^w \\ 0.85 + 0.15\rho^w/\rho, & \rho \leq \rho^w. \end{cases} \tag{2.38}$$

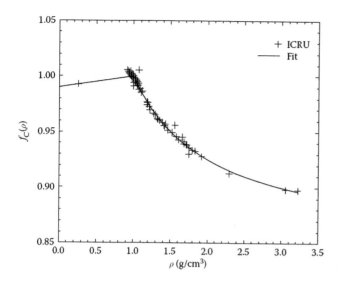

FIGURE 2.7 Ratios of mass Compton interaction coefficients of all materials from ICRU Report 46 to the Compton interaction coefficient of water (crosses). The line represents the fit according to Equation 2.38.

The relative uncertainty of function (2.38) is better than 1% for almost all materials of ICRU Report 46. The only exception is gallstone with an error of 1.6%. Similar fits exist for the other interaction types. Thus, it is sufficient to determine only the mass density in a given region and to use equations like (2.38) to calculate the corresponding cross sections.

2.2.3.2 Charged Particle Stopping and Scattering Powers

For the MC simulation of charged particle transport using the CH technique according to Section 2.2.2.2, knowledge of the linear restricted collision and radiation stopping powers $L_{col}(\vec{r}, E, E_c)$ and $L_{rad}(\vec{r}, E, k_c)$ is necessary. These quantities can be calculated if the unrestricted collision and radiation stopping powers $S_{col}(\vec{r}, E)$ and $S_{rad}(\vec{r}, E)$ for the medium at point $\vec{r}$ are known. They provide the average energy loss dE per step ds without restricting the energy loss by some threshold. The total unrestricted linear stopping power is defined by

$$S(\vec{r}, E) \equiv -\left(\frac{dE}{ds}\right) = S_{col}(\vec{r}, E) + S_{rad}(\vec{r}, E). \quad (2.39)$$

The unrestricted linear collision and radiation stopping powers are calculated using the collision cross section $\sigma_{col}(\vec{r}, E, E')$ and the bremsstrahlung production cross section $\sigma_{rad}(\vec{r}, E, k')$ by

$$S_{col}(\vec{r}, E) = N(\vec{r}) \int_0^E dE' E' \sigma_{col}(\vec{r}, E, E') \quad (2.40)$$

$$S_{rad}(\vec{r}, E) = N(\vec{r}) \int_0^E dk' k' \sigma_{rad}(\vec{r}, E, E'). $$

where

$N(\vec{r})$ is the number of scattering targets per unit volume at point $\vec{r}$ and can be determined by Equation 2.35.

The mixed CH technique as in Section 2.2.2.2 requires the knowledge of the hard inelastic collision and bremsstrahlung cross sections $\sigma_{col}(\vec{r}, E, E')$ and $\sigma_{rad}(\vec{r}, E, k')$ for energy transfer values $E' > E_c$ and $k' > k_c$, respectively. With Equations 2.28 and 2.40, this allows the calculation of the restricted stopping powers according to

$$L_{col}(\vec{r}, E, E_c) = S_{col}(\vec{r}, E) - N(\vec{r}) \int_{E_c}^E dE' E' \sigma_{col}(\vec{r}, E, E') \quad (2.41)$$

$$L_{rad}(\vec{r}, E, k_c) = S_{rad}(\vec{r}, E) - N(\vec{r}) \int_{k_c}^E dk' k' \sigma_{rad}(\vec{r}, E, k'). $$

Stopping power tables suitable for MC simulations in radiation therapy have been published, for example, by ICRU for electrons and positrons in Report 37 (*Stopping Powers for Electrons and Positrons*, 1984), for electrons and protons in Report 46 (*Photon, Electron, Proton and Neutron Interaction Data for Body Tissues*, 1992), as well as for protons and alpha particles in Report 49 (*Stopping Powers and Ranges for Protons and Alpha Particles*, 1993). Online databases like ESTAR, PSTAR, and ASTAR (Berger, 1993) can also be used to determine stopping powers for different elements, compounds, and mixtures.

Comparable to Equation 2.37, the stopping powers can also be determined more directly using the mass stopping power ratios $f_k(E, \rho)$ and the mass stopping power of water $S_k^w(E)/\rho^w$ ($k = $ tot, col, rad,...):

$$S_k(E, \rho) = \frac{\rho}{\rho^w} f_k(E, \rho) S_k^w(E). \quad (2.42)$$

This approach has been adopted, for example, to develop an MC dose calculation algorithm for proton therapy (Fippel and Soukup, 2004). Figure 2.8 shows the total proton stopping power ratio $f_s(E, \rho)$ for two proton energies (10 and 100 MeV) and for all materials of ICRU Report 46 (*Photon, Electron,*

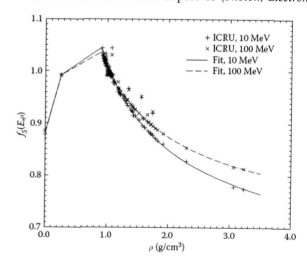

FIGURE 2.8 Mass stopping power ratios of all materials from ICRU Report 46 to the mass stopping power of water for 10 and 100 MeV protons (crosses). The lines represent the fit for both energies.

Proton and Neutron Interaction Data for Body Tissues, 1992). After analyzing the tabulations and calculations of ICRU Report 46 *(Photon, Electron, Proton and Neutron Interaction Data for Body Tissues, 1992)*, ICRU Report 49 *(Stopping Powers and Ranges for Protons and Alpha Particles, 1993)*, and the online database PSTAR (Berger, 1993), the following fit formula has been published (Fippel and Soukup, 2004):

$$f_s(E, \rho) = \begin{cases} 1.0123 - 3.386 \cdot 10^{-5} E + 0.291\left(1 + E^{-0.3421}\right) \\ \left(\rho^{-0.7} - 1\right) \text{ for } \rho \geq 0.9 \\ 0.9925 \quad \text{for } \rho = 0.26 \text{(lung)} \\ 0.8815 \quad \text{for } \rho = 0.0012 \text{(air)} \\ \text{interpolate for all other } \rho \leq 0.9 \end{cases} \quad (2.43)$$

with the kinetic proton energy E in MeV (rest mass not included) and mass density ρ in g/cm^3. Figure 2.8 shows this function for the two proton energies as solid and dashed lines. Besides some outliers (gallstone, protein, carbohydrate, and urinary stone), Equation 2.43 provides an accuracy of better than 1%, that is, knowledge of the atomic composition in each voxel is not necessary. Equation 2.43 and Figure 2.8 also show that an energy dependence of the stopping power ratio can be taken into account by the fit formulas.

Similar formulas exist for electron and positron collision and radiation stopping powers. The material and mass density dependence of the charged particle multiple scattering distributions can also be modeled this way.

References

Agostinelli, S. et al. 2003. GEANT4—A simulation toolkit, *Nucl. Instrum. Methods A* **506**: 250–303.

Berger, M. J. 1963. Monte Carlo calculation of the penetration and diffusion of fast charged particles, *Methods in Computational Physics*, Vol. I, Academic Press, New York, pp. 135–215. eds. B. Alder, S. Fernbach, and M. Rotenberg.

Berger, M. J. 1993. *ESTAR, PSTAR, and ASTAR: Computer Programs for Calculating Stopping-Power and Range Tables for Electrons, Protons, and Helium Ions*, Technical Report NBSIR 4999, National Institute of Standards and Technology, Gaithersburg, MD.

Berger, M. J. and Hubbell, J. H. 1987. *XCOM: Photon Cross Sections on a Personal Computer*, Technical Report NBSIR 87-3597, National Institute of Standards and Technology, Gaithersburg, MD.

Bielajew, A. F. and Rogers, D. W. O. 1987. PRESTA: The parameter reduced electron-step transport algorithm for electron Monte Carlo transport, *Nucl. Instrum. Methods B* **18**: 165–181.

Bohr, N. 1948. The penetration of atomic particles through matter, *Det Kongelige Danske Videnskabernes Selskab Matematisk-Fysiske Meddelelser* **18**(8): 1–144.

Briesmeister, J. F. 1997. *MCNP—A General Monte Carlo N-Particle Transport Code*, Report No. LA-12625-M, Los Alamos National Laboratory.

Chetty, I. J. et al. 2007. Report of the AAPM Task Group No. 105: Issues associated with clinical implementation of Monte Carlo-based photon and electron external beam treatment planning, *Med. Phys.* **34**(12): 4818–4853.

Cullen, D. E., Hubbell, J. H., and Kissel, L. 1997. *EPDL97: The Evaluated Photon Data Library, '97 Version*, Technical Report UCRL-50400, Vol 6, Rev 5, Lawrence Livermore National Laboratory, Livermore, CA.

Eyges, L. 1948. Multiple scattering with energy loss, *Phys. Rev.* **74**: 1534–1535.

Fippel, M. 1999. Fast Monte Carlo dose calculation for photon beams based on the VMC electron algorithm, *Med. Phys.* **26**(8): 1466–1475.

Fippel, M. and Soukup, M. 2004. A Monte Carlo dose calculation algorithm for proton therapy, *Med. Phys.* **31**(8): 2263–2273.

Goudsmit, S. and Saunderson, J. L. 1940a. Multiple scattering of electrons, *Phys. Rev.* **57**: 24–29.

Goudsmit, S. and Saunderson, J. L. 1940b. Multiple scattering of electrons II, *Phys. Rev.* **58**: 36–42.

James, F. 1980. Monte Carlo theory and practice, *Rep. Prog. Phys.* **43**(9): 1145–1189.

Jenkins, T. M., Nelson, W. R., and Rindi, A. (eds) 1988. *Monte Carlo Transport of Electrons and Photons*, Plenum Press, New York and London.

Kawrakow, I. 2000a. Accurate condensed history Monte Carlo simulation of electron transport, I. EGSnrc, the new EGS4 version, *Med. Phys.* **27**: 485–498.

Kawrakow, I. 2000b. Accurate condensed history Monte Carlo simulation of electron transport, II. Application to ion chamber response simulations, *Med. Phys.* **27**: 499–513.

Kawrakow, I. and Bielajew, A. F. 1998. On the condensed history technique for electron transport, *Nucl. Instrum. Methods B* **142**: 253–280.

Kawrakow, I. and Fippel, M. 2000. Investigation of variance reduction techniques for Monte Carlo photon dose calculation using XVMC, *Phys. Med. Biol.* **45**: 2163–2183.

Kawrakow, I., Fippel, M., and Friedrich, K. 1996. 3D electron dose calculation using a Voxel based Monte Carlo algorithm (VMC), *Med. Phys.* **23**(4): 445–457.

Kawrakow, I., Mainegra-Hing, E., Rogers, D. W. O., Tessier, F., and Walters, B. 2011. *The EGSnrc Code System: Monte Carlo Simulation of Electron and Photon Transport*, NRCC Report PIRS-701, National Research Council Canada, Ottawa.

Landau, L. 1944. On the energy loss of fast particles by ionization, *J. Exp. Phys. USSR* **8**: 201–205.

Lüscher, M. 1994. A portable high-quality random number generator for lattice field theory simulations, *Comp. Phys. Commun.* **79**(1): 100–110.

Marsaglia, G. and Zaman, A. 1991. A new class of random number generators, *Ann. Appl. Probab.* **1**(3): 462–480.

Molière, G. Z. 1948. Theorie der Streuung schneller gelad-
ener Teilchen. 2. Mehrfach- und Vielfachstreuung, *Z.
Naturforschung A* **3**: 78–97.

Nakamura, K. et al. 2010. (Particle Data Group) The review
of particle physics, *J. Phys. G* **37**: 075021. Available on
Particle Data Group WWW pages (http://pdg.lbl.gov).

Nelson, W. R., Hirayama, H., and Rogers, D. W. O. 1985.
The EGS4 Code System, SLAC Report No. SLAC-265,
Stanford Linear Accelerator Center.

*Photon, Electron, Proton, and Neutron Interaction Data
for Body Tissues* 1992. ICRU Report 46, International
Commission on Radiation Units and Measurements.

Press, W. H., Teukolsky, S. A., Vetterling, W. T., and Flannery,
B. P. (eds) 1992. *Numerical Recipes in C*, Cambridge
University Press, Cambridge, New York, Port Chester,
Melbourne, Sydney.

Reynaert, N. et al. 2007. Monte Carlo treatment planning for
photon and electron beams, *Rad. Phys. Chem.* **76**(4):
643–686.

Salvat, F., Fernández-Varea, J.M., and Sempau, J. 2011.
PENELOPE-2011: A code system for Monte Carlo simulation of electron and photon transport, nuclear energy
agency. NEA/NSC/DOC20115, *Workshop Proceedings*,
Barcelona, Spain, 4–7 July 2011.

Stopping Powers and Ranges for Protons and Alpha Particles
1993. ICRU Report 49, International Commission on
Radiation Units and Measurements.

Stopping Powers for Electrons and Positrons 1984. ICRU Report
37, International Commission on Radiation Units and
Measurements.

Vavilov, P. V. 1957. Ionization losses of high-energy heavy
particles, *Soviet Phys. JETP* **5**: 749–751.

3

Variance Reduction Techniques

Matthias Fippel

Brainlab AG

3.1 Introduction

Monte Carlo (MC) calculations can be time consuming, especially for applications in radio therapy (RT). Therefore, the algorithmic techniques to speed up the simulations are essential. These techniques are called variance reduction techniques (VRTs). In this chapter, VRTs with a special focus on RT are introduced and explained. For clarity, it is written from the point of view of photon–electron interactions. However, the same concepts can be applied for heavy charged and neutral particles (protons, neutrons, etc.).

Additional information about VRTs is available, for example, in the book chapters by Bielajew and Rogers (1988) or Sheikh-Bagheri et al. (2006), in an article by Kawrakow and Fippel (2000a), and in two task group reports (Chetty et al., 2007; Reynaert et al., 2007). This literature also contains references of the original publications for most of the VRTs presented in the following sections.

3.1.1 Calculation Efficiency

Depending on the number of histories N, the accuracy of any MC calculated mean value $\langle f(N) \rangle$ of quantity f is limited by its statistical uncertainty. This uncertainty is given by the variance $\sigma(N)$ and provides a measure of the statistical fluctuations of the calculated mean value $\langle f(N) \rangle$ around the true value f of that quantity. It is obvious that $\sigma(N)$ decreases with increasing number of histories N and it becomes zero if N approaches infinity. In general, $\sigma(N)$ cannot be calculated because the true value f is unknown. On the other hand, an estimated variance $s(N)$ can be calculated during an MC simulation by

$$s(N) = \sqrt{\frac{\langle f^2(N) \rangle - \langle f(N) \rangle^2}{N-1}}, \tag{3.1}$$

with $\langle f^2(N) \rangle$ being the MC calculated mean of f^2. The best estimate of the variance is obtained if $\langle f(N) \rangle$ and $\langle f^2(N) \rangle$ are calculated using the history-by-history method; that is, they are calculated by averaging over all histories (Salvat et al., 2011). $\langle f(N) \rangle$ and $\langle f^2(N) \rangle$ tend to become constant for large numbers of N. Therefore, Equation 3.1 provides a simple method to reduce the variance just by increasing the number of histories N, that is, increasing the calculation time $T(N)$. However, this is not considered to be a VRT. The purpose of variance reduction is to decrease the time of MC simulations by modifying the algorithm while maintaining an unbiased estimate of the variance $s(N)$. Unbiased means that for any realistic history number N, the result of MC, including VRT,

DOI: 10.1201/9781003211846-4

must not deviate systematically from the corresponding result without VRT.

Instead of VRT, the methods pointed out in this book should be called efficiency enhancement techniques because they improve the efficiency of MC simulations. The calculation efficiency ε is defined by

$$\varepsilon = \frac{1}{\left[s(N)\right]^2 T(N)}. \tag{3.2}$$

From Equation 3.1, it follows that $[s(N)]^2$ becomes proportional to $1/N$ for large N. $T(N)$ is proportional to N. Therefore, the efficiency ε is almost independent of N. The calculation efficiency can be improved by reducing the variance $s(N)$ for a given number of histories N, by decreasing the calculation time $T(N)$ for a given number of histories N, or by doing both. For historical reasons, we will continue to call these techniques VRT throughout this chapter.

3.1.2 Hardware Performance Improvements

The calculation time can be decreased simply by using faster computers or by implementing parallel calculation processes on multicore workstations and computing clusters. However, these methods are not called VRT because they do not make the underlying MC algorithm faster. They just use a given software on a hardware with better performance. Especially, the parallelization of MC calculations is straightforward. Therefore, it is expected that any serious MC algorithm will fully exploit the advantages of present-day computing hardware.

3.1.3 Approximate Methods

In the majority of cases, the calculation time per history is decreased by making approximations. Even if the final result is not affected in a significant way, in fact, the approximate methods do not belong to VRTs. According to its original definition, VRTs must not influence the expected result of an infinitely long MC simulation. However, in literature, approximate methods are often denoted as VRT. Some of them, for example, the condensed history technique (CHT), form the basis of almost all MC calculations in radiation therapy today. Therefore, in this chapter, the approximate methods will also be discussed. It is possible to distinguish between real variance reduction techniques (RVRT) and approximate variance reduction techniques (AVRT).

It is obvious that some techniques are AVRTs, for example, the continuous slowing down approximation (CSDA) of the electron transport. On the other hand, it can be difficult to decide whether a given technique belongs to the RVRTs or to the AVRTs. Sometimes, this decision is purely based on intuition or it is shown numerically that some technique is an RVRT. However, numerical experiments are debatable with this respect because the results of both, RVRT and AVRT, must not be influenced significantly. Only by mathematical

proof, it can be shown that a definite technique is an RVRT. These proofs can be complex; that is, they would cover a significant amount of space in this chapter. Hence, they will not be shown here. This book focuses on the use of MC techniques in radiation therapy. For both types of VRTs, this means that they have to speed up the simulations without significant loss of numerical accuracy. Therefore, it makes sense in the following sections not to demonstrate for all cases that a specific VRT is an AVRT or an RVRT.

3.1.4 Condensed History Electron Transport

The majority of MC algorithms in RT perform electron transport using the CHT. The CHT is an AVRT, and it includes approximations and speeds up MC electron transport significantly compared to analog simulation. Because of its fundamental nature, an in-depth explanation of the CHT is provided in Chapter 2.

3.2 Basic Variance Reduction Techniques

In this section, the basic VRTs are outlined. They provide an introduction into the elemental methods of how MC simulation times can be reduced efficiently and what has to be considered to avoid bias of the results. Some of the basic VRTs can be combined to form a more advanced VRT (see Section 3.3).

3.2.1 Uniform Particle Splitting

A common example to explain variance reduction is the particle splitting technique. It can be applied to photons as well as to charged particles. Figure 3.1 demonstrates the splitting technique applied to bremsstrahlung photons. The left drawing represents the MC simulation of an electron hitting the target of a medical linear accelerator (LINAC) and producing one bremsstrahlung photon. For simplicity, we consider the case of only one bremsstrahlung photon because, in general,

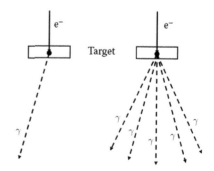

FIGURE 3.1 Schematic representation of an MC simulation of an electron producing one bremsstrahlung photon in the target of a medical linear accelerator (left). On the right-hand side, five bremsstrahlung photons instead of one are created using the splitting technique, each with a statistical weight of $w = 1/5$.

several photons can be produced by one electron in the target. In a "normal" (also called analog) MC simulation, the photon carries a statistical weight of $w = 1$. This means that one realistic photon is represented by one photon in the simulation. The right drawing represents the case with a particle splitting factor of $N_{split} = 5$; that is, instead of one, five independent bremsstrahlung photons are sampled from the same bremsstrahlung production distribution of the given electron. The VRT is called uniform bremsstrahlung splitting (UBS) if all photons are created with the same probability independent of their energy and direction. Then, each of the five photons has to carry a statistical weight of $w = 1/N_{split} = 0.2$; that is, one realistic photon is represented by five photons in the simulation to preserve the total weight.

The reason for bremsstrahlung splitting is given by the purpose of the MC simulation. Two examples are as follows: (1) the bremsstrahlung spectrum of the LINAC target shall be calculated or (2) the target component is part of a full LINAC head simulation with the final goal of dose calculation in a phantom or patient. In both cases, we are interested in a large number of photons to increase the statistical accuracy by lowering the variance. The splitting technique reduces the time required for the creation of five photons in the example of Figure 3.1 because the transport simulation of four additional electrons is saved. In other words, during an analog MC simulation, a lot of time would be "wasted" for the electron transport. However, almost all electrons are absorbed within the target; that is, besides the creation of bremsstrahlung, they do not really contribute to the final result.

Particle splitting is known to be an RVRT, although a proof is not shown here. In the limit of infinite history numbers (number of electrons hitting the target), the results of the analog MC and simulations with photon splitting are identical. This fact is not obvious and it can be wrong if not implemented correctly. An incorrect implementation would arise, for example, if the energy of the electron is reduced by the weighted sum of all split photon energies (five energies in Figure 3.1). Such an approach would bias the shape of the energy spectrum (also known as straggling) of the electrons after the first bremsstrahlung event. Since these electrons can produce secondary and more bremsstrahlung photons, all further results will be influenced by this mistake. To avoid this effect, usually the energy of the electron is reduced just by the energy of one photon arbitrarily selected from all splitting photons. Obviously, the energy is not conserved within one history using this method. On the other hand, the energy conservation law is still fulfilled on an average for large history numbers.

It is important to note that one should not mix analog MC and splitting or splitting with very different splitting numbers thoughtlessly in one simulation. This could create photons of very different statistical weight, that is, "fat" photons with high weight and "meager" photons with low weight. Only the "fat" photons would contribute to the statistical accuracy of the final result, and the time for simulating the "meager" photons would be wasted. A smart implementation of any VRT can produce particles of a very different statistical weight in an intermediate state; however, when the particle properties for the final calculation result are analyzed or processed in a further simulation, it is most efficient if the statistical weight of all the particles is similar.

3.2.2 Russian Roulette

Russian roulette can be considered as the opposite of particle splitting. Very often, both techniques are used in combination. In a Russian roulette technique, for a definite particle type (photon or charged particle), a survival probability $p_{survive}$ with $p_{survive} \ll 1$ is defined. If a particle of this type is created in an MC simulation, a random number ξ is sampled from a uniform distribution in interval [0,1]. The particle survives if $\xi < p_{survive}$; otherwise, it is killed, that is, the simulation of this particle stops. To stay in correspondence with reality, the statistical weight of the surviving particles must be increased by the factor $w = 1/p_{survive}$.

Russian roulette can be applied efficiently, for example, during the simulation of a LINAC head geometry in combination with the splitting technique described in Section 3.2.1. The bremsstrahlung photons with weight $w = 1/N_{split}$ created within the target can hit the secondary collimator of the LINAC. One possible interaction is Compton scattering in the high-Z absorbing material leading to the creation of Compton electrons with weight $w = 1/N_{split}$. Most of these electrons are absorbed without producing any secondary (bremsstrahlung) radiation, and the calculation time to transport these electrons would be wasted if all of them are simulated. Therefore, it makes sense to kill these electrons with a probability of $1 - p_{survive} = 1 - 1/N_{split}$. The weight of the surviving electrons must be increased by the factor $1/p_{survive} = N_{split}$; that is, finally they carry a weight of $w = 1$ like the original electrons hitting the bremsstrahlung target. Afterward, it can be useful to again apply splitting if the Compton electrons produce bremsstrahlung. However, in this case, the simulation of these photons makes sense only if they reach the region of interest (ROI). Therefore, some more advanced techniques other than uniform splitting should be the method of choice (see, e.g., Section 3.3.2).

3.2.3 Range Rejection

The Compton electrons created in the example of Section 3.2.2 can also be omitted using range rejection instead of Russian roulette. For this VRT, the shortest distance from the present charged particle position to the region boundary must be calculated and compared to the maximum range of that particle in the regions material. Therefore, this technique can only be applied to particles with a definite maximum range depending on the energy, that is, to charged particles. If this range is smaller than the shortest distance to the region boundary, then the charged particle can never leave the present region and it is useful to stop the simulation here.

Obviously, the range rejection is an AVRT because a possible production of bremsstrahlung photons is neglected with this approach and photons have a finite probability of leaving the present region. It is of course possible to correctively take these photons into account by adding them via sampling from an approximate distribution.

3.2.4 Cross-Section Enhancement

For the simulation of an ion chamber or an air cavity in water with the focus on the energy absorbed in air, it is beneficial to artificially increase the total photon cross section by some factor $N_{enhance}$ in a predefined region around the chamber or cavity. Therefore, the number of photon interactions increases by factor $N_{enhance}$ in that region causing a correspondingly increased electron fluence. To maintain an unbiased simulation, the weight of all secondary particles produced in these interactions has to be reduced by multiplication with factor $w = 1/N_{enhance}$.

3.2.5 Interaction Forcing

Comparable to cross-section enhancement and also applicable to photons only is the interaction-forcing method, schematically represented in Figure 3.2. This is a method that can be used for a photon MC dose calculation engine in RT treatment planning. In Figure 3.2, an incoming photon hits the calculation grid surface at the point denoted as **Start**. Then,

it is traced along the line denoted with **s** until the interaction point. In a worst-case scenario, this point can be outside the calculation grid, that is, behind the **Stop** position and the photon does not contribute to the tally. Thus, the time spent on sourcing the photon and transporting it through the calculation grid is wasted. To avoid wasting of calculation time, the photon can be forced to interact between **Start** and **Stop**.

Interaction forcing is possible if the number of mean free photon path lengths $\Lambda = \sum_{Start}^{Stop} \mu_i s_i$ between **Start** and **Stop** can be calculated easily and fast for all crossed voxels i. Here, μ_i is the linear attenuation coefficient in voxel i and s_i is the photon step length in voxel i. Then, the number of mean free photon path lengths λ can be selected from the distribution function

$$p(\lambda)d\lambda = \frac{1}{1-e^{-\Lambda}}e^{-\lambda}d\lambda, \qquad (3.3)$$

with λ restricted to the interval $[0,\Lambda]$. That is, with a uniformly distributed random number ξ from interval $[0,1]$, the number of mean free photon path length λ to the interaction point is calculated by

$$\lambda = -\ln\left[1 - \xi\left(1 - e^{-\Lambda}\right)\right]. \qquad (3.4)$$

The distance to the interaction point P is then determined by tracing the photon along the line s until $\lambda = \sum_{Start}^{P} \mu_i s_i$. Photon forcing requires a weight change of the photon using the factor

$$w = 1 - e^{-\Lambda}, \qquad (3.5)$$

that is, the photon weight decreases.

In general, each voxel contains a different medium with a different attenuation coefficient μ_i. Therefore, the interaction forcing is useful only if Λ can be calculated fast enough. Otherwise, the MC calculation efficiency can decrease rather than increase.

3.2.6 Exponential Transform

The exponential transform works in a similar manner as interaction forcing. Here, the exponential depth distribution of photons is stretched or shortened by some factor $F > 0$; that is, the number of mean free photon path lengths λ is sampled from the distribution

$$p(\lambda)d\lambda = \frac{1}{F}e^{-\lambda/F}d\lambda. \qquad (3.6)$$

A uniformly distributed random number ξ from interval $[0,1)$ provides the number of mean free photon path length λ to the interaction point by

$$\lambda = -F\ln(1 - \xi), \qquad (3.7)$$

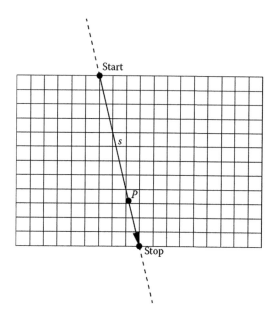

FIGURE 3.2 Forcing the interaction of one simulated photon within the region of interest, that is, between the **Start** and **Stop** positions. The number of mean free photon path lengths $\Lambda = \sum_{Start}^{Stop} \mu_i s_i$ between **Start** and **Stop** must be calculated for all crossed voxels i in advance. This can be time consuming because, in general, each voxel contains a different medium with a different attenuation coefficient μ_i.

that is, for $F<1$, λ is shortened, and for $F>1$, it is stretched. Exponential transform requires a weight change of the photon using the factor

$$w = Fe^{-\lambda(F-1)/F} = F(1-\xi)^{(F-1)/F}, \qquad (3.8)$$

that is, the photon weight is changed depending on the result λ or the random number ξ. This VRT is useful to either increase the calculation efficiency close to the surface (photon entrance point) at the cost of a decreased efficiency in larger depth ($F<1$) or vice versa ($F>1$).

3.2.7 Woodcock Tracking

The basic idea of Woodcock tracking (also called fictitious interaction method) is to make an inhomogeneous simulation geometry homogeneous by adding a fictitious interaction cross section to the total cross section in each region. The fictitious cross sections are chosen in a way to achieve a constant cross section sum for the whole geometry.

For example, to track photons of energy E in an inhomogeneous material grid, we have to determine the maximum total cross section $\mu_{\max}(E)$ for the whole simulation geometry. Then, in each region or each voxel with index i, a fictitious interaction cross section is determined by

$$\mu_{\mathrm{fict}}^i(E) = \mu_{\max}(E) - \mu_{\mathrm{tot}}^i(E). \qquad (3.9)$$

This allows tracking of photons without ray tracing because of the cross section $\mu_{\max}(E)$ being independent of the position within the geometry. As soon as the photon has reached the interaction site, a random number can be used to determine the interaction type, real or fictitious. The probabilities for real and fictitious interactions in each voxel are given by

$$P_{\mathrm{real}}^i(E) = \frac{\mu_{\mathrm{tot}}^i(E)}{\mu_{\max}(E)} \qquad (3.10)$$

$$P_{\mathrm{fict}}^i(E) = \frac{\mu_{\mathrm{fict}}^i(E)}{\mu_{\max}(E)}. \qquad (3.11)$$

If a real interaction is sampled, the simulation is continued with determining the photon interaction type, for example, Compton scattering or pair production. That is, there is no difference to the conventional MC tracking. However, if a fictitious interaction is sampled, the original photon tracking is continued starting from the present position without changing the photon energy and direction. That is, the fictitious interaction has no effect.

Woodcock tracking implemented in this way does not require weight adjustments of primary and secondary particles. However, it is possible to modify this VRT by introducing weight changes depending on the purpose of the simulation. For example, instead of sampling both real and fictitious interactions, only real interactions are sampled with

probability $P=1$. In this case, the weights of all secondary particles produced during these interactions have to be reduced by the factor

$$w = P_{\mathrm{real}}^i(E) = \frac{\mu_{\mathrm{tot}}^i(E)}{\mu_{\max}(E)}. \qquad (3.12)$$

In combination with Russian roulette, further weight modifications are possible to ensure that the statistical weight of all final particles is comparable.

3.2.8 Correlated Sampling

Applied to the problem of dose calculation in a heterogeneous geometry, correlated sampling starts with a known dose distribution $D_{\mathrm{hom}}(\vec{r})$ in a homogeneous phantom, for example, in water. This can be a measured dose distribution, but it can also be a smooth MC dose distribution calculated with very high statistical accuracy. This dose distribution can, for example, be stored in the computer memory. The algorithm then performs a simultaneous calculation of two correlated MC dose distributions, one distribution $D_{\mathrm{hom}}^c(\vec{r})$ in the homogeneous water phantom and another distribution $D_{\mathrm{het}}^c(\vec{r})$ in the heterogeneous geometry. Correlated means that the two simulations are performed using the same sequence of random numbers. To make this calculation fast, the two simulations are performed with low statistical accuracy, that is, with a small number of histories.

The simultaneous simulations are used to determine the distribution of correction factors in each voxel, given by the dose ratios

$$C(\vec{r}) = \frac{D_{\mathrm{hom}}(\vec{r})}{D_{\mathrm{hom}}^c(\vec{r})}. \qquad (3.13)$$

The final dose distribution in the heterogeneous geometry is then calculated by

$$D_{\mathrm{het}}(\vec{r}) = C(\vec{r}) D_{\mathrm{het}}^c(\vec{r}). \qquad (3.14)$$

It has been shown that this technique is useful only if the heterogeneous geometry is not too different from the corresponding homogeneous geometry because strong inhomogeneities destroy the correlation between the two simulations.

This technique can also be generalized to two heterogeneous geometries if an accurate solution for one of the geometries is known. An example is the simultaneous calculation of dose to an air cavity volume in water with and without the wall material present.

3.2.9 Initial Calculation of the Primary Interaction Density

For dose calculations with a simple model of the LINAC head, for example, a point source, the density of primary photon

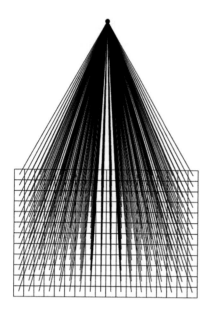

FIGURE 3.3 Precalculation of the primary photon interaction density per voxel of the calculation grid by performing an initial ray tracing between each voxel and the photon source. MC starts with simulating the corresponding number of secondary photons and electrons emitted in each voxel.

interactions can be precalculated in each voxel by performing an initial photon ray tracing. This is schematically shown in Figure 3.3. The MC simulation then starts after this precalculation step with simulating the corresponding number of secondary photons and charged particles in each voxel. The variance is reduced because the primary photon interaction sites are optimally distributed. Investigations have shown that the efficiency can be improved by a factor of about 2 using this method.

The main disadvantage of this method is that it limits the applicability of the code to relatively simple sources. Advanced MC dose calculation engines in RT, however, model the LINAC head using full simulations, using extended virtual source models, or a combination of both. Therefore, an initial calculation of the primary interaction density is not possible for this type of MC applications.

3.2.10 Quasi-Random Numbers

The solution of the transport problem in radiation therapy physics corresponds to an integration in a parameter space with an infinite number of dimensions. If this is performed numerically using the MC method, the dimensionality of the problem has to be limited to a finite number of dimensions, for example, n dimensions. Using a random number generator, it is possible to sample random vectors in this n-dimensional space by generating a sequence of n numbers, forming in this way an n-tuple. In general, one n-tuple corresponds to one particle history of the MC simulation. The simulation of N particle histories corresponds to the generation of a sequence

of N n-tuples. If a normal pseudorandom number generator of good quality is used for that purpose, the distribution of the n-tuples is suboptimal; that is, the statistical error of the result decreases as $1/\sqrt{N}$.

The sequences of n-tuples that fill the n-dimensional space more uniformly than uncorrelated pseudorandom numbers are called quasi-random sequences. These sequences are known to substantially improve the integration efficiency in certain types of problems. Using quasi-random numbers, the statistical error of the result can decrease asymptotically as $1/N$ if the quasi-random n-tuples are optimally distributed. There are different types of quasi-random numbers; examples are Halton's sequence or the Sobol sequence (Press et al., 1992).

Similar to pseudorandom numbers, quasi-random numbers are not really random. However, in contrast to pseudorandom numbers, quasi-random numbers cannot be considered as uncorrelated. For example, the Sobol sequence in one dimension starts with 0.5, 0.75, 0.25, 0.375, …; that is, it tries to fill the interval [0,1] in an optimum way. The method is obvious for the one-dimensional case, but it becomes complex for higher dimensions (Press et al. 1992).

3.3 Advanced Variance Reduction Techniques

Advanced VRTs are formed in general by combining different basic VRTs. Some of them have demonstrated to be very efficient for MC calculations in RT, for example, for dose calculations. They are outlined shortly in this section.

3.3.1 Selective Bremsstrahlung Splitting

With UBS (Section 3.2.1), the splitting factor N_{split} is a constant throughout the whole simulation; that is, it is independent of the direction of the electron producing the bremsstrahlung photons. With selective bremsstrahlung splitting (SBS) (Sheikh-Bagheri et al., 2006), $N_{split}(\theta)$ becomes a function of the variable θ, the angle between the present direction of the electron and the central beam axis. Because bremsstrahlung photons are forward peaked into a direction that differs only slightly from the direction of the initial electron, it makes sense to use smaller splitting factors $N_{split}(\theta)$ with increasing θ. The maximum should be at $\theta = 0$ and the minimum at $\theta = 180°$. Most of the photons transported with large angles are just absorbed within the primary or secondary collimators of the LINAC head, and their calculation time would be wasted. SBS reduces this time waste. Photons with a better chance of reaching the ROI are created with a higher probability using SBS.

There is, however, a serious drawback of this method; it introduces a nonuniform distribution of statistical weights

$$w(\theta) = \frac{1}{N_{split}(\theta)}, \qquad (3.15)$$

whereas θ is *not* the angle of the photon; it is the angle of the original electron. This means that photons in a specific small scoring location can have different weights. This influences the variance in this scoring location, and the final efficiency gain of the method is much smaller than expected. SBS improves the efficiency by a factor of 2–3 compared to UBS (Kawrakow et al., 2004).

3.3.2 Directional Bremsstrahlung Splitting

The disadvantages of SBS are eliminated using the directional bremsstrahlung splitting (DBS) method. Unlike UBS and SBS, DBS is a complex algorithm. Hence, the original and systematic publication of Kawrakow et al. (2004) is referred if the reader is interested in details of the method. Here, only a summary will be provided.

DBS uses (as UBS) a constant user-defined splitting factor N_{split} for the production of bremsstrahlung. Then, the algorithm analyzes the direction of the photons produced. They are transported always, if they are directed into the ROI. If not, Russian roulette (Section 3.2.2) with the survival probability $p_{survive}=1/N_{split}$ is played. Therefore, photons directed into the ROI carry a weight of $w=1/N_{split}$. Photons aiming away from the ROI carry a weight of $w=1$. Additional calculation time can be saved if the directional dependence of the photons is calculated in advance by a smart modification of the bremsstrahlung production cross section (Kawrakow et al., 2004).

By further combining splitting and Russian roulette, by processing "fat" and "meager" particles differently throughout the rest of the simulation as well as by treating photons in gas and higher-density materials differently, it can be ensured that all photons inside the ROI will be "meager," that is, have a weight of $w=1/N_{split}$ and those outside the ROI will be "fat" with a weight of $w=1$.

A drawback of this default DBS implementation is that all electrons are "fat." When they contribute to the result, for example, for dose calculation, the electrons would influence the variance in a negative manner. For these purposes, a DBS with electron splitting is available. This technique also ensures that the electrons reaching the ROI are "meager"; however, the overall calculation efficiency decreases compared to the default DBS.

3.3.3 Macro Monte Carlo

Macro Monte Carlo (MMC) is based on the premise that the simulation time can be decreased by the use of precomputed results. One of these precomputing techniques for the purpose of dose calculation in RT treatment planning of electron beams is called MMC (Neuenschwander and Born, 1992). As shown in Figure 3.4, this technique is based on precalculated fluence distributions on the surface of spheres for electrons hitting the spheres. A general-purpose MC algorithm can be used to calculate the distributions for incoming electrons of different energy as well as spheres with different diameters

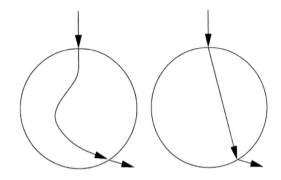

FIGURE 3.4 Principle of macro Monte Carlo. The left drawing shows a "full" MC simulation of an electron path in a homogeneous sphere with a definite diameter. A huge number of electrons hitting those spheres are precalculated, and the distributions of exit points, angles, and energies are stored as a function of sphere radius, material, and initial electron energy. During an MC dose calculation, using the patient geometry, these macroscopic distributions are used to directly jump from the entry point to the exit point (right drawing).

and consisting of different materials. The left drawing in Figure 3.4 shows schematically a "full" MC simulation of an electron path within such a homogeneous sphere. With these presimulations, the distributions of exit points, angles, and energies are calculated and stored as a function of sphere radius, material, and initial electron energy.

Dose calculation using the patient geometry starts with analyzing the computed tomography (CT) information and processing the calculation grid. Each voxel is assigned with a definite material and a definite sphere diameter, corresponding with the distance till the material changes. In homogeneous regions, spheres with larger diameters can be used. In regions with a strong density fluctuation, smaller diameters must be assigned to the voxels. When an electron hits the calculation grid, the sphere radius and material are determined based on the preprocessing result. The parameters of the electron leaving the sphere are randomly selected from the precomputed macroscopic fluence distributions; that is, the electron directly jumps from the entry to the exit point of the sphere (see the right drawing of Figure 3.4). The energy loss within the sphere is used to calculate the absorbed dose in each voxel.

The process continues with determining the new sphere parameters for the electron leaving the old sphere. It is repeated until the whole electron history is simulated and the energy of the initial electron is consumed.

Compared to conventional MC, MMC is faster by about one order of magnitude. One disadvantage is, however, only the electron transport speed can be increased. This means that MMC cannot be applied to photon beams efficiently.

3.3.4 History Repetition

Another technique with recycling of simulation results is called history repetition. This VRT, schematically represented

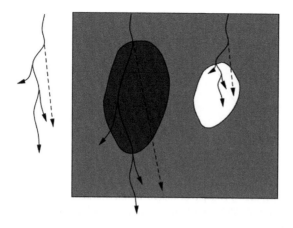

FIGURE 3.5 Electron history repetition. An electron history is simulated in a homogeneous water phantom of infinite size, and all parameters (step lengths, scattering angles, energy losses, and secondary particle parameters) are saved for reuse within the actual simulation (left). In the right drawing, this history is applied to the present patient geometry several times by scaling all parameters depending on the medium and mass density in each voxel.

in Figure 3.5, has been developed originally to increase the speed of MC dose calculations for electron beams in RT. Electron history repetition is based on complete electron histories precalculated in a homogeneous water phantom of infinite size. An example of such a precalculated history is shown in Figure 3.5 left from the box. Precalculated means that all parameters of this history (condensed history step lengths, multiple scattering angles, energy losses, and secondary particle parameters) are saved for reuse.

On the right-hand side of Figure 3.5, the history is applied to the actual patient geometry $N_{repeat}=2$ times, starting at different positions and with different directions at the surface of the calculation matrix. In reality, the optimal N_{repeat} depends on the size of the electron field. The optimum efficiency gain (approximately a factor of 2) is achieved with about one repetition per $5\,cm^2$ of the field; that is, for a $10\times10\,cm^2$ electron field, the repetition number should be about $N_{repeat}=20$. If N_{repeat} is smaller, the potential of history repetition is not fully exploited. If N_{repeat} is too large, the average distance between two neighbor histories becomes small and the variance is influenced by correlations between these neighbor histories.

The precalculations can be performed in advance (e.g., using a general-purpose MC algorithm) for a given large number of electron histories in water. Then, all parameters of all histories have to be stored in a database for use during the dose calculation. In a second approach, one history in water is simulated during the dose calculation immediately before it is applied to the patient geometry N_{repeat} times. After the repetition of this history, the corresponding memory is cleared and the next history in water can be simulated and applied to the patient. This second method allows the simulation of an arbitrary number of particle histories, and the final variance of the dose can become as small as desired. The first method,

on the other hand, is limited by the number of precalculated histories in the database.

Bremsstrahlung photons are generally not included in the repetition. They are simulated on the fly, and they are killed with Russian roulette or they are transported using kinetic energy released per unit mass (KERMA) approximation.

To take the heterogeneities correctly into account, when applied to the patient geometry, the histories in water must be adjusted by scaling the stored parameters depending on the medium, the stopping power, the scattering power, and the mass density in each voxel. Mainly, the electron step lengths and multiple scattering angles are scaled.

An important requirement of history repetition is that the histories must be scalable. Scalability can be assured, for example, if all transport parameters, such as stopping and scattering powers, are approximated as

$$S^T(M, E) = f_c^T(M) f_0^T(E). \tag{3.16}$$

This means that each transport parameter function of type T that depends on the medium M and the electron energy E is approximated as a product of a correction function depending only on the medium and another function depending only on the energy. It has been shown that this is possible for the different types of human tissue, including some phantom materials such as water with an accuracy of 1%–2%. With this factorization and without further loss of generality, a reference medium can be selected. In most cases, this is water; that is, for $f_0^T(E)$, the corresponding transport parameter in water is selected:

$$f_0^T(E) = S^T(H_2O, E). \tag{3.17}$$

Thus, we get

$$S^T(M, E) = f_c^T(M)S^T(H_2O, E). \tag{3.18}$$

If, for example, S^T is the linear collision stopping power $S^{coll}=dE/dx$, then for a given infinitesimal small energy loss dE, the step length dx according to collision loss is scaled by

$$dx(M) = \frac{dE}{f_c^{coll}(M)S^{coll}(H_2O, E)} = \frac{dx(H_2O)}{f_c^{coll}(M)}. \tag{3.19}$$

The factorization of Equation 3.16 is not a necessary condition for electron histories to be scalable. The correction $f_c^T(M)$ can be extended by including a slight energy dependence. However, electron history repetition does not work for arbitrary materials, for example, metals. Therefore, this technique is an AVRT.

3.3.5 Simultaneous Transport of Particle Sets

The limitations of history repetition can be avoided using the simultaneous transport of particle sets (STOPS) technique.

With STOPS, several particles of the same type (electron, positron, or photon) and with the same energy are transported simultaneously as a set. As with history repetition, the initial positions, directions, and weights are different. In contrast to history repetition, the histories in one set are not transformed into each other by path length and scattering angle scaling. However, a variety of parameters are independent of the material, so they are sampled once for all particles in the set.

STOPS can be applied for photons as well as for charged particles. Here, the electrons are used to explain the technique because STOPS has been developed originally in VMC++ (Kawrakow, 2000; Kawrakow and Fippel, 2000b) to replace electron history repetition. Electrons are transported in most of the MC algorithms using a class II CHT (see Chapter 2 for more details) with continuous energy loss to model semielastic collisions as well as discrete Møller (Bhabha for positrons) and bremsstrahlung interactions to model hard inelastic events. The CHT allows one to express distances between discrete interactions as energy losses. For the material of type M, they are sampled using the total discrete interaction cross section (number of interactions) per unit energy loss:

$$\underset{E}{\Sigma}(M, E) = \frac{\Sigma(M, E)}{L(M, E)}. \tag{3.20}$$

$\Sigma(M, E)$ is the total cross section per unit length, and $L(M, E)$ is the restricted stopping power. An advantage of $\underset{E}{\Sigma}(M, E)$ is that it depends only weakly on the material type M as well as on the energy E and a global maximum $\underset{E}{\overset{max}{\Sigma}}$ for all M and E can be efficiently used to perform Woodcock tracking (Section 3.2.7). This means that the geometry becomes homogeneous in terms of number of interactions (fictitious or real) and the energy loss ΔE can be sampled once for all electrons within the set. Thus, at the end of the step, all electrons have the same energy.

The geometric step lengths between the initial and final points depend on ΔE and on the stopping power $L(M, E)$. They are calculated separately for each electron in the set because $L(M, E)$ is different for different materials. Furthermore, the multiple scattering properties are material dependent; that is, the multiple scattering angles are also sampled individually for each particle in the set.

At the end of the step, the interaction type (fictitious, Møller, or bremsstrahlung for electrons) must be determined for each particle separately; however, the same random number can be used for all particles. This causes a very high chance of identical interactions, especially for human tissue because of the weak material dependence of the interaction properties. If the interaction type is identical for all particles, the whole set stays alive. If there are particles with a different interaction type, the set is split into subsets, and henceforward, the new subsets are transported independently.

If a specific (nonfictitious) discrete interaction is sampled, the secondary particle parameters such as the δ-electron energy for Møller interactions or the photon energy for bremsstrahlung production must be determined using the differential cross section. The differential Møller cross section is material independent, and if we assume the same for bremsstrahlung, the secondary particle energies can be sampled once for the entire set. The polar scattering angles are determined uniquely by the kinematics of the processes. Azimuthal scattering angles always follow a uniform distribution in the interval [0,360°]. This distribution is independent of the material type; that is, the azimuthal scattering angle can also be selected once for the particle set. Because identical energies are sampled for equivalent secondary particles within the set, the simulation continues with determining the energy loss to the next interaction by again using the Woodcock scheme. All steps of the simulation are repeated until the entire energy of the set is absorbed or the particles have left the simulation geometry.

The efficiency gain of STOPS is approximately a factor of 2; that is, it is comparable to history repetition. In contrast to history repetition, STOPS can be used for arbitrary media.

3.3.6 Continuous Boundary Crossing

General-purpose MC algorithms usually stop the simulation of charged particles at material interfaces. This is necessary because the underlying physics is valid only for a homogeneous region consisting of one specific material. Especially, the multiple scattering properties are material dependent. Therefore, close to a material boundary, an accurate MC algorithm should switch into a single scattering mode.

However, this is extremely time consuming, and for many applications, this is not necessary. For dose calculation in RT treatment planning, history repetition (Section 3.3.4) can be used without loss of accuracy. The multiple scattering angles and step lengths are precalculated in water and scaled depending on the material properties in each voxel. Therefore, it is possible for charged particles to cross boundaries continuously and to trace them efficiently like photons.

3.3.7 Multiple Photon Transport

In combination with photon splitting and Russian roulette, history repetition (Section 3.3.4) and STOPS (Section 3.3.5) can also be used for photon Monte Carlo. To avoid confusion with the various splitting techniques outlined in Sections 3.2.1, 3.3.1, and 3.3.2, here, we call this technique multiple photon transport (MPT). The method has been introduced in Kawrakow and Fippel (2000a) to speed up MC dose calculation for photon beams in RT treatment planning.

Figure 3.6 schematically represents this VRT. The basic idea comes from interaction forcing (Section 3.2.5); the photon is forced to interact within the calculation grid. However, the time-consuming precalculation of the number of mean free

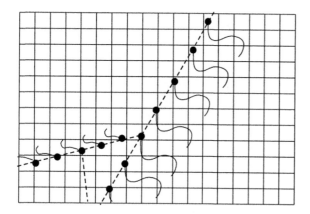

FIGURE 3.6 Multiple photon transport is a combination of photon splitting, Russian roulette, and electron history repetition or STOPS. Instead of one photon, a multiple number of identical photons (dashed lines) are simulated in one ray tracing through the actual geometry. Only the free path length to the interaction point is sampled differently for these photons. Because the interactions are assumed to be identical, for all secondary electrons (solid lines), the history repetition technique (or STOPS) can be applied. Secondary photons such as Compton photons are killed using Russian roulette. The surviving secondary photons are simulated like the primary photons by using the same algorithm.

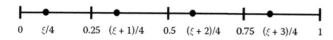

FIGURE 3.7 For the multiple photon transport technique, one uniform random number ξ is used to generate $N_{\text{split}}=4$ (in this example) random numbers, distributed with optimum uniformity in interval [0,1].

photon path lengths Λ between the entry and exit points of the grid is avoided with MPT.

With MPT, instead of one photon, a multiple number of N_{split} identical photons are simulated simultaneously in one ray tracing through the actual geometry. Each of the photons is carrying a weight of $w=1/N_{\text{split}}$. Identical means that the photons have the same energy, direction, and entry position at the surface of the calculation grid. Only the free path lengths λ_i to the first interaction points are sampled differently for the photons. In units of the mean free photon path length, they are calculated using just one random number ξ uniformly distributed in [0,1) by

$$\lambda_i = -\ln\left[1 - \frac{\xi+i}{N_{\text{split}}}\right], \quad i=0,\dots,N_{\text{split}}-1. \tag{3.21}$$

This means that ξ is used to generate the random numbers for the path length determination of N_{split} photons, whereas these numbers are distributed with optimum uniformity in interval [0,1) (see Figure 3.7).

Combined with electron history repetition, the MC simulation is performed by repeating the following steps:

1. If $i=0$, trace primary photons from the cube surface to λ_0 or trace scattered photons from the interaction site to λ_0.
2. If $i=0$, sample one interaction type (Compton scatter, pair production, photoelectric absorption, etc.) for all N_{split} subphotons of the same ray.
3. If $i=0$, sample secondary particle parameters and charged particle histories in water:
   a. For photoelectric absorption, sample the photoelectron direction, create a corresponding history in water, and store the parameters.
   b. For pair production, sample the electron's and positron's energy and direction, create the two histories in water, and store the parameters.
   c. For Compton scattering, sample the scattered particle's energy and direction, create the Compton electron history in water, and store the parameters.
4. If $i>0$, trace photons from λ_{i-1} to λ_i through the inhomogeneous voxel geometry.
5. Apply the charged particle histories to the patient geometry starting at the i-th interaction point.
6. In case of Compton scatter, play Russian roulette with the secondary photon using the survival probability $p_{\text{survive}}=1/N_{\text{split}}$ and store its parameters if it survives.
7. Increase i by 1, that is, take the next subphoton from the ray and go to step 4.

The procedure for the N_{split} primary photons stops when the cube boundary is reached. Then, the remaining Compton photons are simulated in the same way by executing the MPT function recursively. The procedure with STOPS instead of history repetition works correspondingly.

MPT is fast because N_{split} photons are transported with only one single ray tracing through the geometry. Furthermore, many quantities are reused with history repetition or STOPS. MPT also reduces the variance because the photon interaction sites are distributed optimally along the ray. With a splitting number of about $N_{\text{split}}=40$, the MC dose calculation efficiency is increased by a factor of 5–10 using MPT.

3.4 Pure Approximate Variance Reduction Techniques

Some of the AVRTs should not really be considered as VRT if the corresponding technique is dominated by one or more approximations. Here, some of these techniques, called pure AVRTs, are presented.

3.4.1 KERMA Approximation

An approximation based on KERMA is motivated by the fact that the energy released by low-energy photons is absorbed in the direct neighborhood of the photon ray. For these photons, the MC transport of secondary electrons can be switched off. That is, for all voxels along the photon ray, the deposited energy is calculated by

$$\Delta E_{dep} = E\mu_{en}^{M}(E)\Delta s. \tag{3.22}$$

Here, Δs is the length of the photon path within the given voxel, E is the energy of the photon, and $\mu_{en}^{M}(E)$ is the linear energy absorption coefficient for photons of energy E and the material M in the corresponding voxel.

Systematic deviations of the dose distributions can be minimized with this approach if the maximum range of the secondary electrons is smaller than the spatial resolution of the calculation grid. This is possible if the KERMA approximation is applied only to photons with an energy E smaller than some predefined maximum energy K_{cut}. Furthermore, it is useful to apply this technique only for secondary and higher-order scattered photons, but not to primary photons coming directly from the LINAC head.

3.4.2 Continuous Slowing Down Approximation

In CSDA, electrons (or other charged particles) are transported in one step without the creation of secondary particles. The length of the step is calculated from the energy of the electron and the unrestricted total stopping power. This continuous slowing down range is a unique function of the energy of the electron. Therefore, energy straggling is neglected using this approach.

There are versions of CSDA with and without multiple scattering. Without multiple scattering, the charged particle is transported on a straight line. Otherwise, a multiple scattering angle is sampled and included in the simulation.

CSDA should be used only for low-energy charged particles, for example, for electrons within an MC simulation, when their energy drops below some user-defined threshold energy E_{cut}. However, CSDA for subthreshold electrons is more accurate and produces less statistical fluctuations than local absorption of the track-end energy.

3.4.3 Transport Parameter Optimization

Both the accuracy and speed of coupled electron–photon transport MC simulations in RT depend on the selection of various transport parameters such as particle transport cutoff energies, particle production threshold energies, and condensed history step sizes.

MC algorithms usually employ a photon energy cutoff parameter P_{cut}; that is, photons are not transported if they are generated within the simulation with an energy below P_{cut}. The remaining energy can be neglected or it can be deposited locally. It is obvious that the accuracy of the result increases with decreasing P_{cut}; however, the calculation time also increases. On the other hand, if the remaining photon energy is absorbed locally, a large value for P_{cut} can cause additional fluctuations. This can decrease the calculation efficiency, although the calculation time has also decreased.

Another important parameter is the photon production threshold energy P_{min}, for example, for bremsstrahlung. This means that only bremsstrahlung photons with an energy larger than P_{min} can be generated during the MC simulation. The effect of electron stopping due to bremsstrahlung photons below P_{min} is taken into account by the restricted radiative stopping power. The question here is how to deal with the radiative energy released by these electrons? Both local absorption and total disappearance of this energy are approximations. To avoid a significant influence on the result on the one hand and to ensure short calculation times on the other, P_{min} should be selected with care depending on the type of the calculation.

The problem is less complex for the cutoff and production threshold energies of charged particles, E_{cut} and E_{min}. Because of the strong correlation between energy and range, both parameters can be selected depending on the spatial resolution of the calculation geometry. On the other hand, efficiency can be improved by using a higher value of E_{cut} and by simulating track-end electrons with energy below E_{cut} in CSDA (Section 3.4.2).

Another important parameter in condensed history MC electron algorithms is the step size (or the procedure used to determine the actual step size depending on energy, material, geometry, etc.). The step size must be restricted by some maximum if an approximate multiple scattering theory, for example, a small-angle approximation is implemented. In widespread use are approaches with a user-defined maximum energy loss per step parameter E_{step}. The value of E_{step} should be optimized depending on the algorithm and the type of the application.

References

Bielajew, A. F. and Rogers, D. W. O. 1988. Variance-reduction techniques, In: T. M. Jenkins, W. R. Nelson, and A. Rindi (eds), *Proceedings of the International School of Radiation Damage and Protection Eighth Course: Monte Carlo Transport of Electrons and Photons below 50 MeV*, Plenum Press, New York, Chapter 18, pp. 407–419.

Chetty, I. J. et al. 2007. Report of the AAPM Task Group No. 105: Issues associated with clinical implementation of Monte Carlo-based photon and electron external beam treatment planning, *Med. Phys.* **34**(12): 4818–4853.

Kawrakow, I. 2000. VMC++, electron and photon Monte Carlo calculations optimized for radiation treatment planning, In: A. Kling, F. Barao, M. Nakagawa, L. Tavora and P. Vaz (eds), *Advanced Monte Carlo for Radiation Physics, Particle Transport Simulation and Applications, Proceedings of the Monte Carlo 2000 Conference*, Springer-Verlag, Berlin, pp. 229–236.

Kawrakow, I. and Fippel, M. 2000a. Investigation of variance reduction techniques for Monte Carlo photon dose calculation using XVMC, *Phys. Med. Biol.* **45**: 2163–2183.

Kawrakow, I. and Fippel, M. 2000b. VMC++, a MC algorithm optimized for electron and photon beam dose calculations for RTP, *World Congress on Medical Physics and Biomedical Engineering, Med. Phys. 27, Meeting Issue,* Chicago, Illinois, USA.

Kawrakow, I., Rogers, D. W. O., and Walters, B. R. B. 2004. Large efficiency improvements in BEAMnrc using directional bremsstrahlung splitting, *Med. Phys.* **31**(10): 2883–2898.

Neuenschwander, H. and Born, E. 1992. A macro Monte Carlo method for electron beam dose calculations, *Phys. Med. Biol.* **37**: 107–125.

Press, W. H., Teukolsky, S. A., Vetterling, W. T., and Flannery, B. P (eds) 1992. *Numerical Recipes in C*, Cambridge University Press, Cambridge, New York, Port Chester, Melbourne, Sydney.

Reynaert, N. et al. 2007. Monte Carlo treatment planning for photon and electron beams, *Rad. Phys. Chem.* **76**(4): 643–686.

Salvat, F., Fernández-Varea, J. M., and Sempau, J. 2011. PENELOPE-2011: A Code System for Monte Carlo Simulation of Electron and Photon Transport, Nuclear Energy Agency. NEA/NSC/DOC(2011)5, *Workshop Proceedings*, Barcelona, Spain, July 4–7, 2011.

Sheikh-Bagheri, D., Kawrakow, I., Walters, B., and Rogers, D. W. O. 2006. Monte Carlo simulations: Efficiency improvement techniques and statistical considerations, Integrating New Technologies into the Clinic: Monte Carlo and Image-Guided Radiation Therapy, *Proceedings of the 2006 AAPM Summer School*, Medical Physics Publishing, Madison, WI, pp. 71–91.

II

Source Modelling

Monte Carlo Modelling of External Photon Beams in Radiotherapy

Frank Verhaegen
Maastro Clinic

4.1 Introduction

It is widely accepted that Monte Carlo simulation methods offer the most powerful tool for modelling and analyzing radiation transport for radiotherapy applications. One of the most frequent and important uses of Monte Carlo modelling in external beam radiotherapy is the creation of a virtual model of the radiation source. Applications are source design and optimization, studying radiation detector response, and treatment planning. Since the vast majority of radiotherapy treatments are performed with megavolt (MV) photon beams from a linear accelerator (linac), this is also reflected in the research efforts. To a lesser extent, external beam radiotherapy is performed with electron beams, lower-energy photons beams emanating from an X-ray tube, or hadron beams. This chapter will cover Monte Carlo modelling of photon beams and the next one of electron beams, while others will deal with hadron beams.

Historically, Monte Carlo methods provided a simple alternative to measurements to derive photon spectra from radiation sources [1–4]. Monte Carlo techniques were used in the early days of linac design in the seventies to aid in the optimization of the photon beams [2]. The first simplified models of photon beams from linacs were reported from the late seventies till the early nineties [1,5,6]. For linac electron beams, an important early effort was the work by Udale [7,8] in the late eighties. They introduced the modular approach for Monte Carlo modelling of electron beams, which is nowadays very popular since the advent of the BEAM Monte Carlo

user interface [9]. Extensive literature reviews of Monte Carlo modelling of electron beams were provided by Ma and Jiang [10] and of photon beams by Verhaegen and Seuntjens [11]. They cover many details and are in many respects still current.

In this chapter, we will give an overview of the work performed in the field of Monte Carlo modelling of clinical photon beams over the last 40 years or so. We will discuss what is required to derive a Monte Carlo model of such beams, and show some applications. Arguably, the most important clinical application, radiotherapy treatment planning [12], will be covered in other chapters. Monte Carlo models of radiation sources discussed in this chapter will start from the primary electron beam, and hence, we do not cover the extensive work done in the area of accelerator design on, e.g., the beam generation and transport in a linac flight tube.

4.2 Photon Beams from Clinical Linacs

Nowadays, clinically used photon beams in radiotherapy usually are within the energy range of 4–25 MeV. A linac has an essentially modular construction as shown generically in Figure 4.1. Different manufacturers may have the components of a linac in a different order but the ones indicated in Figure 4.1 are usually those that are required in a Monte Carlo model. For an overview of linac technology, the reader is referred to several excellent texts available [13,14]. The Monte Carlo codes employed in this chapter are mostly

DOI: 10.1201/9781003211846-6

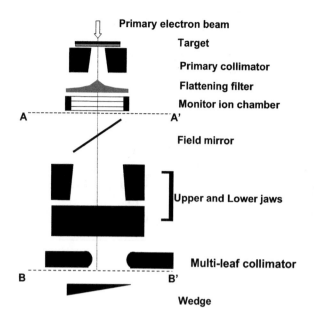

Primary electron beam

Target

Primary collimator

Flattening filter

Monitor ion chamber

A — — — — — — A'

Field mirror

Upper and Lower jaws

Multi-leaf collimator

B — — — — — B'

Wedge

FIGURE 4.1 Schematic presentation of linac components in a Monte Carlo model of a photon linac. Different linac manufacturers may have the linac components in a different order.

EGS4 [15] or EGSnrc [16] (often in conjunction with the user interface BEAM [9] or BEAMnrc [17], GEANT4 [18], MCNP [19 20] and PENELOPE [21,22]). They are all discussed in other chapters. Techniques to speed up simulations, variance reduction techniques, are also discussed elsewhere in this volume. They are heavily relied upon in linac simulations, which can be prohibitively long, even with today's powerful computers. A discussion of variance reduction techniques as they pertain to photon linac modelling can be found in the literature [11].

4.2.1 Components of a Monte Carlo Model of a Linac Photon Beam

The components depicted in Figure 4.1 must be known in detail to build a faithful model of a linac. The composition of materials and alloys, their mass densities, the position, dimensions, and shape of defining surfaces of the components and their motion must all be known in great detail to build an accurate Monte Carlo model. Knowledge of tolerances may help to determine uncertainties in the calculated results. While seemingly trivial, this is a task that has daunted many workers. It usually requires interaction with linac manufacturers to disclose constructional details. Needless to say, errors made at this stage will often translate into systematic errors in the calculated output of the linac. Therefore, it is of the utmost importance that linac blueprints are verified as much as possible and that Monte Carlo linac models are validated against an extensive set of dose measurements of, as a minimum, depth and lateral dose profiles in water, and output factor in water (ratio of dose at a depth in water at the central

axis for a certain field size divided by the same for a reference field size). The validation procedure may comprise comparisons against measurements at various source-to-surface distances of a phantom, should treatment planning be the main application. The literature contains many examples of extensive comparisons of Monte Carlo linac models against measurements [23–27].

In 1995, the BEAM/EGS4 user interface was released [9], since then upgraded to BEAM/EGSnrc (or BEAMnrc [17]). This user interface allows easy modelling of radiotherapy linacs and led to a host of papers presenting photon and electron beam models. It was, and still is, seen as a major step forward in the field. It is currently still a widely used linac simulation package, although the GEANT4 code in packages such as GATE [28,29] is also very popular. Figure 4.2 shows a typical example of a simple photon linac model where linac components and particle tracks can be discerned (Figure 4.2a) and where interactions in the flattening filter can be observed in detail (Figure 4.2b). The BEAM code allows an easy assembly of a linac model (it also enables building models for X-ray tubes and a few other geometries) with a wide choice of building blocks consisting of geometrical shapes such as disks, cones, parallelepipeds, trapezoids etc. The code relies heavily on the fact that linac components don't overlap in the beam direction, which is mostly the case. Separate parts of a linac can, therefore, be built, validated, and studied separately. In addition to linac components which may be defined with many materials (the BEAM package provides an extensive cross-sectional dataset), a wide choice of primary source geometries is available. Particle transport can be done for electrons, photons, and positrons. Particles can be tagged according to interaction types, interaction sites etc, which provides a powerful beam analysis method.

Other user interfaces similarly intended to enable an easy assembly of complex geometries have been introduced, some with great success. An example is PENLINAC for the PENELOPE Monte Carlo code [30].

Around the same time, BEAM was first released. Lovelock et al. [31] also presented a modular Monte Carlo approach to simulate linac photon beams. Both are predated by papers by Udale [7,8] who introduced modular Monte Carlo models for electron beams. In Lovelock et al. [31], the radius of a uniform primary electron beam hitting the target and its energy distribution (a truncated Gaussian) could be taken into account. Information on particles crossing a predefined plane could be stored in a phase-space file. Simulations were done in stages, i.e., phase-space information from one stage was used as an input source for a subsequent stage. Particles could be tagged according to different types of interactions they underwent. They noticed that calculated depth dose profiles in water are relatively insensitive to the primary electron energy, whereas the horns in the lateral dose profiles are a good indicator for the primary electron energy. A lower primary electron energy will result in more pronounced horns. As such, these authors were among the first to introduce a tuning procedure of the

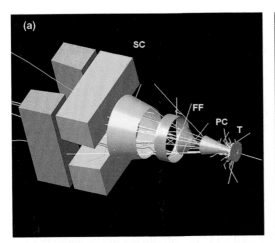
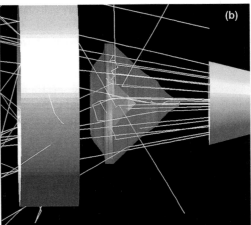

FIGURE 4.2 (a) Simplified BEAM Monte Carlo model of an 18 MV linac photon beam. From right to left, we encounter the primary electron beam hitting the target (T), the primary collimator (PC), the flattening filter (FF), and some secondary collimators (SC). Photon tracks are shown in yellow, electron tracks in blue, and positron tracks in red. In (b), a close-up of the particle interactions in the FF is shown in exquisite detail. By depicting the FF in a semi-transparent fashion, the details of the geometry (notice the cone inside a cone structure) and the interactions are revealed.

Monte Carlo primary electron source characteristics to derive the primary electron energy. This is often needed in case the manufacturer doesn't provide the required information or where simulations of the electron trajectories in the flight tube are not available. This information is rarely at hand.

The photon linac components that are the most important in the model in Figure 4.1 are the target, flattening filter, secondary collimators (T, FF, SC), and wedges, if present. After the accelerated primary electron beam exits the flight tube, it possesses a narrow energy, angular and spatial distribution. This electron beam will hit the target, commonly consisting of a high-Z metal in which the electrons will produce bremsstrahlung photons. It has to be pointed out that in the Monte Carlo models of clinical linacs, usually no modelling is done of the electron beam, prior to exiting the flight tube. This will inevitably entail making assumptions about the primary electron beam, which will be discussed later. The bremsstrahlung photons are then collimated initially by a primary collimator, and the photon fluence is differentially attenuated by the flattening filter to produce a reasonably flat dose distribution in water at a certain depth. Target and flattening filter are the most important sources of contaminant electrons, unless a wedge is present. In the generic model of Figure 4.1, the next components to be considered are the monitor ion chamber and the field mirror. Both present only a small attenuation to the photon beam and are often omitted from Monte Carlo models. However, when backscatter to the monitor chamber is the subject of study (Section 4.2.1.3), some model of the ion chamber should be present. The photon beam is then finally shaped and modulated by secondary collimators and beam modifiers such as jaws, blocks, multileaf collimators (MLCs), and dynamic or static wedges. In the following sections, Monte Carlo modelling of the various linac components will be discussed in more detail.

4.2.1.1 Primary Electron Beam Distribution and Photon Target

4.2.1.1.1 Photon Target Simulation for Radiotherapy Treatment

The bremsstrahlung photons produced in the target by the primary electrons in the energy range [4–25 MeV] are mostly derived from a relatively thin layer on the upstream side. This is because of the quick energy degradation of the electron energy in a high-Z material, the almost linear dependence of the bremsstrahlung cross section on the electron kinetic energy for high energies and electron scattering in the target which also disperses the bremsstrahlung photons. At high electron energies, the average bremsstrahlung photon emission angle is given approximately by m_0c^2/E_0 (m_0c^2 is the electron's rest energy; E_0 its total energy), yielding a strongly forward-peaked angular distribution. Targets in clinical linacs are usually thick enough to stop the primary electrons completely and are for that reason often referred to as 'thick targets'. In such targets, the bremsstrahlung photon angular distribution will be spread out because of electrons undergoing multiple scattering, but the result will nevertheless be a strongly anisotropic photon fluence, while the photon spectrum is fairly isotropic. Due to this complex picture, it is very difficult to obtain the spectral and angular bremsstrahlung photon distribution from thick targets using analytical methods such as the Schiff theory [32,33]. More complex analytical calculation schemes have been developed but the most comprehensive method to generate bremsstrahlung photon distributions from targets in clinical linacs is Monte Carlo simulation. In what follows, we will discuss a few studies related to this topic.

Patau et al. [1] were among the first to present a simple Monte Carlo model, using an in-house code, of a complete photon beam linac. It consisted of a 5.7 MeV electron pencil beam hitting a W/Cu target, followed by a flattening filter of Pb and an unspecified collimator. The authors calculated photon energy fluences, photon transmission through diverse materials, and electron spectra in water. McCall et al. [2] were among the first to attempt to improve the design of a linac by performing Monte Carlo simulations. They used the EGS3 code [34] for the calculation of bremsstrahlung spectra and for the generation of secondary electrons in water. For several combinations of target materials and flattening filters, they noticed a linear correlation between the depth of the dose maximum in water and the average energy of the photon spectrum exiting the linac for 10–25 MV photon beams. They concluded that Monte Carlo simulations are a powerful tool for designing target and flattening filter.

Starting in the late eighties, workers from the National Research Council of Canada (NRCC) in Ottawa devoted several studies to Monte Carlo modelling of bremsstrahlung production in linac targets. Bielajew et al. [35] implemented bremsstrahlung angular sampling from the Koch and Motz distribution [32] in EGS4. The KM angular distribution was found to lead to a significant difference in the degree of self-absorption in the target compared to using a fixed angle of m_0c^2/E_0 (a common approach in many Monte Carlo codes, at one time). This work was complemented by several careful experimental studies (mostly by B Faddegon) to determine the energy and angular distribution of emitted photons by targets [3,4]. They found reasonable, but not perfect, agreement between simulated and measured spectra. The accuracy of cross sections for bremsstrahlung production remains a subject of interest [36]. A study on experimental derivation of linac photon spectra can be found in Ref. [37].

4.2.1.1.2 *Photon Beam Spot Size in Radiotherapy Accelerators*

A point that deserves special attention is the focal spot size of the photon beam or the primary electron beam. The latter is a crucial parameter in Monte Carlo simulations, which influences calculated dose and fluence distributions, as will be demonstrated later. Munro and Rawlinson [38] were among the first to estimate the size of the X-ray source in linacs. A slit camera, in combination with tomography techniques, was used to project an image of the photons that are emitted under small angles with the central beam axis on a diode detector. Determining the size of the photon source this way is also a direct measure for the spot size of the primary electron beam hitting the target. The presence of a flattening filter does not disturb the spot size measurement since only those photons that go through it without interacting can contribute to the image. For 6–25 MV linacs, they found that the size and shape of the X-ray source can differ from machine to machine, with mostly an elliptical shape with the long axis perpendicular to the gantry axis. The full width at half maximum (FWHM) of the measured spots varied between 0.7 and 3.3 mm. Similar studies have been reported [39–42]. Jaffray et al. [43] conducted extensive measurements of focal spot sizes for 6–25 MV photon beams from seven different linacs, using Munro and Rawlinson's tomographic technique [38]. They reported

measured elliptical spot sizes with a FWHM of 0.5–3.4 mm with eccentricities of 1.0–3.1. They concluded from repeated measurements that the long-term stability of the spot size is mostly determined by the linac design and not so much by linac tuning. They noticed that different energies on the same linac can have different source spot centers: a shift of 0.8 mm was noted for 6 and 18 MV beams on a Varian Clinac 2100C. Focal spots also have been reported to move during the start-up phase of irradiations by up to 0.7 mm in the gun-target direction for Elekta linacs [44]. Recently, a method was proposed to estimate the size and the shape of the focal spot from the measured dose profile data [45].

4.2.1.1.3 *Photon Target Simulation for Radiotherapy Imaging*

A final topic we will discuss briefly under this section is the use of Monte Carlo simulations for the optimization of radiotherapy accelerator targets for portal imaging, which is also relevant for MV cone beam CT imaging. Simulation of the imaging detector is of relevance here but a detailed discussion of this is not within the scope of this review. Figure 4.3 shows the complete simulation geometry, including the patient and the portal imager downstream. The figure also shows a simulated versus a recorded portal image of a contrast phantom. A linac target for imaging would typically consist of a lower atomic number material than a therapy target. This results in a photon spectrum that is considerably softer, which is needed for imaging purposes due to the imaging contrast at low photon energies.

In some of the older studies [46,47], detailed experimental work and simulations led to advocating a thin low-Z target (Be or graphite) in a linac with the flattening filter removed to enhance the fraction of photons below 150 keV for the purpose of improving the contrast of portal imaging. These studies also demonstrated that a large fraction of the low-energy photons coming from the special target are absorbed in the phantom, thereby reducing the sought improved imaging contrast. Several others [48,49] used EGS4 simulations of 2–10 MeV electron beam interacting with a range of target materials (Be, C, Al, Ti, Cu). For 4 and 10 MeV electron beams, the optimum Cu target thickness to maximize the integrated photon fluence was found to be 1.5 and 4 mm [49], respectively (these targets cannot be considered 'thick' as defined above). Somewhat at variance with the previously cited studies [46,48], Tsechanski et al. [49] found that there is not much motivation to use imaging targets with an atomic number below Al, mostly due to an increased imaging dose for the low-energy photons. Thin Al or even Cu targets were recommended by these workers.

Building on these works, Flampouri et al. [50,51] presented a comprehensive experimental and simulation study of an optimized portal imaging beam, which was later refined [52]. They used the BEAM/EGS4 MC code for detailed simulations of a complete linac treatment head, including a specially designed contrast phantom and several imaging systems. Optical contrast derived from simulations and experiments was found to be in good agreement (Figure 4.3). Recently, Monte Carlo

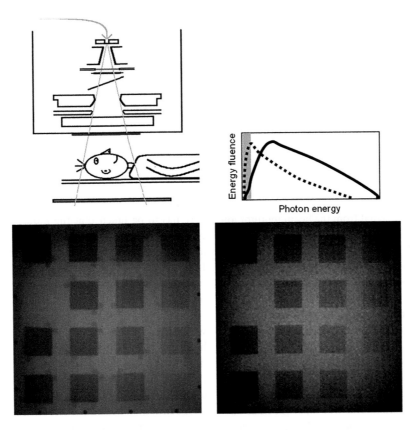

FIGURE 4.3 Schematic representation of a linac simulation including a patient and a portal imager below the patient. The lower-energy photon spectrum (dotted line) used for imaging purposes is compared to the therapy photon spectrum (full line) in the graph. The bottom panels compare a recorded portal image (left) of a contrast phantom to a simulated one (right). (Work by S Flampouri [50], reproduced with kind permission.)

techniques were used for the study of low atomic number targets for MV cone beam CT [53–55].

4.2.1.2 Flattening Filter

In this section, we turn our attention to the next component in a linac downstream from the target that has a major influence on the beam: the flattening filter. Flattening filters are designed to generate flat dose distributions at a certain depth in water. We will highlight a number of Monte Carlo simulation studies that provided an insight in the influence of the usually complex-shaped flattening filter on photon fluence distributions (see Figure 4.2). It has to be pointed out that there are linacs that do not use flattening filters. The Microtron MM50 accelerator, tomotherapy machines, and some modern volumetric arc therapy devices do not employ flattening filters (for examples of Monte Carlo studies on one of these machines, see Refs. [56,57]).

McCall et al. [2] were probably among the first to report the now well-known off-axis differential softening of the photon spectrum, based on Monte Carlo simulations of simplified flattener models made of Al, Ni, and W for a 25 MV photon beam. Since Al was found to cause the largest softening, they recommended that flattening filters for high-energy photon

beams should be made of medium-Z materials such as Cu or steel. Mohan et al. [6] simulated 4–24 MV photon beams from a number of different linacs. They were the first to build accurate models of the flattening filters by developing a special geometry package (another forerunner of BEAM) that allowed easy modelling of the complex geometry. They demonstrated that flattening filters cause significant spectral hardening both on and off the beam axis. For example, a 15 MV beam with no flattening filter present yielded average photon energies at 100 cm from the source of 2.8 and 2.5 MeV, respectively, at the central axis and in an annular scoring region between 10 and 25 cm. These values changed to 4.1 and 3.3 MeV when a flattener was added, increasing the differential off-axis spectral softening significantly. It can be seen from their work that central axis and off-axis spectra differ mostly for energies below 1 MeV. They also reported that the average photon energy from a linac is lower than the one-third of the nominal energy as was commonly assumed.

A generic model of a linac ('McRad') based on the EGS4 code was presented by Lovelock et al. [31] at around the same time BEAM was released. For a 6 MV beam, these authors observed a 10% off-axis softening at 100 cm for an unflattened beam, whereas adding a flattening filter augmented this to about 30%. The fact that all these studies report a relatively small off-axis softening for unflattened beams is related to the observation that the bremsstrahlung spectrum from a thick target is fairly isotropic within the angular range of the photons that can reach the patient, whereas the bremsstrahlung intensity is highly anisotropic, as discussed before (Section 4.2.1.1.1). Faddegon et al. [58] used BEAM simulations to successfully redesign a flattening filter of a clinical accelerator to obtain larger flat 6 MV fields. This involved changing the material of the flattener from steel to brass.

4.2.1.3　Monitor Ion Chamber Backscatter

In some clinical linacs, the signal from the beam monitor ion chamber is affected by the position of the secondary movable beam collimators (jaws). This only occurs in linacs where the distal monitor chamber window is sufficiently thin, where no backscatter plate is present and where the distance between chamber and the upper surface of the collimators is small enough. In that case, particles backscattering from the movable collimators can deposit charge in the monitor chamber, in addition to particles moving in the forward direction in the beam. In a small field, more backscatter from the collimators will occur than in a large field. This means that the monitor chamber will quicker reach its preset number of monitor units (MUs) and terminate the beam. This will cause the linac output to decrease with decreasing field size. The magnitude of this effect is usually limited to a few percent but for some linac types, larger effects have been reported. The decreased output is automatically included in output factor measurements, but when Monte Carlo simulations are used to determine the output factors, the effect has to be taken into account separately. This applies to all cases where outputs from fields of different sizes are combined, so it also plays a role in modelling dynamic (or virtual) wedges and possibly even in modelling of intensity-modulated radiation therapy (IMRT) or volumetric modulated arc therapy (VMAT). This only is possible in certain types of linacs, but one should be aware of this effect, which may cause simulations to disagree with measurements.

A few studies have used Monte Carlo techniques to investigate the backscatter effect. Liu et al. [59,60] performed several detailed simulations of Varian linacs to study the differential effects of the different collimators on monitor backscatter. They concluded that the width of the jaw opening is important to determine the backscatter correction, whereas the actual off-axis position of the field is not important. Verhaegen et al. [61] modeled Varian 6 and 10 MV photon beams, including a detailed model of the monitor ion chamber. By tagging particles and selectively transporting photons and electrons, it was found that electrons cause most of the backscatter effect. A spectral analysis of the forward and backward moving particles was shown. In this study, it was also found that in this linac, the backscatter from MLCs is negligible at the level of the monitor chamber. In the two studies cited where MC simulations and measurements were compared [60,61], good agreement was observed.

4.2.1.4　Wedges

Inserting a physical wedge in a photon beam alters the beam characteristics significantly. Not only is the dose distribution modified by attenuation and scatter from the wedge, but also the photon spectrum is affected. Usually, spectral hardening is seen below some parts of the wedge. When a moving linac jaw is used to form a dynamic (or virtual) wedge, the photon spectrum is much more similar to the open-field spectrum. For very high-energy beams (>20 MV), beam wedges may cause beam softening due to, e.g., pair production and annihilation.

Monte Carlo simulations can be used to model physical or dynamic wedges. When modelling a physical wedge, great care has to be taken that the model for the wedge is exact. Not only the exact shape of the wedge has to be implemented, also the composition and density have to be known exactly. In particular, steel wedges have been known to be difficult to model because of uncertainties in the manufacturer's specified composition.

Liu et al. [62] extended their dual-source model to include a wedge model for a 6–18 MV photon beams of a Varian 2100C linac. Steel and lead wedges with angles from 15° to 45° were modeled. The wedge was seen as part of the patient dose calculation. Special photon dose kernels to take into account the effect of the wedge were added in their convolution/superposition dose calculation method. These kernels were calculated using MC techniques in bimaterial spheres (lead and water). The beam hardening by the wedge was included in the model. Their dose calculations were in agreement with full Monte Carlo simulations, except in the buildup region because of lack of electrons coming from the wedge in their source model. It was found that the wedge-generated photon fluence contributed significantly to the total dose: e.g., a 45° wedge in a $20 \times 20\,\text{cm}^2$ 10 MV photon beam caused 8.5% of the total dose at the central axis at a depth of 5 cm. By distinguishing between annihilation, bremsstrahlung, and Compton-scattered photons from the wedge, it was found that the latter were responsible for the majority of the wedge-generated dose. The wedge was found to produce a near-Gaussian-shaped lateral dose profile.

Schach von Wittenau et al. [63] used crude Monte Carlo wedge models and found that wedge-generated bremsstrahlung photons carried about 20% of the outgoing energy in a 10 MV beam. Li et al. [64] presented their Monte Carlo code MCDOSE, which takes wedges into account as a part of the patient dose calculation. Spezi et al. [65] modified DOSXYZ (now updated to DOSXYZnrc [66]) to include a rectilinear voxel geometry module and the ability to collect phase-space information behind the wedge. They built a model for the lead wedges of their 6 MV photon beam of a Varian 2100CD linac. A step resolution of 1 mm in the wedged and beam direction was found to be sufficient for the dose calculations.

Van der Zee and Welleweerd [67] introduced a new component module in the BEAM code, which allows modelling complex wedges such as the Elekta internal wedge. Figure 4.4 shows the CM WEDGE that was used by the authors to model

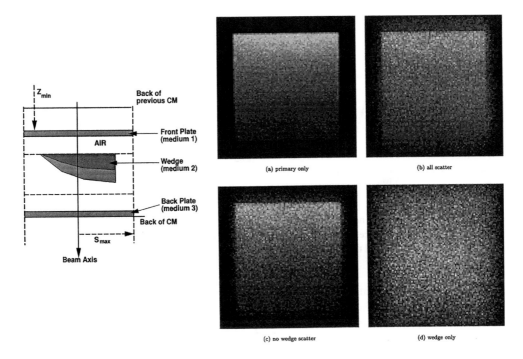

FIGURE 4.4 Left panel: Component Module for the BEAM Monte Carlo code for modelling physical wedges (Van der Zee and Welleweerd [67]). This module is not part of the standard BEAM package. The right panel shows the relative distribution of the energy fluence of primary and scattered photons for a 30×30 cm² wedged field for a 10 MV photon beam. (a) Primary photons only, (b) scattered photons from all scatter sources, (c) scattered photons excluding wedge scatter, (d) scattered photons from the wedge only. The thicker part of the wedge (heel) is at the lower part of the figures. (Reproduced with kind permission.)

the internal wedge in their 6–10 MV photon beams of an Elekta SL linac. They found that the presence of the wedge altered the primary and scattered photon components from the linac significantly: beam hardening by 0.3 and 0.7 MeV was observed for the two components, respectively. They also noted that the wedge-generated photons were mostly generated close to the distal edge of the wedge. Figure 4.4 also shows the photon energy fluence originating from various sources in the linac, including scattered photons from the wedge. Monte Carlo techniques are unique in extracting such information.

In building Monte Carlo models for dynamic wedges, the time-dependent movement of the jaw has to be included. The studies we will discuss here laid the foundations for modelling motion of MLCs in IMRT, to be discussed in the next section. Verhaegen and Das [68] built a BEAM Monte Carlo model for the Virtual Wedge of 6–10 MV photon beams of a Siemens MD2 linac. The wedge was modeled by a discrete sum of a large number of open fields. By comparing calculated photon spectra from the heel, toe, and central regions, they found that the photon spectrum of the Virtual Wedge was only slightly hardened, with no significant difference for the three regions. In contrast, 60° physical wedges of tungsten introduced a significant beam hardening for the 6 MV beam across the whole wedge. Both the transmitted and wedge-generated photons were found to be harder for a tungsten wedge than for an open field. Steel wedges were found not to alter the spectrum significantly.

A full dynamic simulation technique to model a Varian Enhanced Dynamic Wedge was introduced by Verhaegen and Liu [69]. This is the first report of a fully dynamic Monte Carlo simulation of a linac component. They used the BEAM Monte Carlo code to model 6–10 MV dynamically wedged photon beams of a 2100C Varian linac by sampling the position of the moving jaw from the Segmented Treatment Table. More information is given in the chapter on *Dynamic beam delivery and 4D Monte Carlo* in this book. They termed their technique the 'position-probability-sampling method'. The varying backscatter to the monitor chamber during the jaw movement was included in the model. The authors obtained an excellent agreement between measured and calculated wedged dose profiles. Other studies [70,71], e.g., presented a detailed information on photon and electron spectra and angular distributions produced by dynamic and physical wedges, or studied dose distributions.

Monte Carlo models for other beam modifiers such as blocks, compensators etc were discussed in the literature [64,65,72,73].

4.2.1.5 Multileaf Collimators and Dynamic Therapy

MLCs, consisting of banks of tungsten leaves that allow shaping of conformal or intensity-modulated fields in IMRT or VMAT, are among the most challenging geometrical structures in a linac to model. The MLC leaves can have very

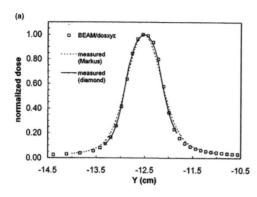
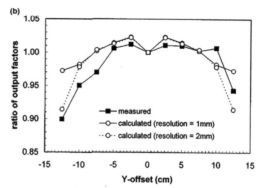

FIGURE 4.5 Comparison between measured and Monte Carlo calculated MLC dose profiles ($10 \times 1\,cm^2$ field, 12.5 cm off-axis, a) and output factors ($10 \times 1\,cm^2$ field at various off-axis distances, b). The agreement is good in general, especially for relative doses but it also shows that absolute dose output for these small fields is hard to calculate correctly. (From De Vlamynck et al. [77]. Reproduced with kind permission.)

complex designs for the leaf ends and the leaf edges. An additional complication is that MLC leaves can move during beam delivery in step-and-shoot and dynamic treatments. The only way that these complex beam-shaping devices can be fully taken into account is by Monte Carlo simulation, but this is a far from trivial task. Furthermore, significant differences in radiation spectrum in different sections of IMRT fields could have significant effects in radiobiology or for film measurements often used in IMRT verification [74].

The Monte Carlo research for modelling MLCs has mainly been focused on two aspects: exactness of the geometrical models and methods to improve calculation efficiency. Because different linac manufacturers have implemented different designs of MLCs, there is no generic MLC model available in codes like BEAM. There are also differences in how IMRT or VMAT beams are delivered in different linacs. All of this has to be taken into account in a Monte Carlo model. We would also like to point out that the mass density of the tungsten used in the MLC leaves usually has to be ascertained for every linac, due to significant variations in different linacs. Backscatter from the MLC leaves to the monitor chamber does not play a significant role due to the large distance of the MLC to the chamber [61], the presence of a backscatter plate [75], or the thickness of the monitor chamber downstream window thickness [68].

An early Monte Carlo study of MLC leaf geometries was done by Küster [76]. She used the GEANT3 Monte Carlo code to investigate the characteristics of two experimental designs of MLCs for a Siemens Primus linac with the purpose of determining the leaf end shape required to optimize dose penumbras. In this feasibility study, leaf leakage was also investigated but good agreement between calculations and measurements was not obtained.

One of the first to show good agreement between measured and simulated MLC dose distributions was De Vlamynck et al. [77]. They built a model for a 6 MV MLC-shaped photon beam from an Elekta SL25 linac. The rounded shape of the MLC leaf ends was approximated by using a stack of MLC

modules in BEAM simulations. Interleaf leakage was not modeled in this study. Figure 4.5 gives an example of the good agreement obtained between calculated and measured dose profiles and output factors for small fields with an off-axis position of up to 12.5 cm. No significant spectral differences were noted when large and small fields were compared. In the same paper, a new component module for the BEAM code, MLCQ, was introduced. This allowed modelling of MLC leaves with curved leaf ends.

Ma et al. [78] modified BEAM/DOSXYZ to allow the simulation of multiple-beam fixed-gantry IMRT treatments. The information in the linac MLC leaf sequence file was used to derive a fluence map from which particles were sampled during the simulations. The statistical weight of the particles was altered using the information in the fluence map. The MLC leakage was incorporated by reducing the particle statistical weight for the MLC-blocked sections. Their model did not include the effects of the leaf shape, tongue-and-groove design, and scattering in the MLC material, but still they reached agreement within 2% for measured and calculated dose distributions. For complex IMRT plans, they showed differences between their Monte Carlo system and a conventional pencil-beam algorithm of up to 20%. The same group of workers [79] later studied the MLC tongue-and-groove effect on IMRT dose distributions. They used a ray-tracing approach to modify the fluence maps so that the presence of the tongue-and-groove geometry was accounted for. For single-field IMRT, dose differences due to the tongue-and-groove effect of 4.5% were noted; when more than five fields were used, the differences fell below 1.6%. They remark that the effect of the tongue-and-groove geometry is probably insignificant in IMRT, especially when organ/patient movement is considered. The same workers [80] introduced a new BEAM component module, VARMLC, that allowed modelling the tongue-and-groove effect of some MLCs.

Work by another group [81–83] introduced an elegant model to take the MLC geometry into account in Monte Carlo simulations without actually having to model any particle transport in the MLC material. In Keall et al. [81], the MLC was compressed in a thin layer in the beam direction. Geometrical paths of photons passing through the MLC were determined for particles sampled from a phase-space file above the MLC. The statistical weights of the particles are modified according to the probability that the particles reach the bottom of the MLC geometry, where a new phase-space file is constructed. The model took into account intraleaf thickness variations and the curved shape of the leaf end but omitted interleaf leakage and charged particle transport, and hence production of bremsstrahlung photons and annihilation photons. First-scattered Compton photons were included but transport across leaves was not considered. This model was extended [83] to take the leaf-edge effect into account. Cross-leaf photon transport and charged particle transport in the leaves were still ignored. They make the interesting observation that it may suffice to model only first Compton scatter and omit electron transport in other linac components upstream from the MLC. The same group also fully simulated MLC geometries [82]. The leaf ends, leaf edges, mounting slots, and holes were included in the study. They investigated radiation leaking through the MLC. Depending on the field size defined by the jaws above the MLC and the photon energy, they found that fully blocked MLC fields had a radiation leakage dose of about 1.5%–2%, compared to the dose in an open field. It was observed that the MLC-generated photons add a broad background dose to the transmitted radiation (Figure 4.6). Electrons emitted from the MLC cause up to 35% of the surface dose [82] in an 18 MV MLC-blocked photon beam. For 6 MV photons, significant hardening of the photon spectrum behind the MLC was observed, resulting in a significant shift of the depth dose curve. No such effect was noted for 18 MV photons.

A method to greatly speed up MLC simulations was reported [84], in which sophisticated variance reduction techniques were used to automatically determine geometrical regions that require detailed transport and others where transport may be performed in an approximate fashion. They also show that modelling only single Compton scatter in the MLCs does not suffice.

Liu et al. [85] were the first to implement fully dynamic MLC motion to allow modelling step-and-shoot or dynamic IMRT delivery. This resulted in a new BEAM component module, DMLCQ. During sampling of particles for transport, the leaf positions were randomized according to the number of MUs in each field segment. This information was derived from the linac leaf sequence file. A correction for the difference between nominal and actual leaf position was required. Figure 4.7 compares Monte Carlo calculations and measurements for an IMRT field delivered in step-and-shoot mode. One of the most detailed MLC models ever published [86], based on Liu et al. [85], included leaf leakage, mounting screws, leaf divergence, support railing groove, leaf tips, and dynamic motion.

Li et al. [87] built a model for intensity-modulated arc therapy (IMAT) delivery, which involves an Elekta linac SL20 arcing around the patient while the MLCs are moving. An additional complication is that the SL20 uses its internal wedge for some fractions of the delivery. They obtained good agreement with measurements but found up to 10% difference between Monte Carlo and conventional dose calculation techniques for clinical IMAT fields. Many more papers on Monte Carlo simulations for dynamic therapy have been published since including flattening filter-free linacs [88–92], more on

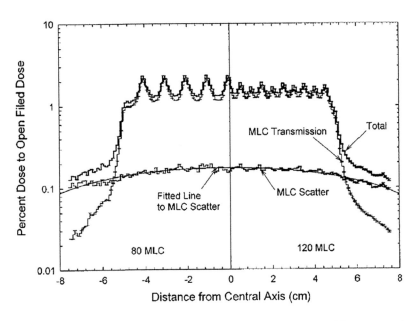

FIGURE 4.6 A dose profile (% of open field dose) perpendicular to the direction of leaf motion of the Monte Carlo computed radiation leakage from a MLC-blocked 6 MV 10×10 cm² field at 5 cm depth in a water phantom for the 80-leaf (left-hand side) and the 120-leaf (right-hand side) MLCs. (From Kim et al. [82]. Reproduced with kind permission.)

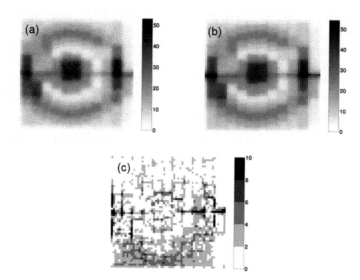

FIGURE 4.7 Monte Carlo-calculated (a) and film-measured (b) dose distributions at 1.5 cm depth for a 6 MV step-and-shoot IMRT field. Panel (c) shows the voxel-to-voxel differences in obtained dose (|calculated−measured| / max(measured)). (From Liu et al. [85]. Reproduced with kind permission.)

this topic in the chapter on *Dynamic beam delivery and 4D Monte Carlo* in this book.

The effect of the MLC on portal dose distributions was studied using Monte Carlo techniques [93,94].

4.2.2 Full Linac Modelling

In the following two sections, we will describe complete linac models for photon beams. The photon beam models can basically be divided into two categories. The first approach is to do complete simulations of linacs and either use the obtained particles directly in calculations of dose and fluence etc in phantoms, or store phase-space information about these particles at possibly several levels in the linac for further use. The phase space can be sampled for further particle transport in the rest of the geometry (linac or phantom). If the phase space is situated at the bottom of the linac, this then effectively replaces the linac and becomes the virtual linac. We will call this the phase-space approach. In Figure 4.1, two levels where phase-space information might be collected, are indicated (levels AA' and BB'). A particularly elegant approach couples BEAMnrc directly to DOSXYZnrc simulations, thereby eliminating the need to store intermediate phase-space files.

The second approach calculates particle distributions differential in energy, position, or angle in any combination of these in one, two, or three-dimensional histograms. This is usually done for several subsources of particles in the linac and is termed a Monte Carlo source model. The linac is then replaced by these subsources, which constitute the virtual linac. The source model approach often starts from information collected in phase spaces in the linac. Source models inevitably have approximations since they condense individual particle information in histograms, whereas all the information on

individual particles is preserved in the phase-space approach. The correlation between angle, energy, and position of particles might be partially or completely lost depending on the degree of complexity of the source model. The disadvantage of working with phase spaces is the large amount of information to be stored and the slower sampling speed during retrieval of all this information. Also, approximations in the linac model will result in uncertainties in the phase-space approach. Source models often involve data smoothing so they usually result in less statistical noise. These models need to be developed, though, which also requires resources. As computer speed and data storage capabilities increase rapidly, there may be less need for source models nowadays.

4.2.2.1 Phase-Space Models

As mentioned before, Patau et al. [1] pioneered Monte Carlo simulation of a more or less complete model of a photon linac. They presented energy and angular distributions of photons before and after a flattening filter but presented no further analysis of the influence of the linac components on the photon beam. In another early study [5], linac components were sequentially made transparent (i.e., no interactions were allowed in them) to study the influence of the components on the particle beam. They found that in a 21 MV beam, the thicker flattening filter combined with the more forward-peaked bremsstrahlung angular distribution caused most of the photon scatter. In a 6 MV beam, the primary collimator contributed most to the scatter, corresponding to the larger angular bremsstrahlung distribution for lower primary electron energies.

Mohan et al. [6] presented a much-quoted EGS3 simulation study of a set of Varian linacs producing 4–24 MV photon beams. The model included a target/backing, a primary collimator, a flattening filter, and a secondary collimator system.

They noted that off-axis photon spectra were softer than at the central axis. The collimating jaws had no significant effect on the energy and angular photon distributions. The beam was mostly determined by the target and the flattening filter. For 15 MV photons, they found that when photons reached a plane at 100 cm from the source, 93.5% had never scattered, whereas 2.8%, 3.5%, and 0.2% suffered scattering interactions in the primary collimator, flattening filter, and secondary collimator system, respectively. The photon spectra presented in this study are a reasonable approximation; they have been frequently used, e.g., for patient dose calculations. Their angular distributions, however, are only crude approximations.

A model of a 6 MV Siemens linac photon beam, including the exit window, target, primary collimator, flattening filter, monitor chamber, and collimating jaws, was published by Chaney et al. [23]. They focused on head scatter contributions. Despite the model being very approximate, the calculated scatter contributions are nevertheless found to be in fair agreement with more detailed studies. They found that about 9% of the photons in a large field were derived from scatter. Figure 4.8 shows their scatterplot of the sites of origin of scattered particles. From this figure, it is evident that Monte Carlo simulations – besides producing aesthetically pleasing pictures – can provide a clear insight into which parts of the linac play an important role as a source of scatter or particle generation.

Libby et al. [95] give a detailed information on how to validate Monte Carlo linac models. As an example of verification of linac geometry, they show how the complex shape

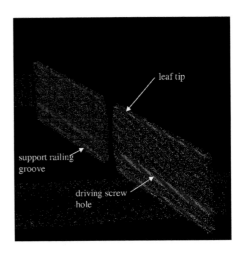

FIGURE 4.9 Photon raytracing an MLC leaf geometry in Monte Carlo simulations. (From Heath and Seuntjens [86]. Reproduced with kind permission.)

of the flattening filter can be reproduced and compared to manufacturer's data by bombarding it with mono-energetic, mono-directional photons, and calculating the attenuation profile. Another approach [86] for verifying complex geometries is raytracing photon tracks in such a way that photon coordinates are stored every time a photon crosses a boundary between MLC leaf material and air. Figure 4.9 gives an example of the geometry that can be visualized in this fashion.

Detailed models of photon linacs were presented by many others [25,27,96–104]. Comprehensive studies on the influence of linac components and the characteristics of the primary electron beam were done by Sheikh-Bagheri et al. [105–107]. They advocated using in-air dose profiles measured with an ion chamber and a buildup cap to derive the energy and spatial characteristics of the primary electron beam. Figure 4.10 illustrates the sensitivity of the off-axis factor (defined as the ratio of measured dose to water with an ion chamber in air at an off-axis position and at the central axis) to the primary electron energy and the width of the radial intensity distribution of the electron beam. By avoiding phantom scatter as a confounding factor, the authors clearly demonstrate the differential effects of the components in their model. Interestingly, the insensitivity of dose calculations to the energy distribution of the primary electron beam for photon beam simulations is in contrast to electron beams where both buildup and build-down regions are highly sensitive to the primary electron energy distribution.

A general finding is that the average photon energy is below one-third of the nominal energy at the central axis (more so for the highest-energy beam) and that the average energy 20 cm off-axis is decreased by 0.5–2.0 MeV (the greatest decrease usually corresponds to the highest nominal energy). Ding [25] noted that the mean energy of the charged particles in a photon beam varies less with off-axis distance than the photon mean energy. A procedure to tune a Monte Carlo linac model was outlined in Verhaegen and Seuntjens [11].

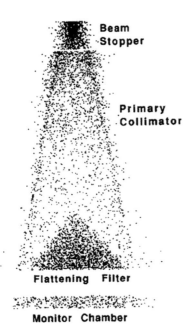

FIGURE 4.8 Distribution of sites of origin of photons for a 6 MV photon beam. Clearly visible are the target/beam stopper, the primary collimator, the flattening filter, and the monitor ion chamber. The center of mass of all the origin sites was found to be inside the flattening filter at 6.2 cm from target. (From Chaney et al. [23]. Reproduced with kind permission.)

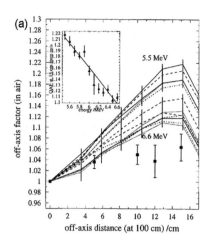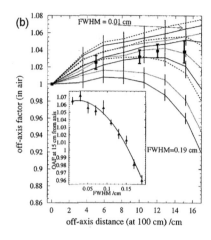

FIGURE 4.10 Linear dependence of off-axis factors at 15 cm off-axis distance (see insert) to the primary electron energy in a 6 MV Siemens photon beam (a). Quadratic dependence of off-axis factors at 15 cm off-axis distance (see insert) to the Gaussian width of the primary electron intensity distribution in an 18 MV Varian photon beam (b). (From Sheikh-Bagheri and Rogers [106]. Reproduced with kind permission.)

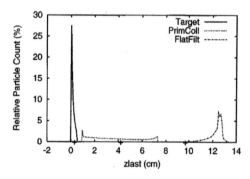

FIGURE 4.11 Distribution of the points of origin of photons in the target, primary collimator, and flattening filter in a 10 MV photon beam. The derived estimates for the source positions are indicated by the crosses on the x-axis. (From Liu et al. [62]. Reproduced with kind permission.)

4.2.2.2 Source Models

Early Monte Carlo-based source models were first introduced for electron beams and then later for photon beams [59,62,108]. Liu et al. [59,62] introduced a dual-photon source model: one source for the primary photons coming directly from the target and a second extra-focal source to describe scatter from mainly the primary collimator and the flattening filter. A third source took the electron contamination into account. Deriving the source model started by performing a full linac Monte Carlo simulation and phase-space analysis. The obtained distribution of points of origin of photons for a 10 MV beam is illustrated in Figure 4.11. Obviously, it is very important in such a model that the geometrical position of the scatter sources is estimated correctly. Inevitably, replacing the real three-dimensional source distributions with two-dimensional ones is an approximation. The primary photon source at the target was modeled with a radius of 0.1 cm.

The effect of the jaws was modeled as eclipsing collimators, thereby ignoring the small amount of scattered radiation the jaws contribute. Backscatter towards the monitor chamber was added explicitly, which is often omitted in later works. Despite the approximations in the model [59,62], good agreement was reached between calculated and measured dose distributions and output factors.

Other complex source models can be found in the literature [24,26,93,109–112]. The PEREGRINE group [109] performed a very detailed study to investigate the influence of linac components on photon fluence, energy spectra, angular distributions etc. They introduced the concept of *correlated histograms* which contain information extracted from phase-space files in such a way that correlations between energy, angles etc are preserved to some degree. Target photons were found to originate from a small disk source with a radial fluence distribution, the primary collimator was found to generate photons in a ring source close to its upper edge, and the flattening filter was found to produce photons uniformly throughout the filter geometry. Figure 4.12 shows the correlated histogram used for the two subsources for a 6 MV beam: target and scattered photons. The authors reached very good agreement between full phase-space-based simulations and their model. A subsource was added to model the electron contamination [26] in the PEREGRINE photon Monte Carlo treatment planning system. The jaws were modeled as a masking collimator, which causes the dose outside the penumbra to be up to 15% too low. Backscatter to the monitor chamber was included as an empirical correction.

One of the most complex source models published up to date was presented by Fix et al. [113]. Using GEANT3 simulations, they derived a model for a 6 MV linac beam with twelve subsources: one for the target, flattening filter, primary collimator, and mirror each, and two each for all four of the secondary collimators. The two subsources for each secondary collimator

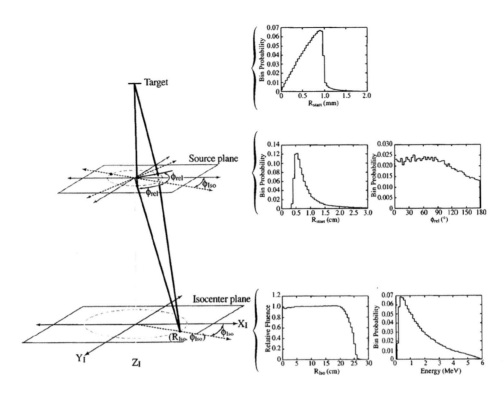

FIGURE 4.12 Correlated histograms of starting position and starting angle for the two subsources in a virtual source model of a 6 MV photon beam. The energy fluence and energy spectrum are described at the isocenter. The origin of photons from the primary collimator or the flattening filter is sampled from the radial and angular distributions. The starting position of the target photons is sampled from a radial distribution, with assumed angular isotropy. (From Schach von Wittenau et al. [109]. Reproduced with kind permission.)

represent radiation emanating from the inner side and the lower side of the collimators. It was found that the inner-side subsource generates most particles close to a 1 cm distance from the top of the collimator, whereas the bottom-side subsource generated most particles close to the field edge. The total contribution from the eight collimator sources was very small and is usually ignored in most linac source models. Data was generated for eight field sizes, which allowed interpolation for arbitrary fields.

The source model approach has to be used with care due to its inherent approximations and potential for systematic error. Its strength lies in the fact that the requirements on data storage are greatly reduced with respect to a full phase-space approach; figures of 400–10,000 have been quoted [24,110,114]. Because of the reduced phase-space reading/writing and because source models inherently involve some degree of fluence smoothing, the calculation time required to achieve a certain specified statistical variation in a dose calculation is also reduced. Monte Carlo-based treatment planning systems often resort to virtual source models.

4.2.2.3 Absolute Dose Calculations (Monitor Unit Calculations)

Dose calculations based on Monte Carlo simulations of linacs (or any other device) usually result in doses expressed in Gy/particle. In the BEAMnrc package, the term 'particle' refers to

the initially simulated particles, so in a linac simulation the calculated doses are in Gy per initial primary electron hitting the target. Even in simulations which are done in several parts, involving multiple phase-space files at different levels in the linac, this information is preserved in the phase-space files. This ensures a correct normalization of the doses per initial particle. Other Monte Carlo codes use different schemes or leave this up to the user. This absolute dose can be related easily to the absolute dose in the real world. By running a simulation that exactly matches an experimental setup to determine absolute dose, one can obtain the ratio of the measured and calculated doses which are expressed in units of [Gy/MU]/[Gy/particle], where MU stands for MU. The latter corresponds to the reading of the monitor ionization chamber in the linac. This ratio may serve as conversion factor for all beams produced by the linac for that particular photon energy. By obtaining these conversion factors for all linac energies, Monte Carlo calculated doses may be converted to absolute doses as used in radiotherapy practice. This approach is also valid for electron beams (see later) and is employed by most users. The reference setup that is frequently modeled is a dose determination in an open large field (e.g., $10 \times 10 \, \text{cm}^2$) in water, but in principle, any setup, not necessarily using open fields, can be modeled to determine the conversion factors for each photon energy. This approach was described in more detail in the literature [115,116] with special attention to IMRT treatments. A different approach which employs the relation between the simulated monitor chamber reading, the initial number of particles impinging on the target, and the field size was introduced by Popescu et al. [117]. This elegant method requires attention to the issue of backscatter of particles to the monitor chamber (Section 2.1.3). It doesn't rely on renormalization to measurements and relative output factors.

4.2.3 Stereotactic Beams

Linac-based stereotactic irradiation of intracranial and extracranial lesions is a frequently employed radiotherapy technique. Monte Carlo simulation of the collimated, narrow photon beams (10 –30 mm at the isocenter) is therefore important in two areas. First, knowledge of detailed beam data allows a proper understanding of detector response which is often affected by the absence of lateral electronic equilibrium. Second, treatment planning systems may employ Monte Carlo techniques to reliably take into account tissue heterogeneities near small targets. This implies that the basic properties of the beam should be accurately calculated using geometric treatment head information.

Several Monte Carlo studies on stereotactic beams [118–129] and a review paper [130] have been published. Studies focused on the accuracy of dose calculations in small fields, photon scatter, photon and electron spectra, and correction factors for measurements. A full simulation [124] of a dedicated 6 MV stereotactic linac with field sizes ranging from 5 to 50 mm showed that photon spectra are influenced by the small beams; mean photon energies at d_{max} in water were found to vary between 2.05 MeV for a 5 mm cone and 1.65 MeV for a 5 cm cone. They also observed that Spencer-Attix stopping power ratios at the central axis were depth independent, which is important information for measurements. Other studies simulated a small-field dynamic radiosurgery unit capable of producing 1.5 mm pencil beams [131,132] and focused on aspects of lateral electronic disequilibrium. Differences of up to 60% were found between Spencer-Attix stopping power ratios and ratios of dose to water and dose to cavity. It has been demonstrated [128,129] that small-field output factors are very sensitive to the dimensions of the primary electron beam, which suggests a way to commission this part of the linac geometry.

Some have suggested that Monte Carlo dose calculations for treatment planning in stereotactic beams may not be needed [133], while others [134] have argued that Monte Carlo simulation is required for reliable small-field dose calculation in regions of tissue heterogeneity for which, in conventional planning systems, often only the primary photon beam component (and not the scatter component) is considered. Full Monte Carlo simulations are preferred for small fields but virtual source models have also been proposed [123].

As an example, Figure 4.13 shows the level of detail that is needed to model a linac with a micro-MLC for stereotactic irradiation. The geometric accuracy of the leaf ends and sides, and the composition of the leaves are very important. The figure also shows the level of agreement with measurements that can be achieved by an accurate Monte Carlo model. Conventional dose calculation was shown to be considerably less accurate.

Monte Carlo studies of a CyberKnife robot-mounted small field linac have also been published [135–138]. A recent study introduced an exotic technique to enhance the simulation efficiency of small fields. The 'ant colony' method [125] was employed whereby geometric cells are overlaid with an importance map which is derived in an analogous fashion in which ants leave pheromone trails to food sources. In their approach, photons that reach regions of interest determine the importance maps so that subsequent photons may be simulated more efficiently.

A recent exotic application in stereotactic irradiation was about using Monte Carlo methods for designing a 'Compton lens' consisting of metal conical slits, with the aim to utilize Compton scattered photons towards the target region, which have the potential to contribute to the overall target dose [139].

4.2.4 Contaminant Particles

4.2.4.1 Electrons

Electron contamination in photon beams has been studied by numerous authors [140–158]. Electrons in the treatment beam are either generated in the high-Z components of the linac or in the air. It is known that electrons in the photon beam are responsible for a significant fraction of the surface dose in a phantom or patient. The dose from contaminant electrons drops off rapidly with depth and is usually insignificant at the depth of maximum dose of the photon beam, d_{max} (for an overview of experimental findings, see Ref. [11]).

Petti et al. [150,159] were the first to dedicate a Monte Carlo study to electron contamination in a 25 MV photon beam. They observed that 70% of the electrons at the phantom surface were generated in the flattening filter and monitor ion chamber, 13% from the collimating jaws, and 17% in the air for an SSD of 80 cm. For an SSD of 400 cm, these fractions changed to 34%, 5%, and 61%, respectively. They concluded from this that at all SSDs the contribution from the flattening filter and monitor chamber is significant, whereas the contribution from air-generated electrons is only important at very large SSD. The former were also found to cause the shift of d_{max} with field size because the collimating jaws act as an eclipsing absorber. At an SSD of 80 cm, these electrons had a most probable energy of 1.5 MeV (with 40% of them having an energy exceeding 5 MeV) and a forward-peaked angular distribution with 60% of the electrons moving at an angle with respect to the beam axis of less than 16°. These findings explain why these electrons still play a significant role at very large SSDs. The electrons generated in the collimating jaws, which are commonly assumed to be important, were found to play a minor role.

Some Monte Carlo studies [113,147] showed energy spectra, energy fluence distributions, and mean energy distributions

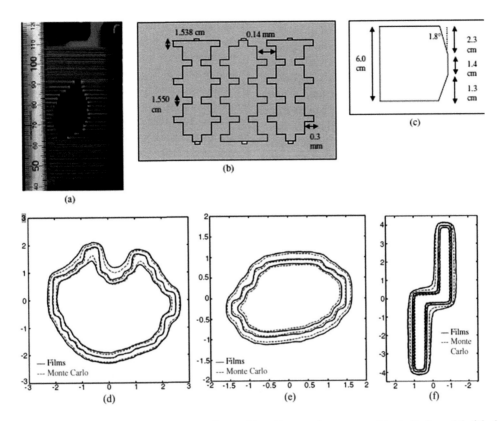

FIGURE 4.13 Top panel; a micro-MLC for stereotactic irradiation (a) and its complex geometry Monte Carlo model of the leaf ends (b) and sides (c). Bottom panel: a few examples of Monte Carlo calculations and film measurements of the isodose distribution in small fields (d-f). (From Belec et al. [118]. Reproduced with kind permission.)

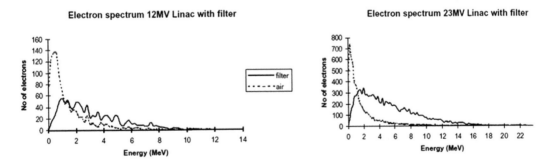

FIGURE 4.14 Monte Carlo generated energy spectra of secondary electrons at 100 cm from the target in a Saturne linac. The electrons originating in the flattening filter and in the air below it are depicted. They exhibit a clear difference in spectral shape. (From Malataras et al. [147]. Reproduced with kind permission.)

for the electrons reaching the phantom or bottom of the linac. They found that the energy spectra of electrons generated in the flattening filter and in air are very different: the former have a broad energy spectrum with an average electron energy comparable to the average photon energy. The latter have a narrow energy distribution, peaked at low energies. These results may be very linac dependent. It was pointed out [147] that the increasing probability of pair production causes the flattening filter to become the dominant electron source at high photon energies (see Figure 4.14).

For quite a long time, Monte Carlo studies of surface dose in photon beams showed significant differences with measurements [26,96,105,106,141]. Only with very detailed linac models and taking into account that radiation detectors may need various response corrections close to an air/phantom interface, a few workers [107,140] succeeded in achieving good agreement. These studies showed that Monte Carlo dose calculations near surfaces are often more reliable than measurements, especially if response corrections are not taken into account.

4.2.4.2 Neutrons

It is known that photons with an energy exceeding about 8 MeV can produce neutrons in the target and collimating structures. For an overview of experimental techniques, we refer to AAPM Report 19 [160]. Several authors used Monte Carlo techniques to simulate 15–50 MV photon beams from a variety of linacs [161–164]. They included methods to estimate (γ, n) production in linac components (mostly the target) and phantom geometries. Monte Carlo codes that can handle photo-neutron production are, e.g., MCNP(X) and GEANT4. It has been determined that photo-neutron production in soft tissue phantoms leads to a neutron dose of about 0.003% of the photon dose inside a 25 MV photon beam [161]. In studies for linac-generated neutrons from 15 to 18 MV photon beams, maximum equivalent neutron doses at the central axis of about 2–5 mSv per Gy absorbed photon dose were found [142,164,165]. Outside the photon field, the linac-generated neutron dose has been observed to decrease by about a factor of ten, meaning the ratio of absorbed dose by the neutrons is higher outside the field compared to inside the field. Most of the linac-generated neutrons depositing energy in a tissue-equivalent phantom had an energy >1 MeV [165]. Recently, the photo-neutron spectrum from a wedged 18 MV photon beam was studied [166]. They found that the neutron fluence decreases with increasing field size for open beams, and increases with increasing field size for wedged beams. Another issue of interest is the calculation of neutron dose in linac mazes for radioprotection [167,168].

4.2.5 Radiotherapy Kilovolt X-Ray Units

Kilovolt X-ray tubes have been used in radiotherapy for many decades but are now being used less frequently for cancer treatment. Linac electron beams and brachytherapy sources have largely replaced the use of kV X-rays for superficial or skin lesions, respectively. This by no means implies that there is no role for Monte Carlo modelling of kV X-ray tubes. In diagnostic imaging or imaging for treatment planning purposes, Monte Carlo methods are commonly used to model X-ray tubes for planar or CT imaging. An example of CT scanner modelling versus Compton scatter spectroscopy is Bazalova and Verhaegen [169]. The literature contains many examples of Monte Carlo modelling of X-ray tubes for diagnostic and mammography applications [170–172] for, e.g., dosimetric purposes. Detailed discussion of these mostly diagnostic applications is outside the scope of this chapter. There are some specialized applications of kV X-rays in radiotherapy, which are under investigation and where Monte Carlo techniques play an important role, such as contrast-enhanced radiotherapy, in which kV X-rays are used to bombard targets laden with dose enhancers such as high atomic number contrast media [173] or gold nanoparticles [174]. To model the interaction of kV X-rays with high atomic number materials accurately, a number of physical processes must be implemented, which makes photon transport in the kV energy

range harder than in the MV energy range where Compton scatter is the dominant interaction. Examples of things that need to be implemented in the transport models and which largely can be ignored at MV energies are electron impact ionization, Rayleigh scattering, photo-electron angular distributions, and fluorescent decay. Another specialized area is the modelling of miniature X-ray tubes which are small enough to be implanted in lesions to irradiate them from within. These devices often consist of very thin transmission X-ray targets made of a composite of various high atomic number materials, stressing the importance of the accuracy of bremsstrahlung cross sections and the modelling of electron impact ionization. Examples are the work of Yanch and Harte [175] on brain applications and Liu et al. [176] on electronic brachytherapy. Another special application is small animal irradiation for preclinical studies, which is usually done with kV X-rays to avoid the buildup regions of MV photon beams in small animals such as mice [177]. Monte Carlo modelling also there plays a role in beam simulation and dose calculation, and this is covered in another chapter in this book.

X-ray spectra are used in many different applications, e.g., in studies in radioprotection, radiobiology, radiation quality, response of radiation detectors etc. Therefore, several authors created methods to generate X-ray spectra as a function of target angle (most efforts focused on tungsten targets), accelerating voltage, and filtration. Whereas earlier work [178,179] employed analytical techniques, the most recent work relies on Monte Carlo methods for a detailed electron penetration in the target [180,181]. The SpekCalc code [182] uses a graphical user interface to perform a very fast, yet accurate, X-ray spectrum calculation for tungsten targets bombarded by electrons in the range 50–300 keV.

In the late nineties, the BEAM code was used to simulate entire radiotherapy X-ray tubes (50–300 kV), including effects such as tube voltage ripple [183]. The results were compared to spectral measurements and another Monte Carlo code, MCNP. Both codes had problems with the characteristic X-ray lines due to limitations in the modelling of electron impact ionization. Compared to spectral measurements, there was a clear overestimation of the low-energy component in the photon spectrum at the expense of contributions in the high-energy part of the spectrum. Other workers modeled X-ray tubes for radiotherapy or diagnostic applications [170,184–187], and special attention was paid to variance reduction techniques [188]. Other studies modeled in great detail the backscatter from electrons at X-ray targets, and the subsequent re-entry of the electrons due to interaction with the electric field inside the X-ray tube [189,190]. This effect causes the well-known extrafocal radiation of which the authors demonstrate that it may influence the spectral shape (therefore, the half-value layer) and the air kerma. Van der Heyden et al. [190] included a model for the complex focal spot shape at the anode from pinhole camera measurements. Models like these can be used to improve, e.g., cone beam CT image reconstruction. Figure 4.15 shows the different generations of photons and electrons (primary, first-generation electron backscatter etc)

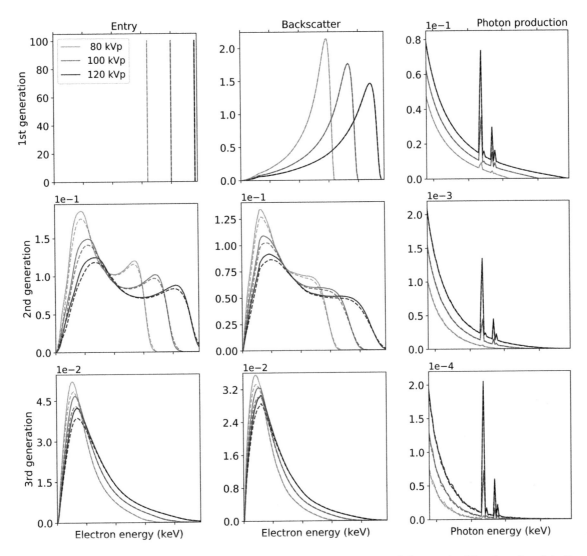

FIGURE 4.15 MC-calculated spectra of the entering electrons (left column), backscattered electrons (middle column), and the X-ray production integrated over all angles (right column) on a 10° angled anode for the small focal spot (0.65 mm; solid lines) and the large focal spot (1.0 mm; dashed lines). (From van der Heyden et al. [190]. Reproduced with kind permission.)

derived from simulations. For applications with a hard filtration and tight beam collimation, the effect of the extrafocal radiation appears minimized.

A final application that deserves brief mention in this section is the simulation of X-ray units for kV cone beam CT imaging, which is a popular technique used in image-guided radiotherapy [184,191–194]. Figure 4.16 gives an overview of a simulated cone beam CT and the distribution of the last interaction sites of the photons in the geometry [193]. These simulation studies may serve to remove scatter from the cone beam images, to optimize the imaging chain, or to derive the patient imaging dose. These topics are covered elsewhere in this book.

4.2.6 ^{60}Co Teletherapy Units

As the relative importance of traditional ^{60}Co teletherapy in comparison with accelerator therapy has dwindled in

industrialized countries, the number of Monte Carlo studies on these modalities is also limited. Although ^{60}Co decay gives rise to only two gamma-ray lines of 1.17 and 1.33 MeV, the spectrum of photons in a teletherapy unit is affected by the source encapsulation as well as by the collimating system. Contaminant electrons are also an issue, especially in calibration setups. Monte Carlo calculations have been used to characterize ^{60}Co beams in clinical as well as standard dosimetry laboratory setups.

Han et al. [195] modeled the Theratron 780 teletherapy machine with a simplified source capsule. Only about 72% of the photons were found to be either 1.17 or 1.33 MeV; the other 28% were single or multiple scattered photons mixed with a small number of unscattered bremsstrahlung and positron annihilation photons. They studied photon fluence spectra, and the effect of field size on output and dose calculations. It was found that an increase in output of the ^{60}Co unit with

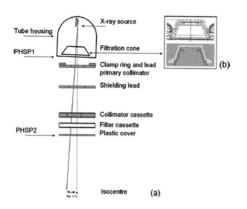
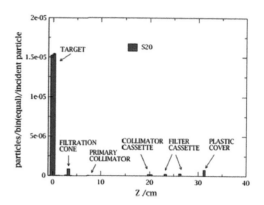

FIGURE 4.16   Left: schematic representation of an Elekta Synergy cone beam CT device. The tube filtration components are shown in the inset (b). Right: distribution of last interaction sites. (From Spezi et al. [193]. Reproduced with kind permission.)

increasing field size is caused by scattered photons from the primary tungsten collimator. The electron contamination of an AECL ^{60}Co teletherapy unit was studied [196], and it was found that electrons originate from the source capsule (most effective at close SSD), the collimators, and the air (most effective at large SSD). At 80 cm SSD, for a field size equivalent to $35 \times 35 \, cm^2$, about 45% of the dose at d_{max} is due to contaminant electrons, of which 22% come from the source capsule and air-generated electrons; 10% and 13% is from electrons emerging from outer and inner collimators, respectively. The major source of scattered photons was the source capsule itself, and scattered photons contribute 18% of the maximum dose. The inner and outer collimators were found to increase the effect of photon scatter by a few percent. More recent studies [197,198] modeled ^{60}Co units from calibration laboratories in great detail and focused on small but important spectral effects. Variance reduction techniques especially for modelling ^{60}Co sources were studied [199].

Besides its use in traditional teletherapy, ^{60}Co is implicated in special treatment techniques such as the GammaKnife. Of special interest for this treatment modality is the study of dose distributions and output factors for the very small field sizes used for which detailed source modelling is important. Detailed Monte Carlo studies can be found in the literature [200,201].

4.3 Summary

Monte Carlo simulations are an extremely powerful tool in modern radiotherapy. That the Monte Carlo community is alive and well can be seen from successful workshops that were held in recent years in various places around the world.

Most of the studies cited in this chapter used EGS/EGSnrc-based codes (including the user interface BEAM/BEAMnrc) but other codes such as GEANT4, MCNP(X), and PENELOPE are popular. We highlighted the capabilities of these codes for modelling and analyzing external radiotherapy photon beams, but they also have proven to be of great value for kV X-ray systems, brachytherapy sources, and patient dose calculations.

In addition, Monte Carlo simulations play an increasingly important role in imaging for diagnostics and patient treatment verification. In this and other chapters, the 4D modelling of linac beam delivery in volumetric arc therapy is mentioned. Elsewhere in this book, the capabilities of Monte Carlo codes to incorporate 4D motion such as in patient/organ movement and deformation is discussed.

Simulation time and storage of large amounts of data, once prohibitive for complex Monte Carlo modelling, is now less of an issue. Technology is aiding here, e.g., with the advent of ultrafast GPU processors.

Monte Carlo-based treatment planning is now rapidly becoming the workhorse of patient planning. Speed and accuracy will be of the essence, and extensive guidelines on commissioning Monte Carlo-based dose calculation systems are only now becoming available [202].

It is clear that Monte Carlo methods will remain very important in the study of new technology in radiotherapy. To name an example, Monte Carlo methods are facing new challenges in, e.g., the design and dosimetry in hybrid MRI-linac systems [203–207] where electrons are perturbed by the strong magnetic fields employed in MRI imaging. Also in exotic beam verification applications such as using Cerenkov radiation from photon beams interacting in a water phantom, Monte Carlo methods have no equal [208,209].

References

1. Patau, J., C. Vernes, M. Terrissol, and M. Malbert. 1978. Calcul des caracteristiques qualitatives (TEL, F.Q., equivalent de dose) d'un faisceau de photons de freinage a usage medical, par simulation de sa creation et de son transport. In *Sixth Symposium on Microdosimetry.* J. Booz and H. Ebert, editors. Harwood Academic Publishing, London. pp. 579–588.

2. McCall, R. C., R. D. McIntyre, and W. G. Turnbull. 1978. Improvement of linear accelerator depth-dose curves. *Med Phys* 5(6):518–524.

3. Faddegon, B. A., C. K. Ross, and D. W. Rogers. 1991. Angular distribution of bremsstrahlung from 15-MeV electrons incident on thick targets of Be, Al, and Pb. *Med Phys* 18(4):727–739.

4. Faddegon, B. A., C. K. Ross, and D. W. Rogers. 1990. Forward-directed bremsstrahlung of 10- to 30-MeV electrons incident on thick targets of Al and Pb. *Med Phys* 17(5):773–785.

5. Nilsson, B., and A. Brahme. 1981. Contamination of high-energy photon beams by scattered photons. *Strahlentherapie* 157(3):181–186.

6. Mohan, R., C. Chui, and L. Lidofsky. 1985. Energy and angular distributions of photons from medical linear accelerators. *Med Phys* 12(5):592–597.

7. Udale, M. 1988. A Monte Carlo investigation of surface doses for broad electron beams. *Phys Med Biol* 33:939–953.

8. Udale-Smith, M. 1992. Monte Carlo calculations of electron beam parameters for three Philips linear accelerators. *Phys Med Biol* 37:85–105.

9. Rogers, D. W., B. A. Faddegon, G. X. Ding, C. M. Ma, J. We, and T. R. Mackie. 1995. BEAM: a Monte Carlo code to simulate radiotherapy treatment units. *Med Phys* 22(5):503–524.

10. Ma, C. M., and S. B. Jiang. 1999. Monte Carlo modelling of electron beams from medical accelerators. *Phys Med Biol* 44(12):R157–189.

11. Verhaegen, F., and J. Seuntjens. 2003. Monte Carlo modelling of external radiotherapy photon beams. *Phys Med Biol* 48(21):R107–164.

12. Brualla, L., M. Rodriguez, and A. M. Lallena. 2017. Monte Carlo systems used for treatment planning and dose verification. *Strahlenther Onkol* 193(4):243–259.

13. Greene, D., and P. Williams. 1997 *Linear Accelerators for Radiation Therapy*. Institute of Physics Publishing, Bristol and Philadelphia.

14. Van Dyk, J. 2005. *Modern Technology of Radiation Oncology*, Volumes 1 and 2 (hardcover). Medical Physics Publishing, Madison, WI.

15. Nelson, W., H. Hirayama, and D. Rogers. 1985. The EGS4 code system. In Stanford Linear Accelerator Center Report SLAC-265.

16. Kawrakow, I. 2000. Accurate condensed history Monte Carlo simulation of electron transport. II. Application to ion chamber response simulations. *Med Phys* 27(3):499–513.

17. Rogers, D. W. O., B. Walters, and I. Kawrakow. 2019. BEAMnrc Users Manual, National Research Council of Canada, NRCC Report PIRS-0509(A)revL.

18. Agostinelli, S. 2003. Geant4-a Simulation Toolkit. *Nucl Instrum Methods Phys Res A* 506(3):250–303.

19. Briesmeister, J. 2000. MCNPTM-A general Monte Carlo N-Particle transport code, Version 4C, LA-13709-M.

20. Werner, C. J., J. S. Bull, C. J. Solomon, F. B. Brown, G. W. McKinney, M. E. Rising, D. A. Dixon, R. L. Martz, H. G. Hughes, L. J. Cox, A. J. Zukaitis, J. C. Armstrong, R. A. Forster, and L. Casswell. 2018. MCNP6.2 Release Notes, LA-UR-18-20808.

21. Sempau, J., E. Acosta, J. Baro, J. Fernandez-Varea, and F. Salvat. 1997. An algorithm for Monte Carlo simulation of coupled electron-photon transport. *Nucl Instrum Methods Phys Res B* 132:377–390.

22. Salvat, F. 2014. PENELOPE: a code system for Monte Carlo simulation of electron and photon transport. Issy-les-Moulineaux: OECD Nuclear Energy Agency.

23. Chaney, E. L., T. J. Cullip, and T. A. Gabriel. 1994. A Monte Carlo study of accelerator head scatter. *Med Phys* 21(9):1383–1390.

24. Deng, J., S. B. Jiang, A. Kapur, J. Li, T. Pawlicki, and C. M. Ma. 2000. Photon beam characterization and modelling for Monte Carlo treatment planning. *Phys Med Biol* 45(2):411–427.

25. Ding, G. X. 2002. Energy spectra, angular spread, fluence profiles and dose distributions of 6 and 18 MV photon beams: results of monte carlo simulations for a varian 2100EX accelerator. *Phys Med Biol* 47(7):1025–1046.

26. Hartmann Siantar, C. L., R. S. Walling, T. P. Daly, B. Faddegon, N. Albright, P. Bergstrom, A. F. Bielajew, C. Chuang, D. Garrett, R. K. House, D. Knapp, D. J. Wieczorek, and L. J. Verhey. 2001. Description and dosimetric verification of the PEREGRINE Monte Carlo dose calculation system for photon beams incident on a water phantom. *Med Phys* 28(7):1322–1337.

27. Mazurier, J., F. Salvat, B. Chauvenet, and J. Barthe. 1999 Simulation of photon beams from a Saturne 43 accelerator using the code PENELOPE. *Physica Medica* XV(3):101–110.

28. Sarrut, D., M. Bardies, N. Boussion, N. Freud, S. Jan, J. M. Letang, G. Loudos, L. Maigne, S. Marcatili, T. Mauxion, P. Papadimitroulas, Y. Perrot, U. Pietrzyk, C. Robert, D. R. Schaart, D. Visvikis, and I. Buvat. 2014. A review of the use and potential of the GATE Monte Carlo simulation code for radiation therapy and dosimetry applications. *Med Phys* 41(6):064301.

29. Jan, S., G. Santin, D. Strul, S. Staelens, K. Assie, D. Autret, S. Avner, R. Barbier, M. Bardies, P. M. Bloomfield, D. Brasse, V. Breton, P. Bruyndonckx, I. Buvat, A. F. Chatziioannou, Y. Choi, Y. H. Chung, C. Comtat, D. Donnarieix, L. Ferrer, S. J. Glick, C. J. Groiselle, D. Guez, P. F. Honore, S. Kerhoas-Cavata, A. S. Kirov, V. Kohli, M. Koole, M. Krieguer, D. J. van der Laan, F. Lamare, G. Largeron, C. Lartizien, D. Lazaro, M. C. Maas, L. Maigne, F. Mayet, F. Melot, C. Merheb, E. Pennacchio, J. Perez, U. Pietrzyk, F. R. Rannou, M. Rey, D. R. Schaart, C. R. Schmidtlein, L. Simon, T. Y. Song, J. M. Vieira, D. Visvikis, R. Van de Walle, E. Wieers, and C. Morel. 2004. GATE: a simulation toolkit for PET and SPECT. *Phys Med Biol* 49(19):4543–4561.

30. Rodriguez, M. L. 2008. PENLINAC: extending the capabilities of the Monte Carlo code PENELOPE for

the simulation of therapeutic beams. *Phys Med Biol* 53(17):4573–4593.

31. Lovelock, D. M., C. S. Chui, and R. Mohan. 1995. A Monte Carlo model of photon beams used in radiation therapy. *Med Phys* 22(9):1387–1394.

32. Koch, H., and J. Motz. 1959. Bremsstrahlung cross section formulas and related data. *Rev Mod Phys* 31:920–955.

33. Desobry, G. E., and A. L. Boyer. 1991. Bremsstrahlung review: an analysis of the Schiff spectrum. *Med Phys* 18(3):497–505.

34. Ford, R., and W. Nelson. 1978. The EGS Code System: computer programs for the Monte Carlo simulation of electromagnetic cascade showers (Version 3). In Stanford Linear Accelerator Center Report SLAC-210.

35. Bielajew, A., R. Mohan, and C.-S. Chui. 1989. Improved bremsstrahlung photon angular sampling in the EGS4 code system. In NRCC Report 1–22.

36. Poon, E., and F. Verhaegen. 2005. Accuracy of the photon and electron physics in GEANT4 for radiotherapy applications. *Med Phys* 32(6):1696–1711.

37. Ali, E. S., M. R. McEwen, and D. W. Rogers. 2012. Unfolding linac photon spectra and incident electron energies from experimental transmission data, with direct independent validation. *Med Phys* 39(11): 6585–6596.

38. Munro, P., J. A. Rawlinson, and A. Fenster. 1988. Therapy imaging: source sizes of radiotherapy beams. *Med Phys* 15(4):517–524.

39. Loewenthal, E., E. Loewinger, E. Bar-Avraham, and G. Barnea. 1992. Measurement of the source size of a 6- and 18-MV radiotherapy linac. *Med Phys* 19(3):687–690.

40. Lutz, W. R., N. Maleki, and B. E. Bjarngard. 1988. Evaluation of a beam-spot camera for megavoltage x rays. *Med Phys* 15(4):614–617.

41. Treuer, H., R. Boesecke, W. Schlegel, G. Hartmann, R. Müller, and V. Sturm. 1993 The source-density function: determination from measured lateral dose distributions and use for convolution dosimetry. *Phys Med Biol* 38:1895–1909.

42. Schach von Wittenau, A. E., C. M. Logan, and R. D. Rikard. 2002. Using a tungsten rollbar to characterize the source spot of a megavoltage bremsstrahlung linac. *Med Phys* 29(8):1797–1806.

43. Jaffray, D. A., J. J. Battista, A. Fenster, and P. Munro. 1993. X-ray sources of medical linear accelerators: focal and extra-focal radiation. *Med Phys* 20(5):1417–1427.

44. Sonke, J. J., B. Brand, and M. van Herk. 2003. Focal spot motion of linear accelerators and its effect on portal image analysis. *Med Phys* 30(6):1067–1075.

45. Wang, L. L., and K. Leszczynski. 2007. Estimation of the focal spot size and shape for a medical linear accelerator by Monte Carlo simulation. *Med Phys* 34(2):485–488.

46. Galbraith, D. M. 1989. Low-energy imaging with high-energy bremsstrahlung beams. *Med Phys* 16(5):734–746.

47. Mah, D. W., D. M. Galbraith, and J. A. Rawlinson. 1993. Low-energy imaging with high-energy bremsstrahlung beams: analysis and scatter reduction. *Med Phys* 20(3):653–665.

48. Ostapiak, O. Z., P. F. O'Brien, and B. A. Faddegon. 1998. Megavoltage imaging with low Z targets: implementation and characterization of an investigational system. *Med Phys* 25(10):1910–1918.

49. Tsechanski, A., A. F. Bielajew, S. Faermann, and Y. Krutman. 1998. A thin target approach for portal imaging in medical accelerators. *Phys Med Biol* 43(8):2221–2236.

50. Flampouri, S., P. M. Evans, F. Verhaegen, A. E. Nahum, E. Spezi, and M. Partridge. 2002. Optimization of accelerator target and detector for portal imaging using Monte Carlo simulation and experiment. *Phys Med Biol* 47(18):3331–3349.

51. Flampouri, S., H. A. McNair, E. M. Donovan, P. M. Evans, M. Partridge, F. Verhaegen, and C. M. Nutting. 2005. Initial patient imaging with an optimised radiotherapy beam for portal imaging. *Radiother Oncol* 76(1):63–71.

52. Roberts, D. A., V. N. Hansen, A. C. Niven, M. G. Thompson, J. Seco, and P. M. Evans. 2008. A low Z linac and flat panel imager: comparison with the conventional imaging approach. *Phys Med Biol* 53(22):6305–6319.

53. Connell, T., and J. L. Robar. Low-Z target optimization for spatial resolution improvement in megavoltage imaging. *Med Phys* 37(1):124–131.

54. Robar, J. L., T. Connell, W. Huang, and R. G. Kelly. 2009. Megavoltage planar and cone-beam imaging with low-Z targets: dependence of image quality improvement on beam energy and patient separation. *Med Phys* 36(9):3955–3963.

55. Faddegon, B. A., M. Aubin, A. Bani-Hashemi, B. Gangadharan, A. R. Gottschalk, O. Morin, D. Sawkey, V. Wu, and S. S. Yom. 2010. Comparison of patient megavoltage cone beam CT images acquired with an unflattened beam from a carbon target and a flattened treatment beam. *Med Phys* 37(4):1737–1741.

56. Kry, S. F., O. N. Vassiliev, and R. Mohan. Out-of-field photon dose following removal of the flattening filter from a medical accelerator. *Phys Med Biol* 55(8):2155–2166.

57. Teke, T., C. Duzenli, A. Bergman, F. Viel, P. Atwal, and E. Gete. 2015. Monte Carlo validation of the TrueBeam 10XFFF phase-space files for applications in lung SABR. *Med Phys* 42(12):6863–6874.

58. Faddegon, B. A., P. O'Brien, and D. L. Mason. 1999. The flatness of Siemens linear accelerator x-ray fields. *Med Phys* 26(2):220–228.

59. Liu, H. H., T. R. Mackie, and E. C. McCullough. 1997. Calculating output factors for photon beam radiotherapy using a convolution/superposition method

based on a dual source photon beam model. *Med Phys* 24(12):1975–1985.

60. Liu, H. H., T. R. Mackie, and E. C. McCullough. 2000. Modelling photon output caused by backscattered radiation into the monitor chamber from collimator jaws using a Monte Carlo technique. *Med Phys* 27(4):737–744.

61. Verhaegen, F., R. Symonds-Tayler, H. H. Liu, and A. E. Nahum. 2000. Backscatter towards the monitor ion chamber in high-energy photon and electron beams: charge integration versus Monte Carlo simulation. *Phys Med Biol* 45(11):3159–3170.

62. Liu, H. H., T. R. Mackie, and E. C. McCullough. 1997. A dual source photon beam model used in convolution/superposition dose calculations for clinical megavoltage x-ray beams. *Med Phys* 24(12):1960–1974.

63. Schach von Wittenau, A. E., P. M. Bergstrom, Jr., and L. J. Cox. 2000. Patient-dependent beam-modifier physics in Monte Carlo photon dose calculations. *Med Phys* 27(5):935–947.

64. Li, J. S., T. Pawlicki, J. Deng, S. B. Jiang, E. Mok, and C. M. Ma. 2000. Validation of a Monte Carlo dose calculation tool for radiotherapy treatment planning. *Phys Med Biol* 45(10):2969–2985.

65. Spezi, E., D. G. Lewis, and C. W. Smith. 2001. Monte Carlo simulation and dosimetric verification of radiotherapy beam modifiers. *Phys Med Biol* 46(11):3007–3029.

66. Walters, B., I. Kawrakow, and D. Rogers. 2009. *DOSXYZnrc Users Manual*. NRCC, Ottawa.

67. van der Zee, W., and J. Welleweerd. 2002. A Monte Carlo study on internal wedges using BEAM. *Med Phys* 29(5):876–885.

68. Verhaegen, F., and I. J. Das. 1999. Monte Carlo modelling of a virtual wedge. *Phys Med Biol* 44(12):N251–259.

69. Verhaegen, F., and H. H. Liu. 2001. Incorporating dynamic collimator motion in Monte Carlo simulations: an application in modelling a dynamic wedge. *Phys Med Biol* 46(2):287–296.

70. Chang, K. P., L. Y. Chen, and Y. H. Chien. 2014. Monte Carlo simulation of linac irradiation with dynamic wedges. *Radiat Prot Dosimetry* 162(1–2):24–28.

71. Shih, R., X. Li, and J. Chu. 2001 Dosimetric characteristics of dynamic wedged fields: a Monte Carlo study. *Phys Med Biol* 46:N281–292.

72. Ma, C. M., J. S. Li, T. Pawlicki, S. B. Jiang, J. Deng, M. C. Lee, T. Koumrian, M. Luxton, and S. Brain. 2002. A Monte Carlo dose calculation tool for radiotherapy treatment planning. *Phys Med Biol* 47(10):1671–1689.

73. Jiang, S. B., and K. M. Ayyangar. 1998. On compensator design for photon beam intensity-modulated conformal therapy. *Med Phys* 25(5):668–675.

74. Liu, H. H., and F. Verhaegen. 2002. An investigation of energy spectrum and lineal energy variations in megavoltage photon beams used for radiotherapy. *Radiat Prot Dosimetry* 99(1–4):425–427.

75. Hounsell, A. R. 1998. Monitor chamber backscatter for intensity modulated radiation therapy using multileaf collimators. *Phys Med Biol* 43(2):445–454.

76. Küster, G. 1999. Monte Carlo Studies for the Optimisation of Hardware Used in Conformal Radiation Therapy (PhD Thesis), University of Heidelberg, Heidelberg.

77. De Vlamynck, K., H. Palmans, F. Verhaegen, C. De Wagter, W. De Neve, and H. Thierens. 1999. Dose measurements compared with Monte Carlo simulations of narrow 6 MV multileaf collimator shaped photon beams. *Med Phys* 26(9):1874–1882.

78. Ma, C. M., T. Pawlicki, S. B. Jiang, J. S. Li, J. Deng, E. Mok, A. Kapur, L. Xing, L. Ma, and A. L. Boyer. 2000. Monte Carlo verification of IMRT dose distributions from a commercial treatment planning optimization system. *Phys Med Biol* 45(9):2483–2495.

79. Deng, J., T. Pawlicki, Y. Chen, J. Li, S. B. Jiang, and C. M. Ma. 2001. The MLC tongue-and-groove effect on IMRT dose distributions. *Phys Med Biol* 46(4):1039–1060.

80. Kapur, A., C. Ma, and A. Boyer. 2000. Monte Carlo Simulations for Multileaf-Collimator Leaves: Design and Dosimetry. In *World Congress on Medical Physics and Biomedical Engineering*, Chicago, Illinois. p. 1410

81. Keall, P. J., J. V. Siebers, M. Arnfield, J. O. Kim, and R. Mohan. 2001. Monte Carlo dose calculations for dynamic IMRT treatments. *Phys Med Biol* 46(4):929–941.

82. Kim, J. O., J. V. Siebers, P. J. Keall, M. R. Arnfield, and R. Mohan. 2001. A Monte Carlo study of radiation transport through multileaf collimators. *Med Phys* 28(12):2497–2506.

83. Siebers, J. V., P. J. Keall, J. O. Kim, and R. Mohan. 2002. A method for photon beam Monte Carlo multileaf collimator particle transport. *Phys Med Biol* 47(17):3225–3249.

84. Brualla, L., F. Salvat, and R. Palanco-Zamora. 2009. Efficient Monte Carlo simulation of multileaf collimators using geometry-related variance-reduction techniques. *Phys Med Biol* 54(13):4131–4149.

85. Liu, H. H., F. Verhaegen, and L. Dong. 2001. A method of simulating dynamic multileaf collimators using Monte Carlo techniques for intensity-modulated radiation therapy. *Phys Med Biol* 46(9):2283–2298.

86. Heath, E., and J. Seuntjens. 2003. Development and validation of a BEAMnrc component module for accurate Monte Carlo modelling of the Varian dynamic Millennium multileaf collimator. *Phys Med Biol* 48(24):4045–4063.

87. Li, X. A., L. Ma, S. Naqvi, R. Shih, and C. Yu. 2001. Monte Carlo dose verification for intensity-modulated arc therapy. *Phys Med Biol* 46(9):2269–2282.

88. Bush, K., R. Townson, and S. Zavgorodni. 2008. Monte Carlo simulation of RapidArc radiotherapy delivery. *Phys Med Biol* 53(19):N359–370.

89. Teke, T., A. M. Bergman, W. Kwa, B. Gill, C. Duzenli, and I. A. Popescu. 2009. Monte Carlo based,

patient-specific RapidArc QA using Linac log files. *Med Phys* 37(1):116–123.

90. Feng, Z., H. Yue, Y. Zhang, H. Wu, J. Cheng, and X. Su. 2016. Monte Carlo simulation of beam characteristics from small fields based on TrueBeam flattening-filter-free mode. *Radiat Oncol* 11:30.

91. Gholampourkashi, S., J. E. Cygler, J. Belec, M. Vujicic, and E. Heath. 2019. Monte Carlo and analytic modelling of an Elekta Infinity linac with Agility MLC: investigating the significance of accurate model parameters for small radiation fields. *J Appl Clin Med Phys* 20(1): 55–67.

92. Almberg, S. S., J. Frengen, and T. Lindmo. 2012. Monte Carlo study of in-field and out-of-field dose distributions from a linear accelerator operating with and without a flattening-filter. *Med Phys* 39(8):5194–5203.

93. Fix, M. K., P. Manser, E. J. Born, R. Mini, and P. Ruegsegger. 2001. Monte Carlo simulation of a dynamic MLC based on a multiple source model. *Phys Med Biol* 46(12):3241–3257.

94. Spezi, E., and D. G. Lewis. 2002. Full forward Monte Carlo calculation of portal dose from MLC collimated treatment beams. *Phys Med Biol* 47(3):377–390.

95. Libby, B., J. Siebers, and R. Mohan. 1999. Validation of Monte Carlo generated phase-space descriptions of medical linear accelerators. *Med Phys* 26(8):1476–1483.

96. van der Zee, W., and J. Welleweerd. 1999. Calculating photon beam characteristics with Monte Carlo techniques. *Med Phys* 26(9):1883–1892.

97. Bednarz, B., and X. G. Xu. 2009. Monte Carlo modelling of a 6 and 18 MV Varian Clinac medical accelerator for in-field and out-of-field dose calculations: development and validation. *Phys Med Biol* 54(4):N43–57.

98. Lin, S. Y., T. C. Chu, and J. P. Lin. 2001. Monte Carlo simulation of a clinical linear accelerator. *Appl Radiat Isot* 55(6):759–765.

99. Mesbahi, A. 2007. Dosimetric characteristics of unflattened 6 MV photon beams of a clinical linear accelerator: a Monte Carlo study. *Appl Radiat Isot* 65(9):1029–1036.

100. Mesbahi, A., M. Fix, M. Allahverdi, E. Grein, and H. Garaati. 2005. Monte Carlo calculation of Varian 2300C/D Linac photon beam characteristics: a comparison between MCNP4C, GEANT3 and measurements. *Appl Radiat Isot* 62(3):469–477.

101. Mesbahi, A., P. Mehnati, A. Keshtkar, and A. Farajollahi. 2007. Dosimetric properties of a flattening filter-free 6-MV photon beam: a Monte Carlo study. *Radiat Med* 25(7):315–324.

102. Reynaert, N., M. Coghe, B. De Smedt, L. Paelinck, B. Vanderstraeten, W. De Gersem, B. Van Duyse, C. De Wagter, W. De Neve, and H. Thierens. 2005. The importance of accurate linear accelerator head modelling for IMRT Monte Carlo calculations. *Phys Med Biol* 50(5):831–846.

103. Titt, U., O. N. Vassiliev, F. Ponisch, S. F. Kry, and R. Mohan. 2006. Monte Carlo study of backscatter in

a flattening filter free clinical accelerator. *Med Phys* 33(9):3270–3273.

104. Vassiliev, O. N., U. Titt, S. F. Kry, F. Ponisch, M. T. Gillin, and R. Mohan. 2006. Monte Carlo study of photon fields from a flattening filter-free clinical accelerator. *Med Phys* 33(4):820–827.

105. Sheikh-Bagheri, D., and D. W. Rogers. 2002. Monte Carlo calculation of nine megavoltage photon beam spectra using the BEAM code. *Med Phys* 29(3):391–402.

106. Sheikh-Bagheri, D., and D. W. Rogers. 2002. Sensitivity of megavoltage photon beam Monte Carlo simulations to electron beam and other parameters. *Med Phys* 29(3):379–390.

107. Sheikh-Bagheri, D., D. W. Rogers, C. K. Ross, and J. P. Seuntjens. 2000. Comparison of measured and Monte Carlo calculated dose distributions from the NRC linac. *Med Phys* 27(10):2256–2266.

108. Ma, C.-M. 1998. Characterisation of computer simulated radiotherapy beams for Monte Carlo treatment planning Radiat. *Phys. Chem.* 35:329–344.

109. Schach von Wittenau, A. E., L. J. Cox, P. M. Bergstrom, Jr., W. P. Chandler, C. L. Hartmann Siantar, and R. Mohan. 1999. Correlated histogram representation of Monte Carlo derived medical accelerator photon-output phase space. *Med Phys* 26(7):1196–1211.

110. Fix, M. K., H. Keller, P. Ruegsegger, and E. J. Born. 2000. Simple beam models for Monte Carlo photon beam dose calculations in radiotherapy. *Med Phys* 27(12):2739–2747.

111. Sikora, M., O. Dohm, and M. Alber. 2007. A virtual photon source model of an Elekta linear accelerator with integrated mini MLC for Monte Carlo based IMRT dose calculation. *Phys Med Biol* 52(15):4449–4463.

112. Tillikainen, L., S. Siljamaki, H. Helminen, J. Alakuijala, and J. Pyyry. 2007. Determination of parameters for a multiple-source model of megavoltage photon beams using optimization methods. *Phys Med Biol* 52(5): 1441–1467.

113. Fix, M. K., M. Stampanoni, P. Manser, E. J. Born, R. Mini, and P. Ruegsegger. 2001. A multiple source model for 6 MV photon beam dose calculations using Monte Carlo. *Phys Med Biol* 46(5):1407–1427.

114. Chetty, I., J. J. DeMarco, and T. D. Solberg. 2000. A virtual source model for Monte Carlo modelling of arbitrary intensity distributions. *Med Phys* 27(1):166–172.

115. Leal, A., F. Sanchez-Doblado, R. Arrans, J. Rosello, E. C. Pavon, and J. I. Lagares. 2003. Routine IMRT verification by means of an automated Monte Carlo simulation system. *Int J Radiat Oncol Biol Phys* 56(1):58–68.

116. Ma, C. M., R. A. Price, Jr., J. S. Li, L. Chen, L. Wang, E. Fourkal, L. Qin, and J. Yang. 2004. Monitor unit calculation for Monte Carlo treatment planning. *Phys Med Biol* 49(9):1671–1687.

117. Popescu, I. A., C. P. Shaw, S. F. Zavgorodni, and W. A. Beckham. 2005. Absolute dose calculations for Monte

Carlo simulations of radiotherapy beams. *Phys Med Biol* 50(14):3375–3392.

118. Belec, J., H. Patrocinio, and F. Verhaegen. 2005. Development of a Monte Carlo model for the Brainlab microMLC. *Phys Med Biol* 50(5):787–799.

119. Kubsad, S. S., T. R. Mackie, M. A. Gehring, D. J. Misisco, B. R. Paliwal, M. P. Mehta, and T. J. Kinsella. 1990. Monte Carlo and convolution dosimetry for stereotactic radiosurgery. *Int J Radiat Oncol Biol Phys* 19(4):1027–1035.

120. Sixel, K. E., and B. A. Faddegon. 1995. Calculation of x-ray spectra for radiosurgical beams. *Med Phys* 22(10):1657–1661.

121. Sixel, K. E., and E. B. Podgorsak. 1993. Buildup region of high-energy x-ray beams in radiosurgery. *Med Phys* 20(3):761–764.

122. Sixel, K. E., and E. B. Podgorsak. 1994. Buildup region and depth of dose maximum of megavoltage x-ray beams. *Med Phys* 21(3):411–416.

123. Chaves, A., M. C. Lopes, C. C. Alves, C. Oliveira, L. Peralta, P. Rodrigues, and A. Trindade. 2004. A Monte Carlo multiple source model applied to radiosurgery narrow photon beams. *Med Phys* 31(8):2192–2204.

124. Verhaegen, F., I. J. Das, and H. Palmans. 1998. Monte Carlo dosimetry study of a 6 MV stereotactic radiosurgery unit. *Phys Med Biol* 43(10):2755–2768.

125. García-Pareja, S., P. Galán, F. Manzano, L. Brualla, and A. M. Lallena. 2010. Ant colony algorithm implementation in electron and photon Monte Carlo transport: application to the commissioning of radiosurgery photon beams. *Med Phys* 37:3782–3790.

126. Ding, G. X., D. M. Duggan, and C. W. Coffey. 2006. Commissioning stereotactic radiosurgery beams using both experimental and theoretical methods. *Phys Med Biol* 51(10):2549–2566.

127. Jones, A. O., and I. J. Das. 2005. Comparison of inhomogeneity correction algorithms in small photon fields. *Med Phys* 32(3):766–776.

128. Scott, A. J., A. E. Nahum, and J. D. Fenwick. 2008. Using a Monte Carlo model to predict dosimetric properties of small radiotherapy photon fields. *Med Phys* 35(10):4671–4684.

129. Scott, A. J., A. E. Nahum, and J. D. Fenwick. 2009. Monte Carlo modelling of small photon fields: quantifying the impact of focal spot size on source occlusion and output factors, and exploring miniphantom design for small-field measurements. *Med Phys* 36(7):3132–3144.

130. Taylor, M. L., T. Kron, and R. D. Franich. 2011. A contemporary review of stereotactic radiotherapy: inherent dosimetric complexities and the potential for detriment. *Acta Oncol* 50(4):483–508.

131. Paskalev, K. A., J. P. Seuntjens, H. J. Patrocinio, and E. B. Podgorsak. 2003. Physical aspects of dynamic stereotactic radiosurgery with very small photon beams (1.5 and 3 mm in diameter). *Med Phys* 30(2):111–118.

132. Paskalev, K., J. Seuntjens, and E. Podgorsak. 2002. *Dosimetry of Ultra Small Photon Fields*. Recent Developments in Accurate Radiation Dosimetry Medical Physics Publishing, Madison, WI, pp. 298–318.

133. Ayyangar, K. M., and S. B. Jiang. 1998. Do we need Monte Carlo treatment planning for linac based radiosurgery? A case study. *Med Dosim* 23(3):161–168.

134. Solberg, T. D., F. E. Holly, A. A. De Salles, R. E. Wallace, and J. B. Smathers. 1995. Implications of tissue heterogeneity for radiosurgery in head and neck tumors. *Int J Radiat Oncol Biol Phys* 32(1):235–239.

135. Yamamoto, T., T. Teshima, S. Miyajima, M. Matsumoto, H. Shiomi, T. Inoue, and H. Hirayama. 2002. Monte Carlo calculation of depth doses for small field of CyberKnife. *Radiat Med* 20(6):305–310.

136. Araki, F. 2006. Monte Carlo study of a Cyberknife stereotactic radiosurgery system. *Med Phys* 33(8):2955–2963.

137. Sharma, S. C., J. T. Ott, J. B. Williams, and D. Dickow. 2010. Clinical implications of adopting Monte Carlo treatment planning for CyberKnife. *J Appl Clin Med Phys* 11(1):3142.

138. Wilcox, E. E., G. M. Daskalov, H. Lincoln, R. C. Shumway, B. M. Kaplan, and J. M. Colasanto. Comparison of planned dose distributions calculated by Monte Carlo and Ray-Trace algorithms for the treatment of lung tumors with cyberknife: a preliminary study in 33 patients. *Int J Radiat Oncol Biol Phys* 77(1):277–284.

139. Shepard, A. J., and E. T. Bender. 2016. Development of Compton lens design for increased dose rate in linear accelerator based SRS. *J Radiosurg SBRT* 4(3):225–234.

140. Abdel-Rahman, W., J. P. Seuntjens, F. Verhaegen, F. Deblois, and E. B. Podgorsak. 2005. Validation of Monte Carlo calculated surface doses for megavoltage photon beams. *Med Phys* 32(1):286–298.

141. Ding, G. X. 2002. Dose discrepancies between Monte Carlo calculations and measurements in the buildup region for a high-energy photon beam. *Med Phys* 29(11):2459–2463.

142. Ding, G. X., C. Duzenli, and N. I. Kalach. 2002. Are neutrons responsible for the dose discrepancies between Monte Carlo calculations and measurements in the build-up region for a high-energy photon beam? *Phys Med Biol* 47(17):3251–3261.

143. Gerbi, B. J., and F. M. Khan. 1990. Measurement of dose in the buildup region using fixed-separation plane-parallel ionization chambers. *Med Phys* 17(1):17–26.

144. Lamb, A., and S. Blake. 1998. Investigation and modelling of the surface dose from linear accelerator produced 6 and 10 MV photon beams. *Phys Med Biol* 43(5):1133–1146.

145. Ling, C. C., M. C. Schell, and S. N. Rustgi. 1982. Magnetic analysis of the radiation components of a 10 MV photon beam. *Med Phys* 9(1):20–26.

146. Mackie, T. R., and J. W. Scrimger. 1982. Contamination of a 15-MV photon beam by electrons and scattered photons. *Radiology* 144(2):403–409.

147. Malataras, G., C. Kappas, and D. M. Lovelock. 2001. A monte carlo approach to electron contamination

sources in the Saturne-25 and -41. *Phys Med Biol* 46(9):2435–2446.

148. Nilsson, B., and A. Brahme. 1979. Absorbed dose from secondary electrons in high energy photon beams. *Phys Med Biol* 24(5):901–912.

149. Padikal, T. N., and J. A. Deye. 1978. Electron contamination of a high-energy X-ray beam. *Phys Med Biol* 23(6):1086–1092.

150. Petti, P. L., M. S. Goodman, J. M. Sisterson, P. J. Biggs, T. A. Gabriel, and R. Mohan. 1983. Sources of electron contamination for the Clinac-35 25-MV photon beam. *Med Phys* 10(6):856–861.

151. Rustgi, S. N., Z. C. Gromadzki, C. C. Ling, and E. D. Yorke. 1983. Contaminant electrons in the build-up region of a 4 MV photon beam. *Phys Med Biol* 28(6): 659–665.

152. Yorke, E. D., C. C. Ling, and S. Rustgi. 1985. Air-generated electron contamination of 4 and 10 MV photon beams: a comparison of theory and experiment. *Phys Med Biol* 30(12):1305–1314.

153. Zhu, T. C., and J. R. Palta. 1998. Electron contamination in 8 and 18 MV photon beams. *Med Phys* 25(1):12–19.

154. Lopez Medina, A., A. Teijeiro, J. Garcia, J. Esperon, J. A. Terron, D. P. Ruiz, and M. C. Carrion. 2005. Characterization of electron contamination in megavoltage photon beams. *Med Phys* 32(5):1281–1292.

155. Mesbahi, A. 2009. A Monte Carlo study on neutron and electron contamination of an unflattened 18-MV photon beam. *Appl Radiat Isot* 67(1):55–60.

156. Sikora, M., and M. Alber. 2009. A virtual source model of electron contamination of a therapeutic photon beam. *Phys Med Biol* 54(24):7329–7344.

157. Ververs, J. D., M. J. Schaefer, I. Kawrakow, and J. V. Siebers. 2009. A method to improve accuracy and precision of water surface identification for photon depth dose measurements. *Med Phys* 36(4):1410–1420.

158. Yang, J., J. S. Li, L. Qin, W. Xiong, and C. M. Ma. 2004. Modelling of electron contamination in clinical photon beams for Monte Carlo dose calculation. *Phys Med Biol* 49(12):2657–2673.

159. Petti, P. L., M. S. Goodman, T. A. Gabriel, and R. Mohan. 1983. Investigation of buildup dose from electron contamination of clinical photon beams. *Med Phys* 10(1):18–24.

160. AAPM. 1986. Neutron measurements around high energy x-ray radiotherapy machines. In AAPM Report 19.

161. Agosteo, S., A. Para, F. Gerardi, M. Silari, A. Torresin, and G. Tosi. 1993 Photoneutron dose in soft tissue phantoms irradiated by 25 MV x-rays. *Phys Med Biol* 38:1509–1528.

162. Gudowska, I., A. Brahme, P. Andreo, W. Gudowski, and J. Kierkegaard. 1999. Calculation of absorbed dose and biological effectiveness from photonuclear reactions in a bremsstrahlung beam of end point 50 MeV. *Phys Med Biol* 44(9):2099–2125.

163. Ing, H., W. R. Nelson, and R. A. Shore. 1982. Unwanted photon and neutron radiation resulting from collimated photon beams interacting with the body of radiotherapy patients. *Med Phys* 9(1):27–33.

164. Ongaro, C., A. Zanini, U. Nastasi, J. Rodenas, G. Ottaviano, C. Manfredotti, and K. W. Burn. 2000. Analysis of photoneutron spectra produced in medical accelerators. *Phys Med Biol* 45(12):L55–61.

165. d'Errico, F., R. Nath, L. Tana, G. Curzio, and W. G. Alberts. 1998. In-phantom dosimetry and spectrometry of photoneutrons from an 18 MV linear accelerator. *Med Phys* 25(9):1717–1724.

166. Ghavami, S. M., A. Mesbahi, and E. Mohammadi. 2010. The impact of automatic wedge filter on photoneutron and photon spectra of an 18-MV photon beam. *Radiat Prot Dosim* 138(2):123–128.

167. Falcao, R., A. Facure, and A. Silva. 2007. Neutron dose calculation at the maze entrance of medical linear accelerator rooms. *Radiat Prot Dosim* 123:283–287.

168. Facure, A., A. X. da Silva, L. A. da Rosa, S. C. Cardoso, and G. F. Rezende. 2008. On the production of neutrons in laminated barriers for 10 MV medical accelerator rooms. *Med Phys* 35(7):3285–3292.

169. Bazalova, M., and F. Verhaegen. 2007. Monte Carlo simulation of a computed tomography x-ray tube. *Phys Med Biol* 52(19):5945–5955.

170. Ay, M. R., M. Shahriari, S. Sarkar, M. Adib, and H. Zaidi. 2004. Monte carlo simulation of x-ray spectra in diagnostic radiology and mammography using MCNP4C. *Phys Med Biol* 49(21):4897–4917.

171. Boone, J. M., M. H. Buonocore, and V. N. Cooper, 3rd. 2000. Monte Carlo validation in diagnostic radiological imaging. *Med Phys* 27(6):1294–1304.

172. Verhaegen, F., and I. A. Castellano. 2002. Microdosimetric characterisation of 28 kVp Mo/Mo, Rh/Rh, Rh/Al, W/Rh and Mo/Rh mammography X ray spectra. *Radiat Prot Dosim* 99(1–4):393–396.

173. Verhaegen, F., B. Reniers, F. Deblois, S. Devic, J. Seuntjens, and D. Hristov. 2005. Dosimetric and microdosimetric study of contrast-enhanced radiotherapy with kilovolt x-rays. *Phys Med Biol* 50(15):3555–3569.

174. Jones, B., S. Krishnan, and S. Cho. 2010. Estimation of microscopic dose enhancement factor around gold nanoparticles by Monte Carlo calculationsB. *Med Phys* 37:3809–3816.

175. Yanch, J. C., and K. J. Harte. 1996. Monte Carlo simulation of a miniature, radiosurgery x-ray tube using the ITS 3.0 coupled electron-photon transport code. *Med Phys* 23(9):1551–1558.

176. Liu, D., E. Poon, M. Bazalova, B. Reniers, M. Evans, T. Rusch, and F. Verhaegen. 2008. Spectroscopic characterization of a novel electronic brachytherapy system. *Phys Med Biol* 53(1):61–75.

177. Tryggestad, E., M. Armour, I. Iordachita, F. Verhaegen, and J. W. Wong. 2009. A comprehensive system for

dosimetric commissioning and Monte Carlo validation for the small animal radiation research platform. *Phys Med Biol* 54(17):5341–5357.

178. Birch, R., and M. Marshall. 1979. Computation of bremsstrahlung X-ray spectra and comparison with spectra measured with a Ge(Li) detector. *Phys Med Biol* 24(3):505–517.

179. Iles, W. 1987. Computation of X-ray bremsstrahlung spectra over an energy range 15 keV – 300 keV. In National Radiological Protection Board Report R204.

180. Poludniowski, G. G. 2007. Calculation of x-ray spectra emerging from an x-ray tube. Part II. X-ray production and filtration in x-ray targets. *Med Phys* 34(6):2175–2186.

181. Poludniowski, G. G., and P. M. Evans. 2007. Calculation of x-ray spectra emerging from an x-ray tube. Part I. electron penetration characteristics in x-ray targets. *Med Phys* 34(6):2164–2174.

182. Poludniowski, G., G. Landry, F. DeBlois, P. M. Evans, and F. Verhaegen. 2009. SpekCalc: a program to calculate photon spectra from tungsten anode x-ray tubes. *Phys Med Biol* 54(19):N433–438.

183. Verhaegen, F., A. E. Nahum, S. Van de Putte, and Y. Namito. 1999. Monte Carlo modelling of radiotherapy kV x-ray units. *Phys Med Biol* 44(7):1767–1789.

184. Ay, M. R., and H. Zaidi. 2005. Development and validation of MCNP4C-based Monte Carlo simulator for fan- and cone-beam x-ray CT. *Phys Med Biol* 50(20):4863–4885.

185. Kakonyi, R., M. Erdelyi, and G. Szabo. 2009. Monte Carlo analysis of energy dependent anisotropy of bremsstrahlung x-ray spectra. *Med Phys* 36(9):3897–3905.

186. Omrane, L. B., F. Verhaegen, N. Chahed, and S. Mtimet. 2003. An investigation of entrance surface dose calculations for diagnostic radiology using Monte Carlo simulations and radiotherapy dosimetry formalisms. *Phys Med Biol* 48(12):1809–1824.

187. Taleei, R., and M. Shahriari. 2009. Monte Carlo simulation of X-ray spectra and evaluation of filter effect using MCNP4C and FLUKA code. *Appl Radiat Isot* 67(2):266–271.

188. Mainegra-Hing, E., and I. Kawrakow. 2006. Efficient x-ray tube simulations. *Med Phys* 33(8):2683–2690.

189. Ali, E. S., and D. W. Rogers. 2008. Quantifying the effect of off-focal radiation on the output of kilovoltage x-ray systems. *Med Phys* 35(9):4149–4160.

190. van der Heyden, B., G. P. Fonseca, M. Podesta, I. Messner, N. Reisz, A. Vaniqui, H. Deutschmann, P. Steininger, and F. Verhaegen. 2020. Modelling of the focal spot intensity distribution and the off-focal spot radiation in kilovoltage x-ray tubes for imaging. *Phys Med Biol* 65(2):025002.

191. Mainegra-Hing, E., and I. Kawrakow. 2008. Fast Monte Carlo calculation of scatter corrections for CBCT images. *J Phys: Conf Ser.* doi:10.1088/1742-6596/102/1/012017.

192. Jarry, G., S. A. Graham, D. J. Moseley, D. J. Jaffray, J. H. Siewerdsen, and F. Verhaegen. 2006. Characterization of scattered radiation in kV CBCT images using Monte Carlo simulations. *Med Phys* 33(11):4320–4329.

193. Spezi, E., P. Downes, E. Radu, and R. Jarvis. 2009. Monte Carlo simulation of an x-ray volume imaging cone beam CT unit. *Med Phys* 36(1):127–136.

194. Downes, P., R. Jarvis, E. Radu, I. Kawrakow, and E. Spezi. 2009. Monte Carlo simulation and patient dosimetry for a kilovoltage cone-beam CT unit. *Med Phys* 36(9):4156–4167.

195. Han, K., D. Ballon, C. Chui, and R. Mohan. 1987. Monte Carlo simulation of a cobalt-60 beam. *Med Phys* 14(3):414–419.

196. Rogers, D., G. Ewart, A. Bielajew, and G. van Dyk. 1988. Calculation of electron contamination in a 60Co therapy beam. Dosimetry in Radiotherapy. International Atomic Energy Agency Vienna, pp. 303–312.

197. Carlsson Tedgren, Å., S. de Luelmo, and J.-E. Jan-Erik Grindborg. 2010. Characterization of a 60Co unit at a secondary standard dosimetry laboratory: Monte Carlo simulations compared to measurements and results from the literature. *Med Phys* 37:2777–2786.

198. Mora, G. M., A. Maio, and D. W. Rogers. 1999. Monte Carlo simulation of a typical 60Co therapy source. *Med Phys* 26(11):2494–2502.

199. Walters, B. R. 2015. Increasing efficiency of BEAMnrc-simulated Co-60 beams using directional source biasing. *Med Phys* 42(10):5817–5827.

200. Cheung, J. Y., and K. N. Yu. 2006. Study of scattered photons from the collimator system of Leksell Gamma Knife using the EGS4 Monte Carlo code. *Med Phys* 33(1):41–45.

201. Moskvin, V., C. DesRosiers, L. Papiez, R. Timmerman, M. Randall, and P. DesRosiers. 2002. Monte Carlo simulation of the Leksell Gamma Knife: I. Source modelling and calculations in homogeneous media. *Phys Med Biol* 47(12):1995–2011.

202. Ma, C. M. C., I. J. Chetty, J. Deng, B. Faddegon, S. B. Jiang, J. Li, J. Seuntjens, J. V. Siebers, and E. Traneus. 2020. Beam modelling and beam model commissioning for Monte Carlo dose calculation-based radiation therapy treatment planning: report of AAPM Task Group 157. *Med Phys* 47(1):e1–e18.

203. Kirkby, C., T. Stanescu, S. Rathee, M. Carlone, B. Murray, and B. G. Fallone. 2008. Patient dosimetry for hybrid MRI-radiotherapy systems. *Med Phys* 35(3):1019–1027.

204. Lagendijk, J. J., B. W. Raaymakers, A. J. Raaijmakers, J. Overweg, K. J. Brown, E. M. Kerkhof, R. W. van der Put, B. Hardemark, M. van Vulpen, and U. A. van der Heide. 2008. MRI/linac integration. *Radiother Oncol* 86(1):25–29.

205. Malkov, V. N., S. L. Hackett, B. van Asselen, B. W. Raaymakers, and J. W. H. Wolthaus. 2019. Monte Carlo simulations of out-of-field skin dose due to spiralling contaminant electrons in a perpendicular magnetic field. *Med Phys* 46(3):1467–1477.

206. Malkov, V. N., S. L. Hackett, J. W. H. Wolthaus, B. W. Raaymakers, and B. van Asselen. 2019. Monte Carlo simulations of out-of-field surface doses due to the electron streaming effect in orthogonal magnetic fields. *Phys Med Biol* 64(11):115029.

207. Ghila, A., B. G. Fallone, and S. Rathee. 2018. Technical Note: experimental verification of EGSnrc calculated depth dose within a parallel magnetic field in a lung phantom. *Med Phys* 45(12):5653–5658.

208. Zlateva, Y., B. R. Muir, I. El Naqa, and J. P. Seuntjens. 2019. Cherenkov emission-based external radiotherapy dosimetry: I. Formalism and feasibility. *Med Phys* 46(5):2370–2382.

209. Zlateva, Y., B. R. Muir, J. P. Seuntjens, and I. El Naqa. 2019. Cherenkov emission-based external radiotherapy dosimetry: II. Electron beam quality specification and uncertainties. *Med Phys* 46(5):2383–2393.

Monte Carlo Modelling of External Electron Beams in Radiotherapy

Frank Verhaegen
Maastro Clinic

5.1 Introduction

The introduction given in the chapter on Monte Carlo modelling of clinical photon beams serves also as an introduction to the present chapter on clinical electron beams. In this chapter, we will give an overview of the work performed in the field of Monte Carlo modelling of clinical electron beams over the last 40 years or so. It is widely believed that to calculate a patient dose correctly in electron beams, Monte Carlo simulations are needed, more so than for radiotherapy photon beams, where there exist alternative dose calculation methods that rival Monte Carlo simulations in accuracy and performance. For electron beams, the situation is different, with many papers exposing flaws in analytical dose calculation methods, even in modern treatment-planning systems. It is therefore not surprising that Monte Carlo simulations are seeing a widespread acceptance in the radiotherapy community.

5.2 Electron Beams from Clinical Linacs

Electron beam radiotherapy is administered much less frequently than photon beam treatment, most commonly as a boost to a small superficial target. The energies used nowadays mostly range from 4 to 20 MeV. The schematics of an electron beam linac for Monte Carlo modelling are represented in Figure 5.1. The most important differences with a photon beam linac are the absence of the photon target, the presence of a scattering foil to broaden the beam, and a multistage collimator to shape the beam close to the irradiated volume. The latter is needed because electrons scatter much more in air than photons and, therefore, require collimation close to the surface of the patient. Monte Carlo modelling of electron beams will be treated more concisely in this chapter than photon beam modelling because of the less frequent clinical use of electron beam radiotherapy. The BEAM Monte Carlo code is again the most heavily used tool in many studies.

5.2.1 Early Work in Electron Beam Modelling

Berger and Seltzer [1] were among the first to model the interaction of electrons with lead scattering foils, which are the linac component that influences the electron beam most significantly and where contaminant bremsstrahlung photon production takes place. The authors found that the intervening air causes a significant energy degradation of the electron beam, while the effect of the air can be ignored in high-energy photon beams. It deserves to be pointed out that the same group that pioneered Monte Carlo photon beam modelling was also one of the first to present a Monte Carlo model for a clinical electron beam. Borrell-Carbonell et al. [2] published in 1980 simplified models of several linacs. They modeled beam collimators as apertures with no interactions in their walls; therefore, no secondary particles were produced. This approximation does not result in a realistic particle fluence for electron beams. Rogers and Bielajew [3] compared, calculated, and measured depth dose curves in water for mono-energetic electrons. They noted that the simulations predicted a less steep dose gradient near the surface and a too steep dose fall-off beyond the depth

DOI: 10.1201/9781003211846-7

of the maximum dose. When the electrons were made to pass through the simulated exit window, scattering foils, and the air, the differences were reduced. The authors pointed out that electron energy straggling has an important influence on the depth dose. Andreo and Fransson [4] studied stopping power ratios in water derived from a variety of simple and more complex electron beam models. They showed that stopping powers are relatively insensitive to the details of the electron spectra. They also indicated that it is important to preserve the correlation between energy and angle, which has implications for virtual source models. In another work [5], they studied the influence of the electron energy and the angular distribution on the depth dependence of stopping power ratios. Figure 5.2 shows that for a variety of simulation geometries, from simple to more complex (depicted in the inset on the left panel), the electron energy spectra are more or less broadened. This causes small differences in water depth doses and, in turn, in water-to-air stopping power ratios. Ebert and coworkers [6–8]

studied simple models of applicators and Cerrobend cutouts. They identified two main processes by which the applicator/cutout may influence the dose distributions: by electron scatter of the inner edges of the aperture and by bremsstrahlung production. Studies like these may lead to an optimized collimator design. Other early studies of simplified Monte Carlo models have been published [9,10].

5.2.2 Monte Carlo Modelling of Complete Electron Linac Beams

5.2.2.1 First Full Linac Models

In general, it is accepted that modelling electron beams is more difficult than modelling photon beams because of the greater sensitivity of particle fluences and absorbed dose distributions to the details of the primary electron beam (energy, spatial, and angular distribution) and the geometry of the linac, in particular the scattering foils, which are present in most linacs, and the applicators/collimators. The pioneering efforts to model complete electron beam geometries with the EGS4 code were made by Udale/Udale-Smith [11,12]. Figure 5.3 shows her models for Philips linacs, which included the exit window, the primary collimator, the scattering foils, the monitor chamber, the mirror, the movable photon jaws, the accessory ring, and the applicator.

The modular approach of building linacs paved the way for the later BEAM code with its "component module" approach. To quantify the effects of various parts of the treatment head, Udale simulated five cases, from a mono-energetic pencil beam in vacuum to the full linac geometry. She used measured depth dose distributions to tune the mono-energetic primary electron source energy by attempting to match the depth of 50% of the maximum dose, R_{50}, and the practical electron range, R_p. Electron range rejection was employed as a variance reduction technique to avoid transporting electrons at a great computational cost that couldn't reach the linac exit anyway. Computers have since then become many orders of magnitude faster, but it is still good advice to invest time in selecting variance reduction techniques wisely. Udale scored phase-space files at the bottom of the linac, and in a second step, these were

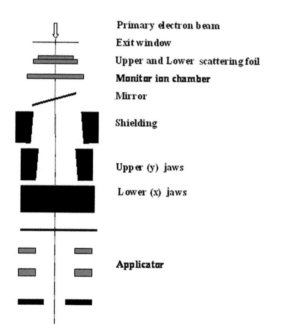

FIGURE 5.1 Schematic representation of an electron beam linac.

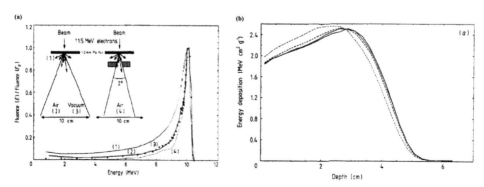

FIGURE 5.2 The influence of the simulation geometry in electron beams on the electron energy spectrum (a). The influence of the spectral variation on water depth dose (b). (From Andreo et al. [5].)

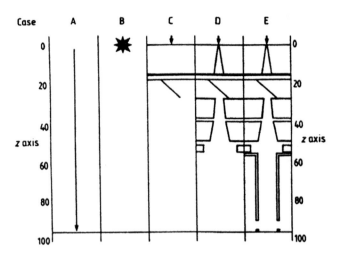

FIGURE 5.3 Five simulated electron beam geometries with increasing complexity from case A to E. In case A, a mono-energetic pencil beam in vacuum impinges directly on the phantom. In case B, a mono-energetic isotropic point source sends electrons through 95 cm of air. Case C adds complexity by modelling the interactions of a mono-energetic pencil beam with an electron window, a scattering foil, a mirror, and air. In case D, movable (photon) jaws were added, and in case E, an electron applicator is present as well. (From Udale [11].)

used for phantom dose calculations, still a common approach today. She also used a second approach, namely, by extracting energy and angular distributions from the phase-space files and using them in a virtual linac model. Correlations between particle position, energy, and angle are ignored this way. She demonstrated that some degree of correlation must be maintained to avoid loss of simulation accuracy. Another approach to reduce calculation time was to calculate dose in water separately for primary electrons, secondary electrons, and photons and to scale their contributions.

Udale-Smith [12] compared models of several linacs and established that some had superior designs, i.e., they led to improved dose distributions, in that they produced fewer contaminant photons, fewer energetic contaminant photons, fewer scattered electrons, and narrower electron angular distributions. Monte Carlo simulation is the ideal tool for these kinds of studies.

Other early works were performed by Kassaee et al. [13] on applicator modelling and by Burns et al. [14] on electron stopping powers and practical ranges in clinical electron beams.

5.2.2.2 Applications of the BEAM Code

As stated in Chapter 4 (photon beam modelling), the advent of the BEAM code in 1995 [15] was a major step forward for Monte Carlo modelling of linacs, including electron beams. In fact, most early results reported with the BEAM code were on electron beams. Also, for electron beams, a wide variety of geometry modules (component modules), source geometries, variance reduction techniques, scoring techniques, and tagging methods are available. It remains a very useful and popular tool nowadays.

Figure 5.4, taken from the original paper on BEAM [15], demonstrates the excellent agreement that can be obtained for dose distributions with BEAM simulations, provided all

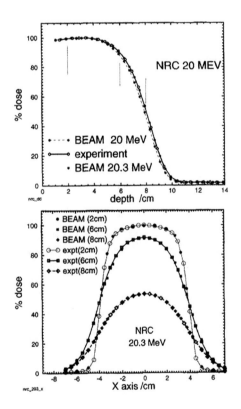

FIGURE 5.4 Measured and calculated Monte Carlo electron dose distributions in water for a 20 MeV electron beam from a research linac with very well-known characteristics. (From Rogers et al. [15].)

the necessary details of the linac are known. This is often a problem such that the user has to resort to "tuning" the model, but in this case, it concerned a research linac with very well-known characteristics. A more complete report, including medical linacs, was given by Ding and Rogers [16]. An

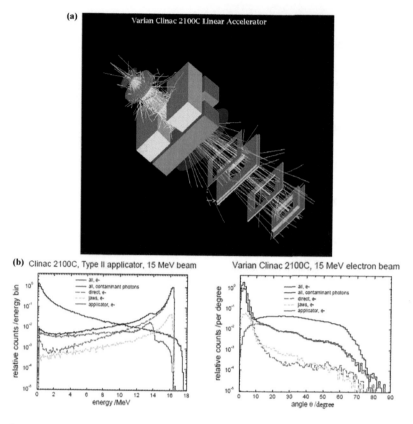

FIGURE 5.5 Monte Carlo model for Varian electron beam (a) and (b) shows particle energy and angular spectra for a 15 MeV electron beam, differentiated according to their origin or interaction site. (From Ding and Rogers [16].)

example is given in Figure 5.5, including again a representation of particle tracks interacting with the linac.

In a series of reports, Ma et al. [17,18] investigated the concept of multiple source models that made use of virtual point source positions due to the diffusivity of electron scatter. This makes deriving source models for electron beams more complicated than for photon beams, which exhibit less scatter. Using realistic linac models, they calculated mean energy in water and stopping power ratios in water [19]. With a similar approach, electron fluence correction factors used in the conversion of dose measured in plastic to dose in water were also reported [20].

Various authors focused on studying components of electron linacs. Verhaegen et al. [21] studied backscatter to the monitor chamber, an effect potentially present in both electron and photon beams in some linacs. The relative increase in backscatter (increased signal of monitor chamber) was found to be 2% when the photon jaws decrease from a square field size of 40 cm to 0 cm in a 6 MeV beam. For higher energies, the effect was smaller. The magnitude of the effect is comparable to photon beams. The contribution to the backscattered fluence from downstream linac components other than the jaws (applicator scrapers, field-defining cutout) was found to be negligible. Figure 5.6 shows the spectra of forward and backward moving particles at the monitor chamber. In contrast to

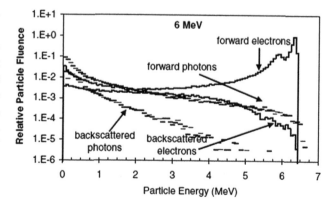

FIGURE 5.6 Monte Carlo simulated forward and backward particles reaching the monitor chamber in a 6 MeV electron beam. (From Verhaegen et al. [21].)

photon beams, in electron beams, a significant difference is obtained in spectral shape between the forward and backscattered electrons. As in photon beams, the backscattered photons were found to contribute insignificantly to the energy deposited in the chamber.

Dose calculations in electron beams are sensitive to the characteristics of the electron spot size hitting the scattering foils. An experimental technique based on a slit camera has been proposed [22] to derive the size of the source of bremsstrahlung photons emerging from the scattering foils, which is equivalent to the size of the electron beam impinging on them. They derived a full width half-maximum (FWHM) size of the elliptically shaped electron primary beams as 1.7–2.2 mm for 6–16 MeV electron beams. They also noted a shift of the primary beam up to 8 mm away from the center of the linac. These authors also studied the influence of primary electron beam divergence on dosimetry in large fields. Introducing a 1.6° primary electron beam divergence changed in-air-measured beam profiles drastically. Primary electron beam information is among the hardest to estimate in clinical electron beams.

It is known that electron beam models are sensitive to the details of the scattering foils, especially for large fields of high-energy electrons. A study [23] considered four parameters of the scattering foils and found that the distance between the foils is critical. This study emphasizes the importance of communication with the manufacturer or verification of the geometry. Large electron fields were also studied in great detail by Faddegon and coworkers [24–28].

5.2.2.3 Studies for Electron Treatment Planning

It is known that conventional treatment planning systems for electron beams have errors in irregular fields and heterogeneous targets [29]. It is generally accepted that Monte Carlo algorithms have unparalleled accuracy for electron dose calculations. Several Monte Carlo–based treatment planning systems are now available. These have been evaluated extensively [30–33], which is covered in Chapter 12 in this book. Much of the present-day work is based on the pioneering efforts of a few groups. Fast Monte Carlo codes with electron treatment planning as their main application were introduced by Neuenschwander et al. [34] and Kawrakow et al. [35] in the mid-1990s.

Ma and coworkers considered various aspects of Monte Carlo electron treatment planning. They simulated clinical electron beams [36,37] and studied beam characterization and modelling for dose calculations [38,39], air gap factors for treatment at extended distances, the commissioning procedure for electron beam treatment planning, stopping power ratios for dose conversions [40], and output factors [37]. The latter aspects were also investigated by others [41,42]. In most of these studies, the BEAM code was used. Virtual source models for electron beams remain an active field of research [43,44].

5.2.2.4 Other Studies with Electron Beams

Several workers developed intensity-modulated electron therapy, in analogy with photon beams. This involves mixing electron beams of various energies, designing or modifying an electron collimator close to the patient, and developing optimization methods for the beam delivery. Al-Yahya et al. [45–47] designed a few-leaf electron collimator, consisting of four motor-driven trimmer bars. Monte Carlo modelling was heavily relied upon during the design stage. The full range

of rectangular fields that can be delivered with the device, in combination with the available electron energies, were used as input for an inverse planning algorithm based on simulated annealing [45]. This system optimizes the beam delivery based on precalculated patient-specific dose kernels. The authors demonstrated that highly conformal treatments can be planned this way.

Modulated electron radiation therapy (MERT), whereby both intensity and energy of the beams are modulated, was studied intensively by Ma and coworkers [48–53]. They first studied this technique for a dedicated electron multileaf collimator [48,51–53] and later by employing the standard photon multileaf collimator to shape the electron beams at short treatment distances [49,50]. They also used inverse Monte Carlo treatment planning.

Combined photon–electron treatment has also been studied by several groups [46,54–60] but is rarely used in clinical practice.

Another application of electron Monte Carlo modelling to highlight in this chapter is the use of portal imaging in electron beams. The bremsstrahlung photon contamination may be used to obtain an image of the patient as was demonstrated by relying on Monte Carlo techniques to study the response of the portal imager to the bremsstrahlung photon fluence. Figure 5.7 compares calculated and recorded portal images in an electron beam. It is clear that here also Monte Carlo techniques play an important role in the optimization of the imaging chain.

A final, very recent application to mention is FLASH radiotherapy. This explores ultrahigh dose rate to create a differential biological response between tumor and normal tissue. At this early stage, prototype linacs are used for these studies. One of the few studies using Monte Carlo simulations with the GATE code for FLASH electron radiotherapy appears to be Lansonneur et al. [62].

5.3 Summary

As in the Chapter 4 on Monte Carlo simulations for photon beams, these techniques are a very powerful tool in modern radiotherapy with electron beams. This is even more so the case here due to the known deficiencies of analytical electron dose calculation techniques. In electron beams, Monte Carlo simulations also play a decisive role in designing complex beam delivery techniques, such as in modulated electron therapy. Also, studies using EGS/EGSnrc based codes (including the user interface BEAM) have dominated this field, but other codes such as GEANT4 are experiencing an increase in use. Since Monte Carlo codes usually differ more in their cross sections and transport methods for electrons than for photons, care has to be taken to properly benchmark simulations for electron beams. Variance reduction techniques for electrons are another point of interest. Several clinical treatment planning systems are offering Monte Carlo modules for electrons, which will be covered elsewhere in this volume. Monte Carlo

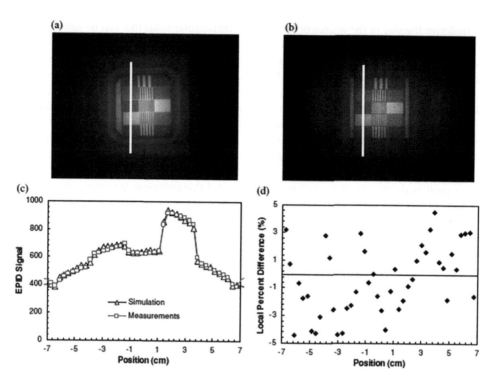

FIGURE 5.7 Measured (a) and simulated (b) images of a test phantom obtained with the bremsstrahlung photons in a 12 MeV electron beam. Profiles (c) and local differences (d) are also shown. The position of the profile is indicated by the white line in (a) and (b). (From Jarry and Verhaegen [61].)

simulations will also play an important role in very novel applications such as FLASH radiotherapy.

References

1. Berger, M., and S. Seltzer. 1978. The influence of scattering foils on absorbed dose distributions from electron beams. Report NBSIR 78-1552 (Gaithersburg: NBS).

2. Borrell-Carbonell, A., J. P. Patau, M. Terrissol, and D. Tronc. 1980. Comparison between experimental measurements and calculated transport simulation for electron dose distributions inside homogeneous phantoms. *Strahlentherapie* 156(3):186–191.

3. Rogers, D. W., and A. F. Bielajew. 1986. Differences in electron depth-dose curves calculated with EGS and ETRAN and improved energy-range relationships. *Med Phys* 13(5):687–694.

4. Andreo, P., and A. Fransson. 1989. Stopping-power ratios and their uncertainties for clinical electron beam dosimetry. *Phys Med Biol.* 34:1847–1861.

5. Andreo, P., A. Brahme, A. Nahum, and O. Mattsson. 1989 Influence of energy and angular spread on stopping-power ratios for electron beams. *Phys Med Biol.* 34:751–768.

6. Ebert, M. A., and P. W. Hoban. 1995. A model for electron-beam applicator scatter. *Med Phys* 22(9):1419–1429.

7. Ebert, M. A., and P. W. Hoban. 1995. A Monte Carlo investigation of electron-beam applicator scatter. *Med Phys* 22(9):1431–1435.

8. Ebert, M. A., and P. W. Hoban. 1996. The energy and angular characteristics of the applicator scattered component of an electron beam. *Australas Phys Eng Sci Med* 19(3):151–159.

9. Keall, P., and P. Hoban. 1994. The angular and energy distribution of the primary electron beam Australas. *Phys Eng Sci Med.* 17 116–123.

10. Manfredotti, C., U. Nastasi, R. Ragona, and S. Anglesio. 1987. Comparison of three dimensional Monte Carlo simulation and the pencil beam algorithm for an electron beam from a linear accelerator. *Nucl Instrum Methods A* 255:355.

11. Udale, M. 1988 A Monte Carlo investigation of surface doses for broad electron beams. *Phys Med Biol* 33:939–953.

12. Udale-Smith, M. 1992 Monte Carlo calculations of electron beam parameters for three Philips linear accelerators. *Phys Med Biol.* 37:85–105.

13. Kassaee, A., M. D. Altschuler, S. Ayyalsomayajula, and P. Bloch. 1994. Influence of cone design on the electron beam characteristics on clinical accelerators. *Med Phys* 21(11):1671–1676.

14. Burns, D. T., S. Duane, and M. R. McEwen. 1995. A new method to determine ratios of electron stopping powers to an improved accuracy. *Phys Med Biol* 40(5): 733–739.

15. Rogers, D. W., B. A. Faddegon, G. X. Ding, C. M. Ma, J. We, and T. R. Mackie. 1995. BEAM: a Monte Carlo code

to simulate radiotherapy treatment units. *Med Phys* 22(5):503–524.

16. Ding, G., and D. Rogers. 1995. Energy spectra, angular spread, and dose distributions of electron beams from various accelerators used in radiotherapy Report PIRS-0439 (Ottawa: NRCC).

17. Ma, C. M., B. A. Faddegon, D. W. Rogers, and T. R. Mackie. 1997. Accurate characterization of Monte Carlo calculated electron beams for radiotherapy. *Med Phys* 24(3):401–416.

18. Ma, C.-M., and D. Rogers. 1995. Beam characterization: a multiple source model Report PIRS-0509(D) (Ottawa: NRCC).

19. Ding, G. X., and D. W. Rogers. 1996. Mean energy, energy-range relationships and depth-scaling factors for clinical electron beams. *Med Phys* 23(3):361–376.

20. Ding, G. X., D. W. Rogers, J. E. Cygler, and T. R. Mackie. 1997. Electron fluence correction factors for conversion of dose in plastic to dose in water. *Med Phys* 24(2):161–176.

21. Verhaegen, F., R. Symonds-Tayler, H. H. Liu, and A. E. Nahum. 2000. Backscatter towards the monitor ion chamber in high-energy photon and electron beams: charge integration versus Monte Carlo simulation. *Phys Med Biol* 45(11):3159–3170.

22. Huang, V. W., J. Seuntjens, S. Devic, and F. Verhaegen. 2005. Experimental determination of electron source parameters for accurate Monte Carlo calculation of large field electron therapy. *Phys Med Biol* 50(5):779–786.

23. Bieda, M. R., J. A. Antolak, and K. R. Hogstrom. 2001. The effect of scattering foil parameters on electron-beam Monte Carlo calculations. *Med Phys* 28(12):2527–2534.

24. Faddegon, B., E. Schreiber, and X. Ding. 2005. Monte Carlo simulation of large electron fields. *Phys Med Biol* 50(5):741–753.

25. Faddegon, B. A., J. Perl, and M. Asai. 2008. Monte Carlo simulation of large electron fields. *Phys Med Biol* 53(5):1497–1510.

26. Faddegon, B. A., D. Sawkey, T. O'Shea, M. McEwen, and C. Ross. 2009. Treatment head disassembly to improve the accuracy of large electron field simulation. *Med Phys* 36(10):4577–4591.

27. O'Shea, T. P., D. L. Sawkey, M. J. Foley, and B. A. Faddegon. 2010. Monte Carlo commissioning of clinical electron beams using large field measurements. *Phys Med Biol* 55(14):4083–4105.

28. Schreiber, E. C., and B. A. Faddegon. 2005. Sensitivity of large-field electron beams to variations in a Monte Carlo accelerator model. *Phys Med Biol* 50(5):769–778.

29. Ding, G. X., J. E. Cygler, C. W. Yu, N. I. Kalach, and G. Daskalov. 2005. A comparison of electron beam dose calculation accuracy between treatment planning systems using either a pencil beam or a Monte Carlo algorithm. *Int J Radiat Oncol Biol Phys* 63(2):622–633.

30. Cygler, J. E., G. M. Daskalov, G. H. Chan, and G. X. Ding. 2004. Evaluation of the first commercial Monte Carlo dose calculation engine for electron beam treatment planning. *Med Phys* 31(1):142–153.

31. Cygler, J. E., C. Lochrin, G. M. Daskalov, M. Howard, R. Zohr, B. Esche, L. Eapen, L. Grimard, and J. M. Caudrelier. 2005. Clinical use of a commercial Monte Carlo treatment planning system for electron beams. *Phys Med Biol* 50(5):1029–1034.

32. Ding, G. X., D. M. Duggan, C. W. Coffey, P. Shokrani, and J. E. Cygler. 2006. First macro Monte Carlo based commercial dose calculation module for electron beam treatment planning--new issues for clinical consideration. *Phys Med Biol* 51(11):2781–2799.

33. Popple, R. A., R. Weinber, J. A. Antolak, S. J. Ye, P. N. Pareek, J. Duan, S. Shen, and I. A. Brezovich. 2006. Comprehensive evaluation of a commercial macro Monte Carlo electron dose calculation implementation using a standard verification data set. *Med Phys* 33(6):1540–1551.

34. Neuenschwander, H., T. R. Mackie, and P. J. Reckwerdt. 1995. MMC--a high-performance Monte Carlo code for electron beam treatment planning. *Phys Med Biol* 40(4):543–574.

35. Kawrakow, I., M. Fippel, and K. Friedrich. 1996. 3D electron dose calculation using a Voxel based Monte Carlo algorithm (VMC). *Med Phys* 23(4):445–457.

36. Ma, C. M., E. Mok, A. Kapur, T. Pawlicki, D. Findley, S. Brain, K. Forster, and A. L. Boyer. 1999. Clinical implementation of a Monte Carlo treatment planning system. *Med Phys* 26(10):2133–2143.

37. Kapur, A., C. M. Ma, E. C. Mok, D. O. Findley, and A. L. Boyer. 1998. Monte Carlo calculations of electron beam output factors for a medical linear accelerator. *Phys Med Biol* 43(12):3479–3494.

38. Ma, C.-M. 1998. Characterization of computer simulated radiotherapy beams for Monte Carlo treatment planning. *Radiat Phys Chem.* 53 329–344.

39. Jiang, S. B., A. Kapur, and C. M. Ma. 2000. Electron beam modelling and commissioning for Monte Carlo treatment planning. *Med Phys* 27(1):180–191.

40. Kapur, A., and C. M. Ma. 1999. Stopping-power ratios for clinical electron beams from a scatter-foil linear accelerator. *Phys Med Biol* 44(9):2321–2341.

41. Verhaegen, F., C. Mubata, J. Pettingell, A. M. Bidmead, I. Rosenberg, D. Mockridge, and A. E. Nahum. 2001. Monte Carlo calculation of output factors for circular, rectangular, and square fields of electron accelerators (6–20 MeV). *Med Phys* 28(6):938–949.

42. Zhang, G. G., D. W. Rogers, J. E. Cygler, and T. R. Mackie. 1999. Monte Carlo investigation of electron beam output factors versus size of square cutout. *Med Phys* 26(5):743–750.

43. Wieslander, E., and T. Knoos. 2006. A virtual-accelerator-based verification of a Monte Carlo dose calculation algorithm for electron beam treatment planning in homogeneous phantoms. *Phys Med Biol* 51(6):1533–1544.

44. Wieslander, E., and T. Knoos. 2007. A virtual-accelerator-based verification of a Monte Carlo dose calculation algorithm for electron beam treatment planning in clinical situations. *Radiother Oncol* 82(2):208–217.

45. Al-Yahya, K., D. Hristov, F. Verhaegen, and J. Seuntjens. 2005. Monte Carlo based modulated electron beam treatment planning using a few-leaf electron collimator-feasibility study. *Phys Med Biol* 50(5):847–857.

46. Al-Yahya, K., M. Schwartz, G. Shenouda, F. Verhaegen, C. Freeman, and J. Seuntjens. 2005. Energy modulated electron therapy using a few leaf electron collimator in combination with IMRT and 3D-CRT: Monte Carlo-based planning and dosimetric evaluation. *Med Phys* 32(9):2976–2986.

47. Al-Yahya, K., F. Verhaegen, and J. Seuntjens. 2007. Design and dosimetry of a few leaf electron collimator for energy modulated electron therapy. *Med Phys* 34(12):4782–4791.

48. Deng, J., M. C. Lee, and C. M. Ma. 2002. A Monte Carlo investigation of fluence profiles collimated by an electron specific MLC during beam delivery for modulated electron radiation therapy. *Med Phys* 29(11):2472–2483.

49. Jin, L., C. M. Ma, J. Fan, A. Eldib, R. A. Price, L. Chen, L. Wang, Z. Chi, Q. Xu, M. Sherif, and J. S. Li. 2008. Dosimetric verification of modulated electron radiotherapy delivered using a photon multileaf collimator for intact breasts. *Phys Med Biol* 53(21):6009–6025.

50. Klein, E. E., M. Vicic, C. M. Ma, D. A. Low, and R. E. Drzymala. 2008. Validation of calculations for electrons modulated with conventional photon multileaf collimators. *Phys Med Biol* 53(5):1183–1208.

51. Lee, M. C., J. Deng, J. Li, S. B. Jiang, and C. M. Ma. 2001. Monte Carlo based treatment planning for modulated electron beam radiation therapy. *Phys Med Biol* 46(8):2177–2199.

52. Lee, M. C., S. B. Jiang, and C. M. Ma. 2000. Monte Carlo and experimental investigations of multileaf collimated electron beams for modulated electron radiation therapy. *Med Phys* 27(12):2708–2718.

53. Ma, C. M., T. Pawlicki, M. C. Lee, S. B. Jiang, J. S. Li, J. Deng, B. Yi, E. Mok, and A. L. Boyer. 2000. Energy- and intensity-modulated electron beams for radiotherapy. *Phys Med Biol* 45(8):2293–2311.

54. Fix, M. K., D. Frei, W. Volken, D. Terribilini, S. Mueller, O. Elicin, H. Hemmatazad, D. M. Aebersold, and P. Manser. 2018. Part 1: optimization and evaluation of dynamic trajectory radiotherapy. *Med Phys*.

55. Mueller, S., M. K. Fix, A. Joosten, D. Henzen, D. Frei, W. Volken, R. Kueng, D. M. Aebersold, M. F. M. Stampanoni, and P. Manser. 2017. Simultaneous optimization of photons and electrons for mixed beam radiotherapy. *Phys Med Biol* 62(14):5840–5860.

56. Mueller, S., P. Manser, W. Volken, D. Frei, R. Kueng, E. Herrmann, O. Elicin, D. M. Aebersold, M. F. M. Stampanoni, and M. K. Fix. 2018. Part 2: dynamic mixed beam radiotherapy (DYMBER): photon dynamic trajectories combined with modulated electron beams. *Med Phys*.

57. Alexander, A., E. Soisson, T. Hijal, A. Sarfehnia, and J. Seuntjens. 2011. Comparison of modulated electron radiotherapy to conventional electron boost irradiation and volumetric modulated photon arc therapy for treatment of tumour bed boost in breast cancer. *Radiother Oncol* 100(2):253–258.

58. Alexander, A., E. Soisson, M. A. Renaud, and J. Seuntjens. 2012. Direct aperture optimization for FLEC-based MERT and its application in mixed beam radiotherapy. *Med Phys* 39(8):4820–4831.

59. Khaledi, N., A. Arbabi, D. Sardari, M. Mohammadi, and A. Ameri. 2015. Simultaneous production of mixed electron--photon beam in a medical LINAC: a feasibility study. *Phys Med* 31(4):391–397.

60. Lloyd, S. A. M., I. M. Gagne, M. Bazalova-Carter, and S. Zavgorodni. 2016. Validation of Varian TrueBeam electron phase-spaces for Monte Carlo simulation of MLC-shaped fields. *Med Phys* 43(6):2894–2903.

61. Jarry, G., and F. Verhaegen. 2005. Electron Beam Treatment Verification Using Measured and Monte Carlo Predicted Portal Images. *Phys Med Biol* 50: 4977–4994.

62. Lansonneur, P., V. Favaudon, S. Heinrich, C. Fouillade, P. Verrelle, and L. De Marzi. 2019. Simulation and experimental validation of a prototype electron beam linear accelerator for preclinical studies. *Phys Med* 60:50–57.

Åsa Carlsson Tedgren
*Department of Health,
Medicine and Caring Sciences,
Linköping University*

Rowan M. Thomson
*Department of Physics,
Carleton University*

Guillaume Landry
*Department of Medical Physics,
Ludwig-Maximilian Universität
München (LMU Munich), Munich*

Gabriel Fonseca
*Department of Radiation Oncology
(MAASTRO), GROW, School for
Oncology and Developmental
Biology, Maastricht University
Medical Center, Maastricht*

Brigitte Reniers
*Research group NuTeC,
Centre for Environmental
Sciences, Hasselt University*

Mark J. Rivard
*Department of Radiation
Oncology, Alpert Medical
School of Brown University*

Jeffrey F. Williamson
*Department of Radiation Oncology,
Washington University School
of Medicine, St Louis, MO*

Frank Verhaegen
*Department of Radiation Oncology
(MAASTRO), GROW, School for
Oncology and Developmental
Biology, Maastricht University
Medical Center, Maastricht*

6

Monte Carlo Techniques in Brachytherapy: Basics and Source and Detector Modelling

6.1 Introduction

This chapter covers core content related to MC techniques in brachytherapy (BT), including simulations of radiation transport, scoring functions, cross-section libraries, and techniques to enhance simulation efficiency. Single-source dosimetry will then be described, including modelling of brachytherapy sources via MC and determination of TG-43 dosimetry parameter reference datasets. Finally, the chapter will consider calculation of photon spectra and their effects and include use of MC to correct for dose perturbations caused by detector/dosimeters in experimental BT.

As for other radiation therapy modalities, Monte Carlo (MC) simulation has become an essential dosimetry tool in modern BT, playing key roles in both clinical practice and research [1–3]. The most established application of MC methods in BT is the determination of dose-rate distributions around individual radiation sources. Modern sources generally contain low-energy radionuclides (mean energies < 50 keV, referred to henceforth as low-energy sources) such as ^{103}Pd, ^{125}I, or ^{131}Cs; contain higher energy radionuclides such as ^{192}Ir, ^{137}Cs, or ^{60}Co (mean energies 355, 662, or 1,250 keV); or may consist of miniature X-ray sources (e.g., 50 kVp bremsstrahlung spectrum).

Source geometries and clinical applications are quite variable. In low-dose-rate (LDR) BT, radioactive material and radio-opaque markers are encapsulated to form permanently implantable seeds, while in high-dose-rate (HDR) BT, e.g., an iridium pellet is encapsulated and welded to the tip of a single-source stepping remote-afterloader cable. Miniature X-ray sources with tungsten anodes also fall in the HDR category. While inverse-square law dependence is the dominating feature of BT dose distributions, photon attenuation and scatter buildup take place in the surrounding medium, and radiation interactions within the source structure give rise to anisotropic dose distributions. The significant modulation of dose distributions must be properly modeled to attain clinically acceptable dosimetric accuracy, and these features are not readily derived from analytical methods such as using the Sievert integral [4]. The complexities of experimentally measuring single-source dose distributions caused by the sharp dose gradients, low and varying photon energies, and dose rates with distance to the source make computational dosimetry techniques such as MC simulations an essential tool in BT [5].

The first computational efforts toward obtaining BT dose distributions are attributed to the 1960s work of Meisberger

DOI: 10.1201/9781003211846-8

who derived 1D tissue-attenuation and scatter buildup factors for [198]Au, ^{192}Ir, ^{137}Cs, [226]Ra, and ^{60}Co point sources [6], while Dale was the first to apply similar techniques to modern ^{125}I sources in 1983 [7]. Although MC modelling of a 3D BT source geometry was performed as early as 1971 by Krishnaswamy for [252]Cf needles [8], it took another decade for the field to fully embrace 3D modelling. Williamson showed in 1983, using 3D MC simulations, that the Sievert integral deviated by 5%–100% from MC results for monoenergetic photons of energies lower than 300 keV emitted from an encapsulated line source, emphasizing the need for accurate computational dosimetry [9]. Burns and Raeside were the first to fully model a commercial ^{125}I BT seed (model 6711), simulating the silver radiomarker, radioactivity distribution, and titanium encapsulation (Figure 6.1) to obtain a 2D dose-rate distribution [10]. Since the range in water of secondary electrons generated by 30 keV photons is less than 20 µm, Burns and Raeside did not transport electrons in their simulations and scored collision kerma using a track-length estimator [11]. Approximating absorbed dose by collision kerma is commonly employed by most MC codes used for BT dosimetry. This means the transport of low-energy secondary electrons is omitted and their vanishingly small number of bremsstrahlung photons is ignored.

While estimating relative 2D dose distributions in a medium around ^{125}I sources from measurements and MC methods was relatively common by the mid-1980s [12], the dose-rate constant Λ (dose rate at the reference position in medium per unit source strength) for low-energy seeds had not been accurately measured or calculated relative to an accurate implementation of the air-kerma strength standard until the late 1980s. Williamson performed simulations in 1988 for models 6711, 6702, and 6701 seeds and the National Institute of Standards and Technology (NIST) Ritz low-energy free-air chamber (Figure 6.2), used as the U.S. air-kerma strength standard, to obtain Λ values [13]. By including the effects of 4.5 keV Ti K-shell characteristic X-rays produced in the source encapsulation, first experimentally observed by Kubo [14], Williamson showed that ^{125}I absolute dose rates obtained from semiempirical methods were overestimated by 10%–14% in good agreement with thermoluminescence dosimetry (TLD) measurements which were being published in the late 1980s; see, e.g., a publication on revisiting the Sievert integral and references therein [4].

The 1990s saw several studies comparing MC simulations of BT sources and thermoluminescence dosimetry (TLD) measurements [15,16]. Williamson observed good agreement

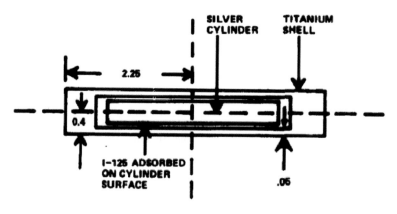

FIGURE 6.1 First 3D model of a 6711 ^{125}I source used in MC simulations by Burns and Raeside [10]. All dimensions are in mm. The silver cylinder is 3 mm in length and 0.5 mm in diameter.

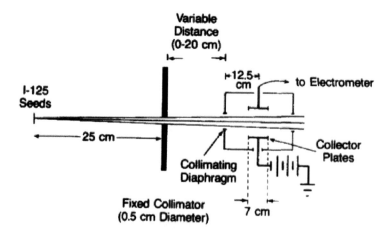

FIGURE 6.2 Williamson's model of the Ritz free-air ion chamber [13].

(1%–5%) between simulations and measurements for ^{125}I when accounting for the measurement phantom medium in the simulations [17]. Excellent agreement (2%–3%) was observed for ^{192}Ir, when the shape and size of the measurement phantom were modeled, although the influence of its composition was found to be less important in this energy range [17]. In a series of articles, Williamson's group performed extensive benchmarking of MC photon-transport calculations against precision diode measurements in water showing that MC accurately (1%–3%) reproduced both relative and absolute dose rates across the entire BT energy range in both homogeneous and heterogeneous phantom geometries [17–22]. These results confirmed that the MC methodology applied to BT dosimetry was mature and sufficiently accurate and robust to support clinical dosimetry. This was reinforced by the American Association of Physicists in Medicine (AAPM) Task Group No. 43 (TG-43) requirement for low-E sources that at least one experimental and one MC determination of dosimetry parameters be published before using a source clinically, making MC source model-specific simulations a de facto dosimetry standard of practice [23]. Investigators tackled ever more complex problems such as the modelling of multicoil radioactive stents used in the treatment of arterial restenosis [24], illustrated in Figure 6.3. However, MC results should not be trusted blindly, as illustrated by the significant dose-estimation differences resulting from using different photon-interaction cross-section databases. This was investigated in the early 2000s by a few groups [25–27], after being first pointed out by Williamson [17]. For BT dose distributions in the <50 keV energy range, where energy deposition is dominated by photoelectric absorption, even 1%–2% errors in the photoelectric cross section can give rise to dose computation errors as large as 10%–15% at 5 cm from a seed. This led to the adoption of modern cross-section libraries derived largely from theoretical quantum mechanical models [28]. The reader is referred to Williamson and Rivard for a more detailed discussion of this complex issue [1].

The rising popularity of LDR prostate seed implantation in the United States, increasing from 5,000 in 1995 to about 50,000 in 2002, fueled a rise in the number of commercially available BT seeds and designs [29]. While the initial 1995 TG-43 report presented consensus dosimetry parameters for one [103]Pd and two ^{125}I seeds [30], its 2004 low-energy seed update (TG-43U1) presented data for 8 seed models [23], while the 2007 supplement presented data for an additional 8 seed models [31]. A second supplement for the remaining commercially available low-energy photon-emitting sources was published in 2017 [32], and this also included a review on later findings on LiF TLD-measured Λ determinations. A joint AAPM/ESTRO report [33] on high-energy photon-emitting BT dosimetry applied the AAPM prerequisite [34] to 21 ^{192}Ir, ^{137}Cs, and ^{60}Co source models. This increasing proliferation of new BT sources and dosimetry datasets was associated with a rapidly growing number of MC-related BT publications in the peer-reviewed literature (Figure 6.4) and resulting in the 2001

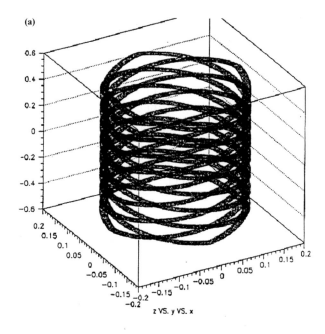

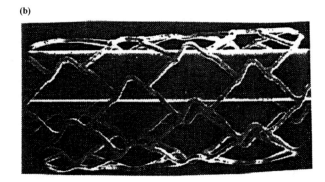

FIGURE 6.3 (a) EGS4 MC model of the ASC Multilink stent. (b) Scanning electron microscope image of the stent. (From Reynaert et al. [24].)

"seed policy" of the *Medical Physics* journal, which limited TG-43 dosimetric parameter papers to technical notes and requested the authors consider other publication venues for TG-43 seed-parameter papers that did not exhibit sufficient scientific novelty to merit *Medical Physics* publication.

6.2 MC Techniques in Brachytherapy

6.2.1 Simulations of Radiation Transport

Radiation transport following radioactive decay is a complex phenomenon, consisting of attenuation and scattering of photons, as well as transport of secondary electrons via elastic collisions with nuclei, elastic and inelastic collisions with orbital electrons, and bremsstrahlung production. Electrons undergo millions of collisions as they slow down, resulting in impractically long computing times, and so electron transport is modeled

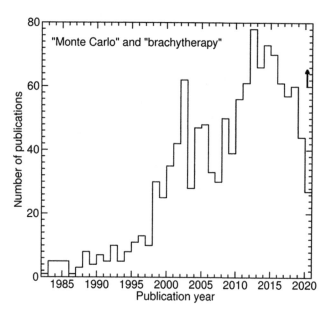

FIGURE 6.4 Results of a PubMed search (1 June 2020) for the terms MC and BT presented per publication year. The arrow indicates the extrapolated 2020 number of publications to a complete year (=65).

via the approximate "condensed history" approach [35,36], if at all. In this approach, a much smaller number of randomly sampled condensed history steps are simulated, each of which represents the collective effects of many discrete collisions by means of multiple scattering distributions and the continuous slowing down approximation (CSDA), with either stochastic or analytical corrections to account for random fluctuations in energy loss about the CSDA energy loss. Despite these approximations, condensed history transport has been demonstrated to accurately characterize absorbed dose distributions over a wide range of circumstances with reasonable computing times.

The simulation of photon interactions may also be simplified. A MC code may simulate K-edge characteristic X-rays, but need not consider L-edge X-rays for low-Z elements, or the atomic relaxation process may be simplified by including only the most probable K- and L-shell energy transitions for high-Z elements. Further, radiological interactions at energies less than a few keV will be subject to molecular binding effects, which are not accounted for in most current MC codes in use, with the possible exception of coherent scattering as described below. Even the impulse approximation, which models the influence of bound orbital electrons on coherent and incoherent (Compton) photon scattering via the form-factor approach, fails to model the influence of orbital electron momentum distributions. This gives rise to a spectrum of scattered photon energies centered about the Compton scattered photon energy, resulting in Compton Doppler broadening [37]. However, even if in the low-energy photon range, the dosimetric differences between the form-factor and consistently implemented free-electron scattering models are small [1,38], the TG-43-U1 recommends use of full form-factor approach at least in the independent atom approximation [23].

By neglecting electron transport altogether and simulating only photon transport, the computation burden in BT dosimetry is substantially reduced. This simplification is exploited by most dosimetry investigators, and all specialized codes intended for BT treatment planning applications. It approximates absorbed dose by collision kerma, which in turn assumes equilibrium of secondary electrons. This approximation is valid everywhere for low-energy BT sources, where electron ranges are < 0.1 mm. However for high-energy ^{192}Ir, ^{137}Cs, and ^{60}Co sources, which have secondary electron ranges of 1–5 mm, the charged-particle equilibrium (CPE) approximation may introduce significant errors near metal-tissue interfaces and near sources. For example, dose errors exceeding 15% at distances less than 1 mm from an HDR ^{192}Ir source have been observed [39,40]. In addition, hot spots due to beta-ray leakage near the ends of LDR ^{192}Ir seeds have been observed experimentally [41]. High-energy photon-emitting sources such as ^{192}Ir and ^{60}Co may not be accurately simulated at distances within a few millimeters [42]. While MC photon transport codes give rise to >5% errors at 1–2 mm from ^{192}Ir and ^{137}Cs sources [33], photon-only transport solutions produce acceptable results for conventional interstitial and intracavitary applications where the clinically relevant distance range is 3–50 mm. Coupled photon-electron calculations and inclusion of the radionuclide beta-ray spectrum may be necessary for applications where near-zone dosimetry is important, e.g., ^{192}Ir-based intravascular BT or estimation of mucosal doses in contact with metal applicators.

6.2.2 Cross Sections

While both low- and high-energy photon-emitting source dose distributions are dominated by inverse-square falloff, radiation interactions within the source capsule and in the surrounding medium significantly alter the dose distribution, requiring Monte Carlo simulation or other linear Boltzmann transport equation solutions to predict doses with clinically acceptable accuracies. Having finalized the source and phantom geometries and chosen the collisional physics model, appropriate photon-interaction cross-section data must be selected and assigned to each material in the simulation geometry. Also to be selected are the detector geometries and estimator, i.e., the mathematical function that estimates the absorbed dose contribution to each detector from each simulated history.

Photon-interaction cross sections are compiled into data libraries and are generally based on quantum mechanical models of each scattering and absorption process which in turn depend upon approximate models of orbital electron wave functions that have previously been validated by comparison to available experimental measurements [43]. These libraries are maintained by organizations such as the National Nuclear Data Center (NNDC) of Brookhaven National Laboratory in New York, the Nuclear Data Section of the International Atomic Energy Agency in Vienna, and the Nuclear Energy

Agency Data Bank of the Organization for Economic Cooperation and Development in Paris. However, the main clearinghouse for evaluated cross-section libraries and public domain transport codes is the Radiation Safety Information Computational Center (RSICC) [http://www-rsicc.ornl.gov] which is located at Oak Ridge National Laboratory. For modest fees, cross-section libraries and many codes for cross-section preprocessing and Monte Carlo simulation can be obtained. Cross-section libraries typically consist of tables of partial cross sections for photoelectric effect, coherent scattering, incoherent scattering, and pair production for each element on a coarse logarithmic energy grid. In addition, atomic form factors and incoherent scattering factors for each element are available. The most modern libraries, e.g., EPDL-97, have subshell-specific photoelectric cross sections and form factors and extensive tables of orbital electron transition and fluorescent yields so that atomic relaxation processes (the cascade of characteristic X-rays and Auger and Coster-König electrons emitted following ejection of an inner shell electron by one of the photon or electron interaction processes) can be properly simulated. For BT, the most important issue is to select a library that contains accurate (post-1983) photoelectric effect and scattering cross sections. Fortunately, all the modern libraries, e.g., EPDL97 from Lawrence Livermore Laboratory, DLC-146 from RSICC, and XCOM from NIST, are based on the same theoretical models, despite having many differences in format and extensiveness of compiled data. Specialized libraries of up-to-date cross sections in EGSnrc, GEANT, and MCNP formats are also available. A more detailed review of both modern and obsolete cross-section libraries and formats is given elsewhere [1]. These cross-section libraries can subtend several hundred megabytes and have complex numerical formats, e.g., ENDF-B. The sparse tables therein are intended for interpolation onto finer grids using specific nonlinear interpolation schemes, e.g., by applying linear interpolation to the logarithmically transformed discrete energy grid and cross-section values. Most libraries come equipped with programs designed to access the database and preprocess the data into user-specified formats. When using the library data, the MC investigator must be careful to employ appropriate interpolation software so as to avoid large errors in the low-energy range [27]. Depending on the estimator type selected, the mass-energy absorption coefficient is another related quantity that may be needed. Tables of mass-energy absorption coefficients, consistent with the NIST XCOM or EPDL97 cross sections, may be obtained from the NIST website [http://www.nist.gov/pml/data/xraycoef/index.cfm], in which elements $Z < 92$ are tabulated over a wide energy range. Tables for user-specified mixtures and compounds may also be obtained. It is essential that the data used for the scoring estimator be derived from the same cross-section data used to transport the photons. For example, if the free-electron scattering model (no coherent scattering or electron-binding corrections to Klein-Nishina Compton scattering) is used for transporting photons, then the NIST mass-energy absorption

coefficients must be recalculated by replacing the term associated with incoherent scattering energy deposition with its Klein-Nishina counterpart.

6.2.3 Scoring Functions

Once cross sections have been assigned, calculation of absorbed dose or collision kerma requires the selection of a scoring grid and a dose estimator. More detailed description of this subject can be found elsewhere [1,11]. In these calculations, the user must be cognizant that results are subject to uncertainties in phantom composition and cross-section data, particularly as distance from the source increases and systematic errors accumulate.

The simplest choice for scoring is the analogue estimator, in which only those simulated collisions that occur within the voxel or volume of interest (referred to generally as "detector volume" in the following) contribute to dose. Invoking CPE, the dose contribution is computed as the difference between energy entering and leaving the detector volume divided by mass.

Another option for scoring is the track-length estimator [11], which approximates dose as collision kerma scored as $(\Delta l \cdot E / \Delta V) \cdot (\mu_{en}/\rho)$, where Δl is the distance within the detector of volume ΔV traversed by incoming photon with energy E. Efficiency is improved since every voxel intersected by a photon flight path produces a non-zero dose score, greatly increasing the information that can be extracted from a finite sample of histories. For example, the track-length estimator can enhance efficiency by a factor (relative to the analogue estimator) of 20–50 for ^{125}I scenarios [44], and by factors of up to 70 (^{103}Pd), 90 (^{125}I), and 300 (^{192}Ir) for different BT treatments [45].

Since the analogue and track-length estimators converge to the same value for collision kerma integrated over the detector volume, the detector grid must be selected carefully to minimize volume-averaging artifacts and to maintain acceptable statistics out to distances of about 7 cm for low-energy BT seeds. For single-source dose estimation, typically a spherical grid is used, with very thin detector elements near the source and thicker ones far from the source to improve statistics. By ignoring azimuthal angle, the detectors become a set of spherical segment shells which can vastly improve efficiency by exploiting the assumed cylindrical symmetry of the source. Since the TG-43 protocol requires specification of dose at geometric points, the MC investigator must correct each MC estimate for volume averaging, typically by using simple inverse-square law–based correction factors [46]. An alternative to volume detectors is to use more complex dose-at-a-point estimators [11,47], in which analytic formulas are used to estimate the contribution from each simulated collision to dose at a geometric point. When using any kind of estimator, the statistical precision of the dose estimates quantified in terms of standard deviation about the mean (67% confidence interval) should be carefully tracked as a function of distance. For volume-detector estimators, detector size is always a trade-off between statistical precision (larger is always better) and spatial resolution, i.e., reduced volume averaging artifact (smaller is always better).

6.2.4 Approaches to Enhance Simulation Efficiency

Scoring via the track-length estimator, simulating only the photon transport (no electron transport), and using the condensed history approach to model electron transport (when it is needed) all enhance simulation efficiency as also overviewed in Section 6.2.3 on estimators [11]. This section describes further approaches to enhance simulation efficiency.

BT treatments may involve many sources within a treatment geometry, and this has the potential to slow calculation times if the geometry modelling approach requires boundary checking of each constituent source geometry for each particle transport step. However, superfluous boundary checks may be avoided by identifying phantom voxels that contain part of a source during geometry initialization [45,48]. Then, the list of phantom voxels containing sources may be used during transport to assess whether the current voxel contains a source: if it does not, then no boundary checks for source geometries occur. This approach, implemented in egs_brachy (via the egs++ geometry class EGS_AutoEnvelope), provides the advantage of achieving calculation times that are effectively independent of the number of seeds modeled (with efficiency gains of up to a factor of 5 reported) [45].

Simulations may begin with particles initialized within sources, e.g., according to the radionuclide decay scheme or the spectrum of accelerating electrons (miniature X-ray tube source), but these *ab initio* simulations involve repeated modelling of particle transport within multiple identical sources and are less efficient than (1) particle recycling and (2) phase-space source [48,49]. For (1), the first source in the simulation acts as a particle generator: particles initialized in this source are tracked until they are absorbed within it or escape source encapsulation. Emitted particles are translated and initiated (optionally rotated) at each source location. For (2), previously generated phase-space data for particles emitted from a source are used to initialize particles on the surface of the source. Both particle recycling and phase-space sources avoid repeated simulation of transport within radionuclide sources, enabling efficiency enhancements of factors of two [45]. The particle recycling offers the advantage of not needing to store large phase-space files and also enables users to simulate more starting particles than there are data for in the phase-space while accounting for correlations between particles of the same primary history.

For simulations involving electronic BT sources (eBT), electron transport is required, but not outside of the source geometry. Standard variance reduction techniques such as bremsstrahlung cross-section enhancement (BCSE), uniform bremsstrahlung splitting (UBS), Russian roulette, and range rejection may be used to enhance electron-transport

simulation efficiency; alternatively, the phase-space approach may be used. For a 50 kV eBT source modeled in egs_brachy, BCSE and UBS increase efficiency by a factor of more than 10^3 (compared with *ab initio* simulation). The phase-space approach offers greater than 10^4 efficiency gain [45]. Thus, BCSE and UBS are useful for research and development purposes where repeated modelling of sources of different models is needed, while the applications of existing X-ray tube models are most efficient with the phase-space source. Correlated sampling has also been investigated to speed up calculations for BT applications [50], [44].

6.3 Single-Source Dosimetry

The application of MC simulations to determine a single-source dose distribution in water for calculation of patient dose distributions using table-lookup methods is a cornerstone of BT treatment planning where it underlies the established TG-43 dosimetry formalism. In addition to describing the general process for source dosimetric characterization, this section will highlight issues unique to MC and other Boltzmann equation numerical solutions, including influence of seed design, phantom size and composition, and other effects not normally encountered in applying experimental and semiempirical dosimetry methods such as the possibility to track the interaction history of radiation particles. The coupling of MC calculated data with traceable measurements to yield absolute dose rates is described in Section 6.3.4. The TG-43 formalism is described as is also the alternative primary and scatter separation formalism (PSS).

6.3.1 TG-43 Dosimetry Parameter Reference Datasets

Current BT treatment planning systems (TPS) utilize the dose calculation formalism of the AAPM TG-43 report [23,30–32]. Databases of approved source data are maintained through the IROC Houston QA Center (http://irochouston.mdanderson.org/RPC/home.htm) and the BRAchytherapy PHYsics Quality assurance System (BRAPHYQS) working group of the European Society for Radiotherapy and Oncology (ESTRO) at the University of Valencia (https://www.uv.es/braphyqs/). The TG-43 formalism assumes a BT source produces a cylindrically symmetric dose distribution and permits the treatment planner to determine patient dose distributions about the source via table lookup in a polar coordinate system using interpolation.

When the source orientation is known, the 2D formalism of the TG-43 report can be used. In this formalism, the contribution of each source to a point of interest (POI) is a function of radial distance r and polar angle θ from the source center to the POI (Figure 6.5). The sources in which the orientation is known include HDR ^{192}Ir or ^{60}Co sources, i.e., those delivered with a remote afterloading unit, LDR ^{137}Cs or ^{192}Ir sources, or low-energy LDR sources such as ^{125}I, ^{103}Pd, and ^{131}Cs. It is important that the imaging modality employed (typically CT

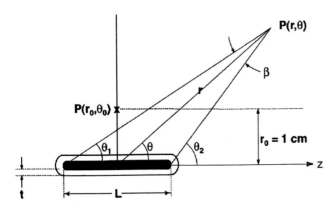

FIGURE 6.5 Coordinate system employed by the TG-43 formalism for a source with active length L. The POI is indicated by $P(r, \theta)$. (From Ref. Rivard et al. [23].)

or three-film projection techniques [51]) permits adequate resolution to discern the capsule orientation. When clinical imaging is inadequate to discern the capsule orientation, the 1D formalism of the TG-43 report is used in which the polar-angle dependence of dose is ignored [52,53].

6.3.2 The TG-43 Dose Calculation Formalism

The 2D dose calculation formalism of the AAPM TG-43 report is given in Equation 6.1.

$$\dot{D}(r,\theta) = S_K \cdot \Lambda \frac{G_L(r,\theta)}{G_L(r_0,\theta_0)} \cdot g_L(r) \cdot F(r,\theta) \qquad (6.1)$$

where the dose rate as a function of r and θ (see Figure 6.5) is equal to the product of the air-kerma strength S_K (a measure of the BT source strength [23]; the dose-rate constant Λ (defined as the ratio of the reference dose rate $\dot{D}(r_0,\theta_0)$ where $r_0 = 1\,\text{cm}$ and $\theta_0 = 90°$ and S_K as $\Lambda = \dot{D}(r_0,\theta_0)/S_K$; the geometry function $G_L(r,\theta)$ at any point r and θ using the line-source (for a source of active length L) approximation divided by its value $G_L(r_0,\theta_0)$ at r_0 and θ_0; the radial dose function $g_L(r)$ based on the line-source approximation; and the 2D anisotropy function $F(r,\theta)$.

The 1D dose calculation formalism of the AAPM TG-43 report is given in Equation 6.2.

$$\dot{D}(r,\theta) = S_K \cdot \Lambda \frac{G_L(r,\theta)}{G_L(r_0,\theta_0)} \cdot g_L(r) \cdot \phi_{an}(r) \qquad (6.2)$$

where the dose rate as a function of r only is equal to the product of S_K, Λ, the geometry function $G_L(r,\theta_0)$ at any distance r using the line-source approximation divided by its value $G_L(r_0,\theta_0)$ at r_0 and θ_0, the radial dose function $g_L(r)$ using the line-source approximation, and the 1D anisotropy function $\phi_{an}(r)$.

These dosimetry parameters for the 2D and 1D dose calculation formalisms are described in great detail within the 2004 AAPM update to the TG-43 report (TG-43U1) [23]. Of note is that all parameters take a value of unity at the reference position (r_0,θ_0) except S_K and Λ such that $\dot{D}(r_0,\theta_0) = S_K \cdot \Lambda$ and $\dot{D}(r_0) = S_K \cdot \Lambda$.

Using equations and recommendations covered in TG-43U1, the MC dosimetry investigator extracts tables of TG-43 parameters from the measured or MC-computed dose distribution, which is typically performed using detectors arranged in polar coordinate grid. In turn, these parameters are then imported into a clinical BT TPS using the same formalism. Recovering the dose distribution in the original MC grid for each source modeled in the TPS is an essential institution-specific quality assurance test [23].

6.3.3 The Primary and Scatter Separation Dose Calculation Formalism, PSS

The PSS was first suggested in 1996 by Russell and Ahnesjö [54] and in a refined version from 2005 where its compatibility with the TG-43 formalism was elucidated [55]. PSS characterization is based on single-source MC simulations, similar to those underlying the TG-43 formalism, with the addition that the dose in water is scored as separated into the primary and the scattered dose fractions. Note how a unique capability of the MC technique, i.e., to keep track of the order of photon scatter, is utilized. Fitted/parameterized MC data are combined with a dose-rate constant Λ to couple with traceable absolute units of dose rate, similar to the TG-43 formalism. A photon is defined as primary until its first interaction outside of the source encapsulation.

The advantage of the PSS source characterization is that the primary dose distribution in water is a fingerprint of the source and independent of phantom size. Since ranges of secondary electrons are short, the primary dose in a heterogeneous environment can be obtained from that in water using one-dimensional attenuation scaling. Additionally, the primary dose can be used as a first collision source term for calculation of the scatter dose component by 3D integral transport or superposition/convolution algorithms. The PSS source characterization is fully compatible with the TG-43 method through adding the initial primary and scatter dose distribution into the total dose. The PSS is used for source characterization for the collapsed cone algorithm "ACE" in the Oncentra treatment planning system (Elekta, Stockholm, Sweden) [56]. The published implementations of the PSS formalism suggest a parameterization of the MC data using exponential functions; however, this is not a necessity and other parameterizations or tabular formats could be used. First-collision source terms have been used in other BT applications including the MC code PTRAN [11] and the discrete ordinates types of analytical solvers of the Boltzmann equation [57,58].

6.3.4 MC Dose Calculation Method for TG-43 and PSS

The process for calculating dose-rate distributions in the vicinity of a BT source has been described in the joint TG-138 report by the AAPM and the European Society of Radiation Oncology (ESTRO) on BT dosimetric uncertainties [42]; the revised TG-43 reports [23,32]; and in a review article [1]. In short, the dosimetry parameters calculated in liquid water $D(r, \theta)$ (where ΔD means absorbed dose in the inherent units of the MC output, usually Gy/simulated primary photon or a multiple thereof) are determined separately from Λ. A typical MC dose-simulation process is described step-by-step below. Source characterization by the PSS method follows the same process, but with the additional requirement to score the primary and scatter dose parts separately.

6.3.4.1 Brachytherapy Source Design

Before initiating MC simulations of BT dose-rate distributions, the dosimetry investigator must have an accurate and complete geometric of model of the BT source. This includes as many as possible of the following: the assembly process, the atomic composition, density, shape, dimensions, and location of each component along with tolerances in any of these features. In addition, the geometry and location of the radionuclide distribution, along with any radioactive contaminants or non-radioactive filler material, must be known. Often this distribution is a very thin layer, the thickness of which can only be known approximately. Typically, the MC investigators start with manufacturer-provided construction drawings of the source, possibly also with proprietary descriptions of the chemical fabrication process, from which additional descriptive details can be inferred. The TG-43U1 report recommends that dosimetry consultants independently validate these specifications. The validation techniques used include, e.g., direct mechanical measurement of non-radioactive seeds and pre-assembly internal components; pin-hole autoradiography; transmission radiography [1]; and electron microscopy [59]. Because the MC simulation process tries to estimate the dose-rate distribution, any differences between the simulated and real geometries may result in erroneous results. While some simulation errors are more forgiving than others (such as for thin-walled, high-energy photon-emitting BT sources), it is not always clear what is the quantitative sensitivity of a design feature to the resultant dose-rate distribution. The TG-43U1 and TG-138 reports recommend that dosimetry investigators perform thorough uncertainty analyses by performing MC simulations over an ensemble of source geometries sampled from the estimated tolerances of geometric parameters whose mean values cannot be accurately determined as illustrated by a study of the model 6711 ^{125}I seed [60]. Among the most error-prone source geometries are thin layers of radioactivity deposited on the surface of an opaque structure with sharp corners as illustrated by a study of the model 6711 ^{125}I seed [59,60].

Some sources have internal components that move under the influence of the gravitational forces, resulting in an orientation-dependent geometry not evident upon external examination of the source [61–64]. This is most prevalent for low-energy LDR sources where the internal components are

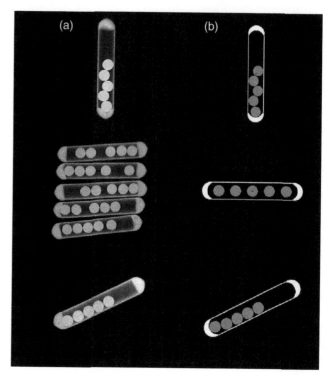

FIGURE 6.6 (a) Radiographs of a BT source with ^{103}Pd adsorbed on Ag spheres in various orientations showing displacement of the Ag spheres. (b) Source geometries simulated in MCNP. (From Rivard et al. [62].)

not secured to the capsule and the radiation emissions are highly attenuated by radio-opaque markers. Figure 6.6 illustrates this phenomenon for ^{103}Pd sources. For these source types, dose rate can vary by more than a factor of two within a few millimeters of the source. In this circumstance, MC simulation is a powerful tool for performing sensitivity analyses to understand how the dose-rate distribution is influenced by dynamic internal source components. Unlike LDR seeds, HDR sources have their capsule attached to a drive cable for translating the source through catheter(s) in a patient with the radioactive component rigidly secured.

6.3.4.2 Brachytherapy Source Radiation Emissions

BT sources generally utilize radionuclides as the radiation source, while eBT sources are an exception. The nuclear disintegration process is well understood with photon energies, emission rates, and branching ratios often known to three or more significant digits for common radionuclides. BT sources are generally designed around a single radionuclide [65,66]. However, radioactivity activation and fabrication processes often result in potentially significant concentrations of contaminant radionuclides even after efforts to improve radiopurity via subsequent radiochemistry. Since radiopurification processes are not perfect, all radionuclide-based BT sources contain radioimpurities. Though the dosimetric influence of radioimpurities has not been sufficiently examined in the literature, the joint AAPM/ESTRO TG-167 report [67] conservatively recommends that radioimpurities be minimized to such levels that their dosimetric contributions over the range of clinically relevant distances in the vicinity of the implant should be <5% of the dosimetric contributions of the primary radionuclide. In addition, the working life of the source must be considered for radioimpurities that are longer-lived than the intended radionuclide. The MC dosimetry investigator is advised to consider the impact of radioimpurities on the recommended dose distribution.

In addition to radioimpurities, all relevant radiation emissions for a given radionuclide should be included in the simulation, where "relevant" means affecting the dose distribution over the range of clinical interest by more than 2%. While neutrino emissions are dosimetrically irrelevant, many common radionuclides disintegrate via beta-decay with subsequent X-ray and gamma-ray emissions. Not all electrons (or positrons) generated within the capsule interior are fully attenuated by the source structure and may deposit dose (or bremsstrahlung X-rays) exterior to the capsule [68]. eBT sources generate characteristic X-rays in addition to bremsstrahlung photons in various internal components, which may also contribute to clinically significant energy deposition outside the source capsule [69]. At a minimum, MC codes used in low-energy dosimetry should accurately model emission of K- and L-shell characteristic X-rays following ejection of inner-shell electrons via knock-on electron collisions, photoelectric effect, or inelastic photon scattering. The MC dosimetry investigator is advised to perform calculations to benchmark the code preceding simulation of a clinical BT source. Only codes that have been well validated for BT or similar applications should be used. Even when using a public-domain code, the investigator should ensure that their implementation accurately reproduces relevant benchmark data.

6.3.4.3 Simulating Phantoms for Reference-Quality or Experimental Dosimetry Measurements

After source radiation originates within the capsule, it may pass into the surrounding medium. A liquid water sphere is the standard phantom for determining reference-quality, single-source BT dose distributions. For low-energy sources, the sphere radius is 15 cm with the BT source located at the center [23]. For high-energy photon-emitting sources (>50 keV photons), the recommended reference phantom radius is 40 cm [33]. Given the ranges of backscattered photons, the phantoms permit acquisition of dose-rate data for low- and high-energy sources within 1% of an unbounded phantom dose distribution for distances up to 10 and 20 cm, respectively [70,71]. In practice, simulating a larger phantom produces dose distributions at the phantom periphery quantitatively closer to that of an infinitely large phantom, with the penalty of longer simulation times for the same number of particle histories. Regardless, the MC dosimetry investigator should state phantom dimensions used and employ phantom-sized correction

factors so that the final recommended dose distribution adheres to the relevant AAPM+ESTRO recommendation.

The AAPM TG-43 dose-calculation formalism ignores patient tissue composition and geometry. It is based upon specification of absorbed dose and transport of radiation outside the source into liquid water. The AAPM TG-43U1 report [23] specifies liquid water to consist of exactly two parts hydrogen and one part oxygen with a mass density of 0.998 g/cm³ at a temperature of 22°C. For low-energy BT dosimetry, cross sections for coherent scattering are sensitive to inter-molecular forces and molecular bonds. Hence, the atomic mixture rule, which generally supports accurate synthesis of photon interaction cross sections for compounds and mixtures, is not applicable to coherent scattering. To accurately model coherent scattering, calculated σ_{coh} values should be recalculated using liquid or molecular water atomic form factors [72] and the water coherent scattering cross sections extracted from NIST XCOM or EPDL97 libraries should be discarded as they are based upon the atomic mixture rule. Secondly, dosimetry investigators utilizing experimental techniques need to be aware that the phantoms used for dose measurement often deviate significantly from liquid water [17,73]. As recommended by TG-43U1, BT experimentalists are expected to use MC simulation to calculate reference-to-experimental phantom correction factors so that the final experimental results represent absorbed dose to liquid water in the spherical liquid water reference phantom. Finally, the reader must remember that the patient tissue compositions, densities, and dimensions deviate significantly from the reference geometry, which can result in large discrepancies between estimated and delivered dose, especially for low-energy sources. The AAPM TG-186 is the first concerted attempt to formulate guidelines for assigning cross sections and densities, other than unit-density water, to organs and tissues based upon patient-specific imaging [74].

6.3.4.4 Dose-Rate Constant

To estimate the value for Λ, the ratio of $\Delta D(r_0, \theta_0)$ to ΔS_K (the air-kerma strength per simulated history) is obtained. Because MC methods generally estimate absorbed dose or collision kerma in terms of energy imparted per mass and per number of particle histories performed or radionuclide disintegrations simulated, the results from the liquid water simulations need to be combined with simulations of the air-kerma strength in a suitable free-air geometry (vacuum or an air sphere). This way, the inherent normalization used by the MC code is irrelevant since both $\Delta D(r_0, \theta_0)$ and ΔS_K are expressed in terms of the same clinically irrelevant normalization. However, the MC normalization constant must be known when absolute dose distribution results are needed in terms of units of air-kerma strength. This process is described by Melhus and Rivard [71] and Williamson and Rivard [1] for calculating dose-rate distributions per unit source strength U, requiring knowledge of the number of photons emitted per disintegration. Thus to compute dose rates in absolute units around an actual BT source, the clinical medical physicist needs only to measure

S_K of the actual source in use (in a traceably calibrated reentrant well-type ionization chamber) and multiply this number by Λ and the other TG-43 dosimetry parameters (i.e., geometry function, radial dose function, and anisotropy function). Since the other TG-43 BT dosimetry parameters are ratios of doses computed in a single simulation run, it is crucial that the same source design and other starting conditions be used between the liquid water and vacuum (or air) simulations or else a scalar offset (percentage systematic dose error) will be present throughout all calculated dose rates.

Vacuum is easy to model while air is more complicated, requiring a mass density (1.196 mg/cm³ at a temperature of 22°C for dry air at 101.325 kPa) with mass compositions of 1.24%, 75.5268%, 23.1781%, and 1.2827% for C, N, O, and Ar, respectively [23]. As for the liquid water description in Section 6.3.4.3, this description for dry air is for standardization purposes since BT source strength (air-kerma strength) is calibrated for these conditions. There are several subtleties MC investigators must appreciate, which have been reviewed by Monroe and Williamson [75]. First, if air is employed, photon attenuation and scatter buildup effects must be corrected to satisfy the AAPM's definition of S_K, which is specified in free space (i.e., without scatter or attenuation). Secondly, care must be taken to filter out or suppress any low-energy contaminant photons below 5 keV as recommended by the TG-43U1 for low-energy sources (<50 keV) and by 10 keV for high-energy sources (>50 keV) [33]. Especially if calibration is simulated in vacuum, failure to suppress Ti characteristic X-rays could inflate ΔS_K by 20% [31]. Finally, care should be taken to simulate any known departures of the U.S. NIST wide-angle free-air chamber (WAFAC) primary standard from the definition of S_K [76]. This was first demonstrated by Williamson in 1988 [13], who found it necessary to simulate the Ritz free-air chamber and the NIST procedure [77] for air-attenuation correction to model the influence of Ti K-shell characteristic X-ray generation on the comparison of theoretical and measured Λ values. Later, it was found that the sharp edges of the palladium-plated carbon pellets of the model 200 ¹⁰³Pd seed induced significant and distance-dependent anisotropy of the polar fluence profile near the transverse axis [75]. This necessitated a detailed simulation of the NIST WAFAC [76], which integrated over this region of polar anisotropy, significantly deviating from the definition of air-kerma strength and altering the dose-rate constant by as much as 12% [62].

6.4 MC in Support of Experimental BT

MC plays an important role in support of experimental BT dosimetry, which presents a difficult measurement scenario characterized by steep dose and dose-rate gradients and photon energy spectra in the keV range that vary considerably with distance to the source and phantom dimensions. MC can be used to derive correction factors for experimental dosimetry, to understand the detector response differences between various beam qualities in use for calibration and measurements, and is a valuable tool in detector design and uncertainty analysis. Experimental BT ranges from setting up primary calibration standards of air kerma strength and verifying TG-43 dosimetry data for low-energy sources to the verification of model-based dose calculation algorithms (MBDCA) and online *in vivo* source tracking. Detectors and dosimeters range from solid-state LiF TLDs, Optically Stimulated Light Dosimeters (OSLDs), electron paramagnetic dosimeters, and radiochromic films to direct-reading detectors such as ion chambers, diode-, diamond-, and fast scintillation detectors. Examples are provided below.

6.4.1 Corrections of Detector Energy Response

MC simulations are excellent tools to generate photon and electron spectra at various distances from BT sources. Detector response depends to a high degree on the photon and/or electron fluence. Spectra outside BT sources depend on the environment of the source (air, water, tissue, phantom material). Information on the photon and / or electron-energy spectrum at the point(s) of interest is important for detector selection with suitable energy-dependent properties. CPE prevails when the detector dimensions are such that it is large in relation to the range of secondary electrons but small compared to the photon mean free paths. Under CPE conditions, the ratios of mass-energy absorption coefficients for water to detector active materials (weighted over the energy-fluence spectrum) may sometimes be enough to predict the energy dependence relative to water. However, as soon as the detector perturbs the initial field cavity theory considerations or better full MC simulation of the detector geometry is needed, see, e.g., [78]. An early example of MC simulations to derive photon energy spectra outside BT sources to correct LiF TLD response was given by Meigooni et al [79], and later examples show differences in the same material, but with different phantom dimensions, e.g., [80]. In materials of low-Z like tissue and water, the spectral variation with distance is larger for ¹⁹²Ir than for low-energy radionuclides due to photon scattering dominating at higher energies and photoelectric effect being more important at the lower energies [18]. When CPE is not fulfilled, coupled photon-electron MC transport is required for high accuracy. The variation of photon energy spectrum with distance depends on photon energy (i.e., radionuclide or eBT design). Use of MC to study the spectra of the measurement geometry reveals information relevant to detector selection. Further, MC can be used to refine the study of detector design and response, taking the detailed geometry into account as exemplified in Section 6.4.3.

6.4.2 Effects of Experimental Phantom

The quantity of interest historically is absorbed dose to water; however, plastic phantoms are commonly used in experiments to mitigate positional uncertainties. Absorbed dose calculations for low-energy photons are highly sensitive to the exact atomic composition of the phantom materials [73]. MC has been used to correct for phantom materials; e.g., Reniers et al. compared two MC codes for dose perturbation by TLDs in various phantom materials for low-energy radionuclides ^{125}I and ^{103}Pd [26]. The consequences of phantom material selection and estimating TLD phantom corrections were discussed in [59,81]. This subject was also reviewed in detail by Williamson and Rivard [1]. A study showing differences between water and plastic phantoms as a function of dimensions and materials (e.g., PMMA, solid water, and polystyrene) for ^{192}Ir was published by Carlsson Tedgren and Carlsson [80]. Schoenfeld et al. have performed simulations of additional phantom materials for both HDR and LDR BT sources [82].

6.4.3 Detector/Dosimeter Response and Perturbation

Another powerful application of MC to BT is the investigation of dosimeter/detector perturbations. An early example of the utility of MC simulations to determine energy response with respect to ^{60}Co radiation of various detector materials embedded in various phantom materials employed in BT is the study by Selvam and Keshavkumar [83]. They used EGSnrc codes and a large cavity approach to calculate the energy response of ^{125}I and ^{169}Yb sources as a function of distance and detector material thickness. Although they did not model detailed detector geometry, their work (and references therein) clearly shows that energy correction factors need to be taken into account in BT and that MC simulation offers a convenient and accurate way to obtain them. Failure to account for these energy response effects may lead to severe over/underestimation of measured dose particularly in low-energy BT but also at higher energies.

The most investigated dosimeter for high-accuracy dosimetry in BT is LiF TLDs, a highly established detector for measuring TG-43 dosimetry parameters. A seemingly systematic difference between MC and LiF TLD measurements was observed and discussed in the early 2000s [44]. A series of TLD experiments, where MC was utilized to determine relative absorbed-dose corrections for low-energy X-ray beam response relative to a high-energy calibration (^{60}Co or 6 MV photon beam), revealed a significant overresponse of the order of 5% in the 22–35 keV energy range compared to ^{60}Co. It was found to depend on differences in the intrinsic efficiency of LiF TLDs at different photon energies at [84–86]. Variation in intrinsic energy-response arise from dependence of the detector signal-formation mechanism on the radiation-field linear energy transfer (LET) and gives rise to an energy-dependent signal per unit dose to the active detector medium. The findings of an intrinsic energy dependence for LiF TLD would not have been possible without complementing experiments with MC simulations to determine the absorbed dose to the active detector volume in the radiation beams investigated. These findings have implications for the experimental determination of Λ as

summarized by Rodriguez and Rogers [81] and accounted for in the 2017 TG-43 recommendations [32]. Similar investigations have been performed for lithium formate electron paramagnetic resonance (EPR) detectors [87], glass dosimeters [88], and microdiamond detectors [89], clearly showing the need for both MC and experiments to resolve the issue of intrinsic efficiency for BT detectors whenever these are calibrated in another beam such as ^{60}Co or 6 MV.

In a study on ion chamber perturbation in BT, Reynaert et al. [90] modeled an NE2571 ionization chamber in detail which was used to perform dosimetry close to a PDR ^{192}Ir source in a water phantom. MC simulation is often the only method to determine reliably these perturbation factors, as discussed in other chapters in this volume. MC calculations were also used to derive corrections factors for ionization chamber measurements to determine absorbed dose to water around an eBT source [91]. Another example is [92] where correction factors for Spencer-Attix cavity theory were investigated for graphite ion chambers used in primary standards for BT.

Adolfsson et al. [93] studied the response of radiochromic film and lithium formate EPR dosimeters at close distances from an eBT source. EGSnrc and GEANT4 was used to calculate several conversion and geometric correction factors for both detector types, after which good dosimetric agreement was obtained, except for ≤1 cm from the source where geometric uncertainties prevented achieving accurate measurements.

In another application using MOSFET detectors, the mean photon energy from an ^{192}Ir source was found to decrease significantly with distance from the source by more than 100 keV over about 10 cm, leading to an increase in dose response by 60% [94].

The important role of MC in deriving dosimeter correction factors for measurements around ^{192}Ir sources was recently applied to a microdiamond detector for obtaining absorbed dose to water measurements in absolute units [78] and the 2D anisotropy function [95]. These works further revealed the importance of detailed electron transport in simulating detector response when dimensions of the active volume were small, and the importance of reliable detector design information. Response and influence on ambient pressure to BT well-type chambers have also been investigated using MC simulations [96–98].

MC simulations have also been successfully used to investigate the response of well-type air-filled chambers, used to determine brachytherapy source strength in terms of air-kerma strength and so couple TG-43 data to absolute units (Equation 6.1). The response of well-type chambers to various types of LDR sources requires calibration to be done per specific source type, while this has hitherto not been considered necessary for HDR sources, where different fabricates are more similar in interior design and also less sensitive to source geometry. However, corrections for source design differences on well-type chamber response between various types of ^{192}Ir HDR source have been shown to increase accuracy [99].

6.5 Future Use of Monte Carlo Methods for Brachytherapy Source Modelling

Over the past decade, there have been significant advances to characterize radiation dose distributions for individual BT sources. These studies have examined the sensitivity of phantom composition on absorbed dose, the influence of dynamic internal components within the source, effects of manufacturing variations on resultant dose distributions, and previously ignored radiations such as beta penetration and characteristic X-ray influence. MC simulations have also been, and persist to be, an important component to correct detector measurements for experimental verifications in BT dosimetry. Forthcoming advances using MC methods for BT may include assessment of dosimetric contributions from radioimpurities, characterization of dissolvable BT sources, design of multi-radionuclide sources, integration of BT dosimetry and patient imaging, and of course evaluation of new radionuclides and miniature eBT sources. MC simulations will also play an important role in treatment planning for novel developments such as anisotropic BT sources [100–102] and intensity-modulated BT (IMBT) using rotating high-energy shields within applicators [103]. Since the focus can range from methodical improvements to existing paradigms to highly creative blue-sky ideas, the possibilities for MC BT dosimetry seem endless.

References

1. Williamson JF, Rivard MJ. Thermoluminescent detector and Monte Carlo techniques for reference-quality brachytherapy dosimetry. In: Rogers DWO, Cygler J, editors. *Clinical Dosimetry Measurements in Radiotherapy (AAPM 2009 Summer School)*. Madison, WI: Medical Physics Publishing; 2009. pp. 437–99.
2. Williamson JF. Brachytherapy technology and physics practice since 1950: a half-century of progress. *Phys Med Biol*. 2006;51(13):R303–25.
3. Rogers DW. Fifty years of Monte Carlo simulations for medical physics. *Phys Med Biol*. 2006;51(13):R287–301.
4. Williamson JF. The Sievert integral revisited: evaluation and extension to ^{125}I,^{169}Yb, and^{192}Ir brachytherapy sources. *Int J Radiat Oncol Biol Phys*. 1996;36(5):1239–50.
5. Williamson JF, Rivard MJ. Quantitative dosimetry methods for brachytherapy. In: Thomadsen BR, Rivard MJ, Butler WM, editors. *Brachytherapy Physics*: Second Edition. Madison, WI: Medical Physics Publishing; 2005. pp. 233–94.
6. Meisberger LL, Keller RJ, Shalek RJ. The effective attenuation in water of the gamma rays of gold 198, iridium 192, cesium 137, radium 226, and cobalt 60. *Radiology*. 1968;90(5):953–7.
7. Dale RG. Some theoretical derivations relating to the tissue dosimetry of brachytherapy nuclides, with particular reference to iodine-125. *Med Phys*. 1983;10(2):176–83.
8. Krishnaswamy V. Calculation of the dose distribution about californium-252 needles in tissue. *Radiology*. 1971;98:155–60.
9. Williamson JF, Morin RL, Khan FM. Monte Carlo evaluation of the Sievert integral for brachytherapy dosimetry. *Phys Med Biol*. 1983;28(9):1021–32.
10. Burns GS, Raeside DE. Two-dimensional dose distribution around a commercial ^{125}I seed. *Med Phys*. 1988;15(1):56–60.
11. Williamson JF. Monte Carlo evaluation of kerma at a point for photon transport problems. *Med Phys*. 1987;14(4):567–76.
12. Williamson JF, Quintero FJ. Theoretical evaluation of dose distributions in water about models 6711 and 6702 ^{125}I seeds. *Med Phys*. 1988;15(6):891–7.
13. Williamson JF. Monte Carlo evaluation of specific dose constants in water for ^{125}I seeds. *Med Phys*. 1988;15(5):686–94.
14. Kubo H. Exposure contribution from Ti K x rays produced in the titanium capsule of the clinical I-125 seed. *Med Phys*. 1985;12(2):215–20.
15. Kirov A, Williamson JF, Meigooni AS, Zhu Y. TLD, diode and Monte Carlo dosimetry of an^{192}Ir source for high dose-rate brachytherapy. *Phys Med Biol*. 1995;40(12):2015–36.
16. Valicenti RK, Kirov AS, Meigooni AS, Mishra V, Das RK, Williamson JF. Experimental validation of Monte Carlo dose calculations about a high-intensity Ir-192 source for pulsed dose-rate brachytherapy. *Med Phys*. 1995;22(6):821–9.
17. Williamson JF. Comparison of measured and calculated dose rates in water near I-125 and Ir-192 seeds. *Med Phys*. 1991;18(4):776–86.
18. Williamson JF, Perera H, Li Z, Lutz WR. Comparison of calculated and measured heterogeneity correction factors for ^{125}I,^{137}Cs, and ^{192}Ir brachytherapy sources near localized heterogeneities. *Med Phys*. 1993;20(1):209–22.
19. Perera H, Williamson JF, Li Z, Mishra V, Meigooni AS. Dosimetric characteristics, air-kerma strength calibration and verification of Monte Carlo simulation for a new Ytterbium-169 brachytherapy source. *Int J Radiat Oncol Biol Phys*. 1994;28(4):953–70.
20. Das RK, Li Z, Perera H, Williamson JF. Accuracy of Monte Carlo photon transport simulation in characterizing brachytherapy dosimeter energy-response artefacts. *Phys Med Biol*. 1996;41(6):995–1006.
21. Das R, Meigooni AS, Mishra V, Langton MA, Williamson JF. Dosimetric characteristics of the type 8 Ytterbium-169 interstitial brachytherapy source. *J Brachytherapy Int*. 1997;13:219–34.
22. Das RK, Keleti D, Zhu Y, Kirov AS, Meigooni AS, Williamson JF. Validation of Monte Carlo dose calculations near ^{125}I sources in the presence of bounded

heterogeneities. *Int J Radiat Oncol Biol Phys*. 1997;38(4): 843–53.

23. Rivard MJ, Coursey BM, DeWerd LA, Hanson WF, Huq MS, Ibbott GS, et al. Update of AAPM Task Group No. 43 Report: a revised AAPM protocol for brachytherapy dose calculations. *Med Phys*. 2004;31(3):633–74.

24. Reynaert N, Verhaegen F, Taeymans Y, Van Eijkeren M, Thierens H. Monte Carlo calculations of dose distributions around ^{32}P and ^{198}Au stents for intravascular brachytherapy. *Med Phys*. 1999;26(8):1484–91.

25. Bohm TD, DeLuca PM, Jr., DeWerd LA. Brachytherapy dosimetry of ^{125}I and ^{103}Pd sources using an updated cross section library for the MCNP Monte Carlo transport code. *Med Phys*. 2003;30(4):701–11.

26. Reniers B, Verhaegen F, Vynckier S. The radial dose function of low-energy brachytherapy seeds in different solid phantoms: comparison between calculations with the EGSnrc and MCNP4C Monte Carlo codes and measurements. *Phys Med Biol*. 2004;49(8):1569–82.

27. Demarco JJ, Wallace RE, Boedeker K. An analysis of MCNP cross-sections and tally methods for low-energy photon emitters. *Phys Med Biol*. 2002;47(8):1321–32.

28. Cullen D, Hubbell JH, Kissel L. EPDL97: the Evaluated Photon Data Library, 97 version. UCRL-50400. 1997;Vol. 6, Rev 5.

29. Grimm P, Sylvester J. Advances in brachytherapy. *Rev Urol*. 2004;6 (Suppl 4):S37–48.

30. Nath R, Anderson LL, Luxton G, Weaver KA, Williamson JF, Meigooni AS. Dosimetry of interstitial brachytherapy sources: recommendations of the AAPM Radiation Therapy Committee Task Group No. 43. American Association of Physicists in Medicine. *Med Phys*. 1995;22(2):209–34.

31. Rivard MJ, Butler WM, DeWerd LA, Huq MS, Ibbott GS, Meigooni AS, et al. Supplement to the 2004 update of the AAPM Task Group No. 43 Report. *Med Phys*. 2007;34(6):2187–205.

32. Rivard MJ, Ballester F, Butler WM, DeWerd LA, Ibbott GS, Meigooni AS, et al. Supplement 2 for the 2004 update of the AAPM Task Group No. 43 Report: joint recommendations by the AAPM and GEC-ESTRO. *Med Phys*. 2017;44(9):e297–e338.

33. Perez-Calatayud J, Ballester F, Das RK, Dewerd LA, Ibbott GS, Meigooni AS, et al. Dose calculation for photon-emitting brachytherapy sources with average energy higher than 50 keV: report of the AAPM and ESTRO. *Med Phys*. 2012;39(5):2904–29.

34. Li Z, Das RK, DeWerd LA, Ibbott GS, Meigooni AS, Perez-Calatayud J, et al. Dosimetric prerequisites for routine clinical use of photon emitting brachytherapy sources with average energy higher than 50 kev. *Med Phys*. 2007;34(1):37–40.

35. Berger MJ. Monte Carlo calculation of the penetration and diffusion of fast charged particles. In: Adler B, Fernbach S, Rotenburg M, editors. *Methods in Computational Physics*, volume 1. New York: Academic Press; 1963. pp. 135–215.

36. Kawrakow I, Bielajew AF. On the condensed history technique for electron transport. *Nucl Instr Meth B*. 1998; 142(3):253–80.

37. Ribberfors R, Carlsson GA. Compton component of the mass-energy absorption coefficient: corrections due to the energy broadening of compton-scattered photons. *Radiat. Res*. 1985;101(1):47–59.

38. Taylor RE, Rogers DW. An EGSnrc Monte Carlo-calculated database of TG-43 parameters. *Med Phys*. 2008;35(9):4228–41.

39. Taylor RE, Rogers DW. EGSnrc Monte Carlo calculated dosimetry parameters for ^{192}Ir and ^{169}Yb brachytherapy sources. *Med Phys*. 2008;35(11):4933–44.

40. Ballester F, Granero D, Perez-Calatayud J, Melhus CS, Rivard MJ. Evaluation of high-energy brachytherapy source electronic disequilibrium and dose from emitted electrons. *Med Phys*. 2009;36(9):4250–6.

41. Chiu-Tsao ST, Duckworth TL, Patel NS, Pisch J, Harrison LB. Verification of Ir-192 near source dosimetry using GAFCHROMIC film. *Med Phys*. 2004;31(2):201–7.

42. DeWerd LA, Ibbott GS, Meigooni AS, Mitch MG, Rivard MJ, Stump KE, et al. A dosimetric uncertainty analysis for photon-emitting brachytherapy sources: report of AAPM Task Group No. 138 and GEC-ESTRO. *Med Phys*. 2011;38(2):782–801.

43. Hubbell JH. Review of photon interaction cross section data in the medical and biological context. *Phys Med Biol*. 1999;44(1):R1–22.

44. Hedtjarn H, Carlsson GA, Williamson JF. Accelerated Monte Carlo based dose calculations for brachytherapy planning using correlated sampling. *Phys Med Biol*. 2002;47(3):351–76.

45. Chamberland MJ, Taylor RE, Rogers DW, Thomson RM. egs_brachy: a versatile and fast Monte Carlo code for brachytherapy. *Phys Med Biol*. 2016;61(23): 8214–31.

46. Ballester F, Hernandez C, Perez-Calatayud J, Lliso F. Monte Carlo calculation of dose rate distributions around ^{192}Ir wires. *Med Phys*. 1997;24(8):1221–8.

47. Li Z, Williamson JF, Perera H. Monte Carlo calculation of kerma to a point in the vicinity of media interfaces. *Phys Med Biol*. 1993;38(12):1825–40.

48. Chibani O, Williamson JF. MCPI: a sub-minute Monte Carlo dose calculation engine for prostate implants. *Med Phys*. 2005;32(12):3688–98.

49. Poon E, Reniers B, Devic S, Vuong T, Verhaegen F. Dosimetric characterization of a novel intracavitary mold applicator for ^{192}Ir high dose rate endorectal brachytherapy treatment. *Med Phys*. 2006;33(12):4515–26.

50. Sampson A, Le Y, Williamson JF. Fast patient-specific Monte Carlo brachytherapy dose calculations via the correlated sampling variance reduction technique. *Med Phys*. 2012;39(2):1058–68.

51. Pokhrel D, Murphy MJ, Todor DA, Weiss E, Williamson JF. Reconstruction of brachytherapy seed positions and orientations from cone-beam CT x-ray projections via a novel iterative forward projection matching method. *Med Phys*. 2011;38(1):474–86.

52. Lindsay P, Battista J, Van Dyk J. The effect of seed anisotrophy on brachytherapy dose distributions using ^{125}I and [103]Pd. *Med Phys*. 2001;28(3):336–45.

53. Corbett JF, Jezioranski JJ, Crook J, Tran T, Yeung IW. The effect of seed orientation deviations on the quality of ^{125}I prostate implants. *Phys Med Biol*. 2001;46(11):2785–800.

54. Russell KR, Ahnesjo A. Dose calculation in brachytherapy for a ^{192}Ir source using a primary and scatter dose separation technique. *Phys Med Biol*. 1996;41:1007–24.

55. Russell KR, Tedgren AK, Ahnesjo A. Brachytherapy source characterization for improved dose calculations using primary and scatter dose separation. *Med Phys*. 2005;32(9):2739–52.

56. Ahnesjo A, van Veelen B, Tedgren AC. Collapsed cone dose calculations for heterogeneous tissues in brachytherapy using primary and scatter separation source data. *Comput Methods Programs Biomed*. 2017;139:17–29.

57. Daskalov GM, Baker RS, Rogers DW, Williamson JF. Multigroup discrete ordinates modelling of [125]I 6702 seed dose distributions using a broad energy-group cross section representation. *Med Phys*. 2002;29(2):113–24.

58. Gifford KA, Horton JL, Jr., Pelloski CE, Jhingran A, Court LE, Mourtada F, et al. A three-dimensional computed tomography-assisted Monte Carlo evaluation of ovoid shielding on the dose to the bladder and rectum in intracavitary radiotherapy for cervical cancer. *Int J Radiat Oncol Biol Phys*. 2005;63(2):615–21.

59. Dolan J, Lia Z, Williamson JF. Monte Carlo and experimental dosimetry of an ^{125}I brachytherapy seed. *Med Phys*. 2006;33(12):4675–84.

60. Rivard MJ, Kirk BL, Leal LC. Impact of radionuclide physical distribution on brachytherapy dosimetry parameters. *Nucl Sci Eng*. 2005;149:101–6.

61. Rivard MJ. Monte Carlo calculations of AAPM Task Group Report No. 43 dosimetry parameters for the MED3631-A/M125I source. *Med Phys*. 2001;28(4):629–37.

62. Rivard MJ, Melhus CS, Kirk BL. Brachytherapy dosimetry parameters calculated for a new[103]Pd source. *Med Phys*. 2004;31(9):2466–70.

63. Rivard MJ. Brachytherapy dosimetry parameters calculated for a [131]Cs source. *Med Phys*. 2007;34(2):754–62.

64. Williamson JF. Dosimetric characteristics of the DRAXIMAGE model LS-1 I-125 interstitial brachytherapy source design: a Monte Carlo investigation. *Med Phys*. 2002;29(4):509–21.

65. Nuttens VE, Lucas S. AAPM TG-43U1 formalism adaptation and monte carlo dosimetry simulations of multiple-radionuclide brachytherapy sources. *Med Phys*. 2006;33(4):1101–7.

66. Nuttens VE, Lucas S. Determination of the prescription dose for biradionuclide permanent prostate brachytherapy. *Med Phys*. 2008;35(12):5451–62.

67. Nath R, Rivard MJ, DeWerd LA, Dezarn WA, Thompson Heaton II H, Ibbott GS, et al. Guidelines by the AAPM and GEC-ESTRO on the use of innovative brachytherapy devices and applications: Report of Task Group 167. *Med Phys*. 2016;43:3178-205.

68. Granero D, Vijande J, Ballester F, Rivard MJ. Dosimetry revisited for the HDR ^{192}Ir brachytherapy source model mHDR-v2. *Med Phys*. 2011;38(1):487–94.

69. Liu D, Poon E, Bazalova M, Reniers B, Evans M, Rusch T, et al. Spectroscopic characterization of a novel electronic brachytherapy system. *Phys Med Biol*. 2008;53(1):61–75.

70. Perez-Calatayud J, Granero D, Ballester F. Phantom size in brachytherapy source dosimetric studies. *Med Phys*. 2004;31(7):2075–81.

71. Melhus CS, Rivard MJ. Approaches to calculating AAPM TG-43 brachytherapy dosimetry parameters for ^{137}Cs,[125]I,[192]Ir,[103]Pd, and [169]Yb sources. *Med Phys*. 2006;33(6):1729–37.

72. Morin LRM. Molecular form factors and photon coherent scattering cross sections of water. *J Phys Chem Ref Data*. 1982;11:1091–8.

73. Patel NS, Chiu-Tsao ST, Williamson JF, Fan P, Duckworth T, Shasha D, et al. Thermoluminescent dosimetry of the Symmetra ^{125}I model I25.S06 interstitial brachytherapy seed. *Med Phys*. 2001;28(8):1761–9.

74. Beaulieu L, Carlsson Tedgren A, Carrier JF, Davis SD, Mourtada F, Rivard MJ, et al. Report TG-186 of the AAPM, ESTRO, and ABG on model-based dose calculation techniques in brachytherapy: status and clinical requirements for implementation beyond the TG-43 formalism. *Med Phys*. 2012;39:6208–36.

75. Monroe JI, Williamson JF. Monte Carlo-aided dosimetry of the theragenics TheraSeed model 200 [103]Pd interstitial brachytherapy seed. *Med Phys*. 2002;29(4):609–21.

76. Seltzer SM, Lamperti PJ, Loevinger R, Mitch MG, Weaver JT, Coursey BM. New national air-kerma-strength standards for ^{125}I and [103]Pd brachytherapy seeds. *J Res Nat Inst Stand Technol*. 2003;108:337–58.

77. Loftus TP. Exposure standardization of Iodine-125 seeds used for brachytherapy. *J Res Nat Bur Stand*. 1984;89:295–303.

78. Kaveckyte V, Malusek A, Benmakhlouf H, Alm Carlsson G, Carlsson Tedgren A. Suitability of microDiamond detectors for the determination of absorbed dose to water around high-dose-rate (192) Ir brachytherapy sources. *Med Phys*. 2018;45(1):429–37.

79. Meigooni AS, Meli JA, Nath R. Influence of the variation of energy spectra with depth in the dosimetry of ^{192}Ir using LiF TLD. *Med Phys*. 1988;33:1159–70.

80. Carlsson Tedgren A, Carlsson GA. Influence of phantom material and dimensions on experimental 192Ir dosimetry. *Med Phys*. 2009;36(6):2228–35.

81. Rodriguez M, Rogers DW. Effect of improved TLD dosimetry on the determination of dose rate constants for (125)I and (103)Pd brachytherapy seeds. *Med Phys.* 2014;41(11):114301.

82. Schoenfeld AA, Thieben M, Harder D, Poppe B, Chofor N. Evaluation of water-mimicking solid phantom materials for use in HDR and LDR brachytherapy dosimetry. *Phys Med Biol.* 2017;62(24):N561–N72.

83. Selvam TP, Keshavkumar B. Monte Carlo investigation of energy response of various detector materials in ^{125}I and ^{169}Yb brachytherapy dosimetry. *J Appl Clin Med Phys.* 2010;11(4):3282.

84. Davis SD, Ross CK, Mobit PN, Van der Zwan L, Chase WJ, Shortt KR. The response of LiF thermoluminescence dosemeters to photon beams in the energy range from 30 kV x rays to ^{60}Co gamma rays. *Radiat Prot Dosimetry.* 2003;106(1):33–43.

85. Nunn AA, Davis SD, Micka JA, DeWerd LA. LiF:Mg,Ti TLD response as a function of photon energy for moderately filtered x-ray spectra in the range of 20-250 kVp relative to ^{60}Co. *Med Phys.* 2008;35(5):1859–69.

86. Tedgren AC, Hedman A, Grindborg JE, Carlsson GA. Response of LiF:Mg,Ti thermoluminescent dosimeters at photon energies relevant to the dosimetry of brachytherapy (<1 MeV). *Med Phys.* 2011;38(10):5539–50.

87. Adolfsson E, Carlsson GA, Grindborg JE, Gustafsson H, Lund E, Carlsson Tedgren A. Response of lithium formate EPR dosimeters at photon energies relevant to the dosimetry of brachytherapy. *Med Phys.* 2010;37(9):4946–59.

88. Hashimoto S, Nakajima Y, Kadoya N, Abe K, Karasawa K. Energy dependence of a radiophotoluminescent glass dosimeter for HDR (192) Ir brachytherapy source. *Med Phys.* 2019;46(2):964–72.

89. Kaveckyte V, Persson L, Malusek A, Benmakhlouf H, Alm Carlsson G, Carlsson Tedgren A. Investigation of a synthetic diamond detector response in kilovoltage photon beams. *Med Phys.* 2020;47(3):1268–79.

90. Reynaert N, Verhaegen F, Thierens H. In-water calibration of PDR ^{192}Ir brachytherapy sources with an NE2571 ionization chamber. *Phys Med Biol.* 1998;43(8):2095–107.

91. Watson PGF, Popovic M, Seuntjens J. Determination of absorbed dose to water from a miniature kilovoltage x-ray source using a parallel-plate ionization chamber. *Phys Med Biol.* 2017;63(1):015016.

92. Borg J, Kawrakow I, Rogers DW, Seuntjens JP. Monte Carlo study of correction factors for Spencer-Attix cavity theory at photon energies at or above 100 keV. *Med Phys.* 2000;27(8):1804–13.

93. Adolfsson E, White S, Landry G, Lund E, Gustafsson H, Verhaegen F, et al. Measurement of absorbed dose to water around an electronic brachytherapy source. Comparison of two dosimetry systems: lithium formate EPR dosimeters and radiochromic EBT2 film. *Phys Med Biol.* 2015;60(9):3869–82.

94. Reniers B, Landry G, Eichner R, Hallil A, Verhaegen F. In vivo dosimetry for gynaecological brachytherapy using a novel position sensitive radiation detector: feasibility study. *Med Phys.* 2012;39(4):1925–35.

95. Rossi G, Gainey M, Kollefrath M, Hofmann E, Baltas D. Suitability of the microDiamond detector for experimental determination of the anisotropy function of High Dose Rate (192) Ir brachytherapy sources. *Med Phys.* 2020; 47(11):5838–51.

96. Bohm TD, Griffin SL, DeLuca PM, Jr., DeWerd LA. The effect of ambient pressure on well chamber response: Monte Carlo calculated results for the HDR 1000 plus. *Med Phys.* 2005;32(4):1103–14.

97. Tornero-Lopez AM, Guirado D, Perez-Calatayud J, Ruiz-Arrebola S, Simancas F, Gazdic-Santic M, et al. Dependence with air density of the response of the PTW SourceCheck ionization chamber for low energy brachytherapy sources. *Med Phys.* 2013;40(12):122103.

98. Torres Del Rio J, Tornero-Lopez AM, Guirado D, Perez-Calatayud J, Lallena AM. Air density dependence of the response of the PTW SourceCheck 4pi ionization chamber for (125)I brachytherapy seeds. *Phys Med.* 2017;38:93–7.

99. Shipley DR, Sander T, Nutbrown RF. Source geometry factors for HDR ^{192}Ir brachytherapy secondary standard well-type ionization chamber calibrations. *Phys Med Biol.* 2015;60:2573–86.

100. Cohen GN, Episcopia K, Lim SB, LoSasso TJ, Rivard MJ, Taggar AS, et al. Intraoperative implantation of a mesh of directional palladium sources (CivaSheet): Dosimetry verification, clinical commissioning, dose specification, and preliminary experience. *Brachytherapy.* 2017;16(6):1257–64.

101. Rivard MJ. A directional (103)Pd brachytherapy device: Dosimetric characterization and practical aspects for clinical use. *Brachytherapy.* 2017;16(2):421–32.

102. Aima M, Reed JL, DeWerd LA, Culberson WS. Air-kerma strength determination of a new directional (103)Pd source. *Med Phys.* 2015;42(12):7144–52.

103. Morcos M, Enger SA. Monte Carlo dosimetry study of novel rotating MRI-compatible shielded tandems for intensity modulated cervix brachytherapy. *Phys Med.* 2020;71:178–84.

7

Monte Carlo Modelling of Scanned Ion Beams in Radiotherapy

Loïc Grevillot
*MedAustron Ion
Therapy Center*

7.1 Introduction

One can divide the ion beam delivery techniques into two main categories, *passive beam delivery techniques* and *dynamic beam delivery techniques*.[1] Hybrid techniques combining a dynamic beam delivery with passive components, such as wobbling, have not been taken into consideration. Instead, this chapter will focus on the Monte Carlo (MC) modelling of scanned ion beam delivery (SIBD) systems in the field of light ion beam therapy (LIBT).[2,3] Light ions are defined as those nuclei with an atomic number lower or equal to 10, i.e., including all ions from protons to neon.[1,4] The terminology "scanned ion beam delivery (SIBD)" is used in this chapter (instead of the widespread terminology "pencil beam scanning") in order to be specific and compliant with the definitions mentioned above. MC is a very useful tool to support medical physics teams in the start-up, design, commissioning, and research activities of LIBT facilities.[5] Several general-purpose MC codes are available, such as GEANT4,[6] FLUKA,[7] PHITS,[8,9] SHIELD-HIT,[10,11] or MCNP.[12–14] In Europe, the FLUKA MC code was originally used at the Heidelberg Ion Beam Therapy Center in Germany and later at the Centro Nazionale di Adroterapia Oncologica (CNAO) in Italy.[5] The GEANT4-based GATE MC code[15,16] is used at the MedAustron Ion Therapy Facility in Austria, in particular the GATE-RTion release.[17] Other GEANT4-based platforms, such as PTSim,[18] GAMOS,[19] and TOPAS[20], are also available. This chapter provides an overview of the work performed in the field of SIBD MC modelling over the past 20 years. After a brief introduction about the SIBD technique, a strong focus is put on describing the modelling methods for SIBD systems. The next two sections address the specificities of modelling nuclear interactions and the applications of MC in LIBT centers. To conclude, a general summary is provided.

7.1.1 Scanned Ion Beam Delivery

SIBD is a time-dependent delivery technique, which takes the advantage of the physical properties of charged particles. Light ion beams are magnetically scanned transversally in two dimensions to conform to the tumor cross-section, and the beam energy is changed to conform to the tumor in depth (Figure 7.1).

Treatment head designs are less standardized than in conventional photon radiation therapy, and variations can be found between different manufacturers and LIBT centers. Typical energy ranges used clinically are 60–250 MeV for protons (30–380 mm range) and 120–430 MeV/u for carbon ions (30–300 mm range). For synchrotron-based delivery systems,

DOI: 10.1201/9781003211846-9

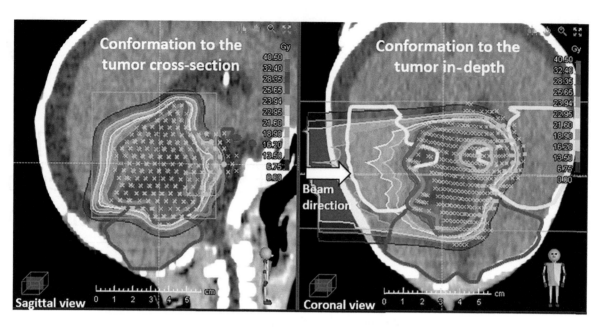

FIGURE 7.1 Illustration of the spot distribution (crosses) for a lateral beam used for a head treatment. Sagittal view (left) and coronal view (right). Treatment plan realized in a horizontal fixed beamline. (Courtesy of MedAustron.)

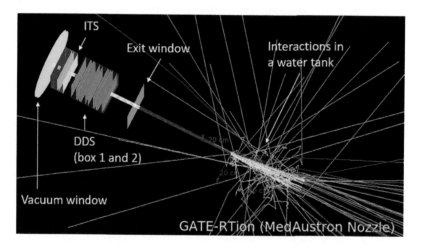

FIGURE 7.2 Monte Carlo simulation of a 400 MeV/u carbon ion pencil beam delivered through a fixed beamline (nozzle) and interacting in a water tank located at isocenter. The nozzle contains a double foil vacuum window, an Independent Termination System (ITS), a redundant beam monitoring system (DDS), a nozzle exit window, and passive elements (not represented in this simulation). Simulation performed using the GATE-RTion release from GATE. (Courtesy of MedAustron.)

the delivered energy is variable; while for cyclotron-based delivery systems, a fixed energy is delivered. Degraders must be used (usually after the cyclotron extraction) to reduce the energy to the desired level before being transported to the treatment nozzle. To further reduce the energy below the minimum energy deliverable by the accelerator (both for synchrotrons and cyclotrons) and to reach proximal tumors down to skin depth, a passive element called range shifter is inserted into the beam at the nozzle. More details about passive elements are given in Section 7.2.3. In some systems, the energy is not adjusted to the desired energy before reaching the treatment nozzle, but it is degraded using additional range shifter

plates inserted directly into the nozzle.[21,22] Overviews of different developments, accelerators, and methods used to apply SIBD treatments can be found elsewhere.[1,21,23] Independently of the technology used and whether the energy is adjusted before reaching the nozzle or in the nozzle using additional range shifter plates, the beam delivery system always requires the same minimum generic equipment: beam monitors, scanning magnets, and passive elements. Without passive elements and additional range shifter plates, the nozzle water equivalent thicknesses (WETs) are typically between 1 and 3 mm. An example of a MC model of a SIBD nozzle is presented in Figure 7.2.

7.2 MC Modelling of SIBD Systems

7.2.1 Description of a Pencil Beam

A pencil beam (or ion beam) can be described by two energy parameters—mean energy and energy spread and three optical parameters—beam size, beam divergence, and beam emittance. In fact, for a given beam size, divergence, and emittance, a beam can converge or diverge; therefore, a fourth optical parameter is required to specify if the beam is converging or diverging. The energy and optical properties are usually approximated by Gaussian distributions. Assuming that the beam is transported along the z-axis, the beam optics parameters can be defined independently from each other in the two orthogonal planes, XoZ and YoZ. In the following, beam optics parameters will be described for one plane, but the same applies to the second plane. Beam optics parameters can be described in a phase space, representing the 2D probability density function of the particles in terms of spatial and angular distributions. The 1-sigma level (i.e., one standard deviation) contour of this 2D probability density function is an ellipse. The ellipse area A corresponds to the beam emittance ε (i.e., $A = \varepsilon$). The ellipse rotates as the beam converges or diverges (Figure 7.3). The smallest spot size is achieved at the so-called waist position when the ellipse is straight. In a vacuum-drift space, the emittance is a constant (due to the absence of scattering). At the waist position, the beam emittance is $\varepsilon = \pi \cdot \sigma_x \cdot \sigma_\theta$, where σ_x and σ_θ correspond to the 1-sigma beam size and beam divergence, respectively.

In accelerator physics, beam optics is frequently described by the so-called Twiss parameters[25]: α, β, γ, and ε, as illustrated in Figure 7.4. The ellipse is described in Equation 7.1:

$$\gamma \cdot x^2 + 2 \cdot \alpha \cdot x \cdot x' + \beta \cdot x'^2 = \varepsilon \qquad (7.1)$$

where x and x' correspond to the beam size and divergence. The parameters α, β, γ, and ε altogether describe the ellipse shape, size, and orientation. These parameters are dependent on each other as described in Equation 7.2:

$$\gamma \beta - \alpha^2 = 1 \qquad (7.2)$$

Different conventions can be used, such as $A = \varepsilon$ as mentioned earlier or $A = \pi \cdot \varepsilon$ or $A = 4 \cdot \pi \cdot \varepsilon$. It is therefore important to verify the convention used to report the beam emittance. If this last

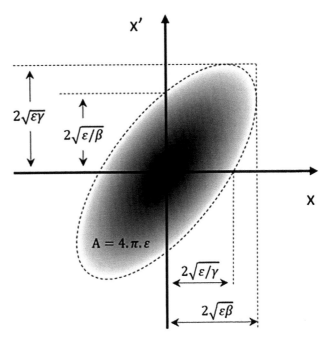

FIGURE 7.4 Phase space description based on Twiss parameters.

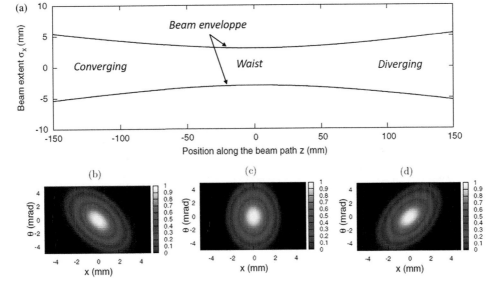

FIGURE 7.3 Illustration of beam size variations during transport in one plane (a), with corresponding phase space representations (b–d). (Reproduced by permission of IOP Publishing[24].)

convention is adopted (i.e., $A = 4 \cdot \pi \cdot \varepsilon$), the beam size is then $\sigma_x = 2 \cdot \sqrt{\varepsilon \cdot \beta}$ and the beam divergence is $\sigma_{x'} = 2 \cdot \sqrt{\varepsilon \cdot \gamma}$.

For a more general description of the transport theory, the reader can refer for instance to the chapters 2 and 3 of this book. A more specific description of the transport theory applied to protons can be found for instance in Ref.[26].

7.2.2 Description of Scanning Properties

The source-to-axis distance (SAD) is the distance from the point where the beam is deflected from the central axis by the scanning magnets to the isocenter. It corresponds in theory to the scanning magnets in the central position (Figure 7.5a). However, in some cases, the SAD may be close to infinite (quasi-parallel SIBD). This is, for example, the case when the scanning magnets are placed before the 90° bending dipole in a vertical beamline or in a gantry (Figure 7.5b). The reader should note that the presented gantry design is specific to gantry 2 from Paul Scherrer Institute (PSI), which is also used at MedAustron. However, in most commercial proton gantries currently available, the scanning magnets are placed in the nozzle after the bending magnets (and not before), hence providing a diverging scanned beam at the isocenter. The virtual SAD (vSAD) can be derived experimentally from spot fluence maps acquired in the air at different air gaps during commissioning, by retroprojecting the beam deflection against the central axis. The vSADs of the two transverse planes (vSADx and vSADy) shall correspond closely to the theoretical SADs (SADx and SADy).

7.2.3 Description of Nozzle Accessories

Geometrical description of nozzle accessories (also called passive elements) is required. For SIBD systems, two main types of accessories are used: accessories for passive shifting of the energy of light ion beams (called "range shifter") and accessories for passive modulation of the energy of light ion beams (called "ripple filter"). Other types of accessories, such as multileaf collimators, may be available for some SIBD systems and are described in other chapters of this book. Passive elements can be integrated either directly in the nozzle (fixed position) or in a moving snout (variable position). The purpose of a moving snout is to bring the passive elements as close to the patient as possible in order to reduce the beam penumbra in the patient (mainly due to multiple Coulomb scatterings in the range shifter). Alternatively, when passive elements are fixed in the nozzle, the patient can be moved toward the nozzle exit to reduce the air gap between the patient and the passive elements[3]. The geometrical description of a range shifter (e.g., a slab of plastic) is simple but critical, and therefore a beam model starting before this passive element allows to provide an accurate model with and without a range shifter. Range shifter modelling is one of the major issues of the current pencil beam algorithms (PBA) used clinically, and this topic will be presented in more detail in Section 7.4.2. In case passive elements such as ripple filters are used, the geometrical description is usually more complex. Different generations of ripple filters have been proposed (1D and 2D), and the optimal design depends on several factors and considerations, such as particle type, energy modulation, manufacturing process, or treatment setup (isocentric or non-isocentric). For carbon ions, ripple filter design accuracy of the order of 10 μm is required. In case the ripple filter geometry varies from the blueprints, some specific correction methods may be required.[27] Therefore, special care should be taken to validate the MC modelling of the passive element geometries. For more details, the reader is referred to a list of relevant publications on the topic.[28-32] Examples of ripple filter and range shifter designs are presented in Figure 7.6.

7.2.4 Beam Modelling Concept

Beam modelling is an attempt to describe the physical properties of the delivered beam as accurately as possible. There may

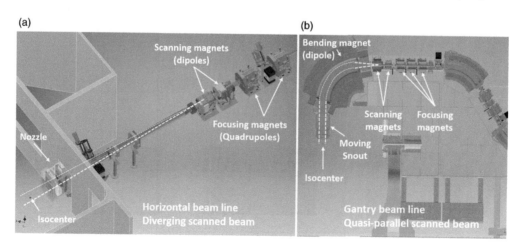

FIGURE 7.5 Illustration of differences between the horizontal beamline design (a) and the gantry beamline design (b) from MedAustron. The horizontal fixed beamline provides a diverging scanned beam at isocenter, while the gantry beamline provides a quasi-parallel scanned beam at isocenter. (Courtesy of MedAustron.)

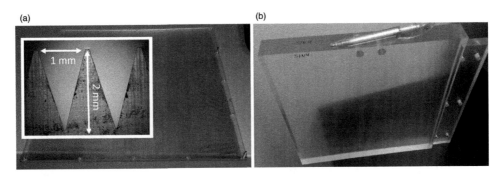

FIGURE 7.6 Illustration of a ripple filter frame (a) with a close-up on the ripple filter structure (image obtained from an optical analysis) and a range shifter of 3 cm WET (b). (Courtesy of MedAustron.)

be different ways of defining and understanding the meaning of *beam modelling* and differences may arise partly from the ion beam delivery technique (passive or dynamic). One possibility is to parameterize a phase space placed at the nozzle exit and derive a model of the beam. This technique is often used for passive delivery systems. For SIBD systems, however, the delivered beam usually interacts very little in the nozzle and is often considered as "pristine Bragg peak." Therefore, it is often more practical to derive the physical properties of the beam directly from measurements in air and in water. In such a case, the beam modelling does not resemble a phase space parametrization, but rather an optimization (or tuning) of MC input parameters.[33] It consists of determining the MC beam model parameters that allow to best match the measurements considered as reference. Ideally, one should adjust the MC parameters according to their uncertainty levels, keeping in mind the uncertainty level of available cross-sections, physics models, and reference measurements.[34] In practical terms, MC experts often refer to the "tuning" of the beam model parameters. Let us consider the example of tuning the simulated beam range to match the measured beam range. Uncertainties in the stopping power of ions in water are estimated to be about 1% or more for clinical energies,[35] which corresponds to 1 mm for a 10 cm beam range. In addition, the energy specified by an accelerator provider does not necessarily correspond to the mean energy of the extracted beam. It can, for instance, correspond to the estimated energy in the synchrotron ring before extraction or other accelerator provider conventions. Moreover, if the beam model does not start from the nozzle entrance, one should additionally estimate the energy loss through all the beamline components crossed by the beam until reaching the position where the beam model starts. Therefore, the uncertainty of the energy at the point where the beam model starts is usually not negligible. To adjust the MC-simulated beam range with the measured beam range in water, one could either adjust the stopping power value of water (usually by adjusting the mean ionization energy I[36,37]) or the MC-simulated energy. Except if specific work was performed to determine the delivered beam energy, one would recommend using the stopping power tables (and I values) provided by international reports such as ICRU and adjust the simulated beam energy (within estimated uncertainties). The

same type of reasoning should be used to tune any other beam model parameters. Beam modelling methods are presented in the following sections.

7.2.5 Generic Modelling Recipe

One can divide the beam modelling methods into two main categories: full-nozzle modelling (i.e., modelling from nozzle entrance) or modelling from the nozzle exit. The advantages and drawbacks of these two methods are described in detail in the next two sections. Independently of the beam modelling method used, a beam model relies on two key pillars: the energy properties and optical properties. Clinically, a third pillar must also be considered: the calibration factor (N/MU) in number of particles (N) per monitor units (MU). According to the previous definition of beam modelling, as "an attempt to describe the physical properties of the delivered beam as accurately as possible," only the first two pillars should be considered as part of the beam model. Instead, the beam model calibration factor should be considered as a separate clinical quantity attached to the beam model. From a clinical point of view, the validation of a beam model in terms of an absolute dose is of paramount importance and relies on this calibration factor. Therefore, when validating a beam model, the beam model calibration factor is often considered as part of the beam model. Since scattering properties vary with beam energy, it is advised to first tune the beam energy properties (using approximate optics parameters) and then tune the beam optics properties. The calibration of the beam model is performed as the last step. The beam modelling should be performed for a subset of clinical energies (subsets from 8 to 27 energies have been reported in the literature) and then interpolated for intermediate energies. The mean energy shall be adjusted to match the measured beam range (usually the range determined at 80% of the maximum dose in the distal dose falloff is used, as it is less sensitive to energy-spread variations) and the energy spread shall be adjusted to match the measured Bragg peak width and distal dose falloff along the beam direction (Figure 7.7). When tuning the energy properties, the limited size of the detector used for measuring integrated radial profiles with depth (IRPD) shall be taken into account in simulation, as illustrated in Figure 7.11 from Section 7.3.

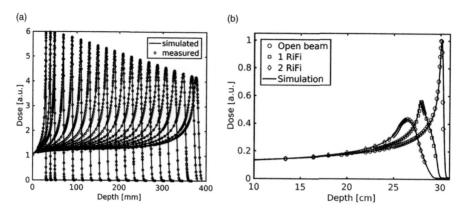

FIGURE 7.7 Example of evaluation of IRPDs for a proton beam model (a, adapted from Ref.[40]) and for a carbon ion beam model including one or two crossed ripple filters (RiFi) to the pristine Bragg peak (b, Reproduced by permission of IOP Publishing[28]). Simulations were performed with the GATE/Geant4 code.

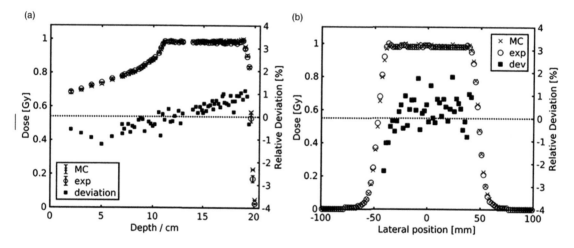

FIGURE 7.8 3D validation of a full-nozzle beam model based on the GATE/Geant4 MC code, considering an 8 cm side cubic target centered at 15 cm depth in water. Evaluation of a depth-dose profile (a) and a transverse profile (b) in absolute dose. (Adapted from Ref.[40].)

The beam optics properties are usually more complex to model than the energy properties, and the modelling recipes are strongly dependent on the beam modelling method used. Therefore, practical details about beam optics modelling are provided in the next two Sections 7.2.6 and 7.2.7.

After the beam optics tuning process, final verification of IRPD in terms of beam range, Bragg peak width, and IRPD shape shall be performed. In case of deviations, a second iteration can be performed to fine-tune the beam energy and optical properties. Once energy and optical parameters are fixed, the beam model can be calibrated in number of particles per monitor unit (N/MU) as a function of energy (see Section 7.2.8). All beam model properties (energy, optical, and calibration) should be modeled as a function of the delivered beam energy to allow predicting beam properties for any clinical energy. Residual differences introduced by interpolation on the spot size, range, and dose shall be evaluated. It is also recommended to evaluate the beam model on intermediate energies which have not been used for beam modelling. Final validation of the beam model should be performed in 3D (e.g.,

using spread-out Bragg peaks (SOBP) delivered during commissioning) and include evaluation of key parameters, such as transverse and longitudinal beam parameters (e.g., field size, range, and penumbras), as well as absolute dose (Figure 7.8). In some cases, validation of the dose in 3D can evidence the need for applying a scaling factor to the beam model[38,39].

Obviously, the accuracy of a beam model does not only depend on the beam modelling procedure. When validating a beam model, one should consider different sources of uncertainties. The uncertainties can be divided between experimental uncertainties and MC simulation uncertainties. Experimental uncertainties are due to measurement uncertainties and beam delivery stability. For more information, the reader may refer, for instance, to the following publications.[34,41,42] MC simulation uncertainties are due to the uncertainties on the cross-sections, physics models (and parameters used), and beam modelling itself. More information on the cross-sections and their uncertainties can be found for instance in Ref.[35,43]. The use of MC simulation for ion beam dosimetry is covered in chapter 2 of the accompanying book.

7.2.6 Modelling Method from Nozzle Exit

Historically, many beam modelling attempts started from the nozzle exit (at least for protons), in contrast to passive scattering delivery systems, where full-nozzle MC simulation is required.[44] Some of the first uses of MC modelling of SIBD were performed at PSI in the early 2000s. An in-house MC code[45,46] was used to evaluate the deficiencies of PBA in complex geometries. In this fast dedicated MC code, the beam model started at the nozzle exit, but the secondary fragments were not explicitly tracked and charged secondaries produced in the nozzle were ignored. The influence of this approximation on the modelling of nuclear processes is described in Section 7.3.1. A key advantage of beam modelling from nozzle exit is that it relies solely on medical physics commissioning data and does not require manufacturer blueprints of the nozzle geometry. Another advantage may be observed in terms of computation speed since the transport of the particles in the nozzle is discarded. Obviously, modelling from the nozzle exit is more accurate as the nozzle is more "transparent" to the beam, i.e., when the energy level is adjusted before being transported to the nozzle (see introductory Section 7.1.1). However, if additional range shifter plates are inserted in the nozzle to adjust the energy, then the spray radiations (see Section 7.3.1) will most probably not be negligible, and a full-nozzle beam model may be more accurate. A first proton source model has been presented in[47] and validated with GEANT3. Later, the first recipe for the modelling of scanned ion beams from nozzle exit solely based on commissioning measurements was proposed in[24], and the open-source code was made available through the OpenGATE platform.[15] In Ref.,[48] the modelling approach from the nozzle exit was validated for protons by comparison against a full-nozzle modelling approach of the Massachusetts General Hospital nozzle, which is a rather "transparent" nozzle. It shows that the fluence of nuclear secondaries produced in the nozzle is typically lower than 0.5% and mainly less than 0.2% of the primary fluence, suggesting that they can be safely neglected for treatment planning system (TPS) validation purposes. However, looking at the low-dose penumbra, the full-nozzle modelling approach shows a larger contribution and may be recommended for studying out-of-field effects, such as radio-induced secondary cancers. Modelling the beam optics consists of describing the beam size variations along the air gap between the nozzle exit and the isocenter. Obviously, the spot size at the nozzle exit can be measured or extrapolated from the commissioning measurements (spot sizes measured at different air gaps between nozzle exit and up to about 20 cm downstream of the isocenter). The beam divergence can also be extrapolated from the measurements, and usually, the unknown parameter remains the beam emittance. In Ref.,[24] the beam size and divergence were determined using the method described above. In addition, the beam divergence at nozzle exit was corrected from the scattering contribution of air in the nozzle to the isocenter air gap, using a quadratic rule (Equation 7.3).

$$\sigma_{\theta,\,\text{nozzle exit}}^2 = \sigma_{\theta,\,\text{air gap}}^2 - \sigma_{\theta,\,\text{air}}^2 \qquad (7.3)$$

The scattering in the air was estimated from GATE/Geant4 simulations. The beam emittance was set empirically to half the beam size (1-sigma) at the nozzle exit times the beam divergence (1-sigma). This choice was made to ensure a purely diverging beam at nozzle exit (i.e., far downstream from the waist position). In case the beam is not purely divergent at nozzle exit or even converging, more complex methods should be used to model the beam optics. In Ref.,[49] the delivered beam at nozzle exit converges to a minimum size and then diverges further. The beam size variation in the air gap was described based on Fermi–Eyges theory using Equation 7.4

$$\sigma_x^2(z) = A_{2,x} + 2A_{1,x}z + A_{0,x}z^2 + \frac{T}{3}z^3 \qquad (7.4)$$

where the $A_{0,x}$, $A_{1,x}$, and $A_{2,x}$ parameters correspond to the variance of the beam angular distribution, the covariance of the beam angular and spatial distributions, and the variance of the spatial distributions at $z=0$, respectively. T corresponds to the scattering power and was derived from GATE/Geant4 simulations as an energy-independent constant. Fitting Equation 7.4 to the measured spot sizes in the air gap between nozzle exit and isocenter allowed deriving the beam size and divergence parameters. The emittance parameter M_x was then determined using Equation 7.5.

$$M_x = \pi\sqrt{A_{0,x}A_{2,x} - A_{1,x}^2} \qquad (7.5)$$

7.2.7 Full-Nozzle Modelling Method

Full-nozzle modelling (i.e., modelling from nozzle entrance) mainly implies the explicit simulation of the beam interactions in the nozzle elements. The explicit transport of the beam through the scanning magnet field, while technically possible, does not increase the beam modelling accuracy.[50] Assuming that detailed and accurate manufacturer's blueprints are available to the user, developing a full-nozzle beam model should provide the most accurate description of the spectra of nuclear secondaries produced in the nozzle. When blueprints are not available, it has been shown that the nozzle elements (mainly the beam monitoring detectors) can be approximated by water slabs having similar WET. However, in some cases and especially if high-density materials (such as tungsten) are used in the beam monitors, additional correction methods may be implemented in order to approximate the scattering and energy modulation properties of the beam monitors.[5] When performing modelling from the nozzle entrance, the manufacturer should provide the user with estimated beam optics and energy parameters at the nozzle entrance. These parameters are usually determined based on other simulation codes and suffer from uncertainties. Therefore, fine-tuning of the beam parameters is required to match the simulations with both the measured IRPDs in water at the isocenter and the measured spot sizes in the air gap between the nozzle exit and the isocenter. In Ref.,[12] a full-nozzle model was developed at the Proton Therapy Center, Houston, in MCNPX (Monte Carlo N-Particle eXtended). The nozzle geometry was based on the detailed blueprints of each component provided by the manufacturer (Hitachi, Ltd). The source was modeled on a subset of eight energies and interpolated over the 94 energies clinically available. Since the beam delivery was based on a synchrotron system, an energy-independent energy spread in percent of the nominal energy was used based on the manufacturer's recommendations. The beam divergence was assumed negligible in comparison to the scattering induced in the nozzle elements and set to 0. This choice simplified the beam optics–tuning process to the only determination of the beam size in *x* and *y* directions. The beam size at the nozzle entrance was then tuned to match the measured spot sizes at different air gaps in the air. At MedAustron, a full-nozzle beam modelling method has also been developed.[38] Despite a purely diverging beam at the nozzle exit, a converging beam was required at the nozzle entrance to predict the spot sizes in the air.[40] Therefore, unlike in,[12] the emittance and divergence parameters could not be set to 0, and additional efforts were required to adjust all optical parameters together in order to match the measured spot sizes in air.

7.2.8 Calibration of the Beam Model

During the commissioning of SIBD systems, MU calibration must be performed for mono-energetic beams in reference conditions. One method consists of delivering a mono-energetic square field (typically within 6–12 cm side) with a constant spot spacing (typically of about one-third of the spot full-width half-maximum) and a constant number of MUs per spot. The dose is measured in a water phantom in the plateau region (typically after the build-up region and up to the midrange). This measurement allows correlating the dose measured in reference conditions with the number of MU per spot delivered, for a given energy. In fact, since a constant spot spacing and MU per spot are used, the measurements can be expressed in dose area product (DAP) per spot.[51] To calibrate the MC beam model, one has to determine the number of particles per MU required to reproduce the measured DAP under reference conditions for the different clinical energies available. The ratio between the measured DAP for a delivered number of MUs per spot and the MC-simulated DAP for a given number of primary particles N simulated per spot allows deriving the calibration factor $\frac{N}{MU}(E)$ for each energy E (Equation 7.6).

$$\frac{N}{MU}(E) = \frac{\dfrac{DAP_{measured}}{MU}(E)}{\dfrac{DAP_{simulated}}{N}(E)} \qquad (7.6)$$

Since the method relies on DAP and due to the reciprocity principle between narrow and broad beams, it is possible to speed up the MC calibration process by relying on the MC simulation of the DAP for laterally integrated central axis ion beams (see Ref.[51] for more details).

7.3 MC Modelling Specificities of Nuclear Models

One of the fundamental differences in LIBT as compared to conventional photon radiation therapy is that the physics models and cross-sections, especially the nuclear models, suffer from significantly larger uncertainties and that nuclear interactions are not negligible. Uncertainties from the nuclear models propagate not only into the physical-dose distribution uncertainties, but also into the modelling uncertainties of relative biological effectiveness (RBE). Considering a 160 MeV proton beam and a 290 MeV/u carbon ion beam having similar ranges in water (about 16–17 cm), about 20% of the protons and 40% of the carbon ions undergo nuclear interactions.[52] This corresponds to a dose contribution from nuclear secondaries up to the Bragg peak depth of about 10% for protons and about 20% for carbon ions at these energies.[52] Several studies benchmarking nuclear models have been performed, but it is generally agreed that the set of available data is very limited and the level of associated uncertainties is substantial. The reader can refer for instance to[53,54] for more details. The following two subsections focus on the MC modelling of the dose contribution from nuclear secondaries for proton and carbon ion beam delivery systems.

7.3.1 Nuclear Halo for Proton Beams

A well-accepted representation of transverse dose distribution from a "pure" primary proton beam impinging a water volume and made of three components has been proposed in[55]: the core consists mainly of primary protons, the halo consists of charged secondaries from hard single scatters, and the aura consists of neutral secondaries (Figure 7.9).

In addition, the term "spray" is used to characterize unwanted doses coming from upstream (mainly nuclear secondaries) and which, in principle, can be reduced by improving nozzle and beamline designs. For example, the spray may arise from additional range shifter plates used to reduce the beam energy in the nozzle,[56] from high Z materials (such as tungsten), which may be used in some beam monitors,[57] or from interactions in the beam pipe prior to the nozzle.

Another common terminology in the literature is the term "low-dose region," which describes the secondary dose deposited away from the beam axis (Figure 7.10). The low-dose region mainly accounts for a mix of halo and spray. In MC simulation, accurate simulation of the core, halo, and aura depends on the beam model and physics models included in the simulation. In case the spray is not negligible, a full-nozzle beam model may be advised. If a beam model from nozzle exit is required, it may be necessary to develop a beam model

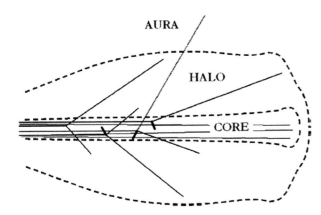

FIGURE 7.9 Description of core, halo, and aura. The dashed lines represent the 10% and 0.01% isodose levels. Recoil nuclei ranges are exaggerated. (Reproduced by permission of IOP Publishing[55].)

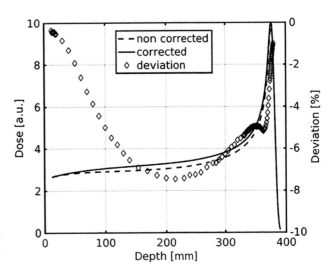

FIGURE 7.11 Illustration of a proton IRPD at 252.7 MeV (38 cm range) measured using a commercial parallel plate ionization chamber of 4.08 cm radius (noncorrected) and corrected based on GATE/Geant4 MC simulations, assuming a 10 cm scoring radius. At midrange, the correction factor is close to 8%. (Adapted from Ref.[62].)

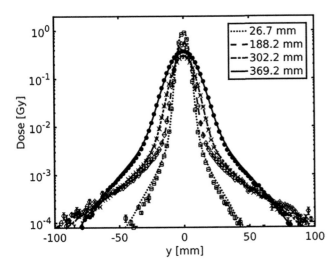

FIGURE 7.10 Illustration of measured (dots) and simulated (lines) transverse dose profiles characterizing the nuclear halo of a 252.7 MeV proton pencil beam in water at 4 depths. The beam has a range of 380 mm. (Adapted from Ref.[58].)

low-dose region. The highest contribution of the "low dose" takes place around midrange (called the "midrange bump"[55]). When measuring IRPDs during commissioning using commercially available large-area parallel-plate chambers, part of the beam escapes the sensitive detector (the low-dose penumbra region). MC simulations can be used to correct measured IRPDs as if they would have been measured with larger detectors,[61,62] which would theoretically encompass the entire halo (Figure 7.11). Since the aura is made of neutral particles, an infinite distance would be required to fully encompass it.

7.3.2 Fragment Spectra for Carbon Ions

For carbon ions, the terminology "nuclear halo" is less employed than for protons, and the scientific community usually refers to the "fragmentation spectra." While for protons the nuclear halo only consists of target nuclei fragments, for carbon ions the projectiles are also fragmented (up to about 50% typically for clinical treatments[63]). The fragmentation spectrum of lighter nuclei is responsible for both a nuclear halo spreading the dose laterally and a tail spreading the dose after the Bragg peak (Figure 7.12). In addition to the issues mentioned previously for protons, for carbon ion therapy the spectra of nuclear secondaries must be described for commissioning the TPS in order to predict the RBE[5]. Parameterization of the transverse dose profiles has been investigated, using FLUKA and GEANT4. Depending on the level of details aimed at describing the spectra of nuclear secondaries, two or three Gaussians may be used.[64,65] If three Gaussians are used and since light fragments scatter with larger angles than heavier fragments, the first, second, and third Gaussians can be assigned to primary carbon ions, fragments with $Z \geq 3$ and

describing not only the primary beam but also the spectra of nuclear secondaries (including energy and optical properties). In the literature, this is often referred to as "multiple Gaussian" beam models: one Gaussian describing the primary pencil beam and a second or even a third Gaussian accounting for the nuclear secondaries.[59] Transverse profile measurements of a single spot, including the low-dose region (as presented in Figure 7.10), are useful to validate nuclear models in MC codes.[58,60] Complementary to transverse profile measurements, output factors (e.g., field size factors) are also valuable measurements to validate the modelling of nuclear processes.[24,57,58] Indeed, as the field size increases, contribution from the far-off axis spots to the dose in the center of the field will mainly be due to the low-dose region from each of these spots. Therefore, good agreement in output factors requires a good description of not only the core but also the

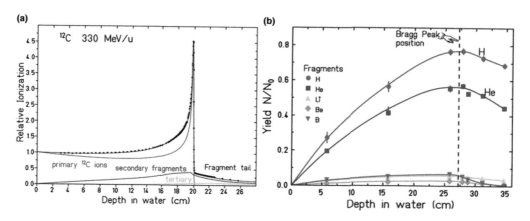

FIGURE 7.12 Illustration of a carbon ion IRPD at 330 MeV/u stopping in water (a) and buildup curves of secondary fragments produced by a 400 MeV/u carbon ion beam in water (b). (Reproduced by permission of IOP Publishing[67].)

fragments with $Z \leq 2$, respectively.[65] Including beam quality variations in the descriptions of transverse dose profiles allows a finer description of RBE variations in the patient geometry. Dose differences up to 2.7% have been reported for small targets using a so-called "trichrome model," as opposed to a previous "monochrome" model for microdosimetric kinetic model–based RBE modelling using GEANT4.[65,66]

7.4 MC Applications

7.4.1 Treatment Planning Workflow

Nowadays, MC has become more and more integrated into the treatment planning workflow to support treatment planning activities. Most TPS manufacturers provide MC algorithms, which are used for routine treatment planning, including dose optimization. Due to their versatility, general-purpose MC codes also found a place of choice in the clinics and are frequently used to support the planning process. For instance, FLUKA has been used for inverse treatment planning[68,69] and for independent dose calculation (IDC) at CNAO.[70] TOPAS has been set up for IDC at the Trento proton therapy center.[71] A new and more efficient MC implementation in the TRiP98 platform has been advertised in ref 72 for IDC and potentially for full inverse optimization of scanned proton and carbon ion beams. Similarly, a dedicated release of GATE, namely GATE-RTion,[17] has been released for clinical applications. GATE-RTion is used for proton IDC at the Christie Hospital in Manchester[73,74] and at the MedAustron LIBT for protons and carbon ions.[38,75] In contrast to general-purpose MC codes, faster and dedicated MC codes have been developed, mainly for protons, such as VMCPRO,[76] MCSQUARE,[77] or FRED.[78] The computation speed depends on the physics implementation details and technical features, such as Central Processing Unit (CPU)- or Graphics Processing Unit (GPU)-based approaches.

7.4.2 Treatment Planning Algorithms

GEANT4- (and formerly GEANT3) and FLUKA-based MC simulation tools have often been used as reference MC codes

to support the development of proton and carbon ion PBA or fast MC algorithms.[16,47,64,65,79,80] For protons, a review of different PBAs clinically available and their typical deficiencies can be found in Ref.[81]. Due to different fragmentation spectra and scattering power, PBA uncertainties are different for protons and carbon ions. MC algorithms show a clear superiority to PBA in complex heterogeneous geometries, especially for protons, due to a significantly larger scattering than carbon ions. As an illustration, dose differences of up to 5% for protons and only 0.5% for carbon ions were observed between MC and PBA in the center of the planning target volume, considering a skull chondrosarcoma and a prostate carcinoma.[72] One of the main issues of PBA in SIBD is to account accurately for the transport of nuclear secondaries in the presence of passive elements, such as a range shifter. This is especially true when large air gaps are used between the treatment nozzle and the patient.[82,83] As an illustration, using a 7.5-cm WET range shifter in combination with an air gap larger than 30 cm resulted in differences between PBA and measurements of up to 10% in the entrance of the SOBP.[82] In contrast, the two MC algorithms tested (RaySearch TPS and GATE/Geant4) were within 3% (Figure 7.13).

In Ref.,[83] comparisons between MC and PBA algorithms for a commercial TPS were performed in various conditions, including critical conditions with superficial targets, range shifter, and various phantoms. In addition, measurements were performed using a 2D array to validate dosimetrically the higher accuracy of MC against PBA. Illustration of the dose and dose volume histogram differences for a beam delivered with 0° gantry angle in a half lamb head are shown in Figure 7.14.

7.4.3 Other Applications

When starting new facilities, MC simulations can be used to estimate beam delivery properties before any beam is available from the accelerator. It is a useful tool to support the definition of accelerator specifications. For instance, MC can be used to estimate the impact of the scanning magnet field fluctuations on the delivered dose[84] or to define beam size and position interlock thresholds based on realistic failure scenarios.[85]

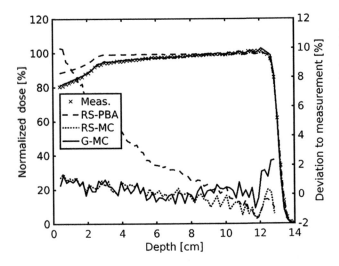

FIGURE 7.13 Comparison of SOBP produced with a range shifter of 7.5 cm WET and an air gap of 31.5 cm, calculated with RayStation PBA (RS-PBA) algorithm, RayStation MC (RS-MC) algorithm, and GATE (G-MC) as independent MC code. Reference measurements (Meas) are also provided for comparison. (Reproduced by permission of IOP Publishing[82].)

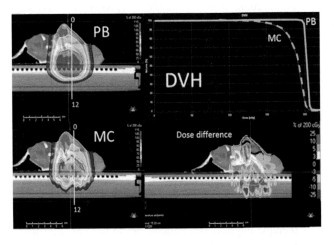

FIGURE 7.14 Comparison of the dose distributions calculated using a PBA and an MC algorithm in the brain target of a half lamb head, using a 0° gantry angle, a range shifter of 4.1 cm WET and an air gap of about 20 cm. (Reproduced by permission of IOP Publishing[83].)

Another application of choice is the design of new beamlines. For example, MC simulations using GATE/Geant4 have been performed at MedAustron to optimize the design of the proton gantry.[86] At CNAO, GEANT4 MC simulations have been conducted to customize one of the existing horizontal fixed beamlines for ocular treatments using protons and carbon ion beams.[87,88]

The treatment of moving targets in radiation therapy requires specific efforts. This is even more complex for SIBD due to interplay effects between the dynamic beam delivery of each spot and patient movement. The impact of the interplay depends on many parameters, such as beam delivery specificities (e.g., accelerator technology, scanning speed, energy switching time, and spot size), patient-specific parameters (e.g., patient condition, 3D motion, and tumor location), and treatment planning considerations (e.g., use of gating/ tracking systems, patient contention, number of beams used, and number of rescanning). Examples of 4D MC applications can be found in Ref.[89,90] The 4D applications and treatment verification techniques based on positron emission tomography or prompt particle imaging are covered in other chapters of this book and of the accompanying book.

The last application to be mentioned relies on the advances of image-guided radiation therapy. While MRI linear accelerator solutions have been proposed in conventional therapy, MRI-guided LIBT concepts are also under investigation. Obviously, the influence of MRI magnetic fields on ion beams is substantial[91] and new PBA must be developed (Figure 7.15).

7.5 Summary

MC simulations play a major role in the field of LIBT and several MC codes are being used. In this chapter, a strong focus was put on describing SIBD modelling methods. Accurate modelling of nuclear processes remains the main source of uncertainty, and it has a direct impact on the physical and biological dose modelling uncertainties. MC is a very useful tool to support the start-up and commissioning of new facilities, as well as new developments and research activities. It is currently being used in clinical routine and, nowadays, most of the TPS vendors are providing MC algorithms. In addition,

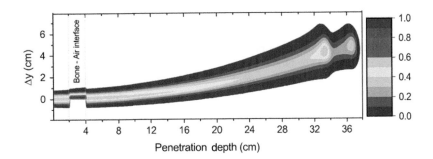

FIGURE 7.15 GATE/Geant4 MC simulation of the relative dose distribution of a 240 MeV proton pencil beam through a 1.5 T magnetic field and crossing an air–bone interface between 2 and 4 cm. (Adapted from Ref.[91].)

general-purpose MC codes are frequently integrated into the clinics to support the treatment planning workflow.

References

1. ICRU. International Commission on Radiation Units and Measurements report 78: Prescribing and Recording and and Reporting Proton-Beam Therapy: Contents. *J ICRU*. 2007;7. doi:10.1093/jicru/ndm021.

2. Moyers MF, Vatnitsky SM. *Practical Implementation of Light Ion BeamTreatments*. Medical Physics Publishing, Madison, WI; 2012. doi:10.1118/1.4851515.

3. Grevillot L, Osorio Moreno J, Letellier V, et al. Clinical implementation and commissioning of the MedAustron Particle Therapy Accelerator for non-isocentric scanned protonbeamtreatments.*MedPhys*.2020;47(2).doi:10.1002/mp.13928.

4. Wambersie A, DeLuca PM, Andreo P, Hendry JH. Light" or Heavy" ions: A debate of terminology? *Radiother Oncol*. 2004;73:iiii.

5. Parodi K, Mairani A, Brons S, et al. Monte Carlo simulations to support start-up and treatment planning of scanned proton and carbon ion therapy at a synchrotron-based facility. *Phys Med Biol*. 2012;57:3759–3784.

6. Allison J, Amako K, Apostolakis J, et al. Recent developments in GEANT4. *Nucl Instrum Methods Phys Res Sect A Accel Spectrom Detect Assoc Equip*. 2016;835(July):186–225. doi:10.1016/j.nima.2016.06.125.

7. Battistoni G, Bauer J, Boehlen TT, et al. The FLUKA Code: An accurate simulation tool for particle therapy. *Front Oncol*. 2016;6(May). doi:10.3389/fonc.2016.00116.

8. Puchalska M, Tessonnier T, Parodi K, Sihver L. Benchmarking of PHITS for carbon ion therapy. *Int J Part Ther*. 2018:IJPT-17-00029.1. doi:10.14338/IJPT-17-00029.1.

9. Sato T, Niita K, Matsuda N, et al. Overview of the PHITS code and its application to medical physics. *Prog Nucl Sci Technol*. 2014;4:879–882. doi:10.15669/pnst.4.879.

10. Hansen DC, Lühr A, Sobolevsky N, Bassler N. Optimizing SHIELD-HIT for carbon ion treatment. *Phys Med Biol*. 2012;57(8):2393–2409. doi:10.1088/0031-9155/57/8/2393.

11. Bassler N, Hansen DC, Lühr A, Thomsen B, Petersen JB, Sobolevsky N. SHIELD-HIT12A - A Monte Carlo particle transport program for ion therapy research. *J Phys Conf Ser*. 2014;489(1):8–13. doi:10.1088/1742-6596/489/1/012004.

12. Sawakuchi GO, Mirkovic D, Perles LA, et al. An MCNPX Monte Carlo model of a discrete spot scanning proton beam therapy nozzle. *Med Phys*. 2010;37(9):4960–4970.

13. Gillin MT, Sahoo N, Bues M, et al. Commissioning of the discrete spot scanning proton beam delivery system at the University of Texas M.D. Anderson Cancer Center, Proton Therapy Center, Houston. *Med Phys*. 2010;37:154–163. doi:10.1118/1.3259742.

14. Ardenfors O, Dasu A, Kopeć M, Gudowska I. Modelling of a proton spot scanning system using MCNP6. *J Phys Conf Ser*. 2017;860(1). doi:10.1088/1742-6596/860/1/012025.

15. Jan S, Benoit D, Becheva E, et al. GATE V6: A major enhancement of the GATE simulation platform enabling modelling of CT and radiotherapy. *Phys Med Biol*. 2011;56(4):881–901. doi:10.1088/0031-9155/56/4/001.

16. Grevillot L, Bertrand D, Dessy F, Freud N, Sarrut D. GATE as a GEANT4-based Monte Carlo platform for the evaluation of proton pencil beam scanning treatment plans. *Phys Med Biol*. 2012;57:4223–4244.

17. Grevillot L, Boersma DJ, Fuchs H, et al. Technical Note: GATE-RTion: A GATE/Geant4 release for clinical applications in scanned ion beam therapy. *Med Phys*. 2020;47(8):3675–3681. doi:10.1002/mp.14242.

18. Akagi T, Aso T, Iwai G, et al. Geant4-based particle therapy simulation framework for verification of dose distributions in proton therapy facilities. *Prog Nucl Sci Technol*. 2014;4:896–900.

19. Arce P, Ignacio Lagares J, Harkness L, et al. Gamos: A framework to do Geant4 simulations in different physics fields with an user-friendly interface. *Nucl Instrum Methods Phys Res Sect A Accel Spectrom Detect Assoc Equip*. 2014;735:304–313. doi:10.1016/J.NIMA.2013.09.036.

20. Perl J, Shin J, Schümann J, Faddegon B, Paganetti H. TOPAS: An innovative proton Monte Carlo platform for research and clinical applications. *Med Phys*. 2012;39(11):6818–6837. doi:10.1118/1.4758060.

21. Noda K. Beam Delivery Method for Carbon-ion Radiotherapy with the Heavy-ion Medical Accelerator in Chiba. *Int J Part Ther*. 2016;2(4):481–489. doi:10.14338/IJPT-15-00041.1.

22. Pedroni E, Bacher R, Blattmann H, et al. The 200-Mev proton therapy project at the Paul Scherrer Institute: Conceptual design and practical realization. *Med Phys*. 1995;22(1):37–53. doi:10.1118/1.597522.

23. Schardt D, Elsässer T, Schulz-Ertner D. Heavy-ion tumor therapy: Physical and radiobiological benefits. *Rev Mod Phys*. 2010;82(1):383–425. doi:10.1103/RevModPhys.82.383.

24. Grevillot L, Bertrand D, Dessy F, Freud N, Sarrut D. A Monte Carlo pencil beam scanning model for proton treatment plan simulation using GATE/GEANT4. *Phys Med Biol*. 2011;56(16):5203–5219. http://www.ncbi.nlm.nih.gov/pubmed/21791731.

25. Wiedemann H. *Particle Accelerator Physics*. Third edit. Springer-Verlag, Berlin Heidelberg; 2007. doi:10.1007/978-3-540-49045-6.

26. Gottschalk B. Techniques of Proton Radiotherapy: Transport Theory. 2012:43. http://arxiv.org/abs/1204.4470.

27. Kurz C, Mairani A, Parodi K. First experimental-based characterization of oxygen ion beam depth dose distributions at the Heidelberg Ion-Beam Therapy Center. *Phys Med Biol*. 2012;57(15):5017–5034. doi:10.1088/0031-9155/57/15/5017.

28. Grevillot L, Stock M, Vatnitsky S. Evaluation of beam delivery and ripple filter design for non-isocentric proton

and carbon ion therapy. *Phys Med Biol.* 2015;60(20):7985–8005. http://www.ncbi.nlm.nih.gov/pubmed/26418366.

29. Weber U, Kraft G. Design and construction of a ripple filter for a smoothed depth dose distribution in conformal particle therapy. *Phys Med Biol.* 1999;44:2765–2775.

30. Bourhaleb F, Attili A, Cirio R, et al. Monte Carlo simulations of ripple filters designed for proton and carbon ion beams in hadrontherapy with active scanning technique. *J Phys Conf Ser.* 2008;102.

31. Courneyea L, Beltran C, Tseung HSWC, Yu J, Herman MG. Optimizing mini-ridge filter thickness to reduce proton treatment times in a spot-scanning synchrotron system. *Med Phys.* 2014;41(6):061713. doi:10.1118/1.4876276.

32. Ringbaek TP, Weber U, Petersen JB, Thomsen B, Bassler N. Monte Carlo simulations of new 2D ripple filters for particle therapy facilities. *Acta Oncol (Madr).* 2014;53(-1):40–49. doi:10.3109/0284186X.2013.832834.

33. Fuchs H, Elia A, Resch AF, et al. Computer-assisted beam modelling for particle therapy. *Med Phys.* 2020. doi:10.1002/mp.14647.

34. Grevillot L, Stock M, Palmans H, et al. Implementation of dosimetry equipment and phantoms at the MedAustron light ion beam therapy facility. *Med Phys.* 2018;45(1):352–369. doi:10.1002/mp.12653.

35. Seltzer SM, Fernández-Varea JM, Andreo P, et al. Key data for ionizing-radiation dosimetry: Measurement standards and applications, ICRU Report 90. 2016.

36. Andreo P. On the clinical spatial resolution achievable with protons and heavier charged particle radiotherapy beams. *Phys Med Biol.* 2009;54(11):N205--N215. doi:10.1088/0031-9155/54/11/N01.

37. Paul H. The stopping power of matter for positive ions. *Mod Pract Radiat Ther.* 2012:113–132.

38. Elia A, Resch AF, Carlino A, et al. A GATE/Geant4 beam model for the MedAustron non-isocentric proton treatment plans quality assurance. *Phys Medica.* 2020;71(August 2019):115–123. doi:10.1016/j.ejmp.2020.02.006.

39. Carlino A, Böhlen T, Vatnitsky S, et al. Commissioning of pencil beam and Monte Carlo dose engines for non-isocentric treatments in scanned proton beam therapy. *Phys Med Biol.* 2019;64(17). doi:10.1088/1361-6560/ab3557.

40. Elia A. Characterization of the GATE Monte Carlo platform for nonisocentric treatments and patient specific treatment plan verification at MedAustron (PhD thesis, INSA Lyon, 2019LYSE002). 2019.

41. Karger CP, Jäkel O, Palmans H, Kanai T. Dosimetry for ion beam radiotherapy. *Phys Med Biol.* 2010;55:193–234. doi:10.1088/0031-9155/55/21/R01.

42. Giordanengo S, Manganaro L, Vignati A. Review of technologies and procedures of clinical dosimetry for scanned ion beam radiotherapy. *Phys Medica.* 2017;43(October):79–99. doi:10.1016/j.ejmp.2017.10.013.

43. ICRU. International Commission on Radiation Units and Measurements report 63: Nuclear Data for Neutron and Proton Radiotherapy and for Radiation Protection. *J ICRU.* 2000. http://jicru.oxfordjournals.org.

44. Paganetti H, Jiang H, Lee SY, Kooy HM. Accurate Monte Carlo simulations for nozzle design and commissioning and quality assurance for a proton radiation therapy facility. *Med Phys.* 2004;31(7):2107–2118.

45. Kohno R, Sakae T, Takada Y, et al. Simplified Monte Carlo Dose Calculation for Therapeutic Proton Beams. *Jpn J Appl Phys.* 2002;41(Part 2, No. 3A):L294–L297. doi:10.1143/JJAP.41.L294.

46. Tourovsky A, Lomax AJ, Schneider U, Pedroni E. Monte Carlo dose calculations for spot scanned proton therapy. *Phys Med Biol.* 2005;50(5):971–981. doi:10.1088/0031-9155/50/5/019.

47. Kimstrand P, Traneus E, Ahnesjö A, Grusell E, Glimelius B, Tilly N. A beam source model for scanned proton beams. *Phys Med Biol.* 2007;52(11):3151–3168. doi:10.1088/0031-9155/52/11/015.

48. Grassberger C, Lomax A, Paganetti H. Characterizing a proton beam scanning system for Monte Carlo dose calculation in patients. *Phys Med Biol.* 2014;60(2):633–645. doi:10.1088/0031-9155/60/2/633.

49. Almhagen E, Boersma DJ, Nyström H, Ahnesjö A. A beam model for focused proton pencil beams. *Phys Medica.* 2018;52:27–32. doi:10.1016/j.ejmp.2018.06.007.

50. Peterson SW, Polf J, Bues M, et al. Experimental validation of a Monte Carlo proton therapy nozzle model incorporating magnetically steered protons. *Phys Med Biol.* 2009;54(10):3217–3229. doi:10.1088/0031-9155/54/10/017.

51. Palmans H, Vatnitsky SM. Beam monitor calibration in scanned light-ion beams Beam monitor calibration in scanned light-ion beams. *Med Phys.* 2016;43(11):5835–5847. doi:10.1118/1.4963808.

52. Dedes G, Parodi K. Monte Carlo simulations of particle interactions with tissue in carbon ion therapy. *Int J Part Ther.* 2015;2(3):447–458. doi:10.14338/IJPT-15-00021.

53. Böhlen TT, Cerutti F, Dosanjh M, et al. Benchmarking nuclear models of FLUKA and GEANT4 for carbon ion therapy. *Phys Med Biol.* 2010;55(19):5833–5847. doi:10.1088/0031-9155/55/19/014.

54. Bolst D, Cirrone GAP, Cuttone G, et al. Validation of Geant4 fragmentation for Heavy Ion Therapy. *Nucl Instrum Methods Phys Res Sect A Accel Spectrom Detect Assoc Equip.* 2017;869:68–75. doi:10.1016/j.nima.2017.06.046.

55. Gottschalk B, Cascio EW, Daartz J, Wagner MS. On the nuclear halo of a proton pencil beam stopping in water. *Phys Med Biol.* 2015;60(14):5627–5654. doi:10.1088/0031-9155/60/14/5627.

56. Pedroni E, Scheib S, Böhringer T, et al. Experimental characterization and physical modelling of the dose distribution of scanned proton pencil beams. *Phys Med Biol.* 2005;50:541–561. doi:10.1088/0031-9155/50/3/011.

57. Sawakuchi GO, Titt U, Mirkovic D, et al. Monte Carlo investigation of the low-dose envelope from scanned

proton pencil beams. *Phys Med Biol.* 2010;55(3):711–721. doi:10.1088/0031-9155/55/3/011.

58. Resch AF, Elia A, Fuchs H, et al. Evaluation of electromagnetic and nuclear scattering models in GATE/Geant4 for proton therapy. *Med Phys.* 2019;46(5):2444–2456. doi:10.1002/mp.13472.

59. Lin L, Kang M, Solberg TD, Ainsley CG, McDonough JE. Experimentally validated pencil beam scanning source model in TOPAS. *Phys Med Biol.* 2014;59(22):6859–6873. doi:10.1088/0031-9155/59/22/6859.

60. Hall DC, Makarova A, Paganetti H, Gottschalk B. Validation of nuclear models in Geant4 using the dose distribution of a 177 MeV proton pencil beam. *Phys Med Biol.* 2016;61(1):N1–N10. doi:10.1088/0031-9155/61/1/N1.

61. Anand A, Sahoo N, Zhu XR, et al. A procedure to determine the planar integral spot dose values of proton pencil beam spots. *Med Phys.* 2012;39(2):891. doi:10.1118/1.3671891.

62. Carlino A. Implementation of advanced methodologies in the commissioning of a Light Ion Beam Therapy facility (PhD thesis, Department of Physics and Chemistry, University of Palermo, Italy). 2017.

63. Durante M, Paganetti H. Nuclear physics in particle therapy: A review. *Rep Prog Phys.* 2016;79(9):096702. doi:10.1088/0034-4885/79/9/096702.

64. Parodi K, Mairani A, Sommerer F. Monte Carlo-based parametrization of the lateral dose spread for clinical treatment planning of scanned proton and carbon ion beams. *J Radiat Res.* 2013;54(SUPPL.1):91–96. doi:10.1093/jrr/rrt051.

65. Inaniwa T, Kanematsu N. A trichrome beam model for biological dose calculation in scanned carbon-ion radiotherapy treatment planning. *Phys Med Biol.* 2014;60(1):437–451. doi:10.1088/0031-9155/60/1/437.

66. Inaniwa T, Furukawa T, Kase Y, et al. Treatment planning for a scanned carbon beam with a modified microdosimetric kinetic model. *Phys Med Biol.* 2010;55(22):6721–6737. doi:10.1088/0031-9155/55/22/008.

67. Haettner E, Iwase H, Krämer M, Kraft G, Schardt D. Experimental study of nuclear fragmentation of 200 and 400 MeV/u (12)C ions in water for applications in particle therapy. *Phys Med Biol.* 2013;58:8265–8279. doi:10.1088/0031-9155/58/23/8265.

68. Mairani A, Böhlen TT, Schiavi A, et al. A Monte Carlo-based treatment planning tool for proton therapy. *Phys Med Biol.* 2013;58(8):2471–2490. doi:10.1088/0031-9155/58/8/2471.

69. Böhlen TT, Bauer J, Dosanjh M, et al. A Monte Carlo-based treatment-planning tool for ion beam therapy. *J Radiat Res.* 2013;54(SUPPL.1):77–81. doi:10.1093/jrr/rrt050.

70. Molinelli S, Mairani A, Mirandola A, et al. Dosimetric accuracy assessment of a treatment plan verification system for scanned proton beam radiotherapy: One-year experimental results and Monte Carlo analysis of the involved uncertainties. *Phys Med Biol.* 2013;58:3837–3847. doi:10.1088/0031-9155/58/11/3837.

71. Fracchiolla F, Lorentini S, Widesott L, Schwarz M. Characterization and validation of a Monte Carlo code for independent dose calculation in proton therapy treatments with pencil beam scanning. *Phys Med Biol.* 2015;60(21):8601–8619. doi:10.1088/0031-9155/60/21/8601.

72. Iancu G, Kraemer M, Zink K, Durante M, Weber U. Implementation of an Efficient Monte Carlo Algorithm in TRiP: Physical Dose Calculation. *Int J Part Ther.* 2015;2(2):415–425. doi:10.14338/IJPT-14-00030.1.

73. Aitkenhead A, Sitch P, Jenny R, Winterhalter C, Patel I, Ranald M. Automated Monte-Carlo re-calculation of proton therapy plans using Geant4/Gate: Implementation and comparison to plan-specific quality assurance measurements. *Br J Radiol.* 2020;93(1114).

74. Winterhalter C, Taylor M, Boersma D, et al. Evaluation of GATE-RTion (GATE/Geant4) Monte Carlo simulation settings for proton pencil beam scanning quality assurance. *Med Phys.* 2020. doi:10.1002/mp.14481.

75. Bolsa M, Palmans H, Boersma DJ, Stock M, Grevillot L. Monte Carlo computation of 3D distributions of stopping power ratios in Light Ion Beam Therapy using GATE-RTion. *Med Phys.* 2021;(accepted).

76. Fippel M, Soukup M. A Monte Carlo dose calculation algorithm for proton therapy. *Med Phys.* 2004;31(8):2263–2273.

77. Souris K, Lee JA, Sterpin E. Fast multipurpose Monte Carlo simulation for proton therapy using multi- and many-core CPU architectures. *Med Phys.* 2016;43(4):1700–1712. doi:10.1118/1.4943377.

78. Schiavi A, Senzacqua M, Pioli S, et al. Fred: A GPU-accelerated fast-Monte Carlo code for rapid treatment plan recalculation in ion beam therapy. *Phys Med Biol.* 2017;62(18):7482–7504. doi:10.1088/1361-6560/aa8134.

79. Soukup M, Fippel M, Alber M. A pencil beam algorithm for intensity modulated proton therapy derived from Monte Carlo simulations. *Phys Med Biol.* 2005;50(21):5089–5104. doi:10.1088/0031-9155/50/21/010.

80. Schaffner B, Pedroni E, Lomax A. Dose calculation models for proton treatment planning using a dynamic beam delivery system: An attempt to include density heterogeneity effects in the analytical dose calculation. *Phys Med Biol.* 1999;44(1):27–41.

81. Saini J, Traneus E, Maes D, et al. Advanced Proton Beam Dosimetry Part I: Review and performance evaluation of dose calculation algorithms. *Transl Lung Cancer Res.* 2018;7(2):171–179. doi:10.21037/tlcr.2018.04.05.

82. Saini J, Maes D, Egan A, et al. Dosimetric evaluation of a commercial proton spot scanning Monte-Carlo dose algorithm: Comparisons against measurements and simulations. *Phys Med Biol.* 2017;62(19):7659–7681. doi:10.1088/1361-6560/aa82a5.

83. Widesott L, Lorentini S, Fracchiolla F, Farace P, Schwarz M. Improvements in pencil beam scanning proton therapy dose calculation accuracy in brain tumor cases with

a commercial Monte Carlo algorithm. *Phys Med Biol.* 2018;63(14). doi:10.1088/1361-6560/aac279.

84. Peterson S, Polf J, Ciangaru G, Frank SJ, Bues M, Smith A. Variations in proton scanned beam dose delivery due to uncertainties in magnetic beam steering. *Med Phys.* 2009;36(8):3693–3702. doi:10.1118/1.3175796.

85. Parodi K, Mairani A, Brons S, et al. The influence of lateral beam profile modifications in scanned proton and carbon ion therapy: A Monte Carlo study. *Phys Med Biol.* 2010;55(17):5169–5187. doi:10.1088/0031-9155/55/17/018.

86. Fuchs H, Grevillot L, Carlino A, et al. Optimizing the MedAustron proton gantry beam delivery: Providing nozzle design recommendations based on Gate/Geant4 Monte Carlo simulation. In: *PTCOG* 55; 2016.

87. Piersimoni P, Rimoldi A, Riccardi C, Pirola M, Molinelli S, Ciocca M. Optimization of a general-purpose, actively scanned proton beamline for ocular treatments: Geant4 simulations. *J Appl Clin Med Phys.* 2015;16(2):261–278. doi:10.1120/jacmp.v16i2.5227.

88. Farina E, Piersimoni P, Riccardi C, Rimoldi A, Tamborini A, Ciocca M. Geant4 simulation for a study of a possible use of carbon ion pencil beams for the treatment of ocular melanomas with the active scanning system at CNAO. *J Phys Conf Ser.* 2015;664(7). doi:10.1088/1742-6596/664/7/072048.

89. Dowdell S, Grassberger C, Sharp GC, Paganetti H. Interplay effects in proton scanning for lung: A 4D Monte Carlo study assessing the impact of tumor and beam delivery parameters. *Phys Med Biol.* 2013;58(12): 4137–4156. doi:10.1088/0031-9155/58/12/4137.

90. Shin J, Perl J, Schümann J, Paganetti H, Faddegon BA. A modular method to handle multiple time-dependent quantities in Monte Carlo simulations. *Phys Med Biol.* 2012;57(11):3295–3308. doi:10.1088/0031-9155/57/11/3295.

91. Padilla-Cabal F, Georg D, Fuchs H. A pencil beam algorithm for magnetic resonance image guided proton therapy. *Med Phys.* 2018:2195–2204. doi:10.1002/mp.12854.

8

Monte Carlo Simulations for Treatment Device Design

Bruce A. Faddegon
*University of California,
San Francisco*

8.1 Introduction

Monte Carlo simulation has broad application in the field of radiotherapy, beyond the well-known applications in dosimetry and treatment planning. The Monte Carlo method has proven to be a necessary component of many developments in radiotherapy, and the tools available today make such virtual design relatively straightforward for a medical physicist. This chapter provides a survey of applications of Monte Carlo simulation in the design of treatment devices for conventional linear accelerators used for X-ray and electron therapy. Examples selected from the research, clinical, and consulting experience of the author show the power of the method and elucidate the general principles of treatment device design. The examples are loosely ordered from the exit window of the accelerator, progressing through the different components of the treatment head, through the patient and into the flat panel detector (Figure 8.1).

The EGS4 Monte Carlo system [1] was used for the earlier work, EGSnrc [2] once it was available. The BEAM user code [3] was used for treatment head simulation, sometimes with modification. The MCRTP user code [4] was used for dose calculation. The results include experimental validation, showing the accuracy of simulation for practical applications. The devices designed have for the most part been fabricated and tested (with the exception of the secondary scattering foils). Many were used in radiotherapy with commercially available accelerators.

8.2 Target for Therapy

The accelerated electron beam incident on X-ray targets generates considerable heat. The targets are generally cooled with circulating, chilled water. One design used a gold target cooled by water channeled beneath the target, with water in the beam path (Figure 8.2). When some of these targets developed hairline cracks, pressurized water would leak out from the cooling channel and drip on components such as the monitor chamber, potentially shorting them out.

Monte Carlo simulation aided in the design of a new target. The primary concern was to eliminate the water from the beam path. In addition, the X-ray beam generated in the new target needed to conform closely to the beam from the original target, not just for one beam energy, but for the full range of X-ray beam energies available for treatment on the different accelerator models that used this target. In order to maintain a comparable dose rate for the same beam current, tungsten was chosen for the target material, being close in atomic number to gold. The tungsten was brazed to copper for heat dissipation. Since X-rays would be generated in both the

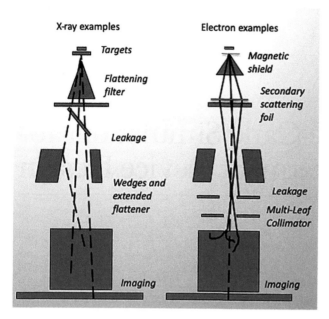

FIGURE 8.1 A generic linear accelerator treatment head configured for X-ray and electron therapy, showing the regions of greatest relevance to the different treatment device design examples (not to scale).

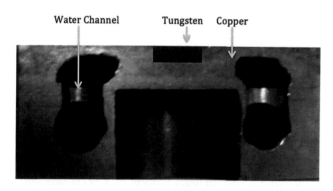

FIGURE 8.2 X-ray target used for radiotherapy, cut in half to show cross section. The water channel for target cooling runs around the outside of the target. The purpose of the cavity below the target is to hold a carbon absorber.

tungsten and copper, this gave a degree of freedom to match X-ray beam characteristics for the two targets. The question to be answered with Monte Carlo simulation was whether there was a combination of thicknesses of the two target materials that would result in matched beams over the full energy range.

The Monte Carlo method proved of great help in answering this question. The approach taken was to match angular distributions of energy fluence and spectral distributions on the beam axis for both targets, covering the full range of beam energies. Although angular distributions of energy fluence could be measured relatively easily by measuring dose profiles, measuring spectral distributions is rather difficult (see, e.g., Ref. [5]). Monte Carlo simulation avoided the need to construct a set of targets with different thicknesses of tungsten

and copper and then measure output and full dose distributions covering the clinical range of field size, including those with beam modifiers (physical wedges, in particular). The simulation details and resulting target design were published [6]. It turned out that a specific thickness of tungsten and copper did yield beam details that were a close enough match to the original target to permit replacement with the new target while avoiding the tedious process of recommissioning.

8.3 Target for Imaging

Patients are traditionally set up on the treatment couch prior to irradiation by aligning marks placed on the skin with lasers mounted on the wall and ceiling. The volume in the patient targeted with radiation may move in relation to the surface markers. Modern radiotherapy often involves imaging the patient immediately prior to treatment. A relatively straightforward means to do so is to generate the imaging beam with the same accelerator used to generate the treatment beam. In this way, the radiation therapists setting up the patient can observe the position of the internal anatomy of the patient in relation to the beam portal (Figure 8.3). A CT image may also be obtained with this beam.

The energy range of photons in the X-ray beam that is most suitable for imaging is 10–150 keV. The treatment beams have energies in the range 0.5–20 MeV. Lower energy photons are actually produced in the target used to generate the treatment beam, but these are largely absorbed in the target and never reach the patient. The resulting beam has a reduced surface dose, generally preferred in radiotherapy. A low atomic number target may be used to retain these photons for the purpose of imaging.

Monte Carlo simulation has proven instrumental in the design of such a target for planar imaging [7] and cone-beam CT [8]. The beam energy is lowered, to fully stop the primary electron beam in the target, as those electrons would otherwise leak through and reach the monitor chamber, undesirable for

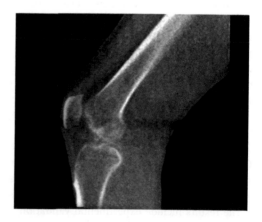

FIGURE 8.3 A sagittal slice of a cone-beam CT of a knee taken with a 4 MV X-ray beam from a carbon target, imaged with a flat panel detector. The dose to the patient was 3 cGy.

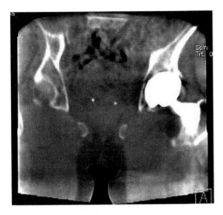

FIGURE 8.4 A coronal slice of a 4 MV carbon target cone-beam CT image taken of a patient immediately prior to radiotherapy.

monitoring dose delivery, and even reach the patient, needlessly adding to the skin dose.

Patient images have been successfully taken with this target for patient alignment with dose in the range 0.3–3 cGy (maximum dose to water at 100 cm SSD for a 10 cm×10 cm field given the same monitor units used to image the patient), a low dose compared to the daily treatment dose, typically around 200 cGy. An image obtained with a 4 MV X-ray beam generated in a carbon target is shown in Figure 8.4. Two of the three seeds implanted in the prostate to aid in patient positioning are visible on the slice shown. Note that the metallic implant in the left hip results in minimal artifact in this image, a strength of imaging with MV X-ray beams.

8.4 Electron Beam Magnetic Shield

Many modern linear accelerators are equipped with a powerful magnet at the end of the beam line to bend the accelerated electron beam 270°. The electrons pass in a circle through the envelope at the end of the evacuated waveguide with a narrow range of energies. The beam emerges through the exit window of the beam line at a right angle to the axis of the waveguide, directed toward the patient on the treatment couch. For the major Siemens linear accelerator models, the resulting full widths at half maxima of the energy distribution peaks are in the range of 6%–14% of the beam energy [9]. The standard approach for adjusting the beam energy is to adjust the bending magnet current and then adjust the RF power in the waveguide while monitoring the dose rate to maintain a constant beam pulse frequency. The bending magnet current gives the desired depth penetration of the electron beam in water, and the RF power that gives the peak dose rate gives the correct beam energy.

At least for one vendor's design, the bending magnet field is known to encroach on the path of the beam between the exit window and monitor chamber. This has no effect on X-rays, since they are uncharged and are not deflected by a magnetic field. However, the fringe magnetic field does deflect electron

FIGURE 8.5 Shield designed to eliminate the fringe magnetic field from the bending magnet that caused the electron field asymmetry in the inplane direction.

beams. This field must be included in the simulation in order to accurately match the known asymmetry in the fluence and dose distributions of electron fields. A study of this effect required modification of the BEAMnrc user code to incorporate asymmetry and spatially varying magnetic fields [10]. The knowledge that this fringe magnetic field fully accounted for the asymmetry leads to the possibility that a magnetic field could be shielded to produce symmetric beams. A hollow cone machined from ferromagnetic stainless steel (Figure 8.5) was inserted into the treatment head with the linear accelerator operated in electron mode. Dose distributions measured once this magnetic field shield was in place were symmetric, as predicted with the Monte Carlo method.

8.5 X-Ray Standard and Extended Flattening Filters

The X-ray beam in radiotherapy is produced by highly relativistic electrons moving within 1% of the speed of light. By conservation of momentum, most of the photons generated in the target travel in a tight cone about the beam axis. A cone-shaped flattening filter is often inserted in the line of the X-ray beam to cut the fluence in the beam center down to match the fluence at the beam edge.

Monte Carlo simulation is an established means to design these flattening filters. In one unique application, the technique was used to develop a beam that can be used to image the patient for setup verification and then treat the patient using the exact same beam [11,12]. This would permit image-guided radiotherapy using the same beam for imaging and treatment, simplifying the design of the treatment head. The beams generated in low atomic number materials are more useful for imaging, having a high proportion of diagnostic energy photons. These photons are low enough in energy to exhibit high contrast to different biological tissues and organs, due to large differences in the photoelectric effect. These imaging beams could not be fully flattened with a flattener of the same material as the target, since there was insufficient space in the treatment head to accommodate such a flattener. Nevertheless, at the time the study was done it was felt that partially flattened beams could prove as effective as fully flattened beams in radiotherapy. This has proved to be the case, as unflattened beams are now commonly used for the treatment of even large target volumes, taking advantage of intensity-modulated radiation therapy (IMRT) and volume-modulated arc therapy (VMAT) techniques.

Monte Carlo simulation was used to determine the trade-off of the compromise beams for imaging and therapy by calculating the response of the flat panel detector used for patient imaging (Figure 8.6) and the spatial distribution of photon fluence at the face of the detector as a function of energy. These were used to calculate contrast for the different target-flattener combinations and relate image quality with field flatness for the different beams. Subsequently, one manufacturer incorporated a graphite target into one of the electron foil slots of a multi-energy machine (see above), slowing the impetus to further develop a single-beam, imaging-treatment machine.

In a separate application, the existing flattener of a clinical machine was redesigned with the help of Monte Carlo simulation, to fully flatten the X-ray beam out along the diagonals of the largest field [13]. Later, a beveled brass plate was designed as an add-on, placed in the accessory tray slot normally used for wedges, to accomplish the same thing. The plate, designed with Monte Carlo simulation, was used to extend the flat region of the field to cover the largest X-ray field available on the machine: a 40 cm × 40 cm field at the machine isocenter. The beveled brass plate did produce a uniform, flat beam over the whole field, out to 23 cm along the diagonals (Figure 8.7). This plate has been used for total body irradiation (TBI) of patients seated 3 m from the source, as a backup to TBI in a larger room where patients stand up to 4.5 m from the source.

8.6 X-Ray Leakage

Radiation from the linear accelerator target is most intense in the direction along the beam axis, where the region in the patient to be treated is generally positioned for treatment. Still, X-rays emanate in all directions from the target, some outside of the area of the treatment beam. Standard practice is for manufacturers to add shielding, usually lead and/or tungsten, to reduce the dose 1 m from the target in all directions to 0.1% of the infield dose at that same 1-m distance.

The Siemens Artiste linear accelerator has a different treatment head than the earlier Oncor model. The distance from the target to the secondary collimators (jaws) was increased, in part to accommodate full overtravel of the movable jaws that collimate the beam in the Y-direction. Monte Carlo simulation was used to determine the thickness and material of shielding to incorporate in the treatment head. Dose profiles calculated at the depth of maximum dose in the direction of jaw travel with the jaw positioned at full overtravel are shown in Figure 8.8. The results show the magnitude of the leakage from several sources (transmission and scatter from different parts of the treatment head). Note that the study was for a prototype shielding design and does not represent the commercial product.

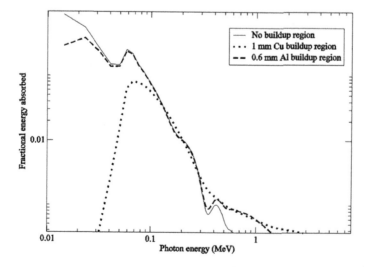

FIGURE 8.6 Calculated response functions for a bare sheet of scintillator and with either 1 mm Cu or 0.6 mm Al on top of the sheet.

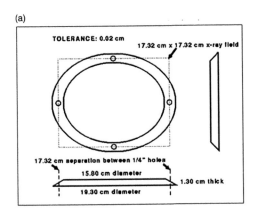

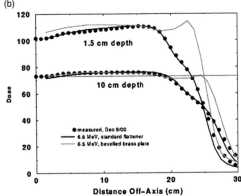

FIGURE 8.7 Brass plate (a), designed to be mounted on the downstream side of a 0.8 cm thick Al tray at 41.3 cm from the source. Diagonal dose profile for the 6 MV MXE 40 cm×40 cm field with and without the brass plate in the beam (b).

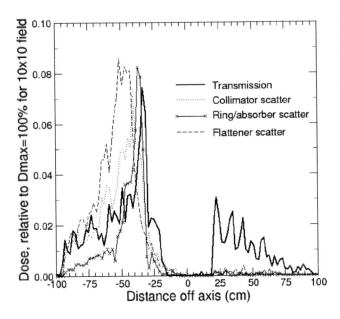

FIGURE 8.8 Calculated leakage profile through isocenter for Artiste 23 MV X-ray beam with full overtravel of jaw and added prototype shielding. Results show contributions from different sources of primary and scattered radiation. The field in this case is completely collimated and would not be used to treat patients.

8.7 Electron Secondary Scattering Foils

Monte Carlo simulation was used to consider what changes in the treatment head were needed to enable precise electron beam collimation with the X-ray MLC [14]. The idea was that by bringing the secondary scattering foil closer to the primary foil, and thus reducing the size of the electron field at the secondary foil, one would obtain a sharper penumbra. The replacement of the air downstream of the secondary foil with the much lower scattering helium atmosphere (at the same pressure as air) was also considered. The secondary foil was positioned at its clinical position, 10.6 cm from the primary foil, and at half that distance, 5.6 cm from the primary foil.

Scattering foils were designed for treatment with one of three choices of final collimation: the conventional electron applicator, with the distal surface of the electron applicator at a distance of 95 cm from the nominal source position (close to the primary scattering foil); an add-on electron MLC at 65 cm (eMLC); and the X-ray MLC at 35 cm (xMLC).

Monte Carlo simulation was used to design the secondary scattering foil. The design constraints are as follows:

1. Similar penetration depth in water, R_{50} (0.2 MeV less energy loss when air is replaced with He at room temperature and pressure).
2. Similar bremsstrahlung tail (it proved sufficient to use the same primary scattering foils).
3. Similar flatness over 30 cm at isocenter without tertiary collimation (40 cm×40 cm field size setting) to produce beams with flatness equivalent to the clinical beams with tertiary collimation, as shown in Figure 8.9 for the 12 MeV beam.

Dose distributions in water were calculated with the Monte Carlo method for the different atmospheres and foil positions and covering the available energy range (6–21 MeV). The three electron collimators were set to the full range of field sizes. The distance from the collimator to the patient, often referred to as standoff, was the same when comparing the penumbral widths. A standoff of 25 cm was used to allow gantry rotation with the patient in treatment position for X-ray IMRT without retracting the tertiary collimator. The width of the penumbra from the calculated dose distributions was largely independent of leaf position and field size. Representative results are shown in Figure 8.10. The penumbral width from the simulated dose distribution was in reasonable agreement for the two configurations measured, limited to the clinically available foil position of 10.6 cm.

The following conclusions were drawn from these results. Replacement of air with helium sharpened the penumbra for all MLC positions (X-ray MLC, extended electron MLC, and applicator). Halving the distance between the foils was only useful when collimating using the X-ray MLC. With the helium atmosphere, the penumbral width under the X-ray MLC was reduced 35%–40%. Of particular interest, collimation of the electron beam with the X-ray MLC provides a clinically reasonable penumbral width when halving the distance between the scattering foils and using helium downstream of the secondary scattering foil. Further experimental validation would be prudent; however, this would require machining the redesigned secondary foil, placing it 5.6 cm from the primary foil, replacing air with helium in the treatment head, then measuring profiles for the different beam energies at different standoffs while using the X-ray MLC to collimate the beam.

8.8 X-Ray Wedges

In radiotherapy, a uniform dose distribution is generally sought throughout the clinical target volume. An X-ray field

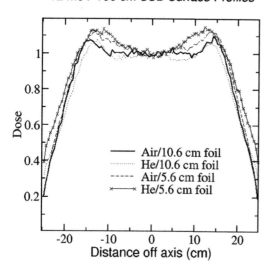

FIGURE 8.9 Surface profiles for the secondary scattering foils designed with Monte Carlo simulation for the 12 MeV electron beam, with air or helium atmosphere, with the foil 5.6 cm and 10.6 cm from the primary foil.

from a single direction produces a uniform dose across the irradiation field with a dose distribution that falls off at a rate of approximately 4% per cm. Diametrically opposed fields, known as a parallel-opposed pair, produce a more uniform distribution with depth. A number of situations arise where a wedge-shaped dose distribution is of value, for example, to reduce the dose to structures distal of the tumor by treating at right angles or to adjust for a slope on the patient surface such as in breast treatment (Figure 8.11). The wedge-shaped dose distribution may be produced by placing a wedge made of a dense material such as iron or lead in the beam, although this may also be done with beam modulation. The wedge is ideally placed as far away from the patient as possible, to reduce the proportion of dose to the patient from secondary radiation emanating from the wedge. Monte Carlo simulation is ideal for evaluating wedge design and placement.

The wedges for the Artiste were designed with the help of Monte Carlo simulation. The objective was to match the profile measured at 10 cm depth from wedges on the earlier Oncor model. The wedge on the Artiste was positioned further from the target than on the Oncor model, requiring larger wedges to cover the same field size. The Oncor wedge thickness was used for the preliminary design, using the same wedge thickness, scaled to the increased distance of the wedge from the target. A unique design was used to keep the material in the wedge closer to the target, primarily so that the weight of the wedge could be minimized. Results for the thickest wedges are shown in Figure 8.12. The scatter from the Artiste wedges was different enough from the Oncor wedges that some design iteration (with simulation) was needed to match the profiles. The final design met manufacturer specification and was adopted for patient treatment.

8.9 Electron Applicator Leakage

Linear accelerators used for radiotherapy often are equipped with scattering foils to provide for treatment with electron beams. Electron beams scatter widely, even in air. Fields are generally collimated close to the patient to sharpen the field edges. Monte Carlo simulation has been used effectively to aid in the design of electron applicators [15]. Electrons that

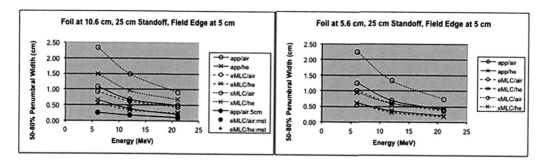

FIGURE 8.10 Calculated penumbral width for the different foil position/atmosphere/electron collimation combinations. The conventional clinical configuration of the foil at 10.6 cm with a 5 cm standoff using an electron applicator is also shown (app/air:5 cm), along with two points where the penumbral width was measured for the 12 MeV beam (mst).

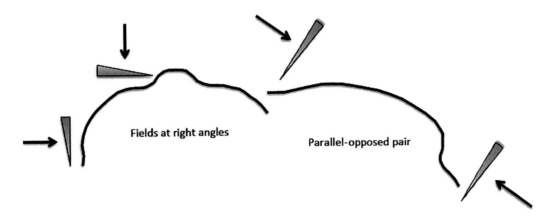

FIGURE 8.11 Examples of treatments using wedged fields: (a) Head and neck treatment with beams at right angles, and (b) breast parallel-opposed pair.

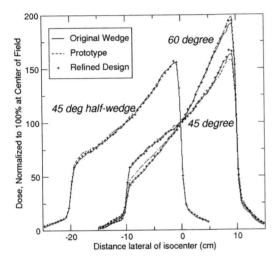

FIGURE 8.12 Prototype wedges based on geometrical divergence alone are compared to the refined design from successive simulations, adjusted to match the wedge profiles for the wedges on the original machine. The calculations are for $0.5 \times 2 \times 0.5$ cm voxels.

hit the applicator can be scattered toward the patient or produce X-rays that can reach the patient. This leakage dose needs to be kept to a low level. The International Electrotechnical Commission document IEC 60621-1 (1998) specifies a limit of 1%–2% of the maximum dose in the field, the larger value for beam energies exceeding 10 MeV, with the leakage dose average over the area 4 cm beyond the field edge and out to the area shielded by the primary collimator.

An investigation using Monte Carlo simulation was conducted to design a retrofit to an existing applicator [16]. The objective was to reduce electron leakage without affecting the field used to treat the patient. The BEAMnrc user code was extended to provide for fluence scoring at the sides of the applicator and to accommodate simulation of the retrofit. A reduced dose (to air) corresponds to reduced fluence, especially in the low-energy electron tail where the stopping power is higher. A retrofit was designed that indeed reduced the fluence of electrons that leaked through the applicator (Figure 8.13) without affecting the relative output factors or dose distributions used to treat patients. Thus, this retrofit could be used without recommissioning the 6 electron beams, a very onerous process.

8.10 Electron Multi-Leaf Collimator

Electrons are charged particles that essentially lose their energy evenly as they penetrate tissue, stopping once they exhaust their energy, with little dose delivered to tissue beyond the electron range. They are widely used in radiotherapy to treat more superficial lesions. They may be used alone or in combination with X-ray therapy, often to boost the dose to the region where there may be residual disease following surgical resection of the tumor, a common approach in treating breast cancer. Electrons, being relatively light compared to the atoms in the tissue, present a challenge for treatment planning as they are scattered widely and as the amount of scattering depends on the material traversed, being very different in air, lung, soft tissue, and bone. Monte Carlo simulation, known to accurately account for electron scatter [17], provides an accurate means for treatment planning in even the most complex radiotherapy geometries.

The availability of accurate treatment planning with Monte Carlo simulation now extends opportunities for electron therapy. Improved conformation to the tumor with significant reduction in harmful energy deposition to healthy tissue may now be achieved by combining X-ray IMRT with a set of adjacent electron fields of different energies [18]. Monte Carlo simulation has been used to aid in the design of the add-on multi-leaf collimator (MLC) to shape the electron fields for this mixed beam radiotherapy [19–20].

8.11 X-Ray Image Detection and Processing

X-rays in both X-ray and electron beams provide useful sources for imaging patients. Imaging is useful for positioning

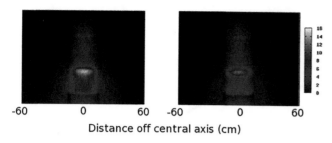

Distance off central axis (cm)

FIGURE 8.13 Dose distribution calculated in air just outside the applicator wall from the isocenter up to 45 cm above the isocenter, without (left) and with (right) additional shielding. The isocenter is situated on the central axis at the bottom of the figure.

the patient prior to treatment and checking and even adjusting for motion during patient treatment.

Patient images, keeping patient comfort and motion (breathing) in mind, are best taken in seconds rather than minutes. Linear accelerator gantry rotation speeds place a lower limit on the time taken for a tomographic CT scan. The time for a single rotation is limited to half a minute or more by regulation [21] and as a consequence of the huge weight they support. Current scanning technologies rely on large fields to cover the region of interest in the patient with fewer rotations of the gantry, using cone-beam CT. The large scatter fractions in kilovoltage cone-beam CT scans reduce image contrast and introduce artifacts, complicating the detection and processing of patient images.

Monte Carlo simulation has proven ideal to characterize the primary beam transmitted through the patient and to characterize the radiation scattered from the patient at the point where that radiation reaches the detector. The technique has provided helpful insight for the design, processing, and interpretation of scatter measurements and for the development of scatter correction algorithms [22–24].

It is also possible to take images of the patient while being treated with electron beams, using the bremsstrahlung generated in the scattering foils [25–26]. These X-rays, along with a smaller portion of bremsstrahlung generated in the patient, result in a slow falloff of dose beyond the electron practical range. The X-rays transmitted through the patient have been used for imaging. The bremsstrahlung dose in the open field, just beyond the practical range, is 0.4% of the maximum dose at 6 MeV, increasing to 1.3% at 12 MeV, and 5% at 21 MeV. Thus, lower energy electron beams have less bremsstrahlung for imaging. Lower energies also produce a flatter bremsstrahlung intensity profile. Patient images were successfully acquired while undergoing electron therapy on a Siemens Primus machine.

8.12 Conclusion

The range of examples shown demonstrates the value of Monte Carlo simulation in treatment device design. A word of caution: Monte Carlo simulation, of great advantage in situations involving complex source and geometry configurations, is necessarily complex and therefore prone to error. Experimental validation should be a general rule. Achieving an accurate match, within the combined experimental uncertainty and statistical precision of the calculation, is both gratifying and simplifies the design process, giving confidence in the results for the newly designed device. However, the effort required can dramatically slow the design process. In practice, device modification can be done without the need to obtain an accurate match between measurement and simulation. An experienced simulator can save considerable time by accepting a match that is deemed sufficient to meet the design objectives. Knowledge of the match (or mismatch) can be enough to redesign the device to a stringent specification, by accounting for this difference as part of the redesign procedure. Experience in simulation is essential to judge whether discrepancies with measured data will unduly affect the geometry and material of the device during the design stage. In any case, once the device is designed, it is prudent to verify the design with measurement. In the end, Monte Carlo simulation can improve the efficiency and cost of the design process by reducing or eliminating the need to machine and make measurements with prototype devices. Indeed, Monte Carlo simulation has become a gold standard for fluence and dose calculation in radiotherapy and is arguably integral to the design process.

References

1. Nelson W. R., Hirayama H., Rogers D.W.O. 1985. The EGS4 Code System. SLAC-Report-265, Stanford Linear Accelerator Center, Stanford, California.

2. Kawrakow I. 2000. Accurate condensed history Monte Carlo simulation of electron transport. I. EGSnrc, the new EGS4 version. *Med. Phys.* 27:485–498.

3. Rogers D. W. O., Faddegon B.A., Ding G. X., Ma C.-M., We J., and Mackie T. R.. 1995. BEAM: A Monte Carlo code to simulate radiotherapy treatment units. *Med. Phys.* 22:503–524.

4. Faddegon B. A., Balogh J., Mackenzie R., Scora D. 1998. Clinical considerations of Monte Carlo for electron radiotherapy treatment planning. *Rad. Phys. and Chem.* 53:217–227.

5. Faddegon B. A., Ross C. K., Rogers D. W. O. 1990. Forward-Directed Bremsstrahlung of 10–30 MeV Electrons Incident on Thick Targets of Al and Pb. *Med. Phys.* 17:773.

6. Faddegon B., Egley B., Steinberg T. 2004. Comparison of Beam Characteristics of a Gold X-ray Target and a Tungsten Replacement Target. *Med. Phys.* 31:91–97.

7. Ostapiak O. Z., O'Brien P. F., Faddegon B. A. 1998. Megavoltage imaging with low Z targets: Implementation and characterization of an investigational system. *Med. Phys.* 25:1910–1918.

8. Faddegon B. A., Aubin M., Bani-hashemi A., Gangadharan B., Gottschalk A. R., Morin O., Wu V.,

Yom S. S. 2010. Comparison of patient megavoltage cone beam CT images acquired with an unflattened beam from a carbon target and a flattened treatment beam. *Med. Phys.* 37:1737–1741.

9. Faddegon B. A., Sawkey D., O'Shea T., McEwen M., Ross C. 2009. Treatment head disassembly to improve the accuracy of large electron field simulation. *Med. Phys.* 36:4577–4591.

10. O'Shea T. P., Foley M. J., Faddegon B. A. 2011. Accounting for the fringe magnetic field from the bending magnet in a Monte Carlo accelerator treatment head simulation. *Med. Phys.* 38:3260–3269.

11. Nishimura K. A. 2005. A Monte Carlo study of low-Z target - flattener combinations for megavoltage imaging. MSc thesis, San Francisco State University.

12. Nishimura K., Svatos M., Zheng Z., Faddegon B. 2005. Target flattener combinations for combined therapy and high contrast megavoltage imaging. *Med. Phys.* 32(6):1909.

13. Faddegon B. A., O'Brien P. F., Mason D. L. D. 1999. The flattened area of Siemens linear accelerator x-ray fields. *Med. Phys.* 26:220–228

14. Faddegon B. A., Svatos M., Karlsson M., Karlsson M., Olofsson L., Antolak J. A. 2002. Treatment head design for mixed beam therapy. *Med. Phys.* 29:1285

15. Janssen R. W. J., Faddegon B. A., Dries W. J. F. 2008. Prototyping a large field size IORT applicator for a mobile linear accelerator. *Phys. Med. Biol.* 53:2089–2102

16. Sawkey D., Faddegon B., 2008. Design of a leakage-reducing electron applicator retrofit. *Med. Phys.* 35:2810.

17. Faddegon B. A., Kawrakow I., Kubyshin Y., Perl J., Sempau J., Urban L. 2009. Accuracy of EGSnrc, Geant4 and PENELOPE Monte Carlo systems for simulation of electron scatter in external beam radiotherapy. *Phys. Med. Biol.* 54:6151–6163.

18. Ge Y., Faddegon B. A. 2011. Study of intensity modulated photon-electron radiotherapy using digital phantoms. *Phys. Med. Biol.* 56:6693–6708.

19. Hogstrom K. R., Boyd R. A., Antolak J. A., Svatos M. M., Faddegon B. A., Rosenman J. G. 2004. Dosimetry of a prototype retractable eMLC for fixed-beam electron therapy. *Med. Phys.* 31:443.

20. O'Shea T. P., Ge Y., Foley M. J., Faddegon B. A. 2011. Characterisation of an Extendable multi-leaf collimator for clinical electron beams. *Phys. Med. Biol.* 56:7621–7638.

21. European standard EN 60601-2-1 medical electrical equipment Part 2-1: Particular requirements for the safety of electron accelerators in the range of 1MeV to 50 MeV (IEC 60601-2-1:1998)

22. Maltz J., Gangadharan B., Vidal M., Paidi A., Bose S., Faddegon B., Aubin M., Morin O., Pouliot J., Zheng Z., Svatos M., Bani-Hashemi A. 2008. Focused beam-stop array for the measurement of scatter in megavoltage portal and cone beam CT imaging. *Med. Phys.* 35:2452–2462.

23. Maltz J., Gangadharan B., Hristov D. H., Faddegon B. A., Paidi A., Bose S., Bani-Hashemi A. R. 2008. Algorithm for X-ray scatter, beam-hardening and beam profile correction in diagnostic (kilovoltage) and treatment (megavoltage) cone beam CT. *IEEE Transactions on Medical Imaging* 227:1791–1810.

24. Bootsma G, Verhaegen F, Jaffray D. 2015. Efficient scatter distribution estimation and correction in cone beam CT using concurrent Monte Carlo fitting. *Med. Phys.* 42, 54–68.

25. Aubin M., Langen K., Faddegon B., Pouliot J. 2003. Electron beam verification with an A-Si EPI. *Med. Phys.* 30(6):1475.

26. Jarry G., Verhaegen F. 2005. Electron Beam Treatment Verification Using Measured and Monte Carlo Predicted Portal Images. *Phys. Med. Biol.* 50:4977–4994.

9

Dynamic Beam Delivery and 4D Monte Carlo

Emily Heath
Department of Physics, Carleton University

Joao Seco
DKFZ, German Cancer Research Center and University of Heidelberg

Tony Popescu
BC Cancer and University of British Columbia

9.1 Introduction

Dynamic delivery techniques such as intensity-modulated radiation therapy (IMRT), tomotherapy, and volumetric-modulated arc therapy (VMAT) are being used on a routine basis in many radiotherapy clinics. In a dynamic beam delivery, the particle fluence is modulated by a beam modifier, such as a multileaf collimator (MLC), whose position within the beam is varied as a function of time. In some techniques, not only the fluence is modified but also the incident direction and energy of the beam may be changed, for example, in an arc therapy delivery or a scanned proton beam. In the absence of Monte Carlo (MC) simulations, the influence of the beam modifiers on the spatial and energy distribution of incident particles is often approximated by the treatment planning system (TPS). Accurate dose calculation methods are required to characterize the dynamic beam modifiers, to test the accuracy of the TPS dose distribution, and to perform independent monitor unit (MU) calculations as well as to reconstruct the patient dose delivery. If properly validated, a dynamic MC model of the beam can serve as a commissioning tool to replace extensive complicated measurements, especially if measurement resolution or accuracy is questionable. This topic comprises the first part of this chapter. The methods for simulation of dynamic beam delivery devices will be discussed followed by examples of specific applications to dynamic radiotherapy techniques.

Over the last two decades, the simulation of dynamic patient geometries has been investigated as the research interest in four-dimensional (4D) radiotherapy techniques for management of respiratory motion has increased. The use of MC-based dose calculation algorithms is a natural choice for such applications owing to their excellent dosimetric accuracy in low-density lung tissue compared to the more standard analytical dose calculation algorithms. Different approaches to dynamic patient dose accumulation using MC methods as well as issues that may limit their practical application will be discussed in the second part of this chapter.

The interaction of dynamic beam delivery methods with the tumor motion, referred to as the "interplay effect" is a concern in intensity-modulated photon and proton therapy, which uses moving MLCs or scanned beams. Numerical methods are best suited for such studies, including approaches that use MC algorithms. The third section of this chapter comprises this topic.

Finally, the last section of this chapter will be devoted to describing the application of MC techniques in development of novel dynamic treatment delivery paradigms or verification strategies.

9.2 Simulations of Dynamic Beam Delivery

9.2.1 Strategies for Simulating Time-Dependent Beam Geometries

In developing MC algorithms to model the IMRT delivery, a very accurate "blueprint" (geometry, atomic composition, etc.) of the beam-modifying devices used to generate the intensity profile is required. The approaches for simulating dynamic geometries in MC simulations range, in the order of increasing complexity, from modification of particle weights to account for attenuation, performing multiple static simulations of "discrete" geometrical states, to treating the beam motion as a probabilistic problem where the beam geometry is sampled on a particle-by-particle basis from probability distributions describing the fraction of the total delivery time for which each geometrical configuration exists.

The simplest approach, particle weighting, arose from techniques used in analytical dose calculation algorithms to model the influence of beam modifiers. This approach has been favored for its relative computational efficiency compared to simulating particle transport in the beam modifiers. Weighting factors may be determined from linear attenuation based on a ray tracing through the beam modifier geometry. Temporal variations of the beam modifier positions are accounted for by scaling the weighting factors by the fraction of the total delivery time that the modifier blocks the beam path. A limitation of the particle weighting approach is that the nonuniformity of particle weights leads to greater statistical variance (Ma et al., 2000). To improve the particle weight uniformity, particle splitting may be applied to particles with large weights, while Russian roulette can be used to reduce the number of low-weight particles.

Simulations that model the different geometrical states of a dynamic beam delivery in detail can be divided into two main approaches. The first uses multiple discrete simulations and will be referred to as the "static component simulation" (SCS) method. This approach is logical when geometry changes occur in discrete steps (e.g., the step-and-shoot IMRT); however, for a continuously variable geometry, a reasonable limit must be imposed on the number of geometry samples and thus the temporal resolution. It should be pointed out that simulating more geometries does not necessarily lengthen the calculation time as the total number of histories to be simulated can be distributed between the individual geometries. Rather, the overhead comes from the input file preparation, initialization, and postprocessing steps. These SCS simulations can be run separately, and the results can be recombined later or by restarting a previous simulation with an updated geometry (Shih et al., 2001). Furthermore, the simulations on different geometries can be performed for equal numbers of particle histories, and the results can be weighted using the fractional MUs to determine the dose. Alternatively, the number of histories to be simulated for each geometry can be calculated based on the fractional MUs. With this latter approach, the statistical variance will be higher for segments that deliver fewer MUs.

The second approach, which is more naturally suited for modelling a continuously changing geometry, is to randomize the sampling of the different geometries. This "position-probability sampling" (PPS) method (Liu et al., 2001) requires cumulative probability distribution functions (CPDFs) for each geometrical parameter that varies. The probabilities are calculated from the fraction of the total delivery time that the particular geometry element (i.e., MLC leaf and jaw) spends at a certain location or configuration. From an operational overhead point of view, the PPS approach may be more efficient than the SCS method. Both approaches should result in the same statistical variance for an equal number of incident particles if the SCS method calculates the number of histories to run for each geometry based on the same CPDF used for the PPS method. It should be noted that the PPS method is not limited to modelling continuously changing geometries. By specifying the CPDF appropriately, a step-and-shoot delivery can be modeled using this approach.

9.2.2 Applications of MC to Model Dynamic Radiotherapy Techniques

9.2.2.1 Dynamic Wedge

Dynamic or virtual wedges are an intensity-modulated delivery approach where a wedge-shaped dose distribution is delivered by the dynamic motion of the collimating jaws. The jaw motion is specified as a function of fractional MUs in the so-called segmented treatment table (STT). The spectrum of particles emerging from a virtual wedge may vary significantly from a static solid wedge, which differentially hardens the beam.

The first reported simulation of a dynamic wedge was performed by Verhaegen and Das (1999). They used the EGS4/BEAM code to simulate delivery with a dynamic wedge on a Siemens linac and to compare the energy spectrum with that obtained with a physical wedge. In the first step, the simulation of transport in the upper section of the treatment head is performed, and a phase space file is scored before the upper jaws. In the next step, 20 discrete simulations of transport through the treatment jaws are performed between which one of the upper jaws moved in 1 cm steps. The resulting 20 phase space files are combined by taking a precalculated number of particles from each phase space file based on a formula for the ratio of dose, or MUs, that must be delivered at each jaw position to obtain a certain wedge angle. This assumption of a one-to-one relationship between the number of delivered MUs and the number of simulated histories is not entirely accurate as it ignores the effect of particles that are backscattered from the jaws into the MU chamber. A correction for this can be applied if the amount of backscatter as a function of jaw position can be characterized for the linac geometry (Liu et al., 2000). In the study by Verhaegen and Das (1999), the dose in the MU chamber resulting from backscattered radiation was less than 1% for all simulated jaw positions. The authors modeled both virtual and physical wedges from 15° to 60° for energies 6–10 MV and obtained a good agreement with measurements, except for the penumbra region of the toe end of the wedge where the discrepancies were up to 4%. They also compared the virtual wedge with the physical wedge in the heel, center, and toe end of the wedges for beam spectrum variations. In comparing physical wedges with virtual wedges for beam hardening effects, no major differences were observed except that the 60° physical wedges produce significantly harder beams across the whole field due to higher absorption by the tungsten.

In a similar fashion, Shih et al. (2001) reproduced the dynamic wedge delivery of a Varian linac, as specified in an STT file, by calculating the number of particles to be simulated for each jaw setting based on the weights in the STT. After the simulation of the first STT entry, the simulation is restarted using the IRESTART feature in BEAM with the updated jaw position and incident histories. In this way, the phase spaces scored for each jaw setting are appended to the main phase space file, and no additional postprocessing is needed.

Verhaegen and Liu (2001) later developed the PPS method to simulate the Varian enhanced dynamic wedge (EDW) delivery. The jaw positions as a function of cumulative MUs in the STT are converted to a CPDF. During initialization of each incident history, the jaw position is randomly sampled from this CPDF. The authors also performed simulations using the SCS method with and without a correction for the reduction in the number of delivered MUs due to backscattering into the MU chamber. The incorporation of this correction was not found to make any difference in the phase space files, thus again confirming the validity of the assumption relating MUs to particle histories. Comparing the results of the PPS and SCS methods, no differences were found in the resulting phase space files created with the two methods.

The use of the particle weighting method to simulate a Varian EDW delivery has been reported by Ahmad et al. (2009) using the method of Ma et al. (2000). Compared to measurements of wedged profiles and output factors, they obtained agreement within 2% and 1 mm.

9.2.2.2 MLC-Based IMRT

With IMRT, nonuniform fluence patterns or intensity-modulated maps that are incident upon a patient from multiple directions are used to generate a uniform dose distribution that covers the target volume. These nonuniform fluence patterns are generated by using a continuous or discrete sequence of apertures formed using an MLC. In step-and-shoot delivery, each MLC-generated subfield is set prior to the beam turning on, and there is no MLC motion during irradiation. In the case of the sliding window (or dynamic) delivery, the leaves move while the beam is on. These MLC leaf sequences, specifying the leaf openings as a function of delivered MU, can be obtained from the TPS or from postdelivery log files. In the TPS, the leaf positions and delivered MU are specified for each control point. In the delivery log files, the machine state (including leaf positions, beam state (on/off), and MU) is output on a fixed time interval (e.g., 10 ms for Varian TrueBeam linac). These log files have been used as a quality assurance method to detect dosimetric errors resulting from errors in positioning of the MLC leaves during treatment delivery (Fan et al., 2006; Tyagi et al., 2006).

MC simulations of step-and-shoot and sliding window IMRT have been used by numerous authors to compare MC dose distributions with those produced by a commercial planning system. For example, Ma et al. (2000) used the particle weighting method to recalculate IMRT treatment plans from the CORVUS TPS. Two-dimensional (2D) intensity maps were calculated from the leaf sequence files by accumulating MUs for unblocked areas and for the blocked areas, where the latter contribution is weighted by the average leaf leakage. This approach ignores the effect of the leaf tongue and groove as well as the scatter contributions from the leaves. To more accurately model the influence of the MLC leaf geometry, Deng et al. (2001) used a ray-tracing algorithm to calculate intensity maps. An example of the fluence maps calculated with and without consideration of the leaf geometry is shown in Figure 9.1.

Keall et al. (2001) included primary and scatter contributions by calculating probability maps for leaves being open, closed, or a leaf tip passing through a grid point. The weighting factors for blocking leaves are calculated from linear attenuation along the path length of the particle through the MLC considering the energy of the incident photon. The scatter contribution is obtained by the integration of attenuation factors obtained for successive positions of the leaf tips by subsampling the intensity map grid. These MLC simulations should be distinguished from the other previously mentioned approaches because they also include modelling of first Compton scattering of each incident photon; thus, a simplified photon transport in the MLC was performed. The weights of these scattered photons are modified by the probability of a leaf blocking the particular grid point.

Leal et al. (2003) and Seco et al. (2005) performed the first full linac and patient simulation (i.e., particle transport through jaws, MLC, and patient) for Siemens and Elekta linacs, respectively, using BEAMnrc, where step-and-shoot delivery was modeled using the SCS method. Each beam consisted of 5–15 segments, which were simulated individually with the number of histories proportional to the number of MUs delivered by the individual segment. Both authors performed their simulations on a personal computer. Scripts were developed to automate generation and distribution of separate files for the simulation of each segment phase space and the resulting dose distribution.

In a study by Liu et al. (2001), the authors simulated both dynamic MLC (DMLC) and step-and-shoot deliveries of IMRT fields using the PPS method. First, the leaf positions as a function of cumulative MUs are used to create a probability function for each leaf (see Figure 9.2). An MC simulation of an IMRT delivery is performed by randomly selecting an MU_{index} value between 0 and 1 on the Y-axis of Figure 9.2. The corresponding MLC subfield is then obtained by converting the MU_{index} (Y-axis) into a segment index (X-axis) using the curve of either step-and-shoot or sweeping leaf curve. If the X-axis value falls between two segment index values, then the leaf positions at this fractional index value will be linearly interpolated from the neighboring segments.

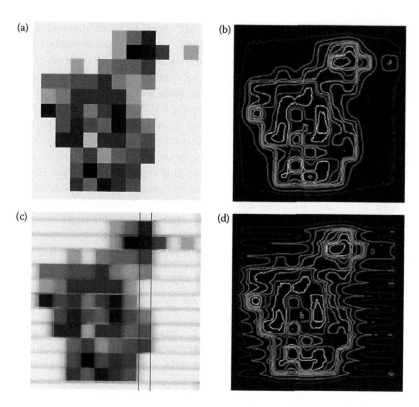

FIGURE 9.1 (a) and (c) Intensity maps used for particle weighting calculated for one field of a step-and-shoot IMRT plan. On the right, (b) and (d) are shown as the corresponding MC dose distributions. The intensity map in (c) is calculated using a ray tracking algorithm, which includes the tongue-and-groove geometry of the MLC leaves. (From Deng J. et al. The MLC tongue-and-groove effect on IMRT dose distributions. *Phys. Med. Biol.* 46, 1039–1060, 2001. With permission.)

Note that the "step-like" appearance of the step-and-shoot curve results from the beam being turned off when the segment index and the leaf positions change. The PPS method has been implemented for the component modules in BEAMnrc which are used to model jaws or MLCs. These "dynamic" component modules are indicated by a "SYNC" prefix.

The complexity of the MLC leaf geometries typically leads to a high number of geometrical regions through which particles must be tracked. When particle transport in these detailed geometries is performed with low cutoff energies, these simulations can become very time consuming. Depending on the quantity being investigated, the contributions of electrons and multiple-scattered photons may not be important. Therefore, a significant improvement in computational efficiency can be realized by simplifying both the particle transport and the simulation geometry. An example of this is the fast MC code VCU-MLC (Siebers et al., 2002), which uses a single-Compton approximation of photon interactions within the MLC. No electron transport in the MLC is performed, and the incident electrons have their weight reduced by the probability of striking an MLC leaf. The leaf geometry is approximated by segmenting each leaf into upper and lower halves, with a correct thickness to model the tongue and groove. Scattered photons are assumed to pass through the current leaf until they exit the MLC; thus, the adjacent leaves are ignored. The VCU-MLC

code correctly predicts the field size dependence of MLC leakage, interleaf and intraleaf leakage, and the tongue-and-groove dose effect. The model is applicable to both step-and-shoot and sweeping window deliveries. For each incident particle, the weight is modified by randomly sampling MLC positions 100 times from the leaf sequence file and averaging the transmission probabilities for each sample. Compared to the SCS and PPS methods that sample an MLC configuration for each incident particle, fewer source particles are required with the transmission probability weighting approach, which results in a gain in calculation efficiency. The VCU-MLC code is thus very useful to patient IMRT dose calculations because of its accuracy and speed of calculation. The MLC model has also been extended to simulate photon transport through both jaws and MLC simultaneously with the addition of multiple Compton interactions (Seco et al., 2008). Tyagi et al. (2007) have also implemented a similar approach in the DPM code to simplify particle transport in a model of the Varian Millennium MLC; however, they adopted a serial approach to simulating IMRT delivery by using the PPS method to determine the leaf positions for each incident particle.

9.2.2.3 Tomotherapy

The challenge of simulating tomotherapy and other delivery techniques that involve dynamic gantry rotation is that the

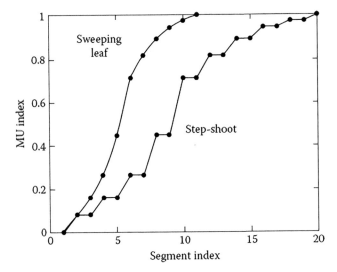

FIGURE 9.2 An example of the IMRT field generated by dynamic leaf sweeping or step-and-shoot curve, where in the Y-axis, MU index represents the CPDF of the MLC motion. (From Liu H.H., Verhaegen F., and L. Dong. A method of simulating DMLCs using MC techniques for IMRT. *Phys. Med. Biol.* 46, 2283–2298, 2001. With permission.)

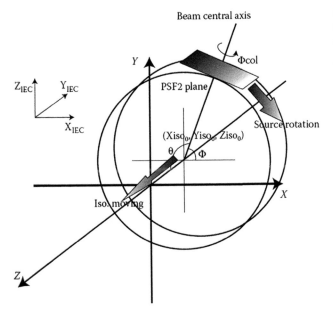

FIGURE 9.3 Geometry of the simulation of helical tomotherapy delivery. Phase space fields (PSF2) are calculated for each projection, which are rotated by the projection angle (θ) and shifted along the Z-axis in accordance with the sinogram file. (From Zhao et al. Monte Carlo calculation of helical tomotherapy dose delivery. *Med. Phys.* 35, 3491–3500, 2008a. With permission.)

gantry angle is continuously changing during the delivery. The geometry of a helical tomotherapy delivery is illustrated in Figure 9.3. Tomotherapy uses a rotating fan beam source whose intensity is modulated by a binary MLC. The helical beam delivery is specified in a sinogram file in which the fractional opening time for each leaf is specified at discrete gantry

positions. Since the delivery is dynamic, leaf motion occurs while the source rotates. Depending on the amount of leaf motion, subsampling of the sinogram may be needed to accurately simulate the dosimetric effects of the combined gantry and leaf motion. The amount by which the couch translates as a function of the gantry rotation is specified by the pitch. Static models of tomotherapy units without MLC modulation have been developed by Jeraj et al. (2004) and Thomas et al. (2005).

Sterpin et al. (2008) developed a PENELOPE user code called TomoPen for simulating helical tomotherapy delivery. First, a phase space file is created for each of the three jaw settings used during delivery. Then, for each entry in the sinogram file, a projection-specific phase space file is created by adjusting the weights of the particles in the phase space file associated with the appropriate jaw setting. To model the continuous delivery, each entry in the sinogram file is further subdivided into 11 subprojections for which the leaf openings are linearly interpolated. It should be noted that the TPS used the sinogram entries to calculate the dose distribution, therefore discretizing the delivery. For all the particles in the jaw-specific phase space, weighting factors are calculated, assuming linear attenuation, by a ray tracing through the MLC leaves. Specifically, the subprojection-specific phase space is calculated by dividing the jaw-specific phase space into 64 regions, one for each MLC leaf, and applying the weighting for leaf attenuation if the current leaf is closed or not adjusting the weight if the leaf is open. Only leaves that are open or adjacent to an open leaf are considered for this step; otherwise, the particles are discarded; thus, the leakage radiation in this situation is not modeled. Each of the resulting subprojection-specific phase space files is then rotated and translated in accordance with the projection angle and couch translation before being applied to the patient geometry. TomoPen has been used to simulate the delivery of commissioning and clinical cases in homogeneous phantoms. A comparison with film and ionization chamber measurements (see Figure 9.4) showed a good agreement within 2% and 1 mm.

Zhao et al. (2008a) also simulated a helical tomotherapy delivery with EGSnrc/BEAMnrc by using the SCS method. They also subdivided each sinogram entry into static MLC subfields. An initial phase space is scored above the MLC. Then, the binary collimator is modeled using the VARMLC component module, and a separate full simulation of particle transport in the MLC is performed for each subprojection, resulting in a second set of phase spaces. The number of histories for each subprojection simulation was proportional to the calculated leaf opening time. The phase spaces are rotated in the XY plane, and the isocenter is modified along the Z direction to simulate the helical delivery as shown in Figure 9.3. This model was used to compare dose distributions calculated by their TPS for clinical treatment plans (Zhao et al., 2008b).

9.2.2.4 VMAT

VMAT is a form of IMRT where the gantry rotates around the patient as the beam apertures are dynamically modified by

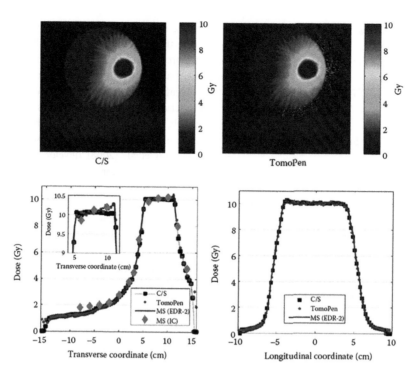

FIGURE 9.4 Comparison of dose distributions calculated with a convolution–superposition algorithm (C/S) and MC (TomoPen) and measured with film (MS EDR-2) and ionization chamber (MS IC) for a commissioning plan in a cylindrical phantom. (From Sterpin E. et al. Monte Carlo simulation of helical tomotherapy with PENELOPE. *Phys. Med. Biol.* 53, 2161–2180, 2008. With permission.)

the MLC. The dose rate is also modulated during the delivery. Similar to MLC-based IMRT, the delivery is specified in the TPS by a sequence of control points at which the gantry angle, MLC leaf positions, and cumulative MU are given. Owing to the complexity of VMAT, there is much interest in using MC simulations for patient-specific quality assurance. However, Boylan et al (2013) cautioned that because of the approximations made in the TPS to model the VMAT delivery, the settings in the plan file may not accurately represent how the plan will be delivered by the linac.

The approach for simulating VMAT delivery is similar to that for tomotherapy. An SCS-based approach to modelling VMAT delivery on an Elekta linac has been described by Li et al. (2001) using EGS4/BEAM. Emulating the dose calculation method used by the TPS, the authors discretize each arc into 5°–10° steps for which static simulations of the MLC apertures defined in the leaf sequence file are performed. A simplified geometry is used for the MLC, without modelling the tongue and groove. For each phase space, a DOSXYZ calculation on the patient geometry is carried out with the appropriate source angle.

A model of a Varian RapidArc delivery has been developed by Bush et al. (2008), which used EGSnrc/BEAMnrc. They also modeled the gantry rotation as a series of discrete static simulations. However, owing to the significant leaf motion that occurs during RapidArc delivery, it is also necessary to model the leaf motion that occurs between the gantry control points. This is addressed by calculating, for each sequential pair of

gantry angles specified in the sequence file, a mean gantry angle. The leaf openings of the adjacent gantry angles are then used to define a pair of control points for a sliding window DMLC delivery at this mean angle, as illustrated in Figure 9.5. The variable dose rate is accounted for by weighting the subsequent dose calculation by the fractional MUs to be delivered by each segment. The authors used the VCU-MLC model (Siebers et al., 2002) to improve the computational efficiency of their simulations. The same group has used the recorded log files to reconstruct the delivered dose from a RapidArc treatment (Teke et al., 2009).

Simulation of tomotherapy and VMAT delivery using the SCS approach requires the calculation of phase space files for a significant number of geometry samples owing to the additional dimension of gantry rotation. Even with automation of the input file preparation and patient dose calculation, the generation of hundreds to thousands of phase space files representing each subprojection is not optimal in terms of storage requirements. Another drawback of using the SCS approach is that the dosimetric influence of the number of subsamples for simulating the continuous delivery must be carefully investigated.

Owing to the continuous motion of the gantry and collimators, the position-probability approach is better suited to the simulation of VMAT and tomotherapy. With the PPS approach, a randomly sampled MU index at the beginning of each history can be used to determine the collimator settings as well as gantry angle and isocenter translation. These delivery settings can be obtained from the sinogram/leaf sequence

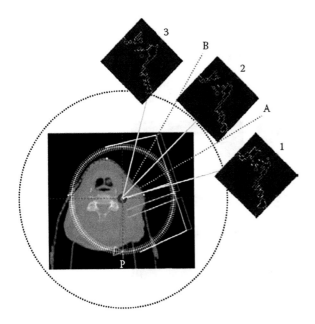

FIGURE 9.5 Illustration of the generation of DMLC segments for the simulation of leaf motion in VMAT delivery. For each consecutive set of gantry positions (1 and 2), a mean gantry angle A is calculated, and the leaf positions from 1 and 2 are used as control points for a dynamic leaf sequence to be simulated at position A. (From Bush K., Townson R., and S. Zavgorodni. Monte Carlo simulation of RapidArc radiotherapy delivery. *Phys. Med. Biol.* 53, N359–N370, 2008. With permission.)

file from the TPS or from a log file recorded during a treatment delivery.

Two PPS-based approaches have been developed for simulating tomotherapy and VMAT using the BEAMnrc code. The first, developed by Lobo and Popescu (2010), is based on the option in DOSXYZnrc to use a BEAMnrc simulation as a particle source (ISOURCE 9). This DOSXYZnrc source has been modified so that the first time a particle is incident on any dynamic component module it is "time stamped" with a randomly sampled fractional MU index. This time stamp allows synchronization between the motion in the BEAMnrc shared library (of the jaws, MLC, etc.) and the motion in the DOSXYZnrc geometry, such as gantry, collimator and couch rotation or translation. These two new sources (ISOURCE 20 and ISOURCE 21) are now distributed with the DOSXYZnrc package. Source 20 uses a phase space scored above the secondary collimators (e.g., jaws) as an input. This is useful for simulations where a detailed model of the linac head is not available and the manufacturer provides a phase space file instead, which is the case for the Varian TrueBeam linac. The randomly sampled MU index is used to sample the MLC configuration for each incident particle. The other source (ISOURCE 21) consists of a full accelerator model simulation and is useful when there is more than one dynamic component. In source 21, the same MU index that is randomly sampled when a particle is incident on the first dynamic component is used when that particle enters the second dynamic

component. This ensures that the motion of the collimators is synchronized, for example, when the jaws track the MLC motion.

A second PPS-based approach was developed by Belec et al. (2011) which used a time variable stored in the phase space file. This time variable replaces the ZLAST variable, which records the z position of the last interaction of the particle. The value of the time variable is actually the same MU index that is used by all PPS approaches and will be randomly sampled when a particle passes through the first dynamic component module. The output of the accelerator simulation can be stored in an intermediate phase space file without losing the temporal information. When the phase space file is read by DOSXYZnrc, the time variable is used to sample from the lookup tables specifying the source configuration. It should be noted that a later modification to the ISOURCE 20 and 21 (Popescu and Lobo, 2013) also implemented storing the MU index associated with each particle in the phase space file, thus creating a so-called "4D" phase space file, in the International Atomic Energy Agency (IAEA) format.

9.2.2.5 Protons

Before the introduction of scanning delivery systems for proton beams, temporal modulation in the form of spinning range modulator wheels was used for modulating the range of an incident proton beam to create a spread-out Bragg peak. In what is perhaps the first published report of a dynamic MC simulation, Palmans and Verhaegen (1998) reported implementing the PPS method in the PTRAN MC code to simulate a range modulator wheel. Paganetti (2004) described a detailed model of the Northwest Proton Therapy Center beam line using GEANT4. Dynamic elements such as the range modulator wheel and beam scanning magnets (see Figure 9.6) were modeled exploiting the feature of GEANT4 to update geometry parameter values during the simulation. The geometry updates were performed in a linear time fashion, with each time step assigned a number of histories. The modulator wheel rotation was simulated in 0.5° steps, whereas the scanning magnet settings were updated in steps of 0.02 T. The authors noted that the calculation time for their dynamic MC simulations were virtually unaffected by the number of times the geometry was updated as this involved only updating a pointer in the memory. More recently, Shin et al. (2012) described the implementation of a framework to model time-dependent geometries into the GEANT4-based proton MC tool TOPAS (Perl et al. 2012). The time-dependent values of a simulation parameter are specified by assigning a Time Feature. The time evolution of the parameter can then be sampled in either a sequential or random fashion which is defined by the Sequence.

9.3 Dynamic Patient Simulations

9.3.1 Patient Motion in Radiotherapy

The need for 4D patient dose calculation methods was initiated by an interest in compensating for the effects of respiratory

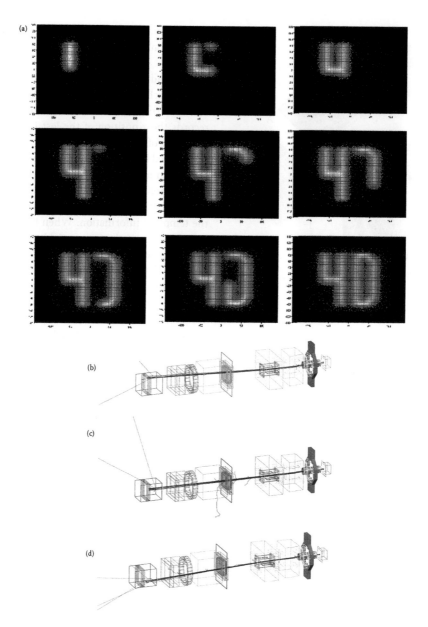

FIGURE 9.6   (a) Fluence distribution from simulation of scanned proton beam delivery, (b–d) simulation of proton trajectories in the treatment nozzle for different settings of the scanning magnets. (From Paganetti H. Four-dimensional Monte Carlo simulation of time-dependent geometries. *Phys. Med. Biol.* 49, N75–N81, 2004. With permission.)

motion during treatment planning and delivery (Keall et al. 2004). 4D dose calculation methods are required to calculate the cumulative dose distribution, which is received by the varying anatomy. The effects of motion on the delivered dose distribution were discussed by Bortfeld et al. (2004). These include (1) blurring of the dose distribution along the path of motion, (2) localized spatial deformations of the dose distribution at deforming organ boundaries or regions of density changes, and (3) interplay between tumor motion and dynamic beam delivery. As will be discussed in the following sections, the blurring effect can be simply modeled with analytical dose calculation algorithms by convolution methods; however, to quantify dose deformation effects, the influence of

individual geometric states on the deposited dose distribution needs to be modeled. The methods discussed here are focused on respiratory motion; however, they can be extended to other examples of patient anatomical variations, such as interfractional geometry variations and applications in adaptive radiotherapy, with appropriate modifications.

9.3.2 Strategies for 4D Patient Simulations

9.3.2.1 Convolution-Based Methods

Similar to approaches used for simulating dynamic beam delivery, the cumulative dose can be calculated either by adjusting the particle fluence or from multiple calculations on different respiratory phases. If the dose distribution is assumed to be shift invariant, the effect of motion on the dose delivered for a large number of fractions can be estimated by convolving the dose distribution with a probability distribution describing the positional variations (Lujan et al., 1999). This models only the blurring effect of motion on the dose distribution, dose deformations, and differential motion that cannot be modeled by convolution. Furthermore, the assumption of spatial invariance of the dose distribution is not valid at tissue interfaces, which can lead to an underestimation of the dose at these locations (Craig et al. 2001; Chetty et al. 2004).

In the reference frame of the patient, patient motion is interpreted as motion of the beam; therefore, the approach of fluence convolution has also been proposed (Beckham et al., 2002; Chetty et al., 2004). This is analogous to the particle weighting approach; however, the (x, y, z) positions and direction cosines of the particles in the phase space file are modified by randomly sampling shifts from a probability distribution function describing the respiratory motion. Fluence convolution attempts to overcome the limitation of the shift invariance assumption in the dose convolution method. However, it is also limited by the approximation of the patient as undergoing rigid body motion.

9.3.2.2 Dose-Mapping Methods: Center-of-Mass and Dose Interpolation

The limitation of the convolution approach is that it does not consider the differential organ motion, deformation, and density changes that cause the shape of the dose distribution to locally "deform" during the beam delivery. To accurately account for these effects, the dose distributions calculated on the different respiratory phases (e.g., exhale, 50% inhale, and 100% inhale) need to be calculated, similar to the SCS method of simulating an IMRT delivery, and weighted by the fraction of the total breathing cycle for which they occur. The number of samples of respiratory states needed to reconstruct the cumulative dose for a full respiratory cycle has been investigated by Rosu et al. (2007). They demonstrated no significant differences in dose volume histograms between cumulative doses calculated from ten respiratory states and calculations based on only the inhale and exhale states. This conclusion is dependent on the treatment plan design; if dose perturbations due to motion are large, nonlinearities of the motion between inhale and exhale can be expected to have a non-negligible influence on the cumulative dose distribution.

Unlike dose deposition from a dynamic beam delivery where the reference frame of the dose deposition remains constant, for a dynamic patient simulation, the position and volume of tissue elements is changing (possibly, in concert with dynamic beam delivery changes). The dose deposited in each geometry needs to be mapped to a reference geometry to determine the cumulative dose deposition from irradiation of the multiple variable geometries. This requires knowledge of the geometrical transformation between tissue locations in

the reference and current geometrical states. Image registration methods are commonly used to estimate these transformations. Briefly, the transformation is calculated by finding a transformation, which maximizes the similarity of the two images. This similarity may be based on image intensity or previously identified landmarks such as points or contours. A wide range of image registration algorithms exist; mainly, they are classified as being rigid or nonrigid. For the types of anatomical changes observed in patient geometries, deformation of the tissue is observed and thus nonrigid or deformable image registration is best suited for this application. A review of deformable image registration algorithms used in medical imaging has been given by Holden (2008), and Xing et al. (2007). It should be obvious to the reader that the accuracy of the mapped dose calculation depends on the accuracy of the image registration. Some discussion of this topic will be made in Section 9.3.2.4.

There are a variety of *dose-mapping methods* that are used to map the dose from the geometry on which it is deposited to a reference geometry. These are applicable to any dose calculation method. Generally, dose mapping is discussed as applying to two single geometry samples, but it could easily be implemented as a "true 4D" simulation if the patient geometry is updated during the simulation and the mapping is applied to accumulate the dose depositions on the current scoring geometry to the reference dose grid.

The simplest approach to mapping the dose delivered on the target geometry to the reference geometry is to use the deformation map to determine the voxel in the target geometry that corresponds with the voxel of interest in the reference geometry. The reference voxel is assigned a dose value equal to the dose computed in the target voxel in which the transformed center of mass of the reference voxel lies (see Figure 9.7a). This "center-of-mass (COM)" remapping method ignores contributions in other voxels, which may be overlapped by the transformed reference voxel. This can easily occur if the transformed reference voxels do not exactly overlap the target voxels. Also, if there are any volume changes, this exact voxel-to-voxel mapping is incorrect. In an application of the COM method, Paganetti et al. (2004) used the UpdateValue method in GEANT4 to update the patient geometry to one of the ten different phases of the patient's breathing from a 4D computed tomography (CT) dataset. To accumulate the dose, pointers in the dose matrix were modified using the COM method to ensure that dose depositions on the different geometries were accumulated in the corresponding voxels of the reference geometry.

To deal with possible deformations and the inexact overlap of transformed and target voxels, a trilinear interpolation of the dose from the local neighboring voxels at the transformed COM point in the target geometry can be used to assign the dose in the reference geometry (Figure 9.7b). Voxel volume changes can also be accounted for by subdividing each reference voxel, estimating the mapped dose for each subvolume, and then interpolating these values to assign the dose in the

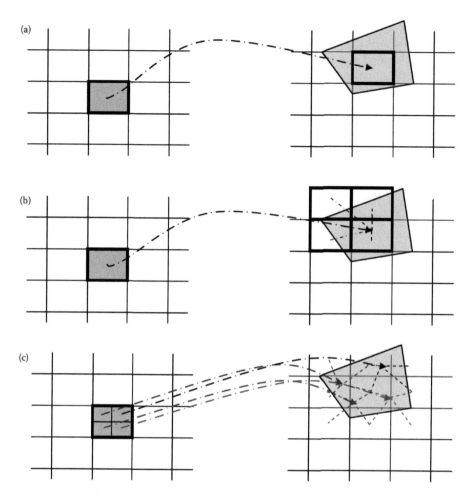

FIGURE 9.7 Illustration of different dose-mapping methods. The left grid represents the reference dose grid. On the right side is the dose grid from the irradiated geometry. The dose deposited on the source irradiated geometry is to be mapped to the reference geometry for accumulation. The shaded polygon represents the boundaries of the reference voxel after the reference-to-target geometry transformation is applied. (Parts of this figure adapted from Rosu M. et al. *Med. Phys.* 32, 2487–2495, 2005.) (a) COM mapping. (b) Dose mapping with trilinear interpolation. (c) Method in (b) with subdivision of the reference voxels.

reference geometry (Figure 9.7c). Rosu et al. (2005) reported the application of these methods to dose distributions calculated using an MC algorithm, although the method is applicable to a dose distribution calculated with any algorithm. Dose interpolation methods are computationally efficient; however, like any interpolation method, they suffer from a reduced accuracy wherever dose gradients exist. Heath and Seuntjens (2006) showed that because of these interpolation errors, the dose interpolation method calculates the dose incorrectly in the regions of large dose and deformation gradients. Siebers and Zhong (2008) also demonstrated that in the situation where voxels of the target geometry merge in the reference geometry (i.e., compression), dose interpolation methods do not conserve the integral dose and thus lead to errors in the regions of dose and density gradients.

9.3.2.3 Voxel Warping Method

Two MC approaches to dose remapping that ensure energy conservation have been developed by Heath et al. (2007, 2011),

and Siebers and Zhong (2008). The first method, called voxel warping, entails deforming the nodes of the reference dose grid voxels using the deformation vectors obtained from registration of the reference to the target image and was first implemented in a modified version of the DOSXYZnrc user code, called defDOSXYZnrc (Heath et al., 2007). Particle transport is performed on this deformed voxel geometry. As the voxel indices do not change between the reference and deformed state, the energy mapped between these states is conserved. The densities of these deformed voxels are adjusted in accordance with the volume change to conserve mass. To describe the deformed voxels, the faces of the rectangular voxels are divided into two subplanes, forming dodecahedrons (Figure 9.8). This voxel face subdivision leads to a twofold increase in the number of distance-to-voxel boundary calculations that need to be performed by the boundary checking algorithm. Furthermore, because of ambiguities in the calculation of particle–plane intersection points, which assume that the plane has an infinite extent, these intersections need to be

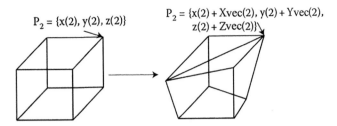

$$P_2 = \{x(2), y(2), z(2)\}$$

$$P_2 = \{x(2) + Xvec(2), y(2) + Yvec(2), z(2) + Zvec(2)\}$$

FIGURE 9.8 Deformed voxel geometry used in defDOSXYZnrc obtained by applying deformation vectors to reference voxels.

further tested to determine if they lie within the boundaries of each plane. These additional computations result in up to a 10-fold increase in computation times with defDOSXYZnrc compared to that in DOSXYZnrc.

A reimplementation of the voxel warping method in VMC++ (Heath and Kawrakow, 2011) achieved a 130-fold improvement in computational efficiency compared to that in defDOSXYZnrc by using a fast MC code and optimized geometry definitions based on tetrahedral elements. Each deformed voxel is divided into six tetrahedrons, which are used for particle tracking while dose deposition is scored in the voxels. The main advantage of using tetrahedral volume elements is the elimination of the plane–particle intersection ambiguity; furthermore, there are fewer planes to be checked in each tetrahedron compared to that in the dodecahedron geometry. Compared to VMC++ calculations in rectilinear geometry, the calculation time with the deformable VMC++ geometry is increased by a factor of 2 for the same level of statistical variance. The deformable geometry was implemented as a new geometry class for VMC++, which was compiled as a dynamic shared library to be loaded at run time. Two alternate definitions of the tetrahedral geometry were implemented, differing by the direction in which the voxel faces fold. Although it could be demonstrated that the geometry definition affects the dose deposition, no differences in patient dose distributions calculated with the two alternate geometries could be found (Heath and Kawrakow, 2011).

A possible limitation of the voxel warping method is that it requires a continuous deformation field, which is not always the physical situation. At the boundary between the lung and the chest wall tissue, sliding occurs, which at the resolution of the dose calculation and image registration, appears as a discontinuous motion. Before using the displacement vector maps obtained from a deformable image registration, they must be checked for discontinuities in the region where the dose will be calculated. This can be performed by calculating the determinant of the Jacobian matrix of the transformations in the neighborhood of each node of the reference voxel grid (Heath et al., 2007; Christensen et al., 1996). A negative value of the determinant (see Equation 9.1) at a particular node indicates a discontinuity in the transformation (u_x, u_y, u_z) at this node. These can be removed by smoothing the transformations either globally or in the vicinity of the discontinuous node.

$$\det\left[J\left(N(\vec{x})\right)\right] = \begin{vmatrix} \dfrac{\partial u_x}{\partial x}+1 & \dfrac{\partial u_x}{\partial y} & \dfrac{\partial u_x}{\partial z} \\[2ex] \dfrac{\partial u_y}{\partial x} & \dfrac{\partial u_y}{\partial y}+1 & \dfrac{\partial u_y}{\partial z} \\[2ex] \dfrac{\partial u_z}{\partial x} & \dfrac{\partial u_z}{\partial y} & \dfrac{\partial u_z}{\partial_z}+1 \end{vmatrix} \quad (9.1)$$

9.3.2.4 Energy-Mapping Methods

Siebers and Zhong (2008) proposed an alternative energy-mapping–based approach in which particles could be transported in the rectilinear target geometry and the energy deposition points are mapped in accordance with the deformation vectors to the target dose grid. To account for the possibility that particle steps in the target geometry may be split over two or more voxels if mapped to the reference geometry, the energy deposition location is randomly sampled along the step. This implementation, termed etmDOSXYZnrc, results in a 10%–50% increase in computation time, depending on how the deformation vectors are interpolated, compared with the standard DOSXYZnrc computations. This approach also does not require a continuous deformation vector field.

However, if the transformation between the reference and target geometry is not exact, the energy mapped to the voxels of the reference phase will not be consistent with the mass assigned to those reference voxels. This means that in the lung, for example, energy could be deposited in a voxel with a density of 1.0 g/cm³, but because of the registration error, that energy deposition is mapped to a reference voxel with a density of 0.3 g/cm³. These inconsistent energy–mass mappings lead to a discontinuous dose distribution on the reference geometry. In deformable registration of patient images, determination of an exact transformation is not possible because of a lack of one-to-one correspondence between images arising from image artifacts, noise, partial volume effects, among other causes (Crum et al. 2003). In a further study, Zhong and Siebers (2009) proposed methods to map the mass along with the energy to ensure that mass is also conserved; thus, the energy and mass distributions used to calculate the final mapped dose distribution are mapped in a consistent manner. This approach is termed the "energy and mass congruent mapping" (EMCM) method. The target voxels are subdivided into 100 subvoxels whose masses are mapped to

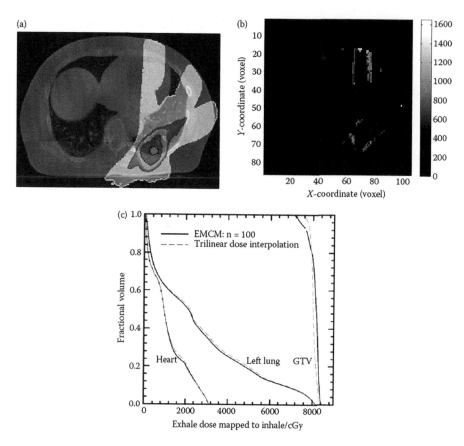

FIGURE 9.9   (a) Dose distribution mapped to Exhale geometry with ETM; (b) difference between dose interpolation and ETM for the same image slice; (c) dose volume histograms. (From Zhong H. and J.V. Siebers. Monte Carlo dose mapping on deforming anatomy. *Phys. Med. Biol.* 54, 5815–5830, 2009. With permission.)

the reference geometry using the same deformation vectors used for mapping the energy deposition points. The precision of the mass mapping was determined to be 99.95% on a simple deformable phantom geometry. Alternatively, a mass "pulling" approach based on the reference–target geometry transformation can be used. This requires a method to invert the deformation field. An energy-mapping method, which performs the energy and mass mapping in the reference–target direction is described below. Dose distributions mapped between the inhale and exhale phases of a lung cancer patient using the EMCM method and dose interpolation are shown in Figure 9.9. The mean difference between the EMCM and trilinear dose interpolation–mapped dose distributions is 7% of the maximum dose.

Heath et al. (2011) proposed an alternative energy-mapping approach that can be performed, after calculation, on a dose distribution. Instead of a point-by-point energy remapping, the energy mapped from the target geometry to each reference voxel is determined by the volume overlap of the deformed reference voxels on the target geometry. Each reference voxel is subdivided into tetrahedrons, which are deformed using the displacement vectors, and the volume intersected by each of these tetrahedral and the target voxels is calculated. The

energy mapped to each reference voxel is simply the sum over all its tetrahedra of the energy deposited in each overlapped target voxel multiplied by the fractional volume overlap. This second method can be applied to a dose distribution calculated with any algorithm; however, information about the spatial distribution of energy depositions within a voxel is lost. This leads to a loss of accuracy when the energy deposited in one voxel is mapped to multiple voxels on the reference geometry.

9.4 Combining Dynamic Beam and Patient Simulations

Combining the simulation of dynamic beam deliveries and variable patient geometries is of interest to study the interplay effects, which arise from the relative motion between the MLC leaves and the moving target region and anatomy upstream from the target, for the case of DMLC delivery. These interplay effects were initially studied by Yu et al. (1998). They demonstrated that when IMRT beams are delivered with dynamic collimation, the problem of intrafraction motion causes large errors in the locally delivered photon dose per fraction due to motion in the penumbra region of the beam. The magnitude of the photon dose variations was shown to be strongly dependent on the speed of the beam aperture relative to the speed of the target motion and on the width of the scanning beam relative to the amplitude of target motion. In the case of proton therapy beams, interplay effects may also occur for proton beam scanning where the dose distribution is affected by (1) patient breathing, (2) proton energy change times, (3) motion amplitude, and (4) proton beam rescanning methodology used (Seco et al., 2009a).

The SCS method can be applied to simulate interplay effects by assigning each aperture to a patient geometry state (e.g., breathing phase) to which it will deliver the dose. The dose distribution from each aperture can be calculated and then accumulated. In the case of a dynamic delivery (sliding windows IMRT or VMAT), apertures at a specific temporal resolution can be interpolated from the control points in a plan file or taken from a delivery log file (Jensen et al. 2012). This approach does not necessitate an MC dose calculation engine.

Modelling of interplay effects using a PPS approach with MC can be carried out if every particle is tagged with information that indicates both the beam delivery (e.g., MLC leaf or beam spot positions) through which it travels and the patient geometry state (e.g., breathing phase) to which it will deliver the dose. For example, in a study by Seco et al. (2009b), each photon was tagged with both the MLC segment and the breathing phase to which it would deliver the dose. Then, dose calculations were performed separately on each CT cube using only the particles that were delivered on that phase. The "time stamp" approach mentioned in Section 9.2.2.4 can also be used if the patient breathing trace can be synchronized to the beam delivery sequence.

An MC method using the PPS method was initially developed by Litzenberg et al. (2007) to study the dosimetric effects of interplay and delivery errors in a phantom geometry. Real-time information about the tracking of the MLC leaf positions is obtained from the Dynalog MLC log file. Real-time target volume position information is measured using wireless electromagnetic transponders (Calypso Medical Technologies, Inc), where three transponders were implanted in the target allowing position updates 10 times per second. The authors termed this method the "synchronized dynamic delivery" approach. In the MC simulation, each particle is transported through a selected MLC field segment and into a specific breathing phase of the dose cube. Therefore, all breathing phase dose cubes would have to reside in the central processing unit (CPU) memory to reduce input/output (I/O) exchanges. Although the study was performed in phantoms, it was a proof of principle that MC could be used to study interplay effects. However, in the case of patients, the method proposed would require large amounts of CPU memory to work efficiently because of the significantly larger patient CT data cubes.

To overcome the limitation of using multiple patient CT data cubes to represent different motion states and to more accurately model the continuous nature of the patient respiratory motion, the "time stamp" approach implemented in

ISOURCE 21 was combined with the voxel warping method implemented in the defDOSXYZnrc user code to create a 4D MC simulation tool, 4DdefDOSXYZnrc (Gholampourkashi et al. 2017). This code uses the full linac model as a particle source. For each incident particle, the MU index that is used to sample the dynamic collimator settings is used to look up the corresponding fractional breathing amplitude from a normalized respiratory trace. This fractional amplitude is used to scale the deformation vectors which are then applied to the patient geometry mesh to recreate the patient anatomy at that current time. Simulations of VMAT delivery were compared with film and MOSTFET dose measurements inside a custom-built programmable deformable lung phantom (Gholampourkashi et al. 2020).

9.5 Novel and Future Applications of MC in Dynamic Beam Delivery

The 4D MC simulation methods that have been summarized in this chapter are capable of simulating dynamic deliveries with many more degrees of freedom than are possible in a conventional linac delivery. Today, such deliveries are possible, for example using the Developer Mode TrueBeam linacs (Varian Medical Systems) or with novel radiotherapy devices such as the Cyberknife or Vero-SBRT (Brainlab and Mitsubishi Heavy Industries). However, no commercial TPS is currently able to calculate dose distributions corresponding to beam trajectories of such complexity. 4D MC simulations have been shown to be capable of continuously modelling variable beam configurations and complex treatment geometry and kinematics with respect to the patient. These 4D MC tools, therefore, have an important role in verification and possibly even planning of these treatments. For example, Teke et al. (2013) developed a BEAMnrc-based MC tool, using ISOURCE 20, to simulate the dose delivery based on the extensible markup language (XML) file which encodes the delivery instructions. Figure 9.10 shows a comparison of film measurements and MC simulations for a delivery with continuous collimator rotation, couch translation, and couch rotation.

The "time stamping" and "4D phase space" methods introduced in Section 9.2.2.4 enable the possibility to encode the temporal information about the dose delivery in a single simulation. For example, Popescu et al. (2015) performed simulations of VMAT delivery using ISOURCE 20 to score a 4D phase space on the boundaries of the patient CT phantom. By applying a filter that rejects particles with a MU index outside a user-defined range, they demonstrated the ability to calculate incremental or cumulative dose distributions in the EPID, simultaneous with the patient dose calculation. The simulated EPID dose distributions showed a good agreement with measured EPID images for two patient cases.

Storing the temporal information about the dose delivery also enables the dose rate to be studied. Podesta et al (2016) were the first to develop an MC technique to generate

(a)

(b)

FIGURE 9.10 A comparison between Gafchromic film (a) and MC simulation, with DOSXYZnrc source 20 (b), of a beam delivery in Developer Mode on a Varian TrueBeam. The dose distribution shown was delivered with continuous and simultaneous collimator rotation, couch translation (lateral and longitudinal), and couch rotation. (Image courtesy of Tony Teke.)

dose rate maps for VMAT treatments using standard and flattening-filter free (FFF) beams. The motivation was to create a research tool in support of radiobiological investigations on the effects of dose rate. It is not clear what time scale is the most critical for these effects. In fact, patient dose rate on various time scales could prove relevant, from the dose per fraction in large-dose single-fraction stereotactice ablative radiotherapy (SABR) treatments, or dose to circulating blood in a low-dose rate total body irradiation (TBI) treatment, to dose per pulse, as in the emerging field of ultra-high dose rate (FLASH) radiotherapy research, where very large doses are delivered in a matter of microseconds.

9.6 Summary and Outlook

Three main approaches exist to simulating dynamic beam and patient geometries: particle weighting/convolution, SCS, and PPS. Dynamic simulations of IMRT deliveries are a well-established technique compared to 4D MC calculations in deforming patient geometries. For the latter, dose interpolation methods have been generally adopted for dose accumulation in radiotherapy planning; however, the development of accurate 4D MC methods that conserve mass and energy have highlighted the inaccuracies of the interpolation approach.

Although particle weighting approaches may appear to be more computationally efficient, comparable performances can be achieved with full "4D" MC simulations using the SCS or PPS method when fast MC codes, with appropriate variance reduction techniques and simplifications of particle transport, are used. The advantage of MC codes compared to analytical algorithms in this regard is that the computation time does not increase as a function of the number of geometries that are simulated as the number of simulated particles may be split between geometries; therefore, a higher temporal resolution can be achieved with MC simulations, which may be of importance for continuous delivery methods such as tomotherapy and VMAT.

The study of temporal interplay effects, which occur when both treatment beam and tumor motion occur, is an ongoing issue particularly in proton radiotherapy. MC methods are well suited to examining these problems although an effort has to be made to reduce the CPU memory requirements. Some initial results have demonstrated that this is possible, which opens up further possibilities of using MC as a method to evaluate the magnitude of these interplay effects for different dynamic beam delivery techniques. Another important application is the ability to track deviations of the beam delivery and patient motion from their planned values through reconstruction of the dynamic dose delivery to the patient based on delivery and patient log files.

In summary, time-dependent treatment planning and delivery are being ever more exploited and scrutinized in radiotherapy. MC codes are well suited to the study of temporal effects and should play an important role in their accurate quantification.

References

Ahmad M., Deng J., Lund M.W., Chen Z., Kimmett J., Moran M.S., and R. Nath. Clinical implementation of enhanced dynamic wedges into the Pinnacle treatment planning system: Monte Carlo validation and patient-specific QA. *Phys. Med. Biol.* 54, 447–465, 2009.

Beckham W.A., Keall P.J., and J.V. Siebers. A fluence-convolution method to calculate radiation therapy dose distributions that incorporate random set-up errors. *Phys. Med. Biol.* 47, 3465–3473, 2002.

Belec J., Ploquin N., La Russa D., and B.G. Clark. Position-probability-sampled Monte Carlo calculations of VMAT, 3DCRT, step-shoot IMRT and helical tomotherapy distributions using BEAMnrc/DOSXYZnrc. *Med. Phys.* 32, 948960, 2011.

Bortfeld T., Jiang S.B., and E. Rietzel. Effects of motion on the total dose distribution. *Semin. Radiat. Oncol.* 14, 41–51, 2004.

Boylan C.J., Aitkenhead A.H., Rowbottom C.G. and R.I. Mackay. Simulation of realistic linac motion improves the accuracy of a Monte Carlo base VMAT plan QA. *Radiother. Oncol.* 109, 377–383, 2013.

Bush K., Townson R., and S. Zavgorodni. Monte Carlo simulation of RapidArc radiotherapy delivery. *Phys. Med. Biol.* 53, N359–N370, 2008.

Chetty I.J., Rosu M., McShan D.L., Fraass B.A., Balter J.M., and R.K. Ten Haken. Accounting for center-of-mass target motion using convolution methods in Monte Carlo-based dose calculations of the lung. *Med. Phys.* 31, 925–932, 2004.

Christensen G.E., Rabbitt R.D., and M.I. Miller. Deformable templates using large deformation kinetics. *IEEE Trans. Image Process.* 5, 1435–1447, 1996.

Craig T., Battista J., and J. Van Dyk. Limitations of a convolution method for modelling geometric uncertainties in radiation therapy I. The effect of shift invariance. *Med. Phys.* 30, 20012011, 2001.

Crum W.R., Griffin L.D., Hill D.L.G., and D.J. Hawkes. Zen and the art of medical image registration: Correspondence, homology and quality. *NeuroImage* 20, 1425–1437, 2003.

Deng J., Pawlicki T., Chen Y., Li J., Jiang S.B., and C-M. Ma. The MLC tongue-and-groove effect on IMRT dose distributions. *Phys. Med. Biol.* 46, 1039–1060, 2001.

Fan J., Li J., Chen L., Stathakis S., Luo W., Du Plessis F., Xiong W., Yang J., and C-M. Ma. A practical Monte Carlo MU verification tool for IMRT quality assurance. *Phys. Med. Biol.* 51, 2503–2515, 2006.

Gholampourkashi S., Vujicic M., Belec J., Cygler J.E. and E. Heath. Experimental verification of 4D Monte Carlo simulations of dose delivery to a moving anatomy. *Med. Phys.* 44, 299–310, 2017.

Gholampourkashi S., Cygler J.E., Lavigne B. and E. Heath. Validation of 4D Monte Carlo dose calculations using a programmable deformable lung phantom. *Physica Medica* 76, 16–27, 2020.

Heath E. and J. Seuntjens. A direct voxel tracking method for fourdimensional Monte Carlo dose calculations in deforming anatomy. *Med. Phys.* 33, 434–445, 2006.

Heath E., Collins D.L., Keall P.J., Dong L., and J. Seuntjens. Quantification of accuracy of the automated nonlinear image matching and anatomical labeling (ANIMAL) nonlinear registration algorithm for 4D CT images of lung. *Med. Phys.* 34, 4409–4421, 2007.

Heath E., Tessier F., Siebers J., and I. Kawrakow. Investigation of voxel warping and energy mapping approaches for fast 4D Monte Carlo dose calculations in deformed geometry using VMC++. *Phys. Med. Biol.* 56, 5187–5202, 2011.

Holden M. A review of geometric transformations for non-rigid body registration. *IEEE Trans. Med. Imaging* 27, 111–128, 2008.

Jensen M.D., Abdellatif A., Chen J. and E. Wong. *Phys. Med. Biol.* 57, N89–N99, 2012.

Jeraj R., Mackie T.R., Balog J., Olivera G., Pearson D., Kapatoes J., Ruchala K., and P. Reckwerdt. Radiation characteristics of helical tomotherapy. *Med. Phys.* 31, 396–404, 2004.

Keall P.J., Siebers J.V., Arnfield M., Kim J.O., and R. Mohan. Monte Carlo dose calculations for dynamic IMRT treatments. *Phys. Med. Biol.* 46, 929–941, 2001.

Keall P.J., Siebers J.V., Joshi S., and R. Mohan. Monte Carlo as a four-dimensional radiotherapy treatment-planning tool to account for respiratory motion. *Phys. Med. Biol.* 49, 3639–3648, 2004.

Leal A., Sanchez-Doblado F., Arran R., Rosello J., Pavon E., and J. Lagares. Routine IMRT verification by means of an automated Monte Carlo simulation system. *Int. J. Radiat. Oncol. Biol. Phys.* 56, 58–68, 2003.

Li X.A., Ma L., Naqvi S., Shih R., and C. Yu. Monte Carlo dose verification for intensity-modulated arc therapy. *Phys. Med. Biol.* 46, 2269–2282, 2001.

Litzenberg D.W, Hadley S.W., Tyagi N, Balter J.M., Ten Haken R., and I.J. Chetty. Synchronized dynamic dose reconstruction. *Med. Phys.* 34(1), 91–102, 2007.

Liu H.H., Mackie T.R., and E.C. McCullough. Modelling photon output caused by backscattered radiation into the monitor chamber from collimator jaws using a Monte Carlo technique. *Med. Phys.* 27(4), 737–744, 2000.

Liu H.H., Verhaegen F., and L. Dong. A method of simulating dynamic multileaf collimators using Monte Carlo techniques for intensity-modulated radiation therapy. *Phys. Med. Biol.* 46, 2283–2298, 2001.

Lobo J. and I. Popescu. Two new DOSXYZnrc sources for 4D Monte Carlo simulations of continuously variable beam configurations, with applications to RapidArc, VMAT, tomotherapy and CyberKnife. *Phys. Med. Biol.* 55, 4431–4443, 2010.

Lujan A.E., Larsen E.W., Balter J.M., and R. Ten Haken. A method for incorporating organ motion due to breathing into 3D dose calculations. *Med. Phys.* 26, 715–720, 1999.

Ma C-M., Pawlicki T., Jiang S.B., Li J.S., Deng J., Mok E., Kapur A., Xing L., Ma L., and A.L. Boyer. Monte Carlo verification of IMRT dose distributions from a commercial treatment planning optimization system. *Phys. Med. Biol.* 45, 2483–2495, 2000.

Paganetti H. Four-dimensional Monte Carlo simulation of time- dependent geometries. *Phys. Med. Biol.* 49, N75–N81, 2004.

Paganetti H., Jiang H., Adams J.A., Chen G.T., and E. Rietzel. Monte Carlo simulations with time-dependent geometries to investigate effects of organ motion with high temporal resolution. *Int. J. Radiat. Oncol. Biol. Phys.* 60, 942–950, 2004.

Palmans H. and F. Verhaegen. Monte Carlo study of fluence perturbation effects on cavity dose response in clinical proton beams. *Phys. Med. Biol.* 43, 65–89, 1998.

Perl J., Shin J., Schumann J., Faddegon B., and H. Paganetti. TOPAS: An innovative proton Monte Carlo platform for research and clinical applications. *Med. Phys.* 39, 6818–6837, 2012.

Podesta M, Popescu I.A., and F Verhaegen. Dose rate mapping of VMAT treatments. *Phys. Med. Biol.* 61 4048–60, 2016.

Popescu I.A. and J. Lobo. Monte Carlo simulations for TrueBeam with jaw tracking, using curved or planar IAEA phase spaces: A Source 20 update. International Conference on the Use of Computers in Radiation Therapy, Melbourne, Australia, 2013.

Popescu I.A., Atwal P., Lobo J., Lucido J. and B.M.C. McCurdy. Patient-specific QA using 4D Monte Carlo phase space predictions and EPID dosimetry. *J. Phys. Conf. Series* 573, 1–10, 2015.

Rosu M., Chetty I.J., Balter J.M., Kessler M.L., McShan D.L., and R.K. Ten Haken. Dose reconstruction in deforming lung anatomy: Dose grid size effects and clinical implications. *Med. Phys.* 32, 2487–2495, 2005.

Rosu M., Balter J.M., Chetty I.J., Kessler M.L., McShan D.L., Balter P., and R. Ten Haken. How extensive of a 4D dataset is needed to estimate cumulative dose distribution plan evaluation metrics in conformal lung therapy? *Med. Phys.* 34, 233–245, 2007.

Seco J., Adams E., Bidmead M., Partridge M., and F. Verhaegen. Head & neck IMRT treatment assessed with a Monte Carlo dose calculation engine. *Phys. Med. Biol.* 50, 817–830, 2005.

Seco J., Sharp G.C., Wu Z., Gierga D., Buettner F., and H. Paganetti. Dosimetric impact of motion in free-breathing and gated lung radiotherapy: A 4D Monte Carlo study of intrafraction and interfraction effects. *Med. Phys.* 35, 356356, 2008.

Seco J., Robertson D., Trofimov A., and H. Paganetti. Breathing interplay effects during proton beam scanning: Simulation and statistical analysis. *Phys. Med. Biol.* 54, N283–N294, 2009a.

Seco J., Sharp G.C., and H. Paganetti. Study of the variability of the dosimetric outcome produced by patient organ-movement and dynamic MLC with focus on intra-fraction effects. *Med. Phys.* 36, 2505, 2009b.

Shih R., Li X.A., and J.C.H. Chu. Dynamic wedge versus physical wedge: A Monte Carlo study. *Med. Phys.* 28, 612–619, 2001.

Shin J., Perl J., Schumann J., Paganetti H., and B.A. Faddegon. A modular method to handle multiple time-dependent quantities in Monte Carlo simulations. *Phys. Med. Biol.* 57, 3295–3308, 2012.

Siebers J.V, Keall P.J., Kim J.O., and R. Mohan. A method for photon beam Monte Carlo multileaf collimator particle transport. *Phys. Med. Biol.* 47, 3225–3249, 2002.

Siebers J.V. and H. Zhong. An energy transfer method for 4D Monte Carlo dose calculation. *Med. Phys.* 35, 4096–4105, 2008.

Sterpin E., Salvat F., Cravens R., Ruchala K., Olivera G.H., and S. Vynckier. Monte Carlo simulation of helical tomotherapy with PENELOPE. *Phys. Med. Biol.* 53, 2161–2180, 2008.

Teke T., Bergman A., Kwa W., Gill B., Duzenli C., and I.A. Popescu. Monte Carlo based RapidArc QA using LINAC log files. *Med. Phys.* 36, 4302, 2009.

Teke T., Gete E., Duzenli C., McAvoy S. and I. Popescu. Monte Carlo simulations for trajectory-based beam delivery in Varian TrueBeam Developer Mode. *Med. Phys.* 40, 87, 2013.

Thomas S.D., Mackenzie M., Rogers D.W.O., and B.G. Fallone. A Monte Carlo derived TG-51, equivalent calibration for helical tomotherapy. *Med. Phys.* 32, 1346–1353, 2005.

Tyagi N., Litzenberg D., Moran J., Fraass B., and I. Chetty. Use of the Monte Carlo method as a comprehensive tool for SMLC and DMLC-based IMRT delivery and quality assurance (QA). *Med. Phys.* 33, 2148, 2006.

Tyagi N., Moran J.M., Litzenberg D.W., Bielajew A.F., Fraass B.A., and I.J. Chetty. Experimental verification of a Monte Carlo- based MLC simulation model for IMRT dose calculation. *Med. Phys.* 34, 651–663, 2007.

Verhaegen F. and I.J. Das. Monte Carlo modelling of a virtual wedge. *Phys. Med. Biol.* 44(12), N251–N259, 1999.

Verhaegen F. and H.H. Liu. Incorporating dynamic collimator motion in Monte Carlo simulations: An application in modelling a dynamic wedge. *Phys. Med. Biol.* 46, 287–296, 2001.

Xing L., Siebers J., and P. Keall. Computational challenges for image-guided radiation therapy: Framework and current research. *Semin. Radiat. Oncol.* 17, 245–257, 2007.

Yu C., Jaffray D.A., and J.W. Wong. The effects of intrafraction organ motion on the delivery of dynamic intensity modulation. *Phys. Med. Biol.* 43, 91–104, 1998.

Zhao Y-L., Mackenzie M., Kirkby C., and B.G. Fallone. Monte Carlo calculation of helical tomotherapy dose delivery. *Med. Phys.* 35, 3491–3500, 2008a.

Zhao Y-L., Mackenzie M., Kirkby C., and B.G. Fallone. Monte Carlo evaluation of a treatment planning system for helical tomo- therapy in an anthropomorphic heterogeneous phantom and for clinical treatment plans. *Med. Phys.* 35, 5366–5374, 2008b.

Zhong H. and J.V. Siebers. Monte Carlo dose mapping on deforming anatomy. *Phys. Med. Biol.* 54, 5815–5830, 2009.

III

Patient Dose Calculation

10

Photons: Clinical Considerations and Applications

Michael K. Fix
*Inselspital – University
Hospital Bern*

10.1 Introduction

The principle aim of radiotherapy is to give a tumoricidal dose to the cancer-bearing tissue and to minimize the dose to the normal healthy tissue. Given the dose-response functions of the tumor and healthy tissue, the requirement stated in the ICRU report 50 (International Commission on Radiation Units and Measurements Report 50. 1993) is to apply the dose to the tumor within -5% and 7% of the prescribed dose. A detailed analysis of uncertainties associated with radiation treatment shows that 3% of accuracy is required in dose calculation to yield ±5% of accuracy in dose delivered to a patient (International Commission on Radiation Units and Measurements Report 24. 1976; Brahme, 1984; Dutreix, 1984; Van Dyk et al., 1993). Some studies have even concluded that for certain types of tumors the uncertainty in dose delivery should be smaller than 3.5% (Brahme, 1984; Mijnheer et al., 1987, 1989), which in turn means that the clinically implemented dose calculation algorithm should be accurate within ±2%. Thus, accurate dose calculation algorithms for treatment planning systems are of critical importance in radiation therapy.

Within the last decades, huge developments have been seen in the field of dose calculation algorithms. The early clinical treatment planning systems used correction-based algorithms (Mackie et al., 1996), whereas nowadays mostly model-based algorithms such as convolution or convolution/superposition algorithms are used. Algorithms using the Monte Carlo (MC) method can also be considered model-based. Although over the last years many applications of MC techniques have been published in the field of medical physics, the clinical impact of MC-calculated patient dose distributions for photon beams of conventional linear accelerators still remains unclear. However, recently linear accelerators combined with magnetic resonance imaging system (MR-LINAC) have been commercially introduced. The challenge of the magnetic field and its influence on dose distribution makes MC techniques currently the only method used for dose calculations (Hissoiny et al., 2011; Ghila et al., 2017; Wang et al., 2017; Kubota et al., 2020). MC in the context of MR-LINACs is also presented in detail in Chapter 12 in the accompanying book.

In contrast to other common techniques, the MC method starts from first principles and tracks individual particle histories; thus, it takes into account the transport of secondary particles. For a long time, a major drawback of using MC methods for photon beam treatment planning in clinical routine was the long calculation time needed to achieve a dose distribution with a reasonable statistical uncertainty. However, recent advances in MC patient dose calculation algorithms coupled with increasing computer processing speed have made MC patient dose calculation speed acceptable for radiotherapy clinics. In principle, the MC method produces accurate results in the regions of tissue heterogeneities, such as lung and surface irregularities, thus providing the most accurate method for the simulation

DOI: 10.1201/9781003211846-13

of patient treatment dose distributions, especially for complex techniques such as intensity-modulated radiotherapy (IMRT), intensity–modulated arc therapy (Andreo, 1991; DeMarco et al., 1998; Wang et al., 1998; Bush et al., 2008; Oliver et al., 2008; Sarkar et al., 2008; Teke et al., 2010; Fix et al., 2011; Gete et al., 2013; Lin et al., 2013; Adam et al., 2020), or when magnetic fields are considered (Hissoiny et al., 2011; Ahmad et al., 2016; Wang et al., 2016; Shortall et al., 2020). The National Cancer Institute recognizes the need for research and development on MC techniques in radiation therapy, and it is anticipated that MC will become a necessary dose calculation tool (Fraass et al., 2003). However, ultimately, the accuracy of the treatment planning dose calculation algorithm depends strongly on the implementation and the accuracy of the input data. For instance, the accuracy does depend on the quality of the anatomical information of patients since this affects the irradiating geometry as well as the tissue cross sections. Another prerequisite is accurate information about the radiation beam incident on the patient. In fact, given the availability of fast radiotherapy-specific MC codes, the major limitation to the widespread implementation of MC dose calculation algorithms is the lack of a general, accurate, and user-specific scalable source model of the accelerator radiation source. More precisely, a user with an arbitrary linear accelerator should be able to commission the source model so that the MC dose calculation algorithm meets the predefined accuracy requirements compared to measurements prior to using the algorithm for patient dose calculation, for example, 2% or 2 mm. For commissioning, often the manufacturer of the treatment planning system provides some support.

10.2 Requirements for Clinical MC Treatment Planning

A clinically usable MC treatment planning (MCTP) system for photon beams is more than just an MC dose calculation algorithm together with a beam model to describe the radiation beam. Additionally, it is important to also have the capability, that is, a tool or a platform, for beam setup, dose display, and dose evaluation. Whereas these platforms are available in commercial treatment planning systems, they are in general missed in the research MC dose calculation packages. If MC dose calculation is used only in some special rare cases, this might be acceptable. However, if MCTP should be used on a large-scale, automation of MC dose calculation is needed. Consequently, some research systems have been extended to provide capabilities for beam setup, dose display, and dose evaluation. The following sections will cover the requirements for clinical MCTP for photon beams.

10.2.1 Beam Setup Capability

For the commercial treatment planning systems, the beam setup capability is the same as for any other dose calculation

algorithm. But research systems, if intended for clinical use, also provide such a platform. Most commonly, this has been realized by interfacing the external MC dose calculation with an already existing commercial treatment planning system either automatically (Fix et al., 2007; Siebers et al., 2000), by another interfacing program (Ma et al., 2002), or by a Digital Imaging and Communications in Medicine-Radiotherapy (DICOM-RT) interface (Alexander et al., 2007; Rodriguez et al., 2013).

10.2.2 Beam Model

Certainly, an accurate characterization of the radiation beam exiting the treatment head is a prerequisite for accurate dose calculations in patients. For MC photon treatment planning, several beam models have been developed and investigated (Ma et al., 2020). They are either measurement- or MC-based (Verhaegen and Seuntjens, 2003). The measurement-based beam models use analytical models for which the parameters are determined from the measurements (Ahnesjo et al., 1992; Ahnesjo, 1994; Ahnesjo et al., 1995; Jiang et al., 2001; Fippel et al., 2003; Faught et al., 2017; Ishizawa et al., 2018). The MC-based beam models can be further divided into those using full MC (Hasenbalg et al., 2008), using phase space (phsp) files (Rogers et al., 1995; Sheikh-Bagheri and Rogers, 2002a, b), using histogram-based beam models (Ma, 1998; Schach von Wittenau et al., 1999; Deng et al., 2000; Fix et al., 2004; Chabert et al., 2016; Aboulbanine and El Khayati, 2018), or using a hybrid of phsp files and histogram beam model (Townson and Zavgorodni, 2014). These MC-based beam models usually model the primary beam, that is, the part of the accelerator head that covers the patient-independent part.

Besides the characterization of the primary beam, the beam model also has to accurately model the clinically used patient-specific beam modifiers. For photon beams, these are blocks, hard wedges, dynamic wedges, multileaf collimators (MLCs), and so on (Magaddino et al., 2011; Hermida-Lopez et al., 2018; Onizuka et al., 2018). See also Chapter 4 on 'MC Modelling of External Photon Beams in Radiotherapy' in this book. For radiation transport through these beam modifiers, a number of transport parameter settings and dedicated variance reduction methods can be applied depending on the treatment case (Schmidhalter et al., 2010). In general, the research systems provide a higher degree of flexibility compared to commercial treatment planning systems with respect to transport options but also on how to use the beam modifiers. For example, reduction in MLC transmission for moving jaws in IMRT has been investigated (Schmidhalter et al., 2007). Those settings need to be considered very carefully before using them for clinical MC dose calculations.

Since errors in the beam model propagate through all subsequent processes for dose calculation, it is very important to extensively verify the performance of the beam model during commissioning and validation process.

10.2.3 Patient Model

Another important issue affecting the accuracy of the MC-calculated dose distribution in patients is the anatomical patient representation, which is the basis for the geometrical and interaction data specification used for dose calculation. Patient computed tomography (CT) scans are used to extract the geometrical information as well as the patient-specific tissue characterization, leading to the interaction data for dose calculation within the patient. Therefore, image artifacts lead to inaccurate patient representation and consequently to inaccurate dose distributions. For non-MC dose calculation algorithms, CT conversion curves from Hounsfield values to electron density or physical density are used. However, for MC algorithms, the interaction data have to be known. Different conversion methods are used, starting from a direct relation between mass density and interaction coefficients (Fippel, 1999) to explicit segmentation of the tissue followed by anatomic composition assignment in order to determine the interaction data (Chetty et al., 2007). The impact of material misassignment on dose distributions is not fully investigated yet, and there are only a few studies in the literature (Verhaegen and Devic, 2005; Vanderstraeten et al., 2007; Ottosson and Behrens, 2011; Demol et al., 2015). More detailed information about this issue can be found in Chapter 11 of this book.

Another issue with CT images is grid resampling since typically a different voxel size is used for the dose calculation than provided by the CT images. The applied interpolation method can introduce errors, which in turn lead to misassignments in the materials and finally the interaction data. A study by Volken et al. (2008) investigating different interpolation algorithms shows that an integral conservative Hermitian curve interpolation improves the interpolation accuracy compared with typically used linear or cubic interpolation functions. Finally, it is worth mentioning that changes occurring between the image acquisition and the dose delivery need to be taken into account. These could be changes to patient, for example, weight loss, accessories only used during CT image acquisition, or different couches used for the CT scan and radiation delivery, respectively. Consequently, it might be necessary to remove that information from the CT images or overwrite the information with more accurate data before using them for dose calculations.

10.2.4 Dose Calculation

The MC dose calculation algorithm needs to be interfaced with the beam model. In principle, there exist several methods. One approach is to store phsp files as the output of the beam model and use these phsp files afterward as input for the dose calculation, that is, the phsp file is used as a beam model in this case. Another approach is to pass the particle from the beam model directly, that is, in memory, to the dose calculation algorithm. The latter has the advantage of being faster

and that no large phsp files have to be stored on hard disks. The fact that MC dose calculation is dealing with random walks offers the possibility to reuse the same particle coming from the beam model. However, by this method, the statistical uncertainty of the calculated dose distribution is affected since those particles are no longer independent. Thus, whereas this method saves some computing time and possibly disc space, the statistical uncertainty of the representation of the radiation beam by the beam model reveals some lower limits. One example is the latent variance of a phsp file (Sempau et al., 2001). Alternatively, modifications of particles from phsp utilizing existing symmetries (Rogers et al., 1995; Fix et al., 2004; Bush et al., 2007; Chabert et al., 2016) or small perturbations (Tyagi et al., 2006) have been investigated.

One of the main advantages of using MC dose calculation is that such an algorithm is not only suitable to calculate the dose for static patient situations but also for dynamic situations, like patient motion. Since MC is able to include a time component—in the beam model as well as in the patient model—MC can be used straightforwardly for dynamic patient dose calculations (Paganetti et al., 2004; Paganetti et al., 2005; Seco et al., 2008). Apart from patient, also the beam delivery system can be dynamic such as for IMRT or volumetric-modulated arc therapy (VMAT), both of which are currently state-of-the-art delivery techniques in clinical routine. Figure 10.1 shows comparisons of dose distributions for a 6 MV VMAT treatment plan of a lung case, using either MC, anisotropic analytical algorithm (AAA) (version 15.6), or Acuros (version 15.6) as dose calculation algorithm. The same number of monitor units (MUs) are used for all calculations. The dose distributions for MC and Acuros are similar. For both, the dose to the target is less homogeneous than predicted using AAA, while for all algorithms the dose-volume histograms (DVHs) to the organs at risk (OARs) are close to each other.

It is worth mentioning that for these applications the dose calculations with MC methods can be performed continuously instead of using a discrete number of situations approximating the dynamic delivery, for example, multiple static beams every 5° to approximate the continuous gantry rotation while the radiation beam is on. Typically, for these techniques also, the dose rate is dynamically changing during dose delivery, for which MC can provide a time-resolved dose rate distribution (Mackeprang et al., 2016; Podesta et al., 2016). More recently, the number of degrees of freedom was even further increased to include dynamic collimator and couch rotations, allowing highly dynamic non-coplanar treatment techniques (Fix et al., 2018; Manser et al., 2019; Smyth et al., 2019). It is also worth mentioning that the calculation time for MC dose calculations does not scale with the number of beams as compared to other conventional algorithms, given that only the statistical uncertainty in the target is considered. However, the calculation time might increase to some extent if a certain statistical uncertainty has to be achieved in OARs.

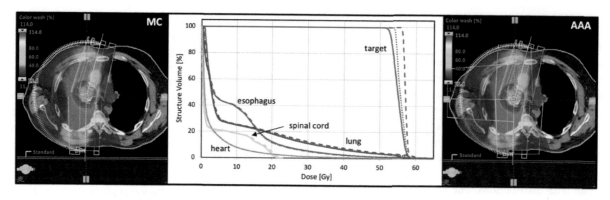

FIGURE 10.1 Comparison of MC-calculated (left and full lines for DVHs) dose distributions with the dose calculated using AAA (right and broken lines for DVHs) and Acuros (broken lines for DVHs) for a lung case (PTV structure shown in red), utilizing VMAT of a 6 MV photon beam. While there are differences in the DVH for the target, the DVH for the OARs is almost the same for all dose calculation algorithms.

10.2.5 Dose Evaluation Capability

Similar to the beam setup, also the dose evaluation can be performed as for any other dose calculation algorithm; thus, the commercial treatment planning systems are able to use their already existing tools for dose display and dose evaluation. This is true also for those research systems that are automatically interfaced with commercial treatment planning systems or those using DICOM files as interfaces. However, additional important information is needed, which is generally not included in commercial dose display and evaluation tools. Since every MC simulation is associated with a statistical uncertainty, documentation of this information for dose evaluation is essential when dealing with MC. Bearing in mind that in the dose evaluation different structures are quantitatively analyzed, it is important to take into account the statistical uncertainty of the dose values within the structure, for example, by displaying the uncertainty distribution. Additionally, it should be stated whether dose to media or dose to water has been calculated independent of which dose calculation algorithm is used. Another possibility to display and evaluate the resulting dose distribution is external viewing and analysis tools, for example, the CERR software development (Deasy et al., 2003).

10.3 Commissioning and Validation

Usually, vendors of commercial treatment planning systems provide a recommendation of the commissioning and validation procedure for their planning systems. Thereby, the commissioning consists of configuration of the treatment planning system. The commissioning and validation are performed by comparing dose measurements with the corresponding calculated dose distributions for different setups. However, especially for MC implementations, it is not fully explored what the least sufficient set of comparisons is for the commissioning and validation process. Of course, this depends strongly on the clinical use foreseen, for example, which type of treatment technique

should be covered: 3D conformal, IMRT, VMAT, stereotactic radiotherapy, and so on. It is further important to determine what kind of beam modifiers has to be provided. Finally, it also depends on the accuracy requested, detector types used for the measurements, and so on. The following sections provide an overview of commissioning and validation of MCTP systems.

10.3.1 Tolerances and Acceptance Criteria

During the commissioning and acceptance procedure, it is always a difficult task to determine the tolerance and acceptance criteria for dose comparisons. Several quantities have been used for comparing two dose distributions, for example, dose difference, distance to agreement, gamma index (Low et al., 1998), or slightly modified gamma index values (Bakai et al., 2003; Jiang et al., 2006; Blanpain and Mercier, 2009; Sumida et al., 2015). For these quantities, a certain tolerance level or acceptance criteria have to be defined. Often 2% or 2 mm is chosen. However, if such criteria are required for patient dose calculation, the estimation for the error due to beam modelling may not be larger than 1% or 1 mm (Keall et al., 2003). It is also questionable to apply the same criteria throughout the comparison, that is, for all field sizes, at every location, and so on. Given the statistical nature of MC, there is a certain probability of dose values with large random errors that might not fulfill the criteria. But there might also be areas in which the accuracy is more important than in others depending on the intended application, for example, buildup region or outside the direct radiation field. Thus, different criteria at different locations and setups might be justified. In summary, it is of great importance to carefully choose the tolerance and acceptance criteria ahead of the commissioning and validation procedure and in relation to the clinical usage of the machine.

10.3.2 CT Conversion

As already mentioned in previous sections, it is of great importance to have accurate anatomical and material information. Depending on the MC algorithm, different conversion methods are needed to finally obtain the interaction data needed for the MC dose calculation. The corresponding input information for the conversion method used has to be provided to the system by the user. Most MCTP systems, including the research systems, use a mapping of Hounsfield values to mass density and material composition. In early MC photon treatment planning, often a default CT to material conversion has been used. This conversion is not appropriate if a highly accurate dose calculation is required. Incorrect assignment of media and/or density can lead to significant dosimetric errors (see Section 10.2.3). A more accurate CT to material conversion requires more segmentation bins compared with the default scheme (Vanderstraeten et al., 2007) and depends on the CT scanner used. Recently developed dual or multi-energy CT scanners have the potential to improve the accuracy of human tissue identification relevant for MC in radiotherapy (Bazalova et al., 2008; Lalonde and Bouchard, 2016). Hence, CT calibration phantoms applied for the CT scanner in use lead to more accurate patient representations and material specifications. However, the benefit of dual-energy CT scanner is larger for protons and kV radiotherapy than for MV photon beams (van Elmpt et al., 2016). In addition, the phantom has to be carefully evaluated in order to determine if it is suitable for such a calibration procedure, for example, Teflon is not an appropriate representation for cortical bone (Verhaegen and Devic, 2005).

10.3.3 Beam Model and Dose Calculation

The specific beam model commissioning depends strongly on the implemented beam model. Most research-based systems have a slightly different implementation, thus an individual beam model commissioning procedure. For beam models based on phsp data, usually the energy and the width of the Gaussian intensity distribution of the initial electron beam are used as parameter. Note that perfect knowledge of the treatment head geometry and material is assumed. Beam models derived from these phsp data usually have settings available for a wide range of different parameterized initial electron beams. During commissioning, the beam model created is determined by interpolation. Analytical or measurement-based beam models have a specified set of measurements needed for commissioning. The scope of these measurements is provided either by the vendors or by the research group.

Beam modifiers typically represent the patient-specific part of the linear accelerator within the beam model. For radiation transport, mainly either MC transport or transmission filters in multiple layers are used. The parameters that are adjusted during the commissioning phase of the beam model depend on the beam modifier considered. Typically, the density is tuned for the secondary collimators and blocks, whereas additionally, the thickness is adjusted for the hard wedges. For the tuning of the MLC, the gap between the leaves is used as a third parameter. As for the open beam, and for the beam modifiers too, a set of measurements is used to determine the settings of the tuning parameters. Since these parameters are not depending on the photon beam energy, the commissioning could in principle be done for only one beam energy. Alternatively and depending on the clinical demands, the commissioning could determine tuning parameters that provide the best fit for all beam energies on average. The validation, of course, has to be done for all beam energies used clinically.

Validation of the beam model is done by comparing dose calculations with measured dose distributions. The comparisons include the measurements that have been used during the commissioning process in order to verify that the treatment planning system is able to reproduce the input dose distributions, that is, a consistency test is performed. For validation, it is important to cover the dosimetric range for which treatment plans are used clinically. Typically, the following measurements are compared:

- Relative depth dose curves and lateral dose profiles for different field sizes and shapes, including off-axis fields. These measurements are needed for open fields as well as for all beam modifiers considered such as wedged, blocked, dynamic wedged, and static and dynamic MLC-shaped fields, including IMRT and VMAT. These measurements are performed in water or water-equivalent phantoms. Since the primary aim is to validate the beam model and the dose calculation, disturbing side effects due to CT conversion or volume averaging and so on should be avoided or minimized. This could be realized by using digital phantoms.
- Since relative depth dose curves and lateral dose profiles are used above, it is necessary to check output factors for the same situations as well as an absolute dose calibration. Alternatively, all dose curves could be compared in absolute units, for example, cGy or Gy per MU.
- Relative measurements using simple inhomogeneous phantoms.
- Clinical treatment plan comparisons between dose distributions calculated with MC and with other available conventional dose calculation algorithms. This should involve simple cases as well as complex cases. Thereby, patient is set to water with density 1 g/cm^3 or with the density resulting from the CT conversion scheme. This provides some consistency checks with conventional dose calculation algorithms.

Some remarks with respect to commissioning and validation:

- The measurement procedure should be consistent with those for conventional treatment planning algorithms, for example, Fraass et al. (1998) or Smilowitz et al. (2015). Appropriate detectors should be used for different situations, for example, measurements at shallow depths. Additionally, the detector type has to be carefully selected to receive a consistent measured data

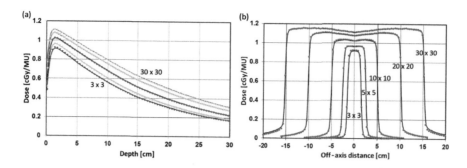

FIGURE 10.2 Comparison of MC calculated (symbols) and measured (lines) dose distributions in units of cGy/MU for different field sizes (in cm²) in water of a 6 MV photon beam: depth dose curves (a), lateral dose profiles (b).

set. Possible dependencies in energy response, effective point of measurement, dose rate, and so on need to be taken into account. Furthermore, the voxel size for the calculated dose distributions should be approximately the same as the sensitive volume of the detector used for the measurements.

- Since MC calculations are always associated with statistical uncertainties, single-point dose comparisons are critical. For example, absolute dose calibration using a single point is inappropriate. A method using multiple points is much more robust (Siebers et al., 1999; Fix et al., 2007). This is also true for dose prescription in clinical treatment plans.
- Generally, it is advisable to perform dose comparisons at shallow depths since these measurements are highly sensitive with respect to the beam model settings. Thus, they are useful to verify the accuracy of the beam model. Additionally, in-air profile measurements can be used for investigating the performance of the beam model since the impact of scatter is reduced.
- Although in the commissioning and validation ultimately the calculated dose distribution has to match measurements, it is important to bear in mind that the measurements themselves are also associated with errors.

Overall, the result of the commissioning and validation process strongly depends on the quality of the measurements. Figure 10.2 shows a typical comparison between MC-calculated and measured absolute depth dose curves and lateral dose profiles.

Using an incompatible voxel size for the calculation compared with the detector used for the measurement leads to dosimetric inaccuracies in the penumbra region. A study by Sahoo et al. (2008) shows that the penumbra width differs by a factor of 2 when different detectors are used. Figure 10.3 illustrates an example of the comparison between calculated and measured absolute dose profiles for 45° hard-wedged beams.

The comparison at different depths provides some indication of whether or not the energy spectrum, including the beam hardening, has been characterized appropriately in the beam model. Validation of MLC has to be done with great care since this complex beam modifier is often used in both static

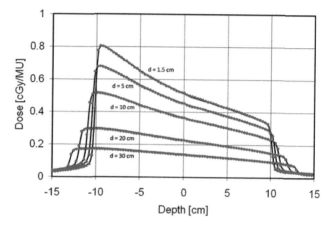

FIGURE 10.3 Comparison of MC calculated (symbols) and measured (lines) dose distributions in units of cGy/MU for a beam with a 45° hard wedge at several depths in water.

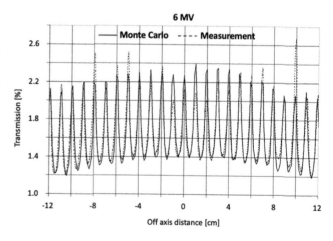

FIGURE 10.4 Comparison of MC calculated (lines) and measured (broken lines) transmission curves for an 80 leaf Varian MLC. (Adapted from Fix et al. 2007. *Phys Med Biol* 52: N425–37, with permission from IOP Publishing.)

and dynamic treatment applications. Transmission, leakage, and various MLC-shaped fields should be included in validation. Figure 10.4 demonstrates calculated and measured transmission and leakage profiles.

If the interleaf leakage is not taken into account for the MLC, the average transmission might be correct; however, the shape of the transmission profile cannot be reproduced. On the other hand, it is not clear if such a detailed modelling of the MLC is clinically needed. Dosimetric comparisons between dose calculation algorithms are important as demonstrated by a study by Reynaert et al. (2005). They encountered a 10% difference in DVHs for the optical chiasm when dose distributions are compared using Peregrine and their in-house MC system and assigned the obtained dose differences found to an inaccurate Elekta MLC modelling. Additional studies demonstrating the importance of accurate MLC modelling are described in the literature (Schwarz et al., 2003; Webster et al., 2007; Fix et al., 2011; Gholampourkashi et al., 2019).

Besides static beam modifiers, there are also dynamic beam modifiers, such as dynamic wedges or MLC. For enhanced dynamic wedges (EDWs) (dynamic wedge for Varian linear accelerators), one jaw of the secondary collimators is traveling from one side to the other while the beam is on, thus generating a wedged beam. Figure 10.5 shows some examples from the validation of EDWs for a 6 MV Varian beam. Measurements with an ionization chamber are taken at some points along the depth dose curves and dose profiles.

Since the MLC can be used for both static and dynamic treatment applications, it is not enough to validate the static characteristics through transmission measurements as mentioned above. Furthermore, dynamic applications have to be validated. Figure 10.6 illustrates the comparison between MC-calculated and measured dose values for one IMRT field from a verification plan using dynamic MLC, that is, one IMRT field is applied with gantry angle zero on a water phantom. For each measurement point using an ionization chamber (CC04), the whole IMRT field has to be delivered.

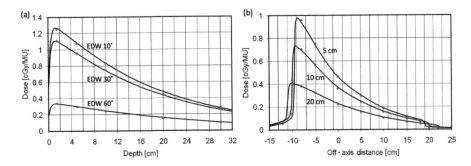

FIGURE 10.5 Comparison of MC calculated (lines) and measured (symbols) dose distributions for a 6 MV beam in units of cGy/MU for different enhanced dynamic wedges (EDWs) in water: depth dose curves (a) and lateral dose profiles for the EDW 60° at several depths (b).

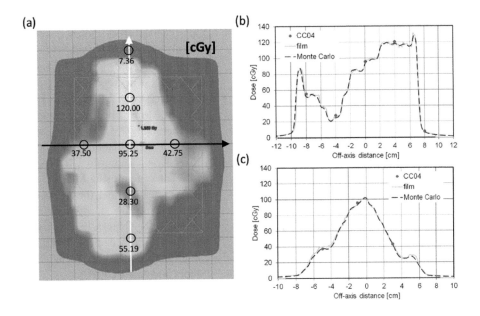

FIGURE 10.6 Comparison of MC calculated and measured dose distributions for one head and neck IMRT field using a 6 MV beam in units of cGy in water: (a) calculated dose distribution and measurement points (with dose values in cGy), (b) calculated dose profile and measurements along the white arrow, and (c) calculated dose profile and measurements along the black arrow.

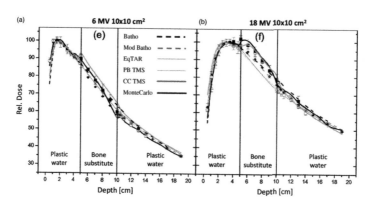

FIGURE 10.7   Comparison of measured and calculated relative depth dose curves in a water-bone-water phantom: 6 MV beam (a), 18 MV beam (b). Measurements are TLD white circles, MOSFET black squares, NACP02 black rhombus, and NE2571 white triangles. (Adapted from Carrasco et al. 2007. *Med Phys* 34: 3323–33, with permission from American Association of Physicists in Medicine.)

Additionally, a corresponding film measurement has been performed in a homogeneous solid water phantom.

Apart from dose comparisons in homogeneous phantoms, it is also important to validate the treatment planning system in inhomogeneous phantoms. This allows validation of transport in non-water materials (density and composition) as well as characterization of materials through CT data sets. The first issue can be verified if digital phantoms are used for the calculation. Figure 10.7 shows an example of an inhomogeneous phantom comparing relative depth dose curves for a slab phantom with water-bone-water interfaces (Carrasco et al., 2007). A 6 and 18 MV beam with a $10 \times 10\,cm^2$ field was used for irradiation. Measurements are performed with thermoluminescent dosimeters (TLDs), MOSFET, and several ionization chambers. These measurements are compared with calculations using different dose calculation algorithms, including MC, pencil beam (PB), and convolution/superposition algorithms.

10.4 Research and Commercial MCTP Systems

10.4.1 Research MCTP Systems

Over the years, many research institutions have implemented MC dose calculation algorithms (Brualla et al., 2017). Those implementations have been coupled with different kinds of source models to allow photon beam MCTP. These research planning systems are described in the literature and have been used to quantify the dosimetric difference between MC-calculated dose distributions and those calculated with traditional treatment planning systems. Research planning systems developed at research institutions include the following:

- University of California Los Angeles, RTMCNP (DeMarco et al., 1998): MCTP based on MCNP4A using a speed-optimized photon particle transport and dose scoring within the standard lattice geometry. The

RTMCNP preprocessor provides a user-friendly interface between the user and the MCNP4A command structure.

- Memorial Sloan Kettering Cancer Center (Wang et al., 1998; Wang et al., 1999): An EGS4-based MCTP environment using a dual-source beam model. The sources describe the primary and the scatter radiation of the treatment head and are based on a full MC simulation of the accelerator.

- Stanford University and Fox Chase Cancer Center, MCDOSE (Ma et al., 2002): This MCTP system for electron and photon beams is coupled with the FOCUS system (Computerized Medical Systems [CMS], Inc., recently purchased by Elekta AB). MCDOSE is based on EGS4 and as a beam model either phsp files or multiple source models can be used.

- Virginia Commonwealth University (Siebers et al., 2000; Siebers et al., 2002): An MCTP system using EGSnrc as transport code. For radiation transport through MLC, a dedicated transport method has been developed (Siebers et al., 2002).

- The University of Michigan, RT_DPM (Chetty et al., 2003): BEAMnrc phsp files coupled with the dose planning method (DPM) MC code (Sempau et al., 2000) are used as dose calculation engine in this MCTP system.

- University of Tübingen (Alber et al., 2003): An MC treatment system using MC also within the optimization. As a beam model, the virtual fluence model (Fippel et al., 2003) is utilized, and the X-ray voxel Monte Carlo (XVMC) code (Fippel, 1999) is used as a dose calculation algorithm. A dedicated source model for electron contamination has been included in the beam model (Sikora and Alber, 2009).

- McGill University, MMCTP (Alexander et al., 2007): The McGill MCTP is a flexible software package with import options, for example, DICOM-RT, tools for contouring, beam editing, visualization, and analysis. MC dose calculations for electron and photon beams are supported using BEAMnrc for simulation of the accelerator head and XVMC for dose calculation.

- Inselspital and University of Bern, SMCP (Fix et al., 2007): The Swiss Monte Carlo Plan (SMCP) is an MCTP system as a registered dose calculation algorithm within Eclipse (Varian Medical Systems, Inc.). Transport methods of different complexity levels for the treatment head combined with MC dose calculation algorithms of EGSnrc or VMC++ allow automatic and flexible MCTP. SMCP was extended to support proton dose calculations (Fix et al., 2013), modulated electron radiotherapy (Henzen et al., 2014), as well as dynamic trajectory radiotherapy (Fix et al., 2018; Manser et al., 2019).

- Universitat Politècnica de Catalunya and Essen University Hospital, PRIMO (Rodriguez et al., 2013): An MCTP system combining a GUI with the MC code PENELOPE for clinical LINAC simulations and absorbed dose estimations of electron or photon beams. For dose calculations, also DPM can be used (Rodriguez et al., 2018). Features such as contouring, DICOM-RT import, dose visualization, and evaluation are available.

- The University of Sevilla, CARMEN (Palma et al., 2012; Ureba et al., 2014): This MCTP system is controlled by a MATLAB˙ interface, supporting a DICOM import of contoured patient data. MC-based inverse treatment planning utilizing fluence maps as well as direct aperture optimization is available. In addition, optimization for mixed electron-photon-modulated radiotherapy is possible. The MC code EGSnrc is used for simulations of photon and electron beams.

10.4.2 Commercial MCTP Systems

In the last decade, there have been only a few commercial treatment planning systems available, offering MC dose calculation for photon beams (Brualla et al., 2017). The following information is focusing on MCTP systems for conventional linacs, while specific information for MR-LINACs is available in Chapter 12 in the accompanying book:

- Peregrine is an MC dose calculation system developed at the Lawrence Livermore National Laboratory. Peregrine was implemented within the commercial treatment planning system Corvus (NOMOS Corporation) and used a four-source beam model based on a phsp simulation of the accelerator head (Schach von Wittenau et al., 1999; Hartmann Siantar et al., 2001). Three photon sub-sources representing the target, the primary collimator, and the flattening filter were combined with the fourth source, representing the electron contamination of the beam utilizing a series of correlated histograms. Validation of the Peregrine system for clinical use is shown by several groups (Hartmann Siantar et al., 2001; Heath et al., 2004; Lehmann et al., 2006;

Rassiah-Szegedi et al., 2007). However, this system is no longer commercially available.

- Monaco is the MCTP option offered by CMS (now Elekta AB). The beam model in Monaco is based on the virtual fluence model for photons developed by Fippel et al. (2003), and for the electron contamination, the approach developed by Sikora and Alber is used (Sikora and Alber, 2009). The inputs for this beam model are depth dose curves, output factors, and lateral dose profiles at several depths. Commissioning uses small field depth dose curves to determine the primary energy spectrum, output factors for a reference field, as well as for the smallest and largest field size to determine the primary source diameter. Finally, cross profiles for the largest field size at several depths are needed to extract the off-axis energy fluence variations. This leads to a beam model which is characterized by 11 parameters. The radiation transport through the MLC is performed by using transmission filters in multiple layers (Sikora et al., 2007). This method results in a speed increase of about a factor of 100 compared with full MC transport and allows simulating the geometry of the MLC but not the scatter radiation produced in MLC. For patient dose calculation, the XVMC code is used (Fippel, 1999). Clinical validations have been reported in the literature (Sikora et al., 2007; Fotina et al., 2009; Sikora and Alber, 2009; Sikora et al., 2009; Narayanasamy et al., 2017; Snyder et al., 2019).

- In December 2008, Brainlab released their MCTP system within iPlan. The beam model is based on a virtual beam model developed by Fippel et al. (2003) and supports linear accelerators, including their MLCs of all major vendors. For the MLC beam modifier, a full MC simulation is used as radiation transport. However, the user is able to select between a speed-optimized (simplified MLC without leakage) and an accuracy-optimized MLC model. Further parameters in iPlan are the voxel size for the dose calculation, the requested mean variance of the MC dose distribution, and whether dose to medium or dose to water is calculated. Currently, the MC dose calculation algorithm requires commissioning of the PB algorithm in iPlan. Commissioning and validation of the iPlan system are done by Brainlab itself using a total of 93 in-air measurements and 97 measurements in water. The in-air measurements include depth profiles, cross profiles, and output factors, whereas the measurements in water include absolute calibration measurements, depth dose curves, and cross profiles, all at a source-to-surface distance (SSD) of 90 and 100 cm as well as output factor measurements at SSD 90. Most of these measurements in water are used for validation of the beam model. The results of the commissioned beam model are finally provided to the customer. Validations of MC implementation in iPlan have been published in the literature (Fotina et al., 2009; Kunzler et al., 2009; Song et al., 2013; Menon et al., 2020).

- In early 2011, the treatment planning system ISOgray (DOSIsoft) received FDA clearance. ISOgray offers MC dose calculations based on the general-purpose MC code PENELOPE (Salvat et al., 2006). The beam model consists of two parts: The first part includes the treatment head components of the linear accelerator which are patient-independent beginning with the target and the second part is the patient-specific part, including the secondary collimator jaws and beam modifiers, such as blocks, wedges, and MLCs. For the patient-independent part, a phsp file is generated using PENELOPE. This computation is performed by DOSIsoft during the commission process in which the incident energy and the spot size of the initial electron beam impinging on the target are tuned to fulfill the acceptance criteria according to the measurements from the user. The phsp file is then used as input for the patient-dependent part of the treatment head where an efficient selective particle tracking method is used (Brualla et al., 2009). Thereby, the geometry of the beam modifier is divided into areas depending on whether or not secondary particles or scatter radiation can emerge from the beam modifier, leading to skin and nonskin areas, respectively. Whereas in skin areas the radiation transport is modeled accurately, in nonskin areas the transport is simplified. In this manner, linear accelerator models for Elekta, Siemens, and Varian are available based on the manufacturer's information. For the dose calculation within the patient, PENFAST is used, which is based on PENELOPE but utilizes particle radiation transport techniques which are optimized with respect to calculation speed (Habib et al., 2010). The work from Habib et al. (2010) also serves as a preclinical validation of this treatment planning system.

- The treatment planning system Precision (Accuray Inc.) also provides an MC dose calculation engine for CyberKnife treatments using fixed and Iris collimators as well as for the MLC. The source model consists of a single source representing the target (Ma et al., 2008). The characteristics were determined by means of phsp data generated by a full MC simulation of the CyberKnife beam defining system. The virtual source is described by probability distributions for the energy spectrum as well as by two 2-dimensional distributions, one representing the origin distribution at the target and another one the fluence distribution of the photons emitted by the target. These distributions are commissioned by internal reference data and specific characteristics of each system acquired by measurements. Before the commissioning for the MC dose calculation algorithm is possible, the beam model for the ray-tracing algorithm has to be commissioned by means of measurements of tissue phantom ratios, lateral dose profiles, and output factors. The MC source model commissioning needs additional measurements of in-air output factors, open field dose profiles (without a collimator), and a percentage depth

dose curve of the 60-mm aperture. Each CT voxel is assigned to one out of the three material types—air, soft tissue, or bone—as well as to mass density. While the mass density is used for photon transport, for electrons only the mass density is used to scale precalculated electron tracks (particle repetition). For photons, additional variance reduction methods including interaction forcing, particle splitting, and Russian roulette are used. Based on the user-defined scoring grid resolution and requested statistical uncertainty for the maximum dose, the number of histories is determined utilizing an empirical equation, including also the patient model. Thereby, the statistical uncertainty is determined by the history-by-history method. This MC algorithm was validated in several studies (Heidorn et al., 2018; Pan et al., 2018; Mackeprang et al., 2019).

- Since Version 8b released in 2019, the treatment planning system RayStation (RaySearch Laboratories) offers a graphical processing unit (GPU)-based MC dose calculation engine developed in-house by RaySearch. The MC code is a class II condensed history algorithm, supporting the transport of photons, electrons, and positrons. The photon interactions included are Compton scattering, photoelectrical effect, and pair production with cross sections taken from the NIST XCOM database. Discrete interactions of Moller scattering and bremsstrahlung are simulated for electrons as well as annihilations for positrons. Multiple scattering is utilizing the random hinge method. The MC dose engine is interfaced with the same beam model as used for the collapsed cone algorithm; however, as the dose engines behave differently, a separate commissioning is needed, that is, different beam model parameters are received for the two algorithms, even though the same measured input data are used. The measurements include depth dose curves and lateral and diagonal dose profiles at several depths (depths of maximum dose, 5, 10, and 20 cm), output factors (including wedge factors), and an absolute calibration point for a reference field size. All measurements have to be performed at the same SSD collimation device (jaws and/or MLC) and should include the field sizes covering the range used for treatment planning. The beam model consists of sources modelling the primary source with a small spot and a high intensity, the scattering source with a large spot and a low intensity, a wedge scattering source, as well as a source for the contamination electrons. These sources are used to calculate the energy fluence in which the collimation devices such as secondary collimator jaws and the MLC are taken into account. The speed-optimized GPU implementation combined with Woodcock tracking offers high performance, for example, the dose calculation time is 11 seconds for a dual arc prostate case using 3 mm^3 voxel (NVIDIA GTX 1080Ti GPU), leading to a statistical uncertainty of 1%.

The radiation transport is performed in medium and reported as dose to medium. The statistical uncertainty is determined online by means of batches. The statistical uncertainty is provided as the mean one standard deviation determined over all dose-scoring voxels with a dose above 50% of the maximum dose. The desired statistical uncertainty is used as a stop criterion for MC dose calculation. Currently, no validation study is published yet; however, some information can be found on www.raysearchlabs.com.

10.5 Clinical Examples and Applications

10.5.1 MC as Treatment Planning Tool

There are many subjects that are worthwhile to be discussed in this section and most of them are described in detail in other chapters of this book. However, there are some issues that have to be carefully considered from a clinical perspective when using MC as a treatment planning tool. The two main issues involving the physicians' point of view are discussed in the following sections.

10.5.1.1 Noise in Dose Distributions

While conventional dose calculation algorithms determine 3D dose distributions deterministically, the MC method leads to a result associated with a statistical uncertainty. Thus, viewing and evaluating MC-calculated dose distributions for the first time within a clinical environment are challenging and need some time to get used to. The statistical uncertainty influences many quantities with respect to dose distributions such as isodose lines, DVHs, dose indices, the convergence of cost functions, and so on. The statistical uncertainty is coupled with the number of independent particle tracks used for the dose calculation and is usually determined by using the history-by-history method (Walters et al., 2002). Roughly, the square number of histories is needed to halve the statistical uncertainty of the dose distribution. Figure 10.8 demonstrates the impact of the statistical uncertainty, that is, the number of independent histories, on the dose distribution for a lung case using three-wedged beams. Thereby, the MC uncertainty is the average of statistical uncertainties (added in quadrature) of all dose values in the dose distribution with more than 50% of its maximum dose. Although the difference in the quality of the isodose line is significant, the differences are less pronounced in the DVHs as shown in the lower right plot for the three cases. Some differences for the planning target volume (PTV) and the spinal cord are visible when the statistical uncertainty is reduced from 10% to 2%. However, in the DVH, there is virtually no difference if the statistical uncertainty is further reduced to 0.5% in the case considered.

From this example, it is clear that for MC dose calculations, all point values, such as D_{max}, D_{min}, and so on, are highly critical since those values might be associated with a

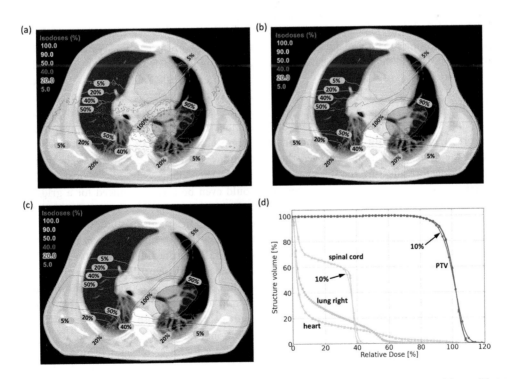

FIGURE 10.8 Comparison of MC-calculated dose distribution with different statistical uncertainties: 10% (a), 2% (b), 0.5% (c), and the impact on the DVHs (d)—10% squares, 2% dots, and 0.5% triangles. For the DVH, there is virtually no difference between the 2% and 0.5% calculations.

high statistical uncertainty in the considered calculation. The uncertainty might be different if the calculation is done the second time using different initial settings for the random number generator. This impact is also important for the dose prescription to a specific point. Thus, quantities relating to a volume or a larger number of dose values, such as D_{median} or D_{mean}, are more reliable quantities when dealing with MC dose distributions. In general, a statistical uncertainty of 2% per beam results in a reasonable precision within the target volume since usually three or more beams are used within a treatment plan. However, this might be no longer the case if the number of beams is below three or if OARs are considered. Since generally the OARs receive a lower particle fluence than the PTV, the statistical uncertainty in OARs can be much higher than in the target volume. Consequently, evaluation for OARs, for example, DVH or normal tissue complication probability values, has to be done very carefully. It might be necessary to increase the number of histories to receive an acceptable statistical uncertainty in the normal tissue.

Other methods influencing the statistical noise are denoising techniques. Several different denoising algorithms have been developed. El Naqa et al. (2005) and the references therein provide a broad overview of existing denoising techniques as well as a detailed comparison between different methods. More recently, also deep learning and neural network methods are investigated in the context of noise reduction in MC-calculated dose distributions with some promising results (Javaid et al., 2019). The main difficulty is to keep the true dose gradients and remove only statistical, that is, random, noise. Additionally, it has to be mentioned that the denoising algorithm introduces some systematic errors. Thus, for clinical applications, denoising can be used during the initial treatment planning process and should only be used with care for final dose calculation.

10.5.1.2 Calculation Time

The statistical uncertainty and the number of histories needed for simulation are directly related to the CPU time needed for the calculation of the dose distribution. However, there are also some additional factors affecting the dose calculation time, for example, the dose voxel size, the dose-scoring volume, the used beam modifiers, and the CPU. For clinical applications, it is important to have fast dose calculation algorithms. A summary of timing results for clinical treatment plans is given in the Task Group Report 105 of AAPM (Chetty et al., 2007). Bearing in mind that this timing comparison has been performed several years ago and that the calculation speed of the computers is still rapidly increasing, nowadays, the calculation time for photon MCTP is acceptable for a clinical routine. More recent data are provided in the review by Brualla et al. (2017). Apart from increased computer speed, the MC dose calculation codes themselves have been optimized using variance reduction methods (Chetty et al., 2007) or due to simplification of the transport code itself. For instance, the radiation transport within the beam modifier

provides some potential for optimizing the calculation time since there is also a trade-off between accuracy and CPU time. For example, the radiation transport through the secondary collimator jaws could be very simple (i.e., fast) if, additionally, the MLC is used below the jaws (Schmidhalter et al., 2010). Recently, MC algorithms have been implemented on GPUs, demonstrating that the efficiency of the MC method could be further increased significantly (Badal and Badano, 2009; Jia et al., 2011; Jahnke et al., 2012; Su et al., 2014; Tian et al., 2015). More information about this issue can be found in Chapter 14 in the accompanying book. Despite all these optimizations, the inverse planning might be the only exception where—given a large number of iterations—the overall calculation time might become unacceptable.

10.5.2 Comparisons of Dose Calculation Engines

10.5.2.1 Lungs

Owing to the difficulties with modelling of electron transport in low densities in the conventional dose calculation algorithms, lungs are potentially the site showing the largest differences between dose distributions calculated using MC and conventional algorithms. There are many studies published in the literature investigating the differences between several dose calculation algorithms. These include investigations using inhomogeneous phantoms (Carrasco et al., 2004; Fogliata et al., 2007; Panettieri et al., 2007; Aarup et al., 2009; Fotina et al., 2009; Bush et al., 2011; Fogliata et al., 2011; Alagar et al., 2016; Elcim et al., 2018) as well as patient treatment planning studies (Wang et al., 2002a; Dobler et al., 2006; Vanderstraeten et al., 2006; Hasenbalg et al., 2007; Madani et al., 2007; Ojala et al., 2014; Tsuruta et al., 2014; Zhao et al., 2014; Galonske et al., 2017; Hoffmann et al., 2018). As an example, for a phantom study, Fogliata et al. (2007) demonstrate that even for simple lung inhomogeneities, algorithms based on PB convolution can lead to dose differences of up to 30% compared with MC. For more advanced algorithms, this difference is reduced to about 8%. This difference could even be further reduced for a grid-based Boltzmann equation solver (Bush et al., 2011; Fogliata et al., 2011; Alagar et al., 2016). In summary, studies with inhomogeneous phantoms show that the higher the energy and the smaller the field sizes, the larger the differences between MC and conventional dose calculation algorithms become. In terms of lung patient studies, Wang et al. (2002a) have shown differences of more than 10% when MC is compared with an algorithm using the equivalent path length inhomogeneity correction method. Additionally, they report that due to the reduced lateral electron range for lower energies, using 6 MV photons instead of 15 MV is advantageous (Wang et al., 2002b). This has been confirmed in a study by Madani et al. (2007). If highly accurate algorithms are available, energy selection depends on the priority ranking of endpoints. Vanderstraeten et al. (2006)

compared MC-calculated dose distributions for lung cancer patients with those calculated with two different convolution/ superposition algorithms—Pinnacle (Philips Electronics N.V.) and Oncentra MasterPlan (Nucletron). Differences above 5% are reported. However, if patient tissue is replaced by water, the differences above 5% disappear, thus demonstrating some limitations of dose calculation algorithms in conventional treatment planning algorithms. Hasenbalg et al. (2007) concluded that using a PB algorithm may be inappropriate for lung cancer patients due to inaccurately calculated dose distributions. The more advanced dose algorithms AAA in Eclipse (Varian Medical Systems, Inc.) and collapsed cone convolution in MasterPlan (Nucletron) show dose differences of 5%. For stereotactic lung lesions, Dobler et al. (2006) show that the PB algorithm overestimates measurements by up to 15%, whereas the collapsed cone and MC underestimate measurements by 8% and 3%, respectively. These findings were confirmed by the study of Ojala et al. (2014) as well as by Zhao et al. (2014), especially when using the IMRT treatment technique. For the CyberKnife system, large differences are reported for lung cases when ray tracing and MC dose calculation algorithms as available in Multiplan are compared (Galonske et al., 2017).

In summary, for lung cases, the discrepancies concern differences in the buildup region and the underestimation of the target coverage and of the isodose lines within the low-density tissue. The latter could also lead to differences in dose distribution within the OARs themselves, which are located tangential to the incident fields. As an example, with respect to the target coverage and isodose lines, Figure 10.9 shows comparison between an MC and PB-calculated dose distribution for a lung case within iPlan. The MC-calculated dose distribution demonstrates that the dose coverage of the target is not achieved when using the PB algorithm for dose calculation. This is also expressed in DVHs as shown in Figure 10.9 if, for both dose calculation algorithms, the same number of MUs is used. Whereas the dose to the PTV is overestimated when using the PB algorithm, the differences for the OARs are small.

10.5.2.2 Head and Neck

Although most comparison studies between MC and conventional dose calculation algorithms have been performed for lungs, there is also a motivation of investigating this comparison for head and neck cancer patients. Since critical structures such as the spinal cord are often located close to the target

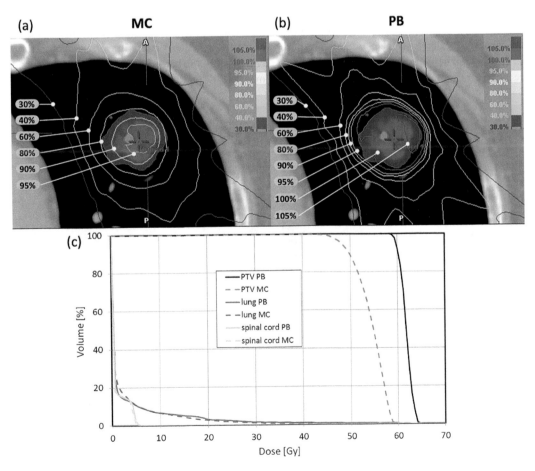

FIGURE 10.9 Comparison of an MC and pencil beam (PB) calculated dose distribution for a lung case within iPlan: (a) MC isodoses, (b) PB isodoses, and (c) DVH comparison.

volumes, dose distributions with high-dose gradients are applied. Additionally, air cavities are in these regions which lead to difficulties for many conventional dose calculation algorithms due to inaccurate dose calculations around air cavities, especially when Batho or equivalent tissue–air ratio correction methods are used (du Plessis et al., 2001). Examples of studies including investigations for head and neck cancer are the following (Wang et al., 2002a; Francescon et al., 2003; Boudreau et al., 2005; Seco et al., 2005; Knoos et al., 2006; Partridge et al., 2006; Sakthi et al., 2006; Mihaylov et al., 2007; Zhao et al., 2008; Pokhrel et al., 2016a; Hoffmann et al., 2018; Yang et al., 2020). The differences are smaller for the head and neck compared with those for lung cancer, and it has not been proven that these differences are clinically relevant. However, in cases with air cavities, differences may have to be carefully considered when choosing the dose calculation algorithm (Paelinck et al., 2006; Smedt et al., 2007). As an example, Figure 10.10 shows a dynamic MLC IMRT larynx case. Comparisons between MC, AAA, and PB (Batho correction method applied) are shown together with the corresponding DVHs using the same number of MUs. The PB calculation results in a higher dose to the PTV compared with AAA and MC algorithms. The MC method predicts a less homogeneous dose distribution within the PTV compared with the dose prediction when using the AAA algorithm.

This observation corresponds with findings from other research groups (Seco et al., 2005; Sakthi et al., 2006). The dose to the OAR is predicted higher when MC is used. Overall, the differences are larger for the more simple PB than for the more accurate AAA algorithm. Those differences need to be carefully considered in the clinical routine.

10.5.3 Inclusion of Time Dependencies

One of the major advantages when using MC methods concerns the time dependencies of irradiation in a 4D dose calculation. Whereas conventional algorithms usually discretize this problem leading to increased calculation time, MC is able to sample the time with almost no increase in calculation time when considering the statistical uncertainty of dose distribution in the target volume. Moreover, several approaches have been investigated to incorporate additionally the temporal patient movement into MC. One approach is to map the time-dependent dose distributions to a reference time point by using some kind of deformable registration between the time steps (Keall et al., 2004; Flampouri et al., 2006). Rosu et al. (2005) applied this method in conjunction with reducing the dose grid size to increase the accuracy of the method. Alternatively, deformation could be taken into account by adjusting the voxel geometry, that is, using deformed voxel instead of a rectilinear dose grid (Heath et al., 2007; Peterhans et al., 2011). A method that separates the particle transport from energy deposition scoring is the energy transfer method. Thereby, the particle transport is done in a rectilinear grid. The scoring, that is, the energy deposition, is done in a reference geometry at a certain time point using deformable image registration (Siebers and Zhong, 2008). Further development of this method is called the energy and mass congruent

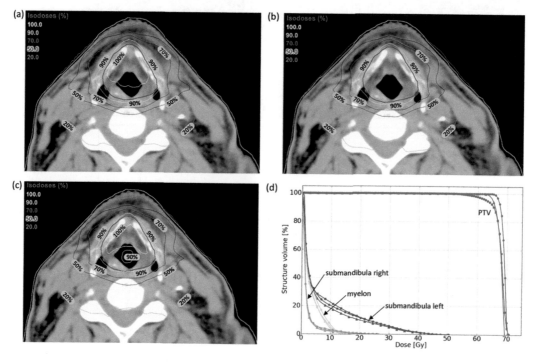

FIGURE 10.10 Comparison of an MC, AAA, and PB-calculated dose distribution for a head and neck case: PB isodoses (a), AAA isodoses (b), MC isodoses (c), DVH comparison with PB (dots), AAA (triangles), and MC (squares) (d). The red lines show PTV.

mapping (Zhong and Siebers, 2009). By this method, not only the energy deposition but also the mass is mapped to a reference geometry by deformable image registration. Apart from using time to resolve patient movements, there are other temporal dependencies that can be handled within the same simulation, for example, dynamic MLCs of an IMRT application and gantry rotation (Paganetti et al., 2004; Manser et al., 2019). This behavior opens the door not only to accurate dose calculation but also to a highly valuable tool for investigations such as interplay effects (Paganetti et al., 2005). Another example is reported in a study by Schmidhalter et al. (2007). They investigated the dosimetric effect in the case where the secondary collimator jaws follow the open window of the MLC during dynamic MLC IMRT to reduce transmission radiation through the MLC.

10.5.4 Reevaluation of Studies

In order to investigate the impact of more accurately calculated dose distributions on the outcome, MC might be the most appropriate method for the reevaluation of clinical studies. There are only a few studies reported in the literature dealing with lung cancer patients (Lindsay et al., 2007; Bibault et al., 2015; Ove et al., 2015; Pokhrel et al., 2016b). For lung patients, the outcome analysis is difficult since the treatment plans for many lung studies are lacking inhomogeneity corrections. However, as already mentioned, the impact of these corrections is large for low-density tissues; thus, it is very important to use accurate dose distributions for investigating the mapping to clinical outcome. These correlations between the dose distribution and effects for tumor and normal tissue gained from retrospective studies may also help define protocols for prospective studies. Additionally, such retrospective studies provide the opportunity to learn how to use MC for routine treatment planning since switching from one dose calculation algorithm to another is always a much-debated issue in the clinical environment. This is mainly because the treatment protocols used are based on clinical experience. Most likely, the change of dose calculation algorithm will also change the dose prescription, that is, a simple scaling, due to changes of the dose distribution in the tumor. However, the situation is more difficult for the normal tissue. Generally, OARs are often located at the field edges, and the impact of the dose calculation algorithm on their dose distribution due to different beam models, implemented lateral electron transport, and so on is not obvious, especially in low-density tissues. Consequently, accurate dose calculations in OARs are of critical importance, particularly if protocols for dose escalation are considered. Thus, many more studies are needed.

10.5.5 MC as QA Tool

Another application of MC dose calculations is its use as a QA tool. This could be done for different purposes. As a first example, MC dose calculations can be used as an independent MU calculation (Seco et al., 2005; Fix et al., 2007; Yamamoto et al., 2007; Pisaturo et al., 2009; Chen et al., 2015; Reynaert et al., 2016; Mackeprang et al., 2017). Thereby, the treatment plan of the conventional algorithm is recalculated with MC, and the resulting MUs are compared. Instead of using MC as an independent MU check, it could be just used to recalculate some complex or unclear treatment plans, especially if it is assumed that for such a plan the conventional algorithm is likely to fail. Bearing in mind that these are some individual cases, not the calculation time but rather the accuracy is highly important. Also, electronic portal imaging systems (EPIDs) are often used for IMRT QA procedures. Thereby, measurements performed with the EPID are compared with a reference. One approach is to use MC to determine the reference image for this comparison (Siebers et al., 2004; Frauchiger et al., 2007; Jarry and Verhaegen, 2007; Beck et al., 2009; Wang et al., 2009; Chabert et al., 2016). Owing to the increased distance of the EPID to the photon source in the linear accelerator head and the high resolution of the EPID detector, the calculation time is very long and thus optimized transport methods have to be used, for example, MC-calculated kernels (Wang et al., 2009). Another purpose for the use of MC dose calculation algorithms is benchmarking since it is assumed that MC is the most accurate method to predict dose distributions (Vanderstraeten et al., 2006; Fix et al., 2007; Hoffmann et al., 2018; Mackeprang et al., 2019). Since for benchmarking the accuracy is most important and there are generally only a few situations to be simulated, the calculation time is usually not very critical. More details can be found in Chapter 14 in this book.

10.5.6 MC in Optimization

In the previous sections, the use of MC dose calculation for forward treatment planning has been described. However, in principle, also the optimization benefits from the accuracy achieved when using MC dose calculations in the inverse treatment planning process for IMRT or VMAT applications. In general, IMRT delivery is performed by a sequence of small fields or a dynamically moving and reshaping aperture. This leads to a great demand for the accuracy of the dose calculation algorithm. Furthermore, conventional dose calculation algorithms may give rise to dose errors since a certain fraction of the dose is deposited by leakage and scatter radiation of the MLC. Most IMRT or VMAT systems use simple but fast dose calculation algorithms, for example, PB, within the optimization process. In many systems, a more sophisticated algorithm is used for final dose calculation. The limited accuracy of these fast algorithms leads to errors in dose prediction and convergence errors in optimization (Mihaylov and Siebers, 2008).

Although the computing time of MC dose calculation for a single field is very long compared with conventional dose calculation algorithms using a PB or convolution/superposition, the MC calculation time is almost independent of the numbers of beams. This is not the case for the conventional dose calculation algorithms. However, the main disadvantage of using MC in optimization remains to be the long calculation time, bearing in mind that many dose calculations during optimization have to be done. As a consequence, one approach is to use MC dose calculation not in each iteration step but always after a certain number of iterations. In all other iteration steps, a fast dose calculation algorithm is used. This approach is called a hybrid scheme for IMRT optimization (Siebers et al., 2007; Siebers, 2008). Another approach is the use of MC-calculated beamlets which are then used in the IMRT or VMAT optimization process or utilizing further speed-optimized MC code implementations on GPUs (Bergman et al., 2006; Sikora et al., 2009; Li et al., 2015; Yang et al., 2016).

A few studies demonstrate that the use of MC in optimization leads to improved dose distributions (Bergman et al., 2006; Mihaylov and Siebers, 2008; Siebers, 2008; Sikora et al., 2009; Dogan et al., 2006). Nevertheless, more detailed studies are needed in order to determine whether these improvements are clinically relevant.

10.6 Conclusions

MC dose calculation offers a great potential in photon treatment planning. The first few commercial treatment planning systems have MC methods for photon treatment planning available, and for MR-LINACs, MC is the only currently applied technique for dose calculations. Thus, MC dose calculation will be of increased importance in the future. Along with the shift to new dose calculation algorithms, there will be adjustments in the clinical routine. It has been demonstrated that when using an MC dose calculation algorithm, not only the algorithm itself is changed, but almost the whole radiotherapy process is affected, for example outlining, dose prescription, beam model, and CT conversion. There are still many open questions and studies required to carefully investigate all the impacts of this new technology and its impacts on outcome. This has to be done in close collaboration between clinical and research physicists.

References

Aarup L. R., Nahum A. E., Zacharatou C. et al. 2009. The effect of different lung densities on the accuracy of various radiotherapy dose calculation methods: implications for tumour coverage. *Radiother Oncol* 91: 405–14.

Aboulbanine Z. and El Khayati N. 2018. Validation of a virtual source model of medical linac for Monte Carlo dose calculation using multi-threaded Geant4. *Phys Med Biol* 63: 085008.

Adam D. P., Liu T., Caracappa P. F., Bednarz B. P. and Xu X. G. 2020. New capabilities of the Monte Carlo dose engine ARCHER-RT: clinical validation of the Varian TrueBeam machine for VMAT external beam radiotherapy. *Med Phys* 47: 2537–49.

Ahmad S. B., Sarfehnia A., Paudel M. R. et al. 2016. Evaluation of a commercial MRI Linac based Monte Carlo dose calculation algorithm with GEANT4. *Med Phys* 43: 894–907.

Ahnesjo A. 1994. Analytic modelling of photon scatter from flattening filters in photon therapy beams. *Med Phys* 21: 1227–35.

Ahnesjo A., Knoos T. and Montelius A. 1992. Application of the convolution method for calculation of output factors for therapy photon beams. *Med Phys* 19: 295–301.

Ahnesjo A., Weber L. and Nilsson P. 1995. Modelling transmission and scatter for photon beam attenuators. *Med Phys* 22: 1711–20.

Alagar A. G., Mani G. K. and Karunakaran K. 2016. Percentage depth dose calculation accuracy of model based algorithms in high energy photon small fields through heterogeneous media and comparison with plastic scintillator dosimetry. *J Appl Clin Med Phys* 17: 132–42.

Alber M., Birkner M., Bakai A. et al. 2003. Routine Use of Monte Carlo Dose Computation for Head and Neck IMRT Optimization. *Int J Radiat Oncol Biol Phys* 57: S208.

Alexander A., Deblois F., Stroian G., Al-Yahya K., Heath E. and Seuntjens J. 2007. MMCTP: a radiotherapy research environment for Monte Carlo and patient-specific treatment planning. *Phys Med Biol* 52: N297–308.

Andreo P. 1991. Monte Carlo techniques in medical radiation physics. *Phys Med Biol* 36: 861–920.

Badal A. and Badano A. 2009. Accelerating Monte Carlo simulations of photon transport in a voxelized geometry using a massively parallel graphics processing unit. *Med Phys* 36: 4878–80.

Bakai A., Alber M. and Nusslin F. 2003. A revision of the gamma-evaluation concept for the comparison of dose distributions. *Phys Med Biol* 48: 3543–53.

Bazalova M., Carrier J. F., Beaulieu L. and Verhaegen F. 2008. Dual-energy CT-based material extraction for tissue segmentation in Monte Carlo dose calculations. *Phys Med Biol* 53: 2439–56.

Beck J. A., Budgell G. J., Roberts D. A. and Evans P. M. 2009. Electron beam quality control using an amorphous silicon EPID. *Med Phys* 36: 1859–66.

Bergman A. M., Bush K., Milette M. P., Popescu I. A., Otto K. and Duzenli C. 2006. Direct aperture optimization for IMRT using Monte Carlo generated beamlets. *Med Phys* 33: 3666–79.

Bibault J. E., Mirabel X., Lacornerie T., Tresch E., Reynaert N. and Lartigau E. 2015. Adapted Prescription Dose for Monte Carlo Algorithm in Lung SBRT: clinical Outcome on 205 Patients. *PLoS One* 10: e0133617.

Blanpain B. and Mercier D. 2009. The delta envelope: a technique for dose distribution comparison. *Med Phys* 36: 797–808.

Boudreau C., Heath E., Seuntjens J., Ballivy O. and Parker W. 2005. IMRT head and neck treatment planning with a commercially available Monte Carlo based planning system. *Phys Med Biol* 50: 879–90.

Brahme A. 1984. Dosimetric precision requirements in radiation therapy. *Acta Radiol Oncol* 23: 379–91.

Brualla L., Rodriguez M. and Lallena A. M. 2017. Monte Carlo systems used for treatment planning and dose verification. *Strahlenther Onkol* 193: 243–59.

Brualla L., Salvat F. and Palanco-Zamora R. 2009. Efficient Monte Carlo simulation of multileaf collimators using geometry-related variance-reduction techniques. *Phys Med Biol* 54: 4131–49.

Bush K., Gagne I. M., Zavgorodni S., Ansbacher W. and Beckham W. 2011. Dosimetric validation of Acuros XB with Monte Carlo methods for photon dose calculations. *Med Phys* 38: 2208–21.

Bush K., Townson R. and Zavgorodni S. 2008. Monte Carlo simulation of RapidArc radiotherapy delivery. *Phys Med Biol* 53: N359–70.

Bush K., Zavgorodni S. F. and Beckham W. A. 2007. Azimuthal particle redistribution for the reduction of latent phase-space variance in Monte Carlo simulations. *Phys Med Biol* 52: 4345–60.

Carrasco P., Jornet N., Duch M. A. et al. 2004. Comparison of dose calculation algorithms in phantoms with lung equivalent heterogeneities under conditions of lateral electronic disequilibrium. *Med Phys* 31: 2899–911.

Carrasco P., Jornet N., Duch M. A. et al. 2007. Comparison of dose calculation algorithms in slab phantoms with cortical bone equivalent heterogeneities. *Med Phys* 34: 3323–33.

Chabert I., Barat E., Dautremer T. et al. 2016. Development and implementation in the Monte Carlo code PENELOPE of a new virtual source model for radiotherapy photon beams and portal image calculation. *Phys Med Biol* 61: 5215–52.

Chen X., Bush K., Ding A. and Xing L. 2015. Independent calculation of monitor units for VMAT and SPORT. *Med Phys* 42: 918–24.

Chetty I. J., Charland P. M., Tyagi N., McShan D. L., Fraass B. A. and Bielajew A. F. 2003. Photon beam relative dose validation of the DPM Monte Carlo code in lung-equivalent media. *Med Phys* 30: 563–73.

Chetty I. J., Curran B., Cygler J. E. et al. 2007. Report of the AAPM Task Group No. 105: issues associated with clinical implementation of Monte Carlo-based photon and electron external beam treatment planning. *Med Phys* 34: 4818–53.

Deasy J. O., Blanco A. I. and Clark V. H. 2003. CERR: a computational environment for radiotherapy research. *Med Phys* 30: 979–85.

DeMarco J. J., Solberg T. D. and Smathers J. B. 1998. A CT-based Monte Carlo simulation tool for dosimetry planning and analysis. *Med Phys* 25: 1–11.

Demol B., Viard R. and Reynaert N. 2015. Monte Carlo calculation based on hydrogen composition of the tissue for MV photon radiotherapy. *J Appl Clin Med Phys* 16: 117–30.

Deng J., Jiang S. B., Kapur A., Li J., Pawlicki T. and Ma C. M. 2000. Photon beam characterization and modelling for Monte Carlo treatment planning. *Phys Med Biol* 45: 411–27.

Dobler B., Walter C., Knopf A. et al. 2006. Optimization of extracranial stereotactic radiation therapy of small lung lesions using accurate dose calculation algorithms. *Radiat Oncol* 1: 45.

Dogan N., Siebers J. V., Keall P. J. et al. 2006. Improving IMRT dose accuracy via deliverable Monte Carlo optimization for the treatment of head and neck cancer patients. *Med Phys* 33: 4033–43.

du Plessis F. C., Willemse C. A., Lotter M. G. and Goedhals L. 2001. Comparison of the Batho, ETAR and Monte Carlo dose calculation methods in CT based patient models. *Med Phys* 28: 582–9.

Dutreix A. 1984. When and how can we improve precision in radiotherapy? *Radiother Oncol* 2: 275–92.

El Naqa I., Kawrakow I., Fippel M. et al. 2005. A comparison of Monte Carlo dose calculation denoising techniques. *Phys Med Biol* 50: 909–22.

Elcim Y., Dirican B. and Yavas O. 2018. Dosimetric comparison of pencil beam and Monte Carlo algorithms in conformal lung radiotherapy. *J Appl Clin Med Phys* 19: 616–24.

Faught A. M., Davidson S. E., Fontenot J. et al. 2017. Development of a Monte Carlo multiple source model for inclusion in a dose calculation auditing tool. *Med Phys* 44: 4943–51.

Fippel M. 1999. Fast Monte Carlo dose calculation for photon beams based on the VMC electron algorithm. *Med Phys* 26: 1466–75.

Fippel M., Haryanto F., Dohm O., Nusslin F. and Kriesen S. 2003. A virtual photon energy fluence model for Monte Carlo dose calculation. *Med Phys* 30: 301–11.

Fix M. K., Frei D., Volken W., Born E. J., Aebersold D. M. and Manser P. 2013. Macro Monte Carlo for dose calculation of proton beams. *Phys Med Biol* 58: 2027–44.

Fix M. K., Frei D., Volken W. et al. 2018. Part 1: optimization and evaluation of dynamic trajectory radiotherapy. *Med Phys* 45: 4201–12.

Fix M. K., Keall P. J., Dawson K. and Siebers J. V. 2004. Monte Carlo source model for photon beam radiotherapy: photon source characteristics. *Med Phys* 31: 3106–21.

Fix M. K., Manser P., Frei D., Volken W., Mini R. and Born E. J. 2007. An efficient framework for photon Monte Carlo treatment planning. *Phys Med Biol* 52: N425–37.

Fix M. K., Volken W., Frei D., Frauchiger D., Born E. J. and Manser P. 2011. Monte Carlo implementation, validation, and characterization of a 120 leaf MLC. *Med Phys* 38: 5311.

Flampouri S., Jiang S. B., Sharp G. C., Wolfgang J., Patel A. A. and Choi N. C. 2006. Estimation of the delivered patient dose in lung IMRT treatment based on deformable registration of 4D-CT data and Monte Carlo simulations. *Phys Med Biol* 51: 2763–79.

Fogliata A., Nicolini G., Clivio A., Vanetti E. and Cozzi L. 2011. Dosimetric evaluation of Acuros XB Advanced Dose Calculation algorithm in heterogeneous media. *Radiat Oncol* 6: 82.

Fogliata A., Vanetti E., Albers D. et al. 2007. On the dosimetric behaviour of photon dose calculation algorithms in the presence of simple geometric heterogeneities: comparison with Monte Carlo calculations. *Phys Med Biol* 52: 1363–85.

Fotina I., Winkler P., Kunzler T., Reiterer J., Simmat I. and Georg D. 2009. Advanced kernel methods vs. Monte Carlo-based dose calculation for high energy photon beams. *Radiother Oncol* 93: 645–53.

Fraass B., Doppke K., Hunt M. et al. 1998. American Association of Physicists in Medicine Radiation Therapy Committee Task Group 53: quality assurance for clinical radiotherapy treatment planning. *Med Phys* 25: 1773–829.

Fraass B. A., Smathers J. and Deye J. 2003. Summary and recommendations of a National Cancer Institute workshop on issues limiting the clinical use of Monte Carlo dose calculation algorithms for megavoltage external beam radiation therapy. *Med Phys* 30: 3206–16.

Francescon P., Cora S. and Chiovati P. 2003. Dose verification of an IMRT treatment planning system with the BEAM EGS4-based Monte Carlo code. *Med Phys* 30: 144–57.

Frauchiger D., Fix M. K., Frei D., Volken W., Mini R. and Manser P. 2007. Optimizing portal dose calculation for an amorphous silicon detector using Swiss Monte Carlo Plan. *J. Phys.: Conf. Ser.* 74: 021005.

Galonske K., Thiele M., Ernst I., Lehrke R. and Zylka W. 2017. Comparison of treatment plans calculated by Ray Tracing and Monte Carlo algorithms for head and thorax radiotherapy with Cyberknife. 3: 647.

Gete E., Duzenli C., Milette M. P. et al. 2013. A Monte Carlo approach to validation of FFF VMAT treatment plans for the TrueBeam linac. *Med Phys* 40: 021707.

Ghila A., Steciw S., Fallone B. G. and Rathee S. 2017. Experimental verification of EGSnrc Monte Carlo calculated depth doses within a realistic parallel magnetic field in a polystyrene phantom. *Med Phys* 44: 4804–15.

Gholampourkashi S., Cygler J. E., Belec J., Vujicic M. and Heath E. 2019. Monte Carlo and analytic modelling of an Elekta Infinity linac with Agility MLC: investigating the significance of accurate model parameters for small radiation fields. *J Appl Clin Med Phys* 20: 55–67.

Habib B., Poumarede B., Tola F. and Barthe J. 2010. Evaluation of PENFAST--a fast Monte Carlo code for dose calculations in photon and electron radiotherapy treatment planning. *Phys Med* 26: 17–25.

Hartmann Siantar C. L., Walling R. S., Daly T. P. et al. 2001. Description and dosimetric verification of the PEREGRINE Monte Carlo dose calculation system for

photon beams incident on a water phantom. *Med Phys* 28: 1322–37.

Hasenbalg F., Fix M. K., Born E. J., Mini R. and Kawrakow I. 2008. VMC++ versus BEAMnrc: a comparison of simulated linear accelerator heads for photon beams. *Med Phys* 35: 1521–31.

Hasenbalg F., Neuenschwander H., Mini R. and Born E. J. 2007. Collapsed cone convolution and analytical anisotropic algorithm dose calculations compared to VMC++ Monte Carlo simulations in clinical cases. *Phys Med Biol* 52: 3679–91.

Heath E., Collins D. L., Keall P. J., Dong L. and Seuntjens J. 2007. Quantification of accuracy of the automated nonlinear image matching and anatomical labeling (ANIMAL) nonlinear registration algorithm for 4D CT images of lung. *Med Phys* 34: 4409–21.

Heath E., Seuntjens J. and Sheikh-Bagheri D. 2004. Dosimetric evaluation of the clinical implementation of the first commercial IMRT Monte Carlo treatment planning system at 6 MV. *Med Phys* 31: 2771–9.

Heidorn S. C., Kilby W. and Furweger C. 2018. Novel Monte Carlo dose calculation algorithm for robotic radiosurgery with multi leaf collimator: dosimetric evaluation. *Phys Med* 55: 25–32.

Henzen D., Manser P., Frei D. et al. 2014. Beamlet based direct aperture optimization for MERT using a photon MLC. *Med Phys* 41: 121711.

Hermida-Lopez M., Sanchez-Artunedo D. and Calvo-Ortega J. F. 2018. PRIMO Monte Carlo software benchmarked against a reference dosimetry dataset for 6 MV photon beams from Varian linacs. *Radiat Oncol* 13: 144.

Hissoiny S., Raaijmakers A. J., Ozell B., Despres P. and Raaymakers B. W. 2011. Fast dose calculation in magnetic fields with GPUMCD. *Phys Med Biol* 56: 5119–29.

Hoffmann L., Alber M., Sohn M. and Elstrom U. V. 2018. Validation of the Acuros XB dose calculation algorithm versus Monte Carlo for clinical treatment plans. *Med Phys* 45: 3509–15.

International Commission on Radiation Units and Measurements Report 24. 1976. *Measurement of Absorbed Dose in a Phantom Irradiation by a Single Beam of X or Gamma Rays.* Bethesda, MD: ICRU.

International Commission on Radiation Units and Measurements Report 50. 1993. *Prescribing, Recording and Report in Photon Beam Therapy.* Bethesda, MD: ICRU.

Ishizawa Y., Dobashi S., Kadoya N. et al. 2018. A photon source model based on particle transport in a parameterized accelerator structure for Monte Carlo dose calculations. *Med Phys* 45: 2937–46.

Jahnke L., Fleckenstein J., Wenz F. and Hesser J. 2012. GMC: a GPU implementation of a Monte Carlo dose calculation based on Geant4. *Phys Med Biol* 57: 1217–29.

Jarry G. and Verhaegen F. 2007. Patient-specific dosimetry of conventional and intensity modulated radiation therapy using a novel full Monte Carlo phase space reconstruction method from electronic portal images. *Phys Med Biol* 52: 2277–99.

Javaid U., Souris K., Dasnoy D., Huang S. and Lee J. A. 2019. Mitigating inherent noise in Monte Carlo dose distributions using dilated U-Net. *Med Phys* 46: 5790–8.

Jia X., Gu X., Graves Y. J., Folkerts M. and Jiang S. B. 2011. GPU-based fast Monte Carlo simulation for radiotherapy dose calculation. *Phys Med Biol* 56: 7017–31.

Jiang S. B., Boyer A. L. and Ma C. M. 2001. Modelling the extrafocal radiation and monitor chamber backscatter for photon beam dose calculation. *Med Phys* 28: 55–66.

Jiang S. B., Sharp G. C., Neicu T., Berbeco R. I., Flampouri S. and Bortfeld T. 2006. On dose distribution comparison. *Phys Med Biol* 51: 759–76.

Keall P. J., Siebers J. V., Joshi S. and Mohan R. 2004. Monte Carlo as a four-dimensional radiotherapy treatment-planning tool to account for respiratory motion. *Phys Med Biol* 49: 3639–48.

Keall P. J., Siebers J. V., Libby B. and Mohan R. 2003. Determining the incident electron fluence for Monte Carlo-based photon treatment planning using a standard measured data set. *Med Phys* 30: 574–82.

Knoos T., Wieslander E., Cozzi L. et al. 2006. Comparison of dose calculation algorithms for treatment planning in external photon beam therapy for clinical situations. *Phys Med Biol* 51: 5785–807.

Kubota T., Araki F. and Ohno T. 2020. Comparison of dose distributions between transverse magnetic fields of 0.35 T and 1.5 T for radiotherapy in lung tumor using Monte Carlo calculation. *Med Dosim* 45: 179–85.

Kunzler T., Fotina I., Stock M. and Georg D. 2009. Experimental verification of a commercial Monte Carlo-based dose calculation module for high-energy photon beams. *Phys Med Biol* 54: 7363–77.

Lalonde A. and Bouchard H. 2016. A general method to derive tissue parameters for Monte Carlo dose calculation with multi-energy CT. *Phys Med Biol* 61: 8044–69.

Lehmann J., Stern R. L., Daly T. P. et al. 2006. Dosimetry for quantitative analysis of the effects of low-dose ionizing radiation in radiation therapy patients. *Radiat Res* 165: 240–7.

Li Y., Tian Z., Shi F. et al. 2015. A new Monte Carlo-based treatment plan optimization approach for intensity modulated radiation therapy. *Phys Med Biol* 60: 2903–19.

Lin M. H., Koren S., Veltchev I. et al. 2013. Measurement comparison and Monte Carlo analysis for volumetric-modulated arc therapy (VMAT) delivery verification using the ArcCHECK dosimetry system. *J Appl Clin Med Phys* 14: 3929.

Lindsay P. E., El Naqa I., Hope A. J. et al. 2007. Retrospective monte carlo dose calculations with limited beam weight information. *Med Phys* 34: 334–46.

Low D. A., Harms W. B., Mutic S. and Purdy J. A. 1998. A technique for the quantitative evaluation of dose distributions. *Med Phys* 25: 656–61.

Ma C. M. 1998. Characterization of computer simulated radiotherapy beams for Monte-Carlo treatment planning. *Radiat Phys Chem* 53: 329–44.

Ma C. M., Li J. S., Deng J. and Fan J. 2008. Implementation of Monte Carlo Dose calculation for CyberKnife treatment planning. *J Phys: Conf Ser* 102: 012016.

Ma C. M., Li J. S., Pawlicki T. et al. 2002. A Monte Carlo dose calculation tool for radiotherapy treatment planning. *Phys Med Biol* 47: 1671–89.

Ma C. M. C., Chetty I. J., Deng J. et al. 2020. Beam modelling and beam model commissioning for Monte Carlo dose calculation-based radiation therapy treatment planning: report of AAPM Task Group 157. *Med Phys* 47: e1–e18.

Mackeprang P. H., Volken W., Terribilini D. et al. 2016. Assessing dose rate distributions in VMAT plans. *Phys Med Biol* 61: 3208–21.

Mackeprang P. H., Vuong D., Volken W. et al. 2017. Independent Monte-Carlo dose calculation for MLC based CyberKnife radiotherapy. *Phys Med Biol* 63: 015015.

Mackeprang P. H., Vuong D., Volken W. et al. 2019. Benchmarking Monte-Carlo dose calculation for MLC CyberKnife treatments. *Radiat Oncol* 14: 172.

Mackie T. R., Reckwerdt P. and McNutt T. 1996 *Photon Beam Dose Computation*. Madison, WI: Advanced Medical Publishing.

Madani I., Vanderstraeten B., Bral S. et al. 2007. Comparison of 6 MV and 18 MV photons for IMRT treatment of lung cancer. *Radiother Oncol* 82: 63–9.

Magaddino V., Manser P., Frei D. et al. 2011. Validation of the Swiss Monte Carlo Plan for a static and dynamic 6 MV photon beam. *Z Med Phys* 21: 124–34.

Manser P., Frauchiger D., Frei D., Volken W., Terribilini D. and Fix M. K. 2019. Dose calculation of dynamic trajectory radiotherapy using Monte Carlo. *Z Med Phys* 29: 31–8.

Menon S. V., Paramu R., Bhasi S., Gopalakrishnan Z., Bhaskaran S. and Nair R. K. 2020. Dosimetric comparison of iPlan() Pencil Beam (PB) and Monte Carlo (MC) algorithms in stereotactic radiosurgery/radiotherapy (SRS/SRT) plans of intracranial arteriovenous malformations. *Med Dosim* 45: 225–34.

Mihaylov I. B., Lerma F. A., Fatyga M. and Siebers J. V. 2007. Quantification of the impact of MLC modelling and tissue heterogeneities on dynamic IMRT dose calculations. *Med Phys* 34: 1244–52.

Mihaylov I. B. and Siebers J. V. 2008. Evaluation of dose prediction errors and optimization convergence errors of deliverable-based head-and-neck IMRT plans computed with a superposition/convolution dose algorithm. *Med Phys* 35: 3722–7.

Mijnheer B. J., Battermann J. J. and Wambersie A. 1987. What degree of accuracy is required and can be achieved in photon and neutron therapy? *Radiother Oncol* 8: 237–52.

Mijnheer B. J., Battermann J. J. and Wambersie A. 1989. Reply to Precision and accuracy in radiotherapy. *Radiother Oncol* 14: 163–7.

Narayanasamy G., Saenz D. L., Defoor D., Papanikolaou N. and Stathakis S. 2017. Dosimetric validation of Monaco treatment planning system on an Elekta VersaHD linear accelerator. *J Appl Clin Med Phys* 18: 123–9.

Ojala J. J., Kapanen M. K., Hyodynmaa S. J., Wigren T. K. and Pitkanen M. A. 2014. Performance of dose calculation algorithms from three generations in lung SBRT: comparison with full Monte Carlo-based dose distributions. *J Appl Clin Med Phys* 15: 4662.

Oliver M., Gladwish A., Staruch R. et al. 2008. Experimental measurements and Monte Carlo simulations for dosimetric evaluations of intrafraction motion for gated and ungated intensity modulated arc therapy deliveries. *Phys Med Biol* 53: 6419–36.

Onizuka R., Araki F. and Ohno T. 2018. Monte Carlo dose verification of VMAT treatment plans using Elekta Agility 160-leaf MLC. *Phys Med* 51: 22–31.

Ottosson R. O. and Behrens C. F. 2011. CTC-ask: a new algorithm for conversion of CT numbers to tissue parameters for Monte Carlo dose calculations applying DICOM RS knowledge. *Phys Med Biol* 56: N263–N74.

Ove R., Parker C. A., Chilukuri M. B. and Russo S. M. 2015. Comparison of Monte Carlo with pencil beam dosimetry for lung CyberKnife SBRT, correlation with local recurrence. *J Radiat Oncol* 4: 257–63.

Paelinck L., Smedt B. D., Reynaert N. et al. 2006. Comparison of dose-volume histograms of IMRT treatment plans for ethmoid sinus cancer computed by advanced treatment planning systems including Monte Carlo. *Radiother Oncol* 81: 250–6.

Paganetti H., Jiang H., Adams J. A., Chen G. T. and Rietzel E. 2004. Monte Carlo simulations with time-dependent geometries to investigate effects of organ motion with high temporal resolution. *Int J Radiat Oncol Biol Phys* 60: 942–50.

Paganetti H., Jiang H. and Trofimov A. 2005. 4D Monte Carlo simulation of proton beam scanning: modelling of variations in time and space to study the interplay between scanning pattern and time-dependent patient geometry. *Phys Med Biol* 50: 983–90.

Palma B. A., Sanchez A. U., Salguero F. J. et al. 2012. Combined modulated electron and photon beams planned by a Monte-Carlo-based optimization procedure for accelerated partial breast irradiation. *Phys Med Biol* 57: 1191–202.

Pan Y., Yang R., Li J., Zhang X., Liu L. and Wang J. 2018. Film-based dose validation of Monte Carlo algorithm for Cyberknife system with a CIRS thorax phantom. *J Appl Clin Med Phys* 19: 142–8.

Panettieri V., Wennberg B., Gagliardi G., Duch M. A., Ginjaume M. and Lax I. 2007. SBRT of lung tumours: Monte Carlo simulation with PENELOPE of dose distributions including respiratory motion and comparison with different treatment planning systems. *Phys Med Biol* 52: 4265–81.

Partridge M., Trapp J. V., Adams E. J., Leach M. O., Webb S. and Seco J. 2006. An investigation of dose calculation accuracy in intensity-modulated radiotherapy of sites in the head & neck. *Phys Med* 22: 97–104.

Peterhans M., Frei D., Manser P., Aguirre M. R. and Fix M. K. 2011. Monte Carlo dose calculation on deforming anatomy. *Z Med Phys* 21: 113–23.

Pisaturo O., Moeckli R., Mirimanoff R. O. and Bochud F. O. 2009. A Monte Carlo-based procedure for independent monitor unit calculation in IMRT treatment plans. *Phys Med Biol* 54: 4299–310.

Podesta M., Popescu I. A. and Verhaegen F. 2016. Dose rate mapping of VMAT treatments. *Phys Med Biol* 61: 4048–60.

Pokhrel D., McClinton C., Sood S. et al. 2016a. Monte Carlo evaluation of tissue heterogeneities corrections in the treatment of head and neck cancer patients using stereotactic radiotherapy. *J Appl Clin Med Phys* 17: 258–70.

Pokhrel D., Sood S., Badkul R. et al. 2016b. Assessment of Monte Carlo algorithm for compliance with RTOG 0915 dosimetric criteria in peripheral lung cancer patients treated with stereotactic body radiotherapy. *J Appl Clin Med Phys* 17: 277–93.

Rassiah-Szegedi P., Fuss M., Sheikh-Bagheri D. et al. 2007. Dosimetric evaluation of a Monte Carlo IMRT treatment planning system incorporating the MIMiC. *Phys Med Biol* 52: 6931–41.

Reynaert N., Coghe M., De Smedt B. et al. 2005. The importance of accurate linear accelerator head modelling for IMRT Monte Carlo calculations. *Phys Med Biol* 50: 831–46.

Reynaert N., Demol B., Charoy M. et al. 2016. Clinical implementation of a Monte Carlo based treatment plan QA platform for validation of Cyberknife and Tomotherapy treatments. *Phys Med* 32: 1225–37.

Rodriguez M., Sempau J., Baumer C., Timmermann B. and Brualla L. 2018. DPM as a radiation transport engine for PRIMO. *Radiat Oncol* 13: 256.

Rodriguez M., Sempau J. and Brualla L. 2013. PRIMO: a graphical environment for the Monte Carlo simulation of Varian and Elekta linacs. *Strahlenther Onkol* 189: 881–6.

Rogers D. W., Faddegon B. A., Ding G. X., Ma C. M., We J. and Mackie T. R. 1995. BEAM: a Monte Carlo code to simulate radiotherapy treatment units. *Med Phys* 22: 503–24.

Rosu M., Chetty I. J., Balter J. M., Kessler M. L., McShan D. L. and Ten Haken R. K. 2005. Dose reconstruction in deforming lung anatomy: dose grid size effects and clinical implications. *Med Phys* 32: 2487–95.

Sahoo N., Kazi A. M. and Hoffman M. 2008. Semi-empirical procedures for correcting detector size effect on clinical MV x-ray beam profiles. *Med Phys* 35: 5124–33.

Sakthi N., Keall P., Mihaylov I. et al. 2006. Monte Carlo-based dosimetry of head-and-neck patients treated with SIB-IMRT. *Int J Radiat Oncol Biol Phys* 64: 968–77.

Salvat F., Fernandez-Varea J. M. and Sempau J. 2006. *PENELOPE-2006: A code System for Monte Carlo Simulation of Electron and Photon Transport.* Issy-les-Moulineaux, France: OECD Nuclear Energy Agency.

Sarkar V., Stathakis S. and Papanikolaou N. 2008. A Monte Carlo model for independent dose verification in serial tomotherapy. *Technol Cancer Res Treat* 7: 385–92.

Schach von Wittenau A. E., Cox L. J., Bergstrom P. M., Jr., Chandler W. P., Hartmann Siantar C. L. and Mohan R. 1999. Correlated histogram representation of Monte Carlo derived medical accelerator photon-output phase space. *Med Phys* 26: 1196–211.

Schmidhalter D., Fix M. K., Niederer P., Mini R. and Manser P. 2007. Leaf transmission reduction using moving jaws for dynamic MLC IMRT. *Med Phys* 34: 3674–87.

Schmidhalter D., Manser P., Frei D., Volken W. and Fix M. K. 2010. Comparison of Monte Carlo collimator transport methods for photon treatment planning in radiotherapy. *Med Phys* 37: 492–504.

Schwarz M., Bos L. J., Mijnheer B. J., Lebesque J. V. and Damen E. M. 2003. Importance of accurate dose calculations outside segment edges in intensity modulated radiotherapy treatment planning. *Radiother Oncol* 69: 305–14.

Seco J., Adams E., Bidmead M., Partridge M. and Verhaegen F. 2005. Head-and-neck IMRT treatments assessed with a Monte Carlo dose calculation engine. *Phys Med Biol* 50: 817–30.

Seco J., Sharp G. C., Wu Z., Gierga D., Buettner F. and Paganetti H. 2008. Dosimetric impact of motion in free-breathing and gated lung radiotherapy: a 4D Monte Carlo study of intrafraction and interfraction effects. *Med Phys* 35: 356–66.

Sempau J., Sanchez-Reyes A., Salvat F., ben Tahar H. O., Jiang S. B. and Fernandez-Varea J. M. 2001. Monte Carlo simulation of electron beams from an accelerator head using PENELOPE. *Phys Med Biol* 46: 1163–86.

Sempau J., Wilderman S. J. and Bielajew A. F. 2000. DPM, a fast, accurate Monte Carlo code optimized for photon and electron radiotherapy treatment planning dose calculations. *Phys Med Biol* 45: 2263–91.

Sheikh-Bagheri D. and Rogers D. W. 2002a. Monte Carlo calculation of nine megavoltage photon beam spectra using the BEAM code. *Med Phys* 29: 391–402.

Sheikh-Bagheri D. and Rogers D. W. 2002b. Sensitivity of megavoltage photon beam Monte Carlo simulations to electron beam and other parameters. *Med Phys* 29: 379–90.

Shortall J., Vasquez Osorio E., Aitkenhead A. et al. 2020. Experimental verification the electron return effect around spherical air cavities for the MR-Linac using Monte Carlo calculation. *Med Phys* 47: 2506–15.

Siebers J., Keall P. J., Kim J. O. and Mohan R. *XIII International Conference on the Use of Computers in Radiation Therapy,* Heidelberg, 2000, vol. Series, ed W Schlegel and T Bortfeld: Springer-Verlag, pp. 129–31.

Siebers J. V. 2008. The effect of statistical noise on IMRT plan quality and convergence for MC-based and MC-correction-based optimized treatment plans. *J Phys* 102: 12020.

Siebers J. V., Kawrakow I. and Ramakrishnan V. 2007. Performance of a hybrid MC dose algorithm for IMRT optimization dose evaluation. *Med Phys* 34: 2853–63.

Siebers J. V., Keall P. J., Kim J. O. and Mohan R. 2002. A method for photon beam Monte Carlo multileaf collimator particle transport. *Phys Med Biol* 47: 3225–49.

Siebers J. V., Keall P. J., Libby B. and Mohan R. 1999. Comparison of EGS4 and MCNP4b Monte Carlo codes for generation of photon phase space distributions for a Varian 2100C. *Phys Med Biol* 44: 3009–26.

Siebers J. V., Kim J. O., Ko L., Keall P. J. and Mohan R. 2004. Monte Carlo computation of dosimetric amorphous silicon electronic portal images. *Med Phys* 31: 2135–46.

Siebers J. V. and Zhong H. 2008. An energy transfer method for 4D Monte Carlo dose calculation. *Med Phys* 35: 4096–105.

Sikora M. and Alber M. 2009. A virtual source model of electron contamination of a therapeutic photon beam. *Phys Med Biol* 54: 7329–44.

Sikora M., Dohm O. and Alber M. 2007. A virtual photon source model of an Elekta linear accelerator with integrated mini MLC for Monte Carlo based IMRT dose calculation. *Phys Med Biol* 52: 4449–63.

Sikora M., Muzik J., Sohn M., Weinmann M. and Alber M. 2009. Monte Carlo vs. pencil beam based optimization of stereotactic lung IMRT. *Radiat Oncol* 4: 64.

Smedt B. D., Vanderstraeten B., Reynaert N., Gersem W. D., Neve W. D. and Thierens H. 2007. The influence of air cavities within the PTV on Monte Carlo-based IMRT optimization. *J Phys: Conf Ser* 74: 021003.

Smilowitz J. B., Das I. J., Feygelman V. et al. 2015. AAPM Medical Physics Practice Guideline 5.a.: Commissioning and QA of Treatment Planning Dose Calculations - Megavoltage Photon and Electron Beams. *J Appl Clin Med Phys* 16: 14–34.

Smyth G., Evans P. M., Bamber J. C. and Bedford J. L. 2019. Recent developments in non-coplanar radiotherapy. *Br J Radiol* 92: 20180908.

Snyder J. E., Hyer D. E., Flynn R. T., Boczkowski A. and Wang D. 2019. The commissioning and validation of Monaco treatment planning system on an Elekta VersaHD linear accelerator. *J Appl Clin Med Phys* 20: 184–93.

Song J. H., Shin H. J., Kay C. S., Chae S. M. and Son S. H. 2013. Comparison of dose calculations between pencil-beam and Monte Carlo algorithms of the iPlan RT in arc therapy using a homogenous phantom with 3DVH software. *Radiat Oncol* 8: 284.

Su L., Yang Y., Bednarz B. et al. 2014. ARCHERRT - a GPU-based and photon-electron coupled Monte Carlo dose computing engine for radiation therapy: software development and application to helical tomotherapy. *Med Phys* 41: 071709.

Sumida I., Yamaguchi H., Kizaki H. et al. 2015. Novel Radiobiological Gamma Index for Evaluation of 3-Dimensional Predicted Dose Distribution. *Int J Radiat Oncol Biol Phys* 92: 779–86.

Teke T., Bergman A. M., Kwa W., Gill B., Duzenli C. and Popescu I. A. 2010. Monte Carlo based, patient-specific RapidArc QA using Linac log files. *Med Phys* 37: 116–23.

Tian Z., Shi F., Folkerts M., Qin N., Jiang S. B. and Jia X. 2015. A GPU OpenCL based cross-platform Monte Carlo dose calculation engine (goMC). *Phys Med Biol* 60: 7419–35.

Townson R. W. and Zavgorodni S. 2014. A hybrid phase-space and histogram source model for GPU-based Monte Carlo radiotherapy dose calculation. *Phys Med Biol* 59: 7919–35.

Tsuruta Y., Nakata M., Nakamura M. et al. 2014. Dosimetric comparison of Acuros XB, AAA, and XVMC in stereotactic body radiotherapy for lung cancer. *Med Phys* 41: 081715.

Tyagi N., Martin W. R., Du J., Bielajew A. F. and Chetty I. J. 2006. A proposed alternative to phase-space recycling using the adaptive kernel density estimator method. *Med Phys* 33: 553–60.

Ureba A., Salguero F. J., Barbeiro A. R. et al. 2014. MCTP system model based on linear programming optimization of apertures obtained from sequencing patient image data maps. *Med Phys* 41: 081719.

Van Dyk J., Barnett R. B., Cygler J. E. and Shragge P. C. 1993. Commissioning and quality assurance of treatment planning computers. *Int J Radiat Oncol Biol Phys* 26: 261–73.

van Elmpt W., Landry G., Das M. and Verhaegen F. 2016. Dual energy CT in radiotherapy: current applications and future outlook. *Radiother Oncol* 119: 137–44.

Vanderstraeten B., Chin P. W., Fix M. et al. 2007. Conversion of CT numbers into tissue parameters for Monte Carlo dose calculations: a multi-centre study. *Phys Med Biol* 52: 539–62.

Vanderstraeten B., Reynaert N., Paelinck L. et al. 2006. Accuracy of patient dose calculation for lung IMRT: a comparison of Monte Carlo, convolution/superposition, and pencil beam computations. *Med Phys* 33: 3149–58.

Verhaegen F. and Devic S. 2005. Sensitivity study for CT image use in Monte Carlo treatment planning. *Phys Med Biol* 50: 937–46.

Verhaegen F. and Seuntjens J. 2003. Monte Carlo modelling of external radiotherapy photon beams. *Phys Med Biol* 48: R107–64.

Volken W., Frei D., Manser P., Mini R., Born E. J. and Fix M. K. 2008. An integral conservative gridding--algorithm using Hermitian curve interpolation. *Phys Med Biol* 53: 6245–63.

Walters B. R., Kawrakow I. and Rogers D. W. 2002. History by history statistical estimators in the BEAM code system. *Med Phys* 29: 2745–52.

Wang L., Chui C. S. and Lovelock M. 1998. A patient-specific Monte Carlo dose-calculation method for photon beams. *Med Phys* 25: 867–78.

Wang L., Lovelock M. and Chui C. S. 1999. Experimental verification of a CT-based Monte Carlo dose-calculation method in heterogeneous phantoms. *Med Phys* 26: 2626–34.

Wang L., Yorke E. and Chui C. S. 2002a. Monte Carlo evaluation of 6 MV intensity modulated radiotherapy plans for head and neck and lung treatments. *Med Phys* 29: 2705–17.

Wang L., Yorke E., Desobry G. and Chui C. S. 2002b. Dosimetric advantage of using 6 MV over 15 MV photons in conformal therapy of lung cancer: Monte Carlo studies in patient geometries. *J Appl Clin Med Phys* 3(1): 51–9.

Wang S., Gardner J. K., Gordon J. J. et al. 2009. Monte Carlo-based adaptive EPID dose kernel accounting for different field size responses of imagers. *Med Phys* 36: 3582–95.

Wang Y., Mazur T. R., Green O. et al. 2016. A GPU-accelerated Monte Carlo dose calculation platform and its application toward validating an MRI-guided radiation therapy beam model. *Med Phys* 43: 4040.

Wang Y., Mazur T. R., Park J. C., Yang D., Mutic S. and Li H. H. 2017. Development of a fast Monte Carlo dose calculation system for online adaptive radiation therapy quality assurance. *Phys Med Biol* 62: 4970–90.

Webster G. J., Rowbottom C. G. and Mackay R. I. 2007. Development of an optimum photon beam model for head and-neck intensity-modulated radiotherapy. *J Appl Clin Med Phys* 8(4): 129–38.

Yamamoto T., Mizowaki T., Miyabe Y. et al. 2007. An integrated Monte Carlo dosimetric verification system for radiotherapy treatment planning. *Phys Med Biol* 52: 1991–2008.

Yang J., Liu G., Liu H. Y. et al. 2020. Influence of CyberKnife Prescription Isodose Line on the Discrepancy of Dose Results Calculated by the Ray Tracing and Monte Carlo Algorithms for Head and Lung Plans: a Phantom Study. *Curr Med Sci* 40: 301–6.

Yang Y. M., Svatos M., Zankowski C. and Bednarz B. 2016. Concurrent Monte Carlo transport and fluence optimization with fluence adjusting scalable transport Monte Carlo. *Med Phys* 43: 3034–48.

Zhao Y., Qi G., Yin G. et al. 2014. A clinical study of lung cancer dose calculation accuracy with Monte Carlo simulation. *Radiat Oncol* 9: 287.

Zhao Y. L., Mackenzie M., Kirkby C. and Fallone B. G. 2008. Monte Carlo evaluation of a treatment planning system for helical tomotherapy in an anthropomorphic heterogeneous phantom and for clinical treatment plans. *Med Phys* 35: 5366–74.

Zhong H. and Siebers J. V. 2009. Monte Carlo dose mapping on deforming anatomy. *Phys Med Biol* 54: 5815–30.

11

Patient Dose Calculation

Joao Seco
DKFZ, German Cancer Research Center and University of Heidelberg

Maggy Fragoso
Alfa-Comunicações

11.1 General Introduction

There is a large variety of Monte Carlo (MC) codes available now in radiation therapy for dose calculation of which the most widely used are EGSnrc (Kawrakow, 2000), MCNP (Brown, 2003), GEANT4 (Agostinelli, 2003), and PENELOPE (Baro et al., 1995; Salvat et al., 2009). All these algorithms are based on the condensed history techniques first developed by Berger (1963), where a large number of electron interactions are condensed into one because electrons lose very little energy in a single electromagnetic interaction. Condensed history implementation was divided into two main classes. In the class I condensed history approach, all collisions are subject to grouping, with the effect of secondary particle creation above a specified threshold energy taken into account after the fact (i.e., independent of the energy actually lost during the step). This is performed by transporting a number of secondary particles. MCNP is an example of a class I type MC code, where the system adopted the electron transport algorithm from ETRAN (Berger and Seltzer, 1973; Seltzer, 1988). This algorithm was developed by Berger and Seltzer at NIST, following the condensed history techniques proposed by Berger (1963). In the class II condensed history approach, interactions are divided into "hard" (also designated as "catastrophic") and "soft" collisions. Soft collisions are grouped in class I scheme, whereas for hard collisions, the particles are tracked individually as independent particles. EGSnrc, GEANT4, and PENELOPE are examples of class II MC simulations.

An important aspect of condensed history is the simulation of a particle crossing a boundary between different regions. A lot of work has gone into MC research of boundary crossing and dependency of calculated results on step size. Step-size artifacts and boundary-crossing issues are now well understood (Larsen, 1992; Kawrakow and Bielajew, 1998; Kawrakow, 2000) and have led to a significant improvement in the accuracy of MC code predictions. For further details of condensed history techniques, consult Chetty et al. (2007) and the literature within.

11.2 Statistical Uncertainties in Patient Dose Calculation

11.2.1 Introduction

The use of MC simulation techniques in radiation therapy planning introduces unavoidable statistical noise in the calculated dose because of the fluctuation of the dose around its mean value in a given volume element (usually termed voxel). This can have a profound effect on the information extracted from the dose distributions, namely, the isodose lines and dose-volume histograms (DVHs). These statistical fluctuations are characterized by the variance, σ^2, on the calculated quantity of interest, which is an estimate of the true variance, σ^2. In an ideal MC simulation, a zero variance is desirable, but it is unachievable within a finite amount of time.

DOI: 10.1201/9781003211846-14

The performance of an MC calculation is characterized by its efficiency, e, which combines the estimated variance with the number of iterations, N, required to achieve that variance, within a given amount of time. In other words, the efficiency of an MC simulation (also called "figure of merit") is an indicator of how "fast" the MC code is achieving a specific σ^2 value. It is defined as follows:

$$\varepsilon = \frac{1}{\sigma^2 T}, \qquad (11.1)$$

where σ^2 is an estimate of the true variance on the quantity of interest, and T is the central processing unit (CPU) time required to achieve this variance. The efficiency of an MC algorithm can be improved by either decreasing the σ^2 value for a given T or by decreasing T for a given N while not changing the variance. Variance reduction techniques (VRTs) are used to reduce the σ^2 value without changing the number of iterations used, N. The VRTs use "clever" physics, mathematics, or numerical "tricks" to accelerate the overall MC algorithm calculation time, thus increasing its efficiency.

11.2.2 Dose-Scoring Geometries and Calculation of Uncertainties

Different types of scoring geometries are used for three-dimensional (3D) dose calculations, one cubic or parallelepiped voxels being the most common. Moreover, the dose is usually scored in the geometrical voxels that are based on the computed tomography (CT) information. One problem that arises is that decreasing the CT resolution enormously increases the memory usage. One alternative that is being currently adopted is to decouple the scoring grid from the geometrical grid. In the following paragraphs, some examples of scoring geometries are presented, followed by some methods of uncertainty calculation.

11.2.2.1 Dosels

In the PEREGRINE MC dose engine, the particle transport is performed in a patient transport mesh, which is a Cartesian map of material composition and density, obtained from the patient's CT images. The dose is scored in an array of overlapping spheres that are independent of the material transport mesh, termed dosels. During the dose calculation, the standard deviation in the dosel receiving the highest dose is tracked, and when it reaches a level specified by the user, the simulation is terminated.

11.2.2.2 Kugels

The electron macro MC (MMC) dose engine calculates the dose distribution from predetermined electron histories that are stored in lookup tables, which were obtained in spherical volume elements, termed kugels, of different materials. The concept was initially introduced by Neuenschwander et al. (1995) for simulating electron beam treatment planning.

The MMC algorithm uses results derived from conventional MC simulations of electron transport through macroscopic spheres of various radii and consisting of a variety of media. On the basis of kugel spheres, the electrons are transported in macroscopic steps through the absorber. The energy loss is scored in the 3D dose matrix. The transport of secondary electrons and bremsstrahlung is also accounted for in the model.

11.2.2.3 Segmented Organs

MC dose scoring can be significantly improved if many of the voxels are combined into a large volume, that is, an organ. In this case, we lose the spatial resolution of the dose distribution, but we gain in the ability to predict organ dose very rapidly and accurately with MC. However, contouring CT volumes into organs can be very subjective and depends significantly on the clinician performing the contouring. Therefore, segment organ dose scoring can only accurately be applied in controlled simulation environments such as the NCAT phantom (Segars, 2001). The NCAT anthropomorphic computational phantom is based on the Visible Human dataset (Spitzer and Whitlock, 1998). The phantom uses nonuniform rational B-spline surfaces to represent 3D human anatomy and also allows the incorporation of four-dimensional (4D) motion of the cardiac and respiratory motions. The anatomical parameters of the phantom are set using an input file so that the anatomy parameters can be scaled to dimensions and values different to that of the default settings based on the Visible Human. Such details include the level of the desired anatomical detail for the blood vessels and lung airway. The dynamic nature of the phantom is invoked by specifying whether respiratory and/or cardiac motion is to be included.

MC simulations using NCAT have been performed allowing both voxel scoring and organ scoring by Riboldi et al. (2008). A 4D MC framework was developed to study dose variations with motion, within a controlled environment of a phantom. Results show that the MC simulation framework can model tumor tracking in deformable anatomy with high accuracy, providing absolute doses for intensity-modulated radiotherapy (IMRT) and conformal radiation therapy. Badal et al. (2009) developed a realistic anatomical phantom based on NCAT on which MC simulation may be performed for imaging applications, from image-guided treatments to portal imaging. In this case, the scoring volumes of the organs were segmented into triangular meshes.

11.2.2.4 Voxel Size Effects

The scoring voxel size has a large impact on the calculation time because of the amount of the time spent in geometry boundary checks and scoring. Cygler et al. (2004) showed that the number of histories to be simulated, per unit area, is linearly proportional to the mass of the scoring voxels.

11.2.2.5 Concept of Latent Variance

There are generally two sources of statistical uncertainty in the MC calculations of patient dose: those resulting from the simulation of the accelerator treatment head and those arising from fluctuations in the phantom/patient dose calculation. Sempau et al. (2001) coined the term "latent variance" to describe the uncertainty due to the statistical fluctuation in the phase space generated from the simulation of the treatment head as opposed to the uncertainty due to the random nature of the dose calculation using MC techniques.

The statistical uncertainty in the dose calculated in a phantom by reusing the particles from the phase space file and assuming that the particles are independent and ignoring the correlations between them will approach the finite, latent variance associated with the phase space data, regardless of the number of times the phase space is used.

More importantly, in estimating the statistical uncertainty in the patient dose calculation, it is necessary to account for the latent variance from the phase space calculation and for the random uncertainty from the patient calculation. Should latent variance be a significant factor in the total uncertainty, more independent phase space particles need to be used in the patient simulations.

11.2.2.6 Batch Method

In the batch method, the estimate of the uncertainty, $s_{\bar{X}}$, of a scored quantity, X, is given as follows:

$$s_{\bar{X}} = \sqrt{\frac{\sum_{i=1}^{N}(X_i - \bar{X})^2}{N(N-1)}}, \qquad (11.2)$$

where N is the number of batches, usually 10; X_i is the value of X in batch I; and $\bar{X}$ is the mean value of X evaluated over all batches. The sample size, N, is thus given by the number of batches.

As pointed out by Walters et al. (2002), three main problems can be identified with this approach for the uncertainty calculation:

1. A large number of batches must be used or else there will be significant fluctuations in the uncertainty itself because the sample size, N, is quite small.
2. Arbitrary grouping of histories into batches ignores any correlations between incident particles.
3. The batch approach adds an extra dimension to the scoring quantities of interest.

11.2.2.7 History-by-History Method

The history-by-history method, implemented in accordance with the Salvat's approach (Salvat et al., 2009), allows the elimination of the above-mentioned problems when using the batch method to estimate the uncertainty in the MC dose calculation. The scored dose data are stored in a very efficient way, while the simulation is running. In this method, X_i now represents the scored quantity in history i rather than in batch i, and N is the number of independent events. Equation 11.2 can then be rewritten as follows:

$$s_{\bar{X}} = \sqrt{\frac{1}{N(N-1)}\left(\frac{\sum_{i=1}^{N}X_i^2}{N} - \left(\frac{\sum_{i=1}^{N}X_i}{N}\right)^2\right)}. \qquad (11.3)$$

When using phase space sources to calculate the dose distribution, one history is defined to be all particle tracks associated with one initial particle. It should also be emphasized that the quantity X_i may be weighted quantities if variance reduction techniques are used.

During the MC calculation, that is, on the fly, a record of the quantities $\sum_{i=1}^{N}X_1^2$ and $\sum_{i=1}^{N}X_i$ is kept. At the end of the calculation, the uncertainty of the calculation is determined in accordance with Equation 11.3, without the need to store the scored quantity in batches. More details on the algorithm implementation can be found in a study by Sempau et al. (2000) and Salvat et al. (2009).

11.3 Denoising and Smoothing Methods

11.3.1 Introduction

As previously mentioned, one possible way to reduce the statistical fluctuations is by performing the MC simulation for a very large number of histories, which is generally not practical without the application of powerful variance reduction techniques. Another possibility is the process of removing or reducing the statistical fluctuations from a noisy calculated data with smoothing or denoising algorithms that are routinely applied in a variety of disciplines that deal with noisy signals (e.g., imaging), thus speeding up the MC calculations.

Analogous to image restoration problems in the field of image processing, an improved estimate of the true image can be produced by smoothing or denoising the raw MC result, produced with fewer source particles than needed with the denoising process, resulting in an accelerated image of the true dose distribution. Obviously, the goal of MC denoising algorithms is to be as aggressive as possible in locally smoothing the raw MC result while attempting to avoid the introduction of systematic errors—bias—especially near sharp features such as beam edges.

The work of Keall et al. (1999 and 2000), Jeraj and Keall (1999), and Buffa et al., (1999) have been seminal in the study of the effect of statistical uncertainty on the evaluation of MC plans and the possibility of removing these effects on MC-calculated DVHs.

Various methods related to image digital filtering, wavelet thresholding, adaptive anisotropic diffusion, and denoising based on a minimization problem have been proposed to solve this problem. These algorithms are an approximate efficiency-enhancing method as they can introduce a systematic bias into the calculation. Nonetheless, denoising techniques are useful as they can reduce the overall (systematic and random) uncertainty when the random component decreases more than the systematic component increases. However, it must be emphasized that the denoising techniques require proper validation under the full range of clinical circumstances before they can be used with MC dose algorithms.

Kawrakow (2002) introduced five accuracy criteria, which have become a standard, to evaluate the performance of smoothing algorithms, where smoothed and unsmoothed MC calculations must be compared to a "benchmark result" that has been obtained through the simulation of a large number of histories:

1. Visual inspection of the isodose lines, where the differences between the benchmark and the smoothed results should be minimal.
2. DVHs. The difference area between two DVHs, one corresponding to the smoothed data and another to the benchmark data, should be quantified. This difference should be small.
3. Maximum dose difference. The maximum dose difference to the benchmark for smoothed and not smoothed MC simulations should be obtained.
4. Root-mean-square difference (RMSD). This is a standard quantitative measure for the degree of agreement between two distributed quantities. The RMSD in an MC calculation without smoothing is solely due to the statistical uncertainties and, after smoothing, it will contain some systematic bias introduced by the denoising process. RMSD should be obtained between the benchmark and the smoothed and unsmoothed dose distribution.
5. $x\%/y$ mm test. The fraction of voxels with a smoothed dose value that differs more than $x\%$ from a benchmark dose value at the same point and there is no point in the benchmark dose distribution that is closer than y mm to the point that has the same dose value (Van Dyk, 1993). The current accuracy recommendation is 2%–3%/2 mm (Fraass et al., 1998).

11.3.2 Denoising Integrated Dose Tallies

11.3.2.1 DVH Denoising Methods

Two groups, Sempau and Bielajew (2000) and Jiang et al. (2000), have introduced techniques for denoising the DVH.

Sempau and Bielajew (2000) used a deconvolution method, where the calculated DVH was considered as the "true" DVH, which would be obtained after an infinite number of histories, thus yielding zero variance, convolved by noise. The implications of this approach would be that decisions based on DVHs could be made quickly, and, more importantly, inverse treatment planning or even optimization methods could be carried out using MC dose calculations at all stages of the iterative process, as the long calculation times at the intermediate calculation steps could be eliminated.

In the work presented by Jiang et al. (2000), the MC-calculated DVH is treated as blurred from the noiseless and "true" DVH. The technique described is similar to image restoration, where a deblurring function is used to obtain the noiseless DVH. An estimate of the smoothed DVH is blurred, and the difference between this image and the blurred and original image is minimized with a least-square minimization method iteratively.

11.3.3 Dose Distributions Denoising Methods

Denoising of DVHs cannot completely address the problem of statistical fluctuations as the dose distributions are frequently assessed and represented by other means (e.g., isodose lines, calculation of the tumor control, and/or normal tissue complication probabilities). This makes the removal of statistical uncertainties from the dose distribution itself more desirable and relevant.

11.3.3.1 Deasy Approach

The first to propose denoising of the 3D dose distribution was Deasy (2000), who suggested that MC results can be obtained from an infinite number of source particles and a noise source due to the statistics of particle counting. He proposed several digital filtering techniques to denoise electron beam MC dose distributions and demonstrated that these techniques improve the visual usability and the clinical reliability of MC dose distributions.

MC-calculated dose distributions with high statistical uncertainty were subjected to various digital filters, and a comparison was then performed between the resulting denoised distributions and calculations with much lower uncertainty using isoline representations. The concluding remarks were that smoothing with digital filters was a promising technique for dealing with the statistical noise of MC simulations. The main objection against the use of denoising for MC-calculated dose distributions results from the concern that smoothing may systematically alter its distribution.

11.3.3.2 Wavelet Approach

Deasy et al. (2002) presented another approach where wavelet threshold denoising was used to accelerate the radiation therapy MC simulations by factors of 2 or more. The dose distribution is separated into a smooth function and noise, represented by an array of dose values that are, subsequently, transformed into discrete wavelet coefficients. Below a positive threshold of the wavelet coefficient values, they were set equal to zero. They showed that with a suitable value of this threshold parameter, the statistical noise was suppressed with little introduction of bias.

11.3.3.3 Savitzky–Golay Method

Kawrakow (2002) presented a 3D generalization of a Savitzky–Golay filter with an adaptive smoothing window size, that is, the number of surrounding voxels that is used, to reduce the probability for systematic bias. The size of the smoothing window is based on the statistical uncertainty in the voxel that is being smoothed. According to the author, this filter decreases the number of particle histories by factors of 2 to 20, concluding that smoothing is extremely valuable for the initial trial and error phase of the treatment planning process.

11.3.3.4 Diffusion Equation Method

Miao et al. (2003) investigated the denoising of MC dose distributions with a 3D adaptive anisotropic diffusion method. The conventional anisotropic diffusion method was extended by adaptively modifying the filtering parameters in accordance with the local statistical noise. The work presented showed that this method can reduce statistical noise significantly, that is, 2–5 times, corresponding to a reduction in the simulation time by a factor of up to 20, while maintaining the characteristics of the gradient dose areas.

11.3.3.5 IRON Method

Fippel and Nusslin (2003) introduced the so-called iterative reduction of noise (IRON) algorithm, which iteratively reduces the statistical noise for MC dose calculations. By varying the dose in each voxel, this algorithm minimizes the second partial derivatives of the dose with respect to the three spatial coordinates. This algorithm requires MC dose distributions with or without known statistical uncertainties as the input.

Smoothing of the MC dose distributions using IRON was proven to lead to an additional reduction of the MC calculation time by factors between 2 and 10. As with the application of the other smoothing techniques, this reduction is particularly useful if the MC dose calculation is part of an inverse treatment planning calculation.

11.3.3.6 Content Adaptive Median Hybrid Filter Method

In this method, linear filters were combined through a weighted sum, with the median operation to produce hybrid median filters (El Naqa et al., 2005). In regions with strong second derivative features, the median operator is used, and in smoother regions, the mean operator is preferred. In general, median filters outperform the moving average and other linear filters in the process of removing impulsive noise (outliers) and in the preservation of edges; however, they fail to provide the same degree of smoothness in homogenous regions. An adaptive combination with mean value linear filters was then chosen in this study.

11.4 CT to Medium Conversion Methods

11.4.1 CT Stoichiometric Conversion Methods

The conversion of CT HUs into material composition and mass density may strongly influence the accuracy of patient dose calculations in MC treatment planning or in any other dose calculation algorithm. To establish an accurate relationship between CT HU and electron density of a tissue, the CT scale is usually divided into a number of subsets. The initial MC dose algorithms used six or fewer materials to define the conversion, for example, air, lung, fat, water, muscle, and bone (DeMarco et al., 1998; Ma et al., 1999). The influence of the differences in tissue composition on MC computation was investigated by du Plessis et al. (1998) for a set of 16 human tissues. They assessed the optimal number of tissue subsets to achieve 1% dose accuracy for millivolt photon beams and concluded that seven basic tissue subsets were sufficient. By varying the physical mass density for the lung and the cortical bone, 57 different tissue types (excluding air) could be constructed, spanning a total CT range of 3,000 HU. The division adopted by du Plessis et al. was 21 types of cortical bone ranging from 1,100 to 3,000 HU and 31 different lung tissues ranging from 20 to 950 HU. The remaining five types were chosen from the basic tissue subsets. The disadvantage of using the method proposed by du Plessis is that it does not use a stoichiometric conversion scheme and therefore is limited to the beam quality under consideration.

Stoichiometric conversion schemes were a large step forward in the accuracy of MC algorithms for patient dose calculations relative to the methods discussed previously. The stoichiometric conversion scheme from CT to electron (or proton stopping power) was initially proposed by Schneider et al. (1996) for proton MC dose algorithms. Initially, a set of materials of well-known atomic composition and density (usually plastic materials of varying composition and density to mimic human tissues) are CT scanned to measure the corresponding CT HU values. Next, the measured CT HU values are fitted by a theoretical equation interrelating the CT HU, mass density, atomic number Z, and atomic weight A of each material. Finally, the fitted parameters can be used to calculate the CT HU values for real tissues using tabulated composition data.

Schneider et al. (2000) proposed a CT conversion based on the stoichiometric conversion scheme, considering 71 human tissues, and they adopted the total cross section parameterization given by Rutherford et al. (1976). To reduce the effort of fitting the CT HU numbers to mass density and elemental weights, the authors created four sections on the CT scale within which the selected tissues were confined. Within each section, both the mass density and elemental weights of the tissues were interpolated.

11.4.1.1 Calculation of CT HU Number for Stoichiometric Conversion Schemes

In a study by Rutherford et al. (1976), a parameterization of the cross section in the diagnostic X-ray energy range is presented:

$$\sigma_i(E) = Z_i K^{KN}(E) + Z_i^{2.86} K^{sca}(E) + Z_i^{4.62} K^{ph}(E), \quad (11.4)$$

where E is the energy; $\sigma_i(E)$ is the total cross section in units of barn/atom; K^{KN} denotes Klein–Nishina coefficient; K^{sca} is the Rayleigh or coherent scattering coefficient; and K^{ph} is the photoelectric coefficient. The linear attenuation coefficient, μ, is now obtained by replacing the total cross section:

$$\mu(E) = \rho N_A \sum_{i=1}^{n} \frac{w_i}{A_i} \sigma_i(E)$$

$$= \rho N_A \sum_{i=1}^{n} \frac{w_i}{A_i} \left[Z_i K^{KN}(E) + Z_i^{2.86} K^{sca}(E) + Z_i^{4.62} K^{ph}(E) \right], \quad (11.5)$$

where ρ (g/cm³) is the mass density; N_A (mol⁻¹) is the Avogadro constant (6.022045×10^{23}); i is the element index; and w_i is the element weight. Recall that the μ values are converted to HU using the below formula:

$$HU = \left(\frac{\mu}{\mu_{H_2O}} - 1 \right) 1,000, \quad (11.6)$$

where the CT HU is defined such that water has the value 0, and air has the value 1,000. The equation then becomes as follows:

$$\frac{\mu}{\mu_{H_2O}} = \frac{\rho}{\rho_{H_2O}} \frac{\sum_{i=1}^{n} \frac{w_i}{A_i} \left(Z_i + Z_i^{2.86} k_1 + Z_i^{4.62} k_2 \right)}{\frac{w_H}{A_H}(1 + k_1 + k_2) + \frac{w_O}{A_O}\left(8 + 8^{2.86} k_1 + 8^{4.62} k_2 \right)}, \quad (11.7)$$

with $k = K^{sca}/K^{KN}$ and $k_2 = K^{ph}/K^{KN}$, and the values of (k_1, k_2) are dependent on the CT kilovolt value used for imaging and are determined experimentally by a least-square fit of the measured HU values to the previous equation, that is, minimizing the following expression:

$$\sum_{i=1}^{n} \left[\left(\frac{\mu}{\mu_{H_2O}} \right)(k_1, k_2) - \left(\frac{H(\text{meas.})}{1,000} + 1 \right) \right]^2. \quad (11.8)$$

Additional predictions of the k_1 and k_2 values are also presented by Vanderstraeten et al. (2007) for a large group of CT scanners of different models and manufacturers for several European radiation oncology centers.

11.4.1.2 CT Hounsfield Units Interpolation Method

In this section, we present a brief description of the method of interpolating within a tissue range for tissue samples composed of only two components by Schneider et al., where the components are denoted as $(\rho_1, w_{1,i}, H_1)$ and $(\rho_2, w_{2,i}, H_2)$ and the composite by (ρ, w_i, H) and where H_1, H_2 represent CT HU values range and where $H_1 < H_2$. In addition, W_1 represents the proportion of the first component, whereas the second is obtained from $W_2 = 1 - W_1$. A brief note on the meaning of W_1 and W_2 and how they differ from $w_{1,i}$ and $w_{2,i}$ can be given as follows: while $w_{1,i}$ and $w_{2,i}$ represent the weighting within a tissue of the chemical elements such as oxygen, hydrogen, and so on, W_1 and W_2 represent the weighting of two reference tissues to form a new "interpolated" tissue such as combining osseous tissue and bone marrow to form different skeletal tissues. The new interpolated medium has the following weight per chemical element (w_i) and mass density (ρ), respectively:

$$w_i = W_1 w_{1,i} + W_2 w_{2,i} = W_1 \left(w_{1,i} - w_{2,i} \right) + w_{2,i} \quad (11.9)$$

$$\rho = \frac{m}{V} = \frac{m}{(m_1/\rho_1) + (m_2/\rho_2)}$$

$$= \frac{1}{(W_1/\rho_1) + (W_2/\rho_2)} = \frac{\rho_1 \rho_2}{W_1(\rho_2 - \rho_1) + \rho_1}. \quad (11.10)$$

The linear attenuation coefficient of the new media can now be generated by substituting Equations 11.9 and 11.10 in Equation 11.5:

$$\mu = \frac{\rho_1 \rho_2}{W_1(\rho_2 - \rho_1) + \rho_1}$$
$$N_A \left[W_1 \sum_i \left(\frac{w_{1,i} - w_{2,i}}{A_i} \sigma_i \right) + \sum_i \left(\frac{w_{2,i}}{A_i} \sigma_i \right) \right] \quad (11.11)$$

$$= \frac{\rho_1 \rho_2}{W_1(\rho_2 - \rho_1) + \rho_1} N_A \left[W_1 \left(\frac{\mu_1}{\rho_1} - \frac{\mu_2}{\rho_2} \right) + \frac{\mu_2}{\rho_2} \right]. \quad (11.12)$$

The composite tissue with H, such that $H_1 < H < H_2$, is then given by the following:

$$W_1 = \frac{\rho_1(H_2 - H)}{(\rho_1 H_2 - \rho_2 H_1) + (\rho_2 - \rho_1)H}. \quad (11.13)$$

The composite medium will then have the following mass density and elemental composition:

$$\rho = \frac{\rho_1 H_2 - \rho_2 H_1 + (\rho_1 - \rho_2)H}{H_2 - H_1} \quad (11.14a)$$

$$w_i = \frac{\rho_1(H_2 - H_1)}{(\rho_1 H_2 - \rho_2 H_1) + (\rho_2 - \rho_1)H}(w_{1,i} - w_{2,i}) + w_{2,i},$$

$$H_1 \le H \le H_2 \quad (11.14b)$$

11.4.1.3 Stoichiometric Conversion Method Based on Dose-Equivalent Tissue Subsets

Vanderstraeten et al. (2007) generalized the Schneider et al. (2000) method to generate subsets of tissue materials that were dosimetrically equivalent to within 1%. Their key assumption was that in the case of tissues that are situated close to the interface between two subsets, the dose differences between them should not exceed 1%. In a previous study by Verhaegen and Devic (2005), the authors had showed the dosimetric importance of the CT calibration curve "tissue-binning" approach and the large dosimetric impact of tissue substitutes containing fluorine. Vanderstraeten et al. (2007) proposed the use of 14 dosimetrically equivalent tissue subsets (bins), where 10 were bone, to significantly reduce tissue-binning effects. The study was performed for a large number of different CT scanners obtained from nine different radiotherapy departments across Europe. These were then compared to the conventional five-bin scheme with only one bone bin. In Figure 11.1, the MC dose prediction for 5 and 14 bins of the CT is presented, where it is possible to see 5% differences at the interface.

The 10 bone tissue subsets used were generated to accurately model both the hydrogen (H) and calcium (Ca) content from 100 HU to beyond 1,450 HU, consisting of cortical bone or denser bone. Both H and Ca contents strongly influence photon attenuation as pointed out by Seco and Evans (2006). In addition, the Ca content at the diagnostic energies will strongly increase the importance of the photoelectric and Rayleigh scattering terms because of their strong Z-dependence. The 14- and 5-bin approaches were then studied for a cohort of nine patients, where dosimetric differences

of up to 5% were observed. However, DVH comparisons did not show any significant differences.

11.4.2 Dual-Energy X-Ray CT Imaging: Improved HU CT to Medium Conversions

Dual-energy X-ray CT imaging involves scanning an object with two significantly different tube voltages to obtain a better estimate of the effective atomic numbers, Z, and the relative electron densities, ρ_e. In a study by Torikoshi et al. (2003), the linear attenuation coefficient of a material at a set X-ray energy, E, is given by the following:

$$\mu(E) = \rho'_e \left(Z^4 F(E, Z) + G(E, Z) \right), \qquad (11.15)$$

Where ρ'_e is the electron density and $\rho'_e Z^4 F(E, Z)$ and $\rho'_e G(E, Z)$ are the photoelectric and combined Rayleigh and Compton scattering terms of the linear attenuation coefficient, respectively. Both F and G are obtained by quadratic fits of the photoelectric and scattering terms of NIST attenuation coefficients (Berger et al., 2005).

A detailed method of performing dual-energy tissue segmentation is presented by Bazalova et al. (2008). Dual-energy CT material extraction was performed using the 100 and 140 kVp CT images with corresponding X-ray spectra. The authors found that the mean errors of extracting ρ'_e and Z are 1.8% and 2.8%, respectively. Dose calculations were then performed using various photon beam energies of 250 kVp, 6 MV, and 18 MV, and 18 MeV electron beams and a solid water phantom with tissue-equivalent inserts. The dose calculation

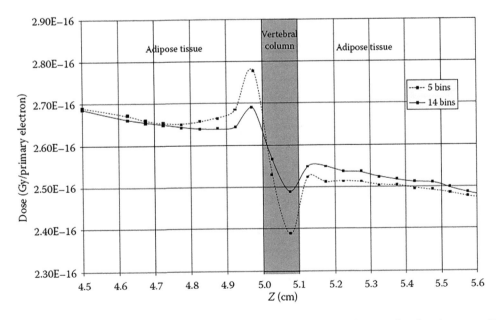

FIGURE 11.1 Depth dose curve with a slab phantom calculated with MC for an interface between the adipose tissue and bone (thickness 1 mm). The results compare the five-bin CT with 14-bin CT proposed by Vanderstraeten et al. (2007). (Courtesy of Vanderstraeten et al. 2007. Conversion of CT numbers into tissue parameters for MC dose calculations: A multicenter study. *Phys. Med. Biol.* 52, 539–62.)

errors were particularly high for the 250 kVp beam, leading to 17% error due to a misassigned soft bone tissue-equivalent cylinder. In the case of the 18 MeV electron beam and 18 MV photon beams, they found 6% and 3% dose calculation errors due to misassignment. In the case of the 6 MV electron beam, dose differences were less than 1%.

11.5 Deformable Image Registration

(A brief overview of deformable image registration is given; for more detailed description, please consult the Heath, Seco and Popescu, Chapter 9).

The use of deformable image registration varies from 4D radiotherapy to contour propagation, treatment adaptation, dosimetric evaluation, and 4D optimization (Keall et al., 2004; Rietzel et al., 2005; Trofimov et al., 2005; Brock et al., 2006). A variety of different approaches exist to the nonrigid registration; Wang et al. (2005) used an accelerated Demons algorithm, and Rietzel and Chen (2006) used a B-spline method that respects discontinuity at the pleural interface. A variety of other methods exist, which use optical flow, thin-plate spline, calculus of variations, and finite element methods with different motion algorithms (Meyer et al., 1997; Guerrero et al., 2004; Lu et al., 2004).

11.5.1 Developing Dose Warping Approaches for 4D MC

Keall et al. (2004) performed one of the first "pseudo" 4D MC dose calculations. The method used MC to calculate dose on each of the N (= 8) individual 3D CT cubes representing an individual breathing phase. The dose distribution of each breathing phase was then mapped back to the end-inhale CT image set, using deformable image registration. The flowchart presented in Figure 11.2 represents the various steps of performing 4D MC dose calculation with deformable image registration. There are several stages to the process: from deforming contours of the anatomy from the reference phase to all CT datasets to performing dose warping from the CT of the various breathing phases back to the reference dataset.

Deformable image registration has mostly been used in lung cancer dosimetric studies with MC (Flampouri et al., 2005; Rosu et al., 2005; Seco et al., 2008). A major issue with deformable registration algorithms is that no true one-to-one correspondence exists between the initial dataset (image or dose) and the registered dataset. This is a consequence of the initial voxels being rearranged (split or merged) to produce a new registered dataset. This occurs because registration algorithms either are statistical in nature or involve interpolation methods. Heath and Seuntjens (2006) developed a method to circumvent errors introduced by deformable image registration mapping dose distributions. They developed the MC code defDOSXYZnrc, which uses a vector field to directly warp the dose grid on the fly while tracking individual particles. The major disadvantage of this method was the increased calculation time of the MC dose calculation by a factor of 2 or more.

In a more recent study by Siebers and Zhong (2008), a different approach was adopted to circumvent dose warping, while maintaining a fast MC dose calculation algorithm. The authors designated their method as the energy transfer method (ETM), where the separation between radiation transport and energy deposition is performed. The authors showed

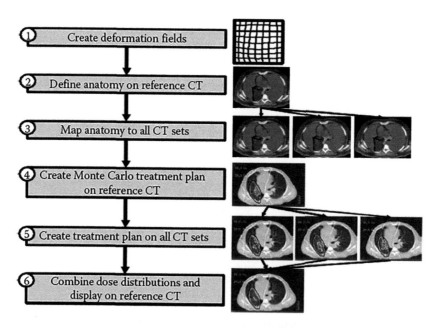

FIGURE 11.2 A flowchart of the 4D radiotherapy planning process using MC. (Courtesy of Keall P. et al. 2004. MC as a 4D radiotherapy treatment-planning tool to account for respiratory motion. *Phys. Med. Biol.* 49, 3639–48.)

that some compensation mechanism must be provided that accounts for the rearrangement of the energy deposited per unit mass, particularly in the dose gradient regions. As an example of the ETM approach, consider two adjacent voxels $A1$ and $A2$ that merge into one voxel $X1$, following image registration. The dose interpolation method calculates the dose, $d(X1)$, from $d(A1)$ and the following:

$$d^{\text{INTERP}}(X1) = \frac{d(A1) + d(A2)}{2}$$
$$= \frac{\left(E(A1)/m(A1)\right) + \left(E(A2)/m(A2)\right)}{2}, \quad (11.16)$$

where $E(A1)$, $E(A2)$, $m(A1)$, and $m(A2)$ are the energy deposited and masses of voxel $A1$ and $A2$, respectively. However, the authors point out that the correct interpretation of the dose should be as follows:

$$d^{\text{ETM}}(X1) = \frac{E(A1) + E(A2)}{m(A1) + m(A2)}, \quad (11.17)$$

where the dose interpolation error is $\varepsilon = \left|d^{\text{INTERP}} - d^{\text{ETM}}\right|$. The ETM approach allows energy mapping from multiple different source anatomies to multiple different "reference" anatomies simultaneously. In addition, the authors indicate that the method could also be useful in 4D IMRT optimization for 4D lung cases. In this case, beam intensities are simultaneously optimized on multiple breathing phases; thus, accurate dose evaluation is desired not only at a reference phase but also at each phase of the breathing cycle.

11.5.2 Comparing Patient Dose Calculation between 3D and 4D

Flampouri et al. (2005) studied the effect of respiratory motion on the delivered dose for lung IMRT plans, where a group of six patients were studied. For each patient, a free-breathing (FB) helical CT and a 10-phase 4D CT scan were acquired. A commercial planning system was then used to generate an IMRT plan. The authors found that conventional planning was sufficient for patients with tumor motion less than 12 mm, where FB helical CT had small or no artifacts relative to the 4D CT dataset. The authors also found that conventional planning was not adequate for patients with larger tumor motion or severe CT artifacts. Their study indicated that CT reconstruction artifacts have far bigger effects than tumor motion. Therefore, a major conclusion was that for accurate 4D CT MC dosimetry, CT datasets with less artifacts and distortions are required. In addition, the authors also evaluated the minimum number of breathing phases required to recreate the 10-phase composite dose. They found that a three-phase composite dose would allow a 3% error relative to a 10-phase dose prediction, whereas five phases would achieve 0.5% error relative to the 10-phase dose. These results are in agreement with predictions by Rosu et al.

(2007), where two phases (inhale and exhale) were shown to be sufficient to predict dose distribution in the respiratory motion. In a study by Seco et al. (2008), the authors compared the dose distributions for 3D CT FB with 4D CT FB and 4D CT-gated treatments to assess if gated treatments provided improved delivered dose distributions. The following question was addressed: do we require gating to mitigate motion effects in delivered dose distributions? The breathing pattern of each individual patient was accounted for while studying interfraction and intrafraction effects of breathing in the delivered dose. The respiratory motion was recorded by the RPM system of Varian (Varian Medical Systems, Palo Alto, CA). The authors showed that the largest dosimetric differences occurred between inhale and the other breathing phases, including the FB scan.

To analyze the spatial differences between two MC dose distributions for different breathing phases as a function of the density, the authors defined a parameter called omega index, $\Omega(\rho)$ or $\Omega(\rho; f, \text{VOI})$, where the density, ρ, is the independent variable, and f is a dose or dose difference distribution, and VOI is the volume of interest for which the omega index is calculated:

$$\Omega(\rho) = \frac{\int f(x) * \delta(\rho(x) - \rho) \mathrm{d}x}{\int \delta(\rho(x) - \rho) \mathrm{d}x}, \quad (11.18)$$

The parameter f can be either a 3D dose cube or a dose difference between two 3D cubes, that is, $f = D_{\text{MC}\varphi 1} - D_{\text{MC}\varphi 2}$, where φ_1 and φ_2 are any breathing phases. The omega index indicates dose differences as a function of the density. In Figure 11.3, a comparison of the dose differences between inhale and FB dose calculations is given.

It is possible to observe dose differences of the order 3–5 Gy between inhale and FB that occur in both the low- and high-density regions of the lung and the bone, respectively. This indicates that 4D MC can improve accurate dose predictions not only within the primary tumor but also for the surrounding tissue, that is, the lung or the bone, which is considered part of the planning target volume (PTV) because of the expanding margin to account for motion and setup errors. The expanding area can encompass a large volume of both the lung and the bone, which may affect the final dose estimate to the PTV due to movement.

In addition, the authors also addressed the question of whether 4D-gated treatments are better than 4D FB treatments in providing a better coverage of the PTV, where gated treatments have a far longer total delivery time and subsequent patient time on the couch. No significant dosimetric differences were observed between 4D CT FB and 4D CT-gated dose distributions; however, only a small group of three patients was assessed, so no generalization can be made. Larger differences (of approximately 3%) were observed between 3D FB and 4D FB, which could be attributed to various factors, such as (1) differences in image reconstruction of serial versus

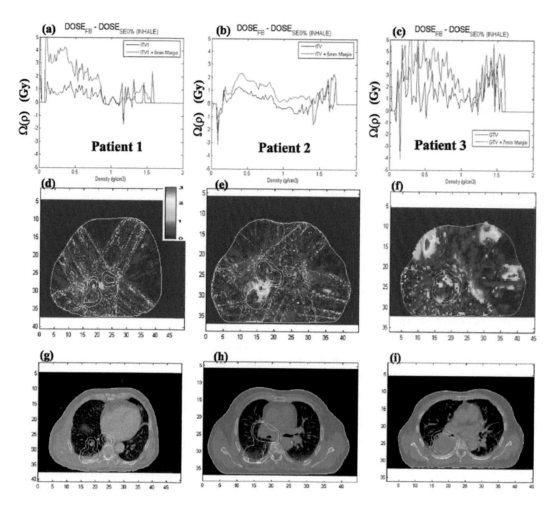

FIGURE 11.3   (a–c) Omega index prediction, (d–f) gamma index predictions, and (g–i) an example of CT slice for patients 1–3 (gamma values >3 are represented as red, the maximum gamma value of the color scale). (Courtesy of Seco J. et al. 2008. Dosimetric impact of motion in FB and gated lung radiotherapy: A 4D MC study of intrafraction and interfraction effects. *Med. Phys.* 356–66.)

helical CT, (2) importance of the inhale phase in the final 3D FB CT, and (3) coughing or breath-hold by the patient during the serial imaging process.

11.5.3 Intercomparison of Dose Warping Techniques for 4D Dose Distributions

In a study by Heath et al. (2008), an intercomparison of the various dose warping methods was performed for 4D MC dose calculations in the lung. The methods compared were center of mass tracking, trilinear interpolation, and the defDOSXYZ method (Heath and Seuntjens, 2006). No clinically significant dose differences were observed with the three methods. However, for the extreme case where motion was not accounted for in the treatment plan, the authors noticed an underestimate of the target volume dose by up to 16% by the remapping techniques. The authors also pointed out that the accuracy of the dose calculation is significantly affected by the continuity of the deformation fields from nonlinear image registration.

11.6 Inverse Planning with MC for Improved Patient Dose Calculations in Both 3D and 4D

Inverse planning for IMRT with any dose algorithm, including MC, involves the calculation of what is usually called the D_{ij} matrix, which allows the conversion of a fluence map into a dose distribution by the following relationship:

$$D_i(x) = \sum_{j \in \text{Beamlet}} D_{ij} x_j, \tag{11.19}$$

where D_{ij} is the respective dose influence, and x_j is a given fluence or beamlet intensity from a set of all beamlets. The D_{ij} value can then be calculated using pencil beam (PB), superposition-convolution, or MC dose algorithms. A major issue with the formulation given in Equation 11.16 is that usually delivery constraints and field size dependencies of output factors are not accounted for in the generation of D_{ij} values.

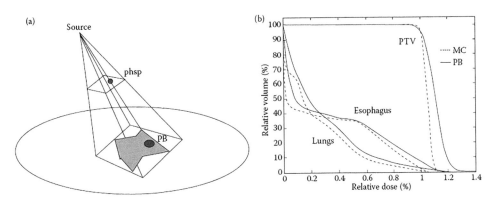

FIGURE 11.4 (a) MC fluence-based PB dose calculation method. (b) Dose-volume histogram of optimized IMRT plan based on MC (full line) and standard PB-like dose algorithm (dashed line) normalized to 95% isodose line. (Courtesy of Nohadani O. et al. 2009. Dosimetry robustness with stochastic optimization. *Phys. Med. Biol.* 54, 3421–32.)

These have to be added after the planning stage, which can lead to suboptimal delivered dose distributions.

One of the initial attempts in using MC to perform 3D inverse planning was performed by Jeraj and Keall (Jeraj and Keall, 1999) using the MCNP code. The authors performed optimization on MC-generated PBs, where an initial back-projection estimate was obtained with MCNP to generate an initial IMRT fluence. EGS4 was then used to calculate dose within the patient and therefore generate the required PBs for optimization of all beams, using a narrow-beam approach to generate each PB. The major issues with the method proposed by Jeraj and Keall (1999) was that each MC-generated PB did exactly model the spectrum exiting the linac, MLC delivery constraints were not modeled, and head scatter variations with field sizes and motion effects were not taken into account. More recently, Siebers et al. (Siebers and Kawrakow, 2007) proposed a hybrid method of 3D MC inverse planning, where an initial dose prediction was performed by the PB algorithm. For subsequent iterations, the PB dose prediction was improved by using the MC algorithm such that the optimized plans were equivalent to an MC-based optimization. This produced a gain of 2.5 times over a full MC-based optimization.

Nohadani et al. (2009) proposed a 4D MC-based inverse planning approach using the broad beam approach, in contrast to the narrow beam approach of Jeraj et al. (1999). For each incident beam, the jaw setting does not change during IMRT delivery for a Varian linac, making it possible to generate an open-field phase space file just after the *X/Y* jaws. Therefore, the MC phase space after the jaws can be split into PB-like pixels from which particles can be sampled to produce a 3D PB dose distribution within the patient. In Figure 11.4a, we present a diagram showing how sampling a pixel in the phase space produces an MC–PB distribution within the patient.

The following are the advantages of the broad beam approach for generating the MC–PB distribution: (1) it will account for heterogeneity effects in the patient IMRT calculation; (2) if all PBs are combined into a homogeneous open field, the output

factor will be that of standard open field, and no further output corrections are required; and (3) when delivering the IMRT profile with, for example, a multileaf collimator, the changes in output, due to field size changes during delivery, will be small and therefore will not affect the planned optimal dose distribution. In Figure 11.4b, the DVH comparing IMRT obtained with MC–PB versus standard PB is given. It shows that an IMRT plan based on PB may appear to be a better plan in terms of PTV coverage, hot spots, and organs at risk sparing. However, it deviates significantly from an MC-based optimized plan that closely mimics the reality of the deposited dose. The figure also illustrates that an otherwise optimal IMRT plan based on PB dose calculation may not correspond to the actual dose distribution because of these dosimetric errors. Therefore, errors in dose calculation cannot be neglected.

References

Agostinelli S. 2003. GEANT4—A simulation toolkit. *Nucl. Instrum. Methods Phys. Res. A* 506, 250–303.

Badal A. et al. 2009. PenMesh-Monte Carlo radiation transport simulation in a triangle mesh geometry. *IEEE Trans. Med. Imaging* 28(12), 1894–901.

Baro J., Sempau J., Fernandez-Varea J.M. and Salvat F. 1995. PENELOPE—An algorithm for Monte Carlo simulation of the penetration and energy loss of electrons and positrons in matter. *Nucl. Instrum. Method Phys. Res. A* 100, 31–46.

Bazalova M. et al. 2008. Dual-energy CT-based material extraction for segmentation in Monte Carlo dose calculations. *Phys. Med. Biol.* 53, 2439–56.

Berger M. and Seltzer S. 1973. ETRAN Monte Carlo code system for electron and photon transport through extended media. Radiation Shielding Information Center (RSIC) Report CCC-107, Oak Ridge National Laboratory, Oak Ridge, TN.

Berger M.J. 1963. *Methods in Computational Physics*. B. Alder, S. Fernbach, and M. Rothenberg (eds.), Academic, New York, Vol. 1, p. 135.

Berger M.J. et al. 2005. XCOM: Photon Cross Section Database NBSIR 87-3597 (web version 1.3 (http://physics.nist.gov/PhysRefData/Xcom/Text/XCOM.html)).

Brock K.K. et al. 2006. Feasibility of a novel deformable image registration technique to facilitate classification, targeting and monitoring of tumor and normal tissue. *Int. J. Radiat. Oncol. Biol. Phys.* 64, 1245–54.

Brown F.B. 2003. MCNP—A general Monte Carlo particle transport code, Version 5. Report LA-UR-03 1987 Los Alamos National Laboratory, Los Alamos, NM.

Buffa F.M., Nahum A.E., and Mubata C., 1999. Influence of statistical fluctuations in Monte Carlo dose calculations on radiobiological modelling and dose volume histograms. *Med. Phys.* 26, 1120.

Chetty I.J. et al. 2007. Report of the AAPM Task Group No. 105: Issues associated with clinical implementation of Monte Carlo-based photon and electron external beam treatment planning. *Med. Phys.* 34(12), 4818–53.

Cygler J.E., Daskalov G.M., Chan G.H., and Ding G.X. 2004. Evaluation of the first commercial Monte Carlo dose calculation engine for electron beam treatment planning. *Med. Phys.* 31, 142–53.

Deasy J.O. 2000. Denoising of electron beam Monte Carlo dose distributions using digital filtering techniques. *Phys. Med. Biol.* 45, 1765–79.

Deasy J.O., Wickerhauser M.V., and Picard M. 2002. Accelerating Monte Carlo simulations of radiation therapy dose distributions using wavelet threshold denoising. *Med. Phys.* 29, 2366–73.

DeMarco J.J., Solberg T.D., and Smathers J.B., 1998. A CT-based Monte Carlo simulation tool for dosimetry planning and analysis. *Med. Phys.* 25, 1–11.

El Naqa I., Kawrakow I., Fippel M., Siebers J.V., Lindsay P.E., Wickerhauser M.V., Vicic M., Zakarian K., Kauffmann N., and Deasy J.O. 2005. A comparison of Monte Carlo dose calculation denoising techniques. *Phys. Med. Biol.* 50, 909–22.

Fippel M. and Nusslin F. 2003. Smoothing Monte Carlo calculated dose distributions by iterative reduction of noise. *Phys. Med. Biol.* 48, 1289–304.

Flampouri S. et al. 2005. Estimation of the delivered patient dose in lung IMRT treatment based on deformagistration of 4D-CT data and Monte Carlo simulations. *Phys. Med. Biol.* 51, 2763–79.

Fraass B., Doppke K., Hunt M., Kutcher G., Starkschall G., Stern R., and Van Dyke J. 1998. American association of physicists in medicine radiation therapy committee task group 53: Quality assurance for clinical radiotherapy treatment planning. *Med. Phys.* 25, 1773–829.

Guerrero T. et al. 2004. Intrathoracic tumor motion estimation from CT imaging using the 3D optical flow method. *Phys. Med. Biol.* 49, 4147–61.

Heath E., Seco J., Wu Z., Sharp G.C., Paganetti H., and Seuntjens J. 2008. A comparison of dose warping methods for 4D Monte Carlo dose calculations in lung. *J. Phys.: Conf Ser.* 102,102.

Heath E. and Seuntjens J. 2006. A direct voxel tracking method for four-dimensional Monte Carlo dose calculations in deforming anatomy. *Med. Phys.* 33, 434–45.

Jeraj R. and Keall P. 1999. Monte Carlo-based inverse treatment planning. *Phys. Med. Biol.* 44, 1885–96.

Jeraj R., Keall P., and Ostwald P.M. 1999. Comparisons between MCNP, EGS4 and experiment for clinical electron beams. *Phy. Med. Biol.* 44, 705–18.

Jiang S.B., Pawlicki T., and Ma C.-M. 2000. Removing the effect of statistical uncertainty on dosevolume histograms from Monte Carlo dose calculations. *Phys. Med. Biol.* 45, 2151–62.

Kawrakow I. 2000. Accurate condensed history Monte Carlo simulation of electron transport, I. EGSnrc, the new EGS4 version. *Med. Phys.* 27, 485–98.

Kawrakow I. 2002. On the denoising of Monte Carlo calculated dose distributions. *Phys. Med. Biol.* 47, 3087–103.

Kawrakow I. and Bielajew A.F. 1998. On the condensed history technique for electron transport. *Nucl. Instrum. Methods Phys. Rev. B* 142, 253–80.

Keall P., Siebers J.V., Libby B., Mohan R., and Jeraj R. 1999. The effect of Monte Carlo noise on radiotherapy treatment plan evaluation. *Med. Phys.* 26, 1149.

Keall P. et al. 2004. Monte Carlo as a four-dimensional radiotherapy treatment-planning tool to account for respiratory motion. *Phys. Med. Biol.* 49, 3639–48.

Keall P.J., Siebers J.V., Jeraj R., and Mohan R. 2000. The effect of dose calculation uncertainty on the evaluation of radiotherapy plans. *Med. Phys.* 27, 478–84.

Larsen E.W. 1992. A theoretical derivation of the condensed history algorithm. *Ann. Nucl. Energy* 19, 701–14.

Lu W. et al. 2004. Fast free-form deformable registration via calculus of variations. *Phys. Med. Biol.* 49, 3067–87.

Ma C.-M., Mok E., Kapur A., Pawlicki T., Findley D., Brain S., Forster K., and Boyer A. L. 1999. Clinical implementation of a Monte Carlo treatment planning system. *Med. Phys.* 26, 2133–43.

Meyer C.R. et al. 1997. Demonstration of accuracy and clinical versatility of mutual information for automatic multimodality image fusion using affine and thin-plate spline warped geometric deformation. *Med. Image Anal.* 1, 195–206.

Miao B., Jeraj R., Bao S., and Mackie T.R. 2003. Adaptive anisotropic diffusion filtering of Monte Carlo dose distributions. *Phys. Med. Biol.* 48, 2767–81.

Neuenschwander H. et al. 1995. MMC—A high performance Monte Carlo code for electron beam treatment planning. *Phys. Med. Biol.* 40, 543–74.

Nohadani O., Seco J., Martin B.C., and Bortfeld T. 2009. Dosimetry robustness with stochastic optimization. *Phys. Med. Biol.* 54, 3421–32.

du Plessis F.C.P. et al. 1998. The direct use of CT numbers to establish material properties needed for Monte Carlo calculation of dose distributions in patients. *Med. Phys.* 25, 1195–201.

Riboldi M. et al. 2008. Design and testing of a simulation framework for dosimetric motion studies integrating an anthropomorphic phantom into four-dimensional Monte Carlo. *Technol. Cancer Res. Treat.* 7(6), 449–56.

Rietzel E. and Chen G.T.Y. 2006. Deformable registration of 4D computed tomography data. *Med. Phys.* 33, 4423–30.

Rietzel E. et al. 2005. Four-dimensional image-based treatment planning: Target volume segmentation and dose calculation in the presence of respiratory motion. *Int. J. Radiat. Oncol. Biol. Phys.* 61, 1535–50.

Rosu M., Chetty I.J., Balter J.M., Kessler M.L., McShan D.L., and Ten Haken R.K. 2005. Dose reconstruction in deforming lung anatomy: Dose grid size effects and clinical implications. *Med. Phys.* 23, 2487–95.

Rosu M. et al. 2007. How extensive of a 4D dataset is needed to estimate cumulative dose distribution plan evaluation metrics in conformal lung therapy? *Med. Phys.* 34, 233–45.

Rutherford R.P., Pullan B.R., and Isherwood I. 1976. Measurement of effective atomic number and electron density using EMI scanner. *Neuroradiology* 11, 15–21.

Salvat F., Fernandez-Varea J.M., and Sempau J. 2009. PENELOPE-2008: A code system for Monte Carlo simulation of electron and photon transport. *Proceedings of a Workshop/Training Course*, OECD.

Schneider W., Bortfeld T., and Schlegel W. 2000. Correlation between CT numbers and tissue parameters needed for Monte Carlo simulations of clinical dose distributions. *Phys. Med. Biol.* 45, 459–78.

Schneider U., Pedroni E., and Lomax A. 1996. The calibration of CT Hounsfield units for radiotherapy treatment planning. *Phys. Med. Biol.* 41, 111–24.

Seco J. and Evans P.M. 2006. Assessing the effect of electron density in photon dose calculations. *Med. Phys.* 33, 540–52.

Seco J. et al. 2008. Dosimetric impact of motion in free-breathing and gated lung radiotherapy: A 4D Monte Carlo study of intrafraction and interfraction effects. *Med. Phys.* 35, 356–66.

Segars W.P. 2001. Development of a new dynamic NURBS-based cardiac-torso (NCAT) phantom. PhD thesis. University of North Carolina, North Carolina.

Seltzer S.M. 1988. An overview of ETRAN Monte Carlo methods in T.M. Jenkins, W.R. Nelson, A. Rindi, A.E. Nahum, and D.W.O. Rogers (eds.) *Monte Carlo Transport of Electrons and Photons Below 50MeV*, Plenum, New York, pp. 153–82.

Sempau J. and Bielajew A.F. 2000. Towards the elimination of Monte Carlo statistical fluctuation from dose volume histograms for radiotherapy planning. *Phys. Med. Biol.* 45, 131–158.

Sempau J., Sanchez-Reyes A, Salvat F., ben Tahar H.O., Jiang S.B., and Fernandéz-Varea J.M. 2001. Monte Carlo simulation of electron beams from an accelerator head using PENELOPE. *Phys. Med. Biol.* 46, 1163–86.

Sempau, J., Wilderman S.J., and Bielajew A.F. 2000. DPM, a fast accurate Monte Carlo code optimized for photon and electron radiotherapy treatment planning dose calculations, *Phys. Med. Biol.* 45, 2263–92.

Siebers J.V. and Kawrakow I. 2007. Performance of a hybrid MC dose algorithm for IMRT optimization dose evaluation. *Med. Phys.* 34, 2853–63.

Siebers J.V. and Zhong H. 2008. An energy transfer method for 4D Monte Carlo dose calculation. *Med. Phys.* 35(9), 4095–105.

Spitzer V. and Whitlock D. 1998. The Visible Human dataset: The anatomical platform for human simulation. *Anat. Rec.* 253, 49–57.

Torikoshi M. et al. 2003. Electron density measurements with dual-energy x-ray CT using synchroton radiation. *Phys. Med. Biol.* 48, 673–85.

Trofimov A. et al. 2005. Temporo-spatial IMRT optimization: Concepts implementation and initial results. *Phys. Med. Biol.* 50, 2779–98.

Van Dyk J., Barnett R.B., Cygler, J.E., and Shragge, P.C. 1993. Commissioning and quality assurance of treatment planning computers. *Int. J. Radiat. Oncol. Biol. Phys.* 26, 261–73.

Vanderstraeten B. et al. 2007. Conversion of CT numbers into tissue parameters for Monte Carlo dose calculations: A multicentre study. *Phys. Med. Biol.* 52, 539–62.

Verhaegen F. and Devic S. 2005. Sensitivity study for CT image use in Monte Carlo treatment planning. *Phys. Med. Biol.* 50, 937–46.

Walters B.R.W., Kawrakow I., and Rogers D.W.O. 2002. History by history statistical estimators in BEAM code system. *Med. Phys.* 34(12), 4818–53.

Wang H. et al. 2005. Validation of an accelerated "demons" algorithm for deformable image registration in radiation therapy. *Phys. Med. Biol.* 50, 2887–905.

12

Electrons: Clinical Considerations and Applications

Michael K. Fix
Inselspital – University Hospital Bern

George X. Ding
Vanderbilt University School of Medicine

Joanna E. Cygler
The Ottawa Hospital Cancer Centre

12.1 Introduction: Rationale for Monte Carlo-Based Treatment Planning Systems for Electron Beams

This chapter covers the practical aspects of Monte Carlo (MC) implementation, the challenges encountered, the main application areas, and comparison with conventional algorithms. The discussion includes implementation of commercial packages and research MC packages.

Lots of effort has been devoted to technological developments toward improving treatment outcomes. As a result, modern radiation therapy is a complex process that consists of multiple steps, with each step containing inherent uncertainties and assumptions as described in the International Commission on Radiation Units and Measurements Report 71 (ICRU-71, 2004). It has been stated that the uncertainty of the dose delivered to the patient should be less than 5% (ICRU-29, 1978; Brahme, 1984; Papanikolaou et al., 2004). This in turn requires that the uncertainty of dose calculation be less than 2% (Papanikolaou et al., 2004). Most commercial treatment planning systems (TPSs) cannot provide

this level of accuracy for all clinically encountered situations. This has been especially true for electron beams for which older dose calculation algorithms failed in many cases (Cygler et al., 1987; Ding et al., 2005; Hogstrom and Almond, 1983; Hogstrom et al., 1989; Hogstrom et al., 1981; Hogstrom et al., 1984). In principle, MC gives an accurate answer to within the statistical uncertainty, as there are no significant approximations (except in beam models) involved in the calculations, no approximate scaling of dose kernels is needed, and the electron transport is fully modeled. Both treatment machines and patient geometries can be modeled accurately. All types of inhomogeneities are properly handled in the calculations. It has been well documented that MC calculations can be very accurate (Ma et al., 1997; Reynaert et al., 2007; Zhang et al., 1999). In spite of that, for many years, MC calculations were available only for research purposes and used in more advanced departments for special projects or as a quality assurance tool for commercial non-MC systems. This has been mostly due to the long calculation times for MC algorithms. That problem has been overcome with the arrival of modern fast computers and efficient calculation algorithms. Finally, at the dawn of the twenty-first century, state-of-the-art MC dose calculations for electron beams were implemented in commercial TPSs. Such systems are now routinely used in clinics for radiotherapy patients. This is truly a major breakthrough in radiotherapy as no other existing algorithm can calculate the dose from electron beams accurately for all treatment situations. There have been several studies describing commissioning and clinical implementation of the MC-based TPS for electron beams (Ali et al., 2011; Cygler et al., 2004; Ding et al., 2005; Ding et al., 2006; Edimo et al., 2009; Fragoso et al., 2008; Popple et al., 2006; Ojala et al., 2016; Huang et al., 2019; Snyder et al., 2019). Recognizing that MC-based TPSs had become a reality, the American Association of the Physicists in Medicine (AAPM) formed a special task group, TG-105, to address issues related to clinical implementation of such systems. The TG-105 report provides a useful framework for commissioning of MC-based TPSs (Chetty et al., 2007). Finally, such accurate dose calculation algorithms allow further use of electron beam radiotherapy and its future applications such as standard electron beams without applicators, modulated electron radiotherapy (MERT), or mixed photon and electron beam radiotherapy (MBRT).

12.1.1 Advantages of MC versus Pencil Beam Algorithm

Before the arrival of MC dose calculations, commercial TPSs for electron beams were based on pencil beam (PB) algorithms. Although at the time of its implementation, the PB approach was a big step forward; it was known to have very serious limitations. PB-based planning systems cannot really handle monitor unit (MU) calculations for arbitrary source-to-surface distance (SSD) values when only one machine of a single SSD is configured in the system. This is due to the fact that unlike in the case of high-energy photon beams, the electron beam source is not a point but rather an extended source as the fraction of electrons scattered from the electron cones is significant. The effective source position depends on the beam energy, the jaw-opening size, and the geometry of the electron cone. Therefore, the simple inverse square law is applicable in case of point source does not work for electron beams (Cygler et al., 1997; Khan et al., 1991). Another limiting factor for PB algorithms for electron beams is that the simple ray tracing is not possible because of electron scattering. To handle MU calculations for extended SSDs in PB systems, one has to configure several virtual machines, one for each SSD. This is not the case in MC TPSs. In such systems, users need to install just a single virtual machine for a standard SSD for each beam energy to achieve MU and dose distributions within the clinically acceptable accuracy not only at standard but also at any extended SSD (>110 cm) (Cygler et al., 2004; Ding et al., 2006).

In addition, the dose distributions in heterogeneous media have large errors for the complex geometries, frequently encountered in human anatomy. This has been well documented in the literature (Cygler et al., 1987; Ding et al., 2005; Fragoso et al., 2008; Hogstrom et al., 1989; Keall and Hoban, 1996; Shiu and Hogstrom, 1991; Starkschall et al., 1991). Figure 12.1 shows one of the most striking examples of the failure of the PB algorithm (Ding et al., 2005).

Figure 12.1 illustrates the "trachea and spine" phantom used in electron beam commissioning by Cygler et al. (1987) and Ding et al. (2005; 2006). The phantom consists of overlying low- and high-density heterogeneities, as frequently encountered in the human body. In a solid water slab with an overall thickness of 6.2 cm, there is an air pipe of 2.5 cm diameter and 10 cm length, and about half centimeter below it, there are four 1 cm thick bone cylinders of 2.5 cm diameter each. Figure 11.1 presents the dose profiles measured and calculated at various depths behind the "trachea and spine" phantom for a 9 MeV beam. The PB cannot even remotely resolve the shape of the hot and cold spots for this complex geometry, whereas MC can do it very well. The MC versus measured disagreement that we see here in the peaks is about 4% and can be attributed to the fact that the measurement resolution was 1 mm, and the MC calculation voxel was 3.9 mm. The measurement and calculation resolutions could not be the same because in the software used (Nucletron), users had no control over the size of the calculation voxel. For a size of 2 mm calculation voxel, one sees excellent agreement with the measured data (Ding *et al.*, 2005; Ding *et al.*, 2006). In addition to their limited accuracy, PB algorithms require relatively long calculation times for multislice anatomies. For example, on a standard Eclipse TPS work station (with 8 Xeon processors 2.8 MHz), the MC calculations (with 2% statistical uncertainty on doses >50% of

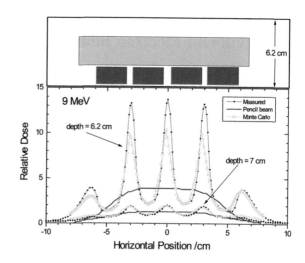

FIGURE 12.1 Comparison between measured and calculated cross-beam dose profiles at various depths for a 9 MeV beam incident on the trachea and spine phantom with a 10×10 cone and SSD=110 cm. (From Ding, G. X. et al. 2005. *Int J Radiat Oncol Biol Phys* 63: 622–33 with permission from Elsevier.)

D_{max} and voxel size of 0.25 cm) took 4 minutes, whereas the PB calculations took 21 minutes for the same single beam of 9 MeV electron beam and a 10×10 cm^2 field size. The calculation speed gain for MC calculations is partially due to more efficient algorithms and using multiple processors of the work station (Eclipse). The PB algorithm calculations for a single beam on Eclipse use a single processor.

12.2 Research and Commercial MC TPSs

12.2.1 Meeting the Challenges: The OMEGA Project

The MC method is regarded as the "gold standard" in dose calculation, as it is capable of accurately predicting the dose distributions under almost all circumstances. It has played a significant role in the electron treatment planning because it provides accurate dose predictions even in complex three-dimensional heterogeneous geometries, where the model-based dose calculations algorithms, such as PB, have shown significant limitations. Recognizing the requirement of accurate patient treatment planning, the National Institutes of Health (NIH) funded the OMEGA research project in 1990. It was a collaborative project between the National Research Council (NRC) of Canada and the University of Wisconsin to develop MC calculation algorithms for use in fully 3D electron beam TPSs to calculate the dose to patients. The result of this project was a user code package, called BEAM (Rogers et al., 1995), using the general-purpose MC code Electron Gamma Shower Version 4 (EGS4) to simulate radiation beams from radiotherapy machines, including high-energy photon and electron beams, Co-60 beams, and kilovoltage units. Since then, the codes have been improved by the NRC, leading to BEAMnrc and EGSnrc, which are currently used worldwide for research. BEAM and BEAMnrc have also been used to generate and verify beam models used in commercial TPSs.

12.2.2 Methods to Speed Up the MC Calculations

Lengthy computing time was considered to be a major drawback of the MC technique. To meet the needs of routine clinical treatment planning, special methods were developed to speed up MC calculations for these applications. More generally, the efficiency of the MC code is of importance here. The efficiency is defined as follows:

$$\varepsilon = \frac{1}{s^2 T}$$

where s is the estimated statistical uncertainty of the MC simulated quantity, and T is the corresponding central processing unit (CPU) time needed for the calculations. Thus,

improvements in efficiency can either be achieved by reducing the variance or by reducing the corresponding calculation time, e.g., simulation time per history. Note that in this context, the usage of faster computer hardware is not considered an efficiency improvement of the MC code.

One of the methods that speed ups MC calculations is called "a voxel-based electron beam MC algorithm", which introduced some simplifications and approximations into the transport algorithm (Fippel et al., 1997; Kawrakow, 1997, 2001). These methods have been implemented into commercial TPSs for electron dose calculations (Nucletron and Elekta Software) (Ali et al., 2011; Cygler et al., 2004).

The other technique developed to speed up electron beam calculations is called the macro MC (MMC) method (Neuenschwander and Born, 1992; Neuenschwander et al., 1995). See also Section 3.3.3 in this book. The following are the key features of the MMC transport model: (1) Conventional MC simulations of electron transport are performed in well-defined local geometries, namely, spheres ("kugels"). The result of these calculations is a library of probability distribution functions (PDFs) of particles emerging from the "kugels". The PDFs used in a commercial TPS (Eclipse) were generated in extensive precalculations by using the EGSnrc (Kawrakow and Rogers, 2002) software to simulate the transport of vertically incident electrons of variable energies through macroscopic spheres of various sizes and five materials (air, lung, water, Lucite, and solid bone). They were calculated only once for a variety of clinically relevant energies and the abovementioned five materials; (2) patient-specific MC calculations are performed in a global geometry. Electrons are transported through the patient in macroscopic steps based on the PDFs generated in the above-described local calculations. As already mentioned, this method has been implemented in a commercial TPS by Varian (Ding et al., 2006; Popple et al., 2006).

Another method to speed up MC dose calculations is to use source (beam) models instead of direct particle transport through the machine head.

12.2.3 EGSnrc-Based MC Research Package

MC particle transport techniques were first introduced in the 1940s and since then have evolved into many different areas of applications, including radiotherapy physics. In the last several decades, the MC technique has become ubiquitous in medical physics (Rogers, 2006). There are many MC codes available for the applications in radiotherapy (Allison et al., 2006; Briesmeister, 1997; Kawrakow, 2000; Nelson et al., 1985; Seltzer, 1991; Sempau et al., 1997). The user code package BEAM with the general-purpose MC code EGS4 for simulation of radiotherapy beams from treatment units was made available in 1995 (Rogers et al., 1995). This code makes it easier to simulate radiotherapy units and has been widely used for the study of photon and electron beams from accelerators. Owing to the easy interface for using the BEAM/EGS4 (now BEAMnrc/EGSnrc) research package, the BEAM and later BEAMnrc code have been used extensively to characterize therapy beams, including megavoltage electron and photon beams from linear accelerators and kilovoltage photon beams from X-ray units (Bazalova and Verhaegen, 2007; Bazalova et al., 2009; Ding, 1995; Ding et al., 2006; Ding and Rogers, 1995; Francescon et al., 2000; Francescon et al., 2008; Francescon et al., 2009; Iaccarino et al., 2011; Jarry et al., 2006; Keall et al., 2003; Pasciuti et al., 2011; Pimpinella et al., 2007; Sheikh-Bagheri and Rogers, 2002a, b; Zhang et al., 1998, 1999). Not only is the BEAMnrc/EGSnrc research package capable of simulating the treatment beams but also it uses the simulated realistic beams to calculate dose to a patient using the information from CT images. MC research packages have also been used in validating the accuracy of dose calculations of commercial TPSs (Ali et al., 2011; Ma et al., 2000; Wieslander and Knoos, 2007).

12.2.4 Commercial MC TPSs

MC calculations of dose distribution in a patient can be divided into three steps (Chetty et al., 2007):

1. Particle transport through the top of machine head (patient-independent step).
2. Particle transport through the patient-specific part of the machine.
3. Dose calculations in the patient.

In research, for particle transport through the machine head, BEAMnrc is frequently used. BEAMnrc generates a particle phase space file, which is used as the input file to steps 2 and 3. In Step 2 and 3, codes such as EGSnrc (only step 2), voxel MC (VMC), XVMC, and VMC++ are used.

Direct simulation of particles in the machine head is not used in commercial TPSs. The main reason is that simulations of the direct particle transport through the machine head strongly depend on knowing accurately the machine head details (materials, dimensions, etc.), which may not be easily available to the users (Schreiber and Faddegon, 2005).

So, typically commercial MC-based TPSs use source models, which describe the beams with several fitting parameters derived from the fully simulated phase space file and/or measured dose profiles (Brualla et al., 2017).

Presently, there are four commercial MC-based TPSs available: Elekta Monaco (VMC++), Varian Eclipse (eMC), Elekta Software XiO* (XVMC), and RaySearch RayStation (VMC++).

12.2.4.1 Elekta Monaco VMC++

In 2002, Nucletron* was the first to release a commercial MC-based TPS for electron beams. The MC option was available in the dose calculation module (DCM) of Theraplan Plus and Oncentra MasterPlan and now in the Monaco TPS

from Elekta. The beam transport is divided into two components: treatment independent and treatment dependent (Oncentra MasterPlan—physics reference manual). The treatment-independent component, called the phase space engine, handles particle transport through the linac head down to the lowest patient-independent collimation level. The patient-dependent component, called the dose engine, is based on the VMC++ algorithm (Kawrakow, 2000). The accelerator beam model in the Nucletron TPS was developed by Traneus et al. (2001). It consists of a coupled multisource beam model. It is based on five parameters and requires a limited set of clinic-measured data to configure the beam model. This beam model is flexible enough to support present collimation types such as applicators with optional inserts, variable trimmer machines, and potential multileaf collimator (MLC) applications in the future. The modelling of a particular beam starts upstream of the uppermost changeable/movable collimating element with a source phase space (SPS). The SPS is propagated through the treatment head to an exit phase space (EPS) plane located at the lowest collimating element, thereby defining the interface between the phase space and the dose engines; see Figure 12.2.

For a treatment unit with fixed applicators with optional patient-specific cutouts, the block collimators and possible MLC have fixed positions per energy/applicator combination. The SPS for these applicators is prepropagated as part of the beam data handling process to an EPS plane located just in front of the insert position. The in-patient dose calculation is then driven by sampling from a parameterized description of this EPS. In this way, the transport through the treatment head does not have to be performed for every dose calculation. For more general treatments that are not using fixed collimation setups (e.g., MLC fields), the dose calculation can be driven by sampling directly from the SPS.

To bypass time-consuming in-collimator transport during dose planning, scattering effects in the applicator are accounted for by using precalculated collimator scatter kernels, describing the coupled angular and energy probability distributions of indirect electrons coming from applicator elements.

Therefore, for the Nucletron planning system, the accelerator model provides the EPS file, which serves as the input file for the in-phantom/patient calculations performed with VMC++ (Kawrakow, 1997, 2001; Kawrakow et al., 1996).

The code can handle fixed applicators with optional, arbitrary inserts of any shape. It also can handle the variable size fields defined by an applicator such as the digital electron variable applicator (DEVA) (Siemens). The code always calculates the absolute dose per MU, cGy/MU.

Theraplan Plus and Oncentra planning systems can handle all major accelerators available on the market. The beam modelling is performed by Nucletron.

The user interface for the Nucletron planning systems does not allow for selection of any calculation parameters, except the number of particle histories/cm^2. The larger this number

---

[1] It actually was MDS Nordion who later sold the treatment planning system to Nucletron. Meanwhile, Nucletron became part of Elekta.

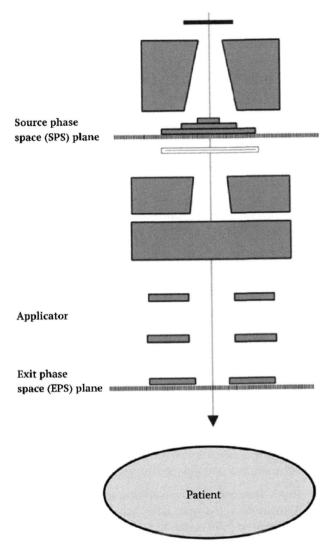

Source phase space (SPS) plane

Applicator

Exit phase space (EPS) plane

Patient

FIGURE 12.2 Linac head and beam model. (Adapted from *Oncentra MasterPlan—Physics Reference Manual* 92.724ENG-00.)

and number of histories per cm^2 (10^3–10^6). That last parameter controls the calculation uncertainty. For example, the number of histories set to $2 \times 10^5/cm^2$ results in about 1% statistical uncertainty in voxels containing doses $> 50\%$ of D_{max}.

12.2.4.2 Varian Electron MC

The electron MC (eMC) algorithm used by Eclipse is an implementation of the MMC method (Neuenschwander et al., 1995) for calculation of dose from high-energy electron beams. In its initial implementation in Eclipse, it was only capable to model the electron beams from Varian Linacs, and its calculation accuracy was reported by Ding et al. (2006). Since version 13.6 of Eclipse (2015), an improved eMC algorithm is available, which uses a new beam model supporting 4 MeV and non-Varian machines as well as providing some improvements in the transport code (Fix et al., 2013; Fix et al., 2010). The algorithm consists of the following: (1) a multiple-source model describing the electrons and photons that emerge from the treatment head of the linear accelerator. This model is based on precalculated data for a machine type and configured using measured beam data; (2) electron transport/dose deposition in the patient is calculated using the MMC method.

The patient's anatomy is based on CT images. The CT image volume is first converted into a mass density volume with a user-defined resolution (0.1–0.5 cm), applying appropriate CT-to-mass density conversion factors. The resulting density volume is then scanned for heterogeneities. To each voxel of the density volume, a sphere index is assigned. A voxel of the density volume is considered to be part of a heterogeneous volume if the density ratio of the voxel and its neighbors exceeds a limit (typically 1.5). If the densities in both voxels are below a threshold (typically 0.05 g/cm³), the ratio is not evaluated. The material assigned to each sphere depends on the average mass density within the sphere. If the average mass density of a sphere is exactly equal to the mass density of one of the preset materials, the preset material is selected for the sphere. If the mass density of a sphere is between two preset materials, the material is randomly selected for these two materials each time a particle enters the sphere. The probability for a material to be selected is proportional to the closeness of a sphere's average mass density to the mass density of the material. There are five different preset materials (air, lung, water, Lucite, and bone) in the MMC calculation database (The Reference Guide for Eclipse Algorithms Version 6.5 (P/N B401653R01M)). For example, if the average density within a sphere is 1.12, there is a 12/19 chance that the material is selected to be Lucite and a 7/19 chance that it is selected to be water. Unlike conventional TPSs, the reported doses are the doses calculated to the material (Ding et al., 2006) and not the dose to water of different densities. However, the user can define other calculation parameters, such as the number of particle histories or required statistical uncertainty of the calculation or voxel size, or whether to apply isodose smoothing.

is, the lower is the statistical uncertainty in the dose distribution. The user has no control over the calculation's voxel size, which is automatically assigned by the system. The voxel size assignment is based on the computed tomography (CT) anatomy volume and the total number of allowed voxels being 800,000. So, for smaller CT volumes, one achieves smaller calculation voxel sizes. The larger overall volumes result in being divided into larger calculation voxels. A database of 21 materials is used in segmentation of patient CT anatomy.

In Theraplan Plus, the user can select to calculate and report dose to medium or dose to water. Unfortunately, the user does not have this choice in Oncentra, which only reports dose to water.

Implementation of VMC++ in the Monaco TPS gives the user more control over calculation parameters. In the Calculation Properties tab, the user can select the voxel size (range, 0.1–0.8 cm), dose to water or dose to medium option,

12.2.4.3 Elekta XiO eMC

The XiO eMC module is based on XVMC (Fippel, 1999; Fippel et al., 1999; Kawrakow et al., 1996). The beam model has been developed by Elekta Software, and it consists of a weighted combination of particle sources (direct and indirect electrons and photons).

Beam modelling based on user measurements is performed by the Elekta software. As mentioned above, the dose calculation in the phantom/patient is based on XVMC. The user interface in XiO allows for full control over the dose calculation parameters. The user can define calculation voxel dimensions and can control the statistical uncertainty in the dose distributions by setting either the maximum number of particle histories or the goal mean relative statistical uncertainty (MRSU) defined as follows:

$$\text{MRSU} = \sqrt{\frac{1}{N} \sum_{D(\vec{r}_i) \geq aD_{\max}}^{N} \left(\frac{\sigma(\vec{r}_i)}{D_{\max}} \right)^2},$$

where D_i is the dose in ith voxel and $\sigma(\vec{r}_i)$ is the statistical uncertainty of dose for the ith voxel. The sum is computed over all N voxels in which the dose is greater than or equal to some fraction, α, of the maximum dose $D_{\max}$. It is commonly accepted that for tumor dose calculations $\alpha = 50\%$. However, for accurate calculation of dose to organs at risk, α should be set to a much lower value, which will of course increase the calculation time. XiO eMC allows the user to calculate and report either dose to water or dose to medium, which makes it compliant with TG-105's recommendations (Chetty et al., 2007). XiO can handle calculations for linacs from all major vendors.

12.2.4.4 RaySearch Electron MC Algorithm

The RaySearch electron MC algorithm in RayStation combines a phase space beam model with the VMC++ MC code for the dose calculation (Kawrakow, 2001). The beam model consists of a SPS at the level of the secondary scattering foil and an EPS at the plane of the cutout. The EPS contains weighted direct and indirect electrons as well as photons. These particles from the EPS are handed over to the dose calculation algorithm upon its request. The radiation transport between SPS and EPS is performed using a dedicated MC code considering the following beam defining components of the linac: jaws, MLC, scrapers, and cutout (Huang et al., 2019).

On the one hand, patient representation is carried out by a HU to density conversion, whereas on the other hand, the material composition is determined by using a mapping of 50 fixed materials. Those were interpolated from ten core materials out of which six are human tissue plus air, two types of aluminum, and iron. The dose calculation in a voxelized phantom or patient is then carried out using the VMC++ MC transport code, resulting in a dose distribution reported as dose to water. The dose grid resolution can be defined by the user. The statistical uncertainty of the dose distribution is indirectly steered by the user's setting of the number of histories per cm² per field

of an electron beam. The performance of the RayStation electron MC algorithm was assessed by several publications such as Huang et al. (2019) and Richmond et al. (2019).

12.3 Commissioning of an MC-Based TPS

Commissioning of an MC-based TPS can be divided into three logical steps:

1. Collection of data to allow beam modelling, as outlined in the user manual.
2. Verification of the beam model.
3. Verification of dose calculations in heterogeneous media and under typical clinical conditions.

12.3.1 Measurements Required for Beam Characterization

To create a beam model, the user has to provide the following information about the treatment unit to the treatment planning vendor:

- Position and thickness of jaw collimators and MLCs.
- For each applicator scraper layer: thickness, position, shape (perimeter and edge), and composition (including density).
- For inserts: thickness, shape, and composition.

The open-field measurements for each energy (without the applicator, with collimator jaws wide open) consist of the following:

- Depth–dose curves in water at SSD = 100 cm.
- Absolute doses (reference dose), expressed in cGy/MU, at a specified point on the depth–dose curve.
- Dose profiles in air at source-to-detector distance = 95 cm (applicator level).

The applicator measurements for each energy/applicator combination are as follows:

- Depth–dose curves in water at SSD = 100 cm.
- Absolute doses (reference dose), expressed in cGy/MU, at a specified point on the depth–dose curve.

The reference dose (in cGy/MU) for the beam defined by an applicator is normally greater than the reference dose of the open beam (no applicator) due to the applicator scatter.

The electron applicator field size is entered in the beam configuration task. The photon jaw settings are set corresponding to beam energy and electron applicator size based on the accelerator manufacturer's specifications.

12.3.2 Additional Beam Data Measurements for Commissioning MC-Based TPSs

To commission an MC-based TPS, additional input data related to the beam characteristics may be needed. For example, to configure the eMC algorithm in the Varian Eclipse system, additional dose profiles in air at a specified source-to-detector distance of 95 cm for open fields (without applicator) but jaws at field sizes used for applicators are required, whereas no dose profiles in water are necessary (Fix et al., 2013). In the MC-based TPS, the user typically needs to measure for each beam energy dose profiles in air and water, both with and without the applicators (Cygler et al., 2004; Ding et al., 2006; Huang et al., 2019). For example, the RayStation electron algorithm requires in-air profiles at two levels being at least 20 cm apart as well as dose profiles in water at two distinctive depths, which are electron beam energy dependent. The open-field (beam without electron applicator) beam data are necessary in order for the TPS to accurately model the incident electron beams for dose calculations. In addition, the in-air profiles provide incident electron beam information (without the effect of an applicator) needed to configure the beam (Cygler et al., 2004; Ding et al., 2006; Huang et al., 2019).

The beam modelling procedure in XiO is based on in-water measurements only. It requires, for each energy and applicator, depth–dose curves and profiles at seven depths for both an SSD of 100 cm and for an extended SSD commonly used in the clinic. The same set of scans is also required for a $15 \times 15\,cm^2$ applicator with a $5 \times 5\,cm^2$ cutout in place. In addition, depth–dose curves and profiles at seven depths are also required for SSD=100 cm with no applicator present and the jaws set to their maximum. The user has to also provide point dose values at a reference depth (typically d_{max}) for each field size and SSD listed above.

12.3.3 Beam Measurements Required for In-Phantom/Patient Dose Calculation Verification

Good clinical practice requires careful commissioning of a TPS. During the commissioning, the user should verify the accuracy of dose calculations of a TPS before such a system is put into clinical use. This can be carried out by performing calculations and comparing the results with the data measured in homogeneous and some heterogeneous phantoms.

12.3.3.1 Homogeneous Phantom: Dose Profiles, MU Calculations at Various SSDs

Water is normally the choice of homogeneous phantom because the dose distributions can be obtained through depth–dose curve or dose profile scanning measurements by using commercially available scanning systems with an ionization chamber, diode, or other type of detector. The measurements of depth–dose curves and dose profiles at several different SSDs that are clinically relevant (e.g., SSD=100 cm, 115 cm) are necessary not only for beam modelling but also for verification of dose calculations at standard and extended SSDs. The validation should also include the accuracy of MU calculations at standard and extended SSDs for open applicators and for a variety of cutout-defined field sizes for each

specified electron applicator. One of the important strengths of an MC-based electron TPS is that it is capable of accurately calculating both dose distributions and MUs not only at a standard SSD but also at any extended SSD (Cygler et al., 2004; Ding et al., 2006). This is a significant advantage because conventional electron dose calculation algorithms are not capable of accurately predicting either the dose distributions or MUs at an extended SSD when the beam is commissioned only at the standard SSD (Ding et al., 2005; Ding et al., 1999).

12.3.3.2 Heterogeneous Phantoms

In addition to the validation of dose calculation in homogeneous phantoms, some tests should be carried out for heterogeneous phantoms before clinical implementation of the TPS. Such tests should include heterogeneous phantoms of various complexities, such as 1D-type or slab, 2D-type, and more complex 3D-type geometries. The density of the heterogeneous materials must include high-density (bone) and low-density (lung) materials that are clinically relevant in the patient treatment planning calculations. The validation involves the comparison of doses calculated by the TPS and the measured ones. Studies by Cygler et al. (2004) and Ding et al. (2006) showed some possible methods of evaluating the accuracy of a commercial electron TPS in a complex phantom containing 3D-type heterogeneities.

12.4 Issues Arising from the Clinical Implementation of MC Dose Calculations

12.4.1 Calculation Normalization, Dose Prescription, and Isodose Lines

As part of the commissioning of an electron TPS for clinical use, one should also validate the accuracy of this system to calculate output factors for various cutout sizes for arbitrary SSDs for all available beam energies. A clinically acceptable agreement between measured and calculated output factors should be obtained. This includes both open applicator and cutout-defined fields. When a calculated dose distribution is not normalized appropriately in a TPS, user applied renormalization may be necessary to obtain accurate MUs. It has been shown that MC methods available in commercial TPSs are capable of calculating output factors accurately even at extended SSDs. This is a significant improvement over the PB algorithm implemented in CadPlan˙ TPS by Varian (Ding et al., 2005).

12.4.2 Statistical Uncertainty, Smoothing, and Calculation Voxel Size

It is noteworthy that any MC calculation includes a statistical uncertainty. It is not easy to have a rigorous method to obtain an accurate statistical uncertainty of the calculated dose distribution. Therefore, if a target prescription dose is based on a point dose calculation, the statistical uncertainty may have a significant effect on the dose prescription. To reduce computational time and to reduce statistical uncertainties of the MC calculated dose distributions, a method called smoothing to smooth the calculated doses between neighboring voxels was introduced. The tools provided for smoothing dose distributions in some commercial TPSs do not discriminate between real dose gradients and statistical variations. When applying the smoothing tools or using large calculation voxels to reduce the computation time, caution must be exercised or the resultant details of the calculated dose distributions may not be accurate, especially in regions of high-dose gradients (Ding et al., 2006). On the other hand, the statistical uncertainty can be misinterpreted as dose variation when the real statistical uncertainty of the calculation is not well known. Figure 12.3 illustrates the effect of using different calculation voxel sizes, smoothing techniques, and statistical uncertainties on dose distributions in an MC-based TPS. More recently, deep learning neural network methods are investigated to reduce the statistical uncertainty (Javaid *et al.*, 2019), which show some promising results (see Chapter 17 in this book).

12.4.3 Dose-to-Medium versus Dose-to-Water Calculations

One of the issues arising when using MC-based TPSs is "dose to water vs. dose to medium" calculations (Ding et al., 2006; Gardner et al., 2007; Keall, 2002; Liu, 2002). The two are conceptually different and the one which is used has a significant impact on the reported doses to different organs, such as the bone and the lung. In conventional TPSs, CT numbers are converted to electron densities of water-like material before calculations are carried out. Unlike conventional TPSs, MC inherently calculates the dose to medium. There can be differences (>10%) in reported doses when dose to water as opposed dose to medium is calculated. This is due to the fact that the stopping-power ratios of water/medium are not likely to be unity for electron beams. There could be significant differences. The study by Ding et al. (2006) provided some examples of such situations. Their investigation found significant differences for the lung and the bone, respectively, when "dose to medium" is reported instead of "dose to water" as done in the conventional planning systems. In addition, the magnitude of this difference is not constant but depends on the beam energy, calculation depth, and medium. This variation is due in part to the fact that water/medium stopping-power ratios are strongly dependent on beam energy unlike the case in photon beams. Figures 12.4 and 12.5 illustrate the magnitude of this difference for hard bone. These differences reflect the fact that the mass stopping power of the bone is lower than that of water while the opposite is true for the lung and water.

This large variation only due to the method of dose reporting may have significant clinical implications. Since MC-based TPSs have entered routine clinical practice, it is imperative that a consistent approach to dose reporting is used. Only

then can the prescription dose and treatment outcomes be compared in a meaningful way between different centers.

12.4.4 Examples of CT-Based Dose Calculations

Figure 12.6 shows a calculated dose distribution for a tumor near the left ear.

It can be seen that the MC calculated dose distributions show significant hot and cold regions due to the presence of low- and high-density media, whereas the PB dose calculation algorithm is not able to accurately predict the dose changes caused by the inhomogeneity. Based on the water/bone stopping-power ratios shown in Figure 12.4, it is worth to note that the lower dose shown in the MC calculated dose profiles is mostly due to the fact that MC calculates dose to medium, whereas the PB calculated dose to water. The ability to accurately predict hot and cold regions in patient treatment planning is essential to provide the best possible choice for treatment options.

12.4.5 Typical Calculation Times

MC dose calculation times depend on the computer hardware, dose calculation algorithm, beam energy, field size, and calculation parameters. The parameters that affect the speed of dose calculation are the number of particle histories or required statistical uncertainty and the voxel size. Users of a commercial TPS do not always have control over all parameters that affect the speed of dose calculations. The Elekta software is the vendor that allows users to set the largest range of calculation parameters. The proper choice of parameters depends on the anatomical site and required statistical uncertainty. Figure 12.7 shows typical calculation times for the trachea and spine phantom using a clinical XiO (version 4.51) workstation (Linux OS, 8 CPU 3 GHz each, 8.29 GB of RAM) for two beam energies and two voxel sizes as a function of MRSU (see Section 12.2.4.3). Typical calculation times for other planning systems can be found in a study by Cygler et al. (2004), Ding et al. (2005), Chetty et al. (2007), Fragoso et al. (2008), Brualla et al., (2017), and Snyder et al. (2019) and are in the range of a few minutes for clinical situations. A concrete example shows calculation times of about 20 and 40 seconds for eMC (2.5 mm^3 voxel, Version 10.0.28) and 200 and 390 seconds for XiO (3 mm^3 voxel, Version 4.60) for a 6 and 16 MeV electron beam, using a 10×10 cm^2 field size, SSD = 100 cm in a homogeneous water phantom, and 2% average standard statistical uncertainty (Brualla et al., 2017).

12.5 Future Clinical Applications

Currently, the potential of electron beam radiotherapy is not fully used; thus, there are several options on how electron beams can improve treatments to patients. In this section, examples of future clinical applications are presented. All

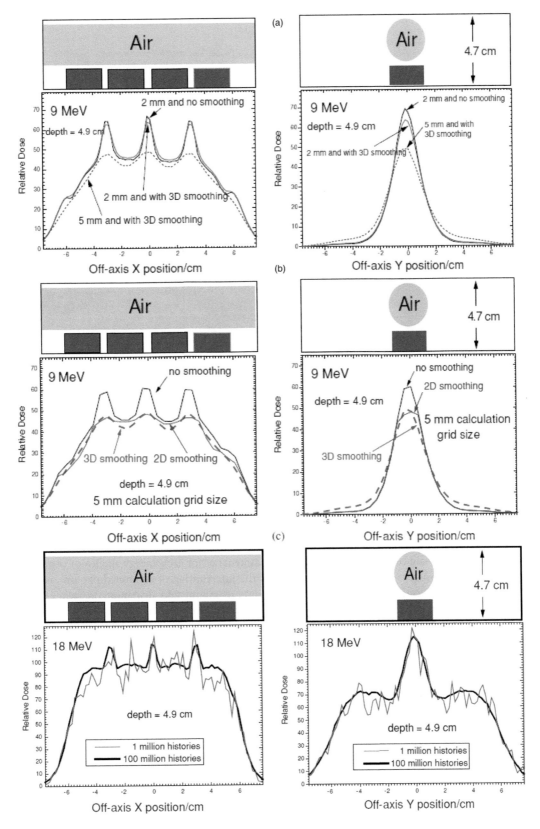

FIGURE 12.3 (a) Effect of using different smoothing techniques with the same calculation voxel sizes on the dose distributions. (b) Effect of using different calculation voxel sizes and smoothing techniques on the depth–dose distributions. (c) Effect of statistical uncertainties on dose distributions in the MC-based TPS. (Reproduced from Ding et al. 2006. *Phys Med Biol* 51(11):2781–99. With permission.)

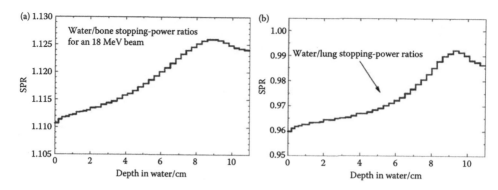

FIGURE 12.4 Illustration of water-to-medium restricted stopping power ratios as a function of depth for an 18 MeV electron beam in water determined using the EGS4 user code to calculate stopping-power ratios (SPRRZ) (cutoff $\Delta = 10$ keV). It is seen that the magnitude of differences between (a) dose to bone or (b) dose to lung and dose to water can be up to 12% or 4%, respectively. (Reproduced from Ding et al. 2006. *Phys Med Biol* 51(11):2781–99. With permission.)

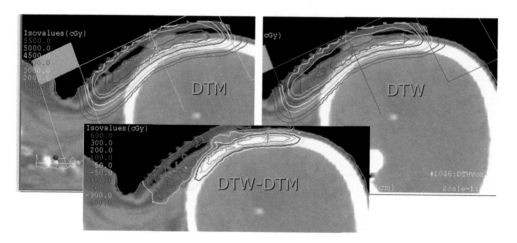

FIGURE 12.5 Clinical example of differences between dose to medium and dose to water calculations for 6 MeV beam (XiO eMC).

these features rely on accurate beam representation and dose calculation algorithms. MC techniques are perfectly suited to fulfill these demands and are essential for promoting these new applications and bring them into clinical routine.

12.5.1 Electron Beams without Applicator

Over the last few decades, an enormous effort was made to replace patient-specific blocks in photon radiotherapy with a photon MLC. The MLC not only substantially improves treatment efficiency and safety but it also has stimulated the development of dynamic delivery techniques such as intensity modulated radiotherapy (IMRT) and volumetric modulated arc therapy (VMAT). A similar effort has not been made for electron beam radiotherapy.

Standard electron treatments currently used in clinical routine are still applied using the extremely cumbersome and inefficient standard or molded patient-specific cutouts placed in dedicated electron applicators for which limited planning features are available. The usage of standard electron treatments needs a huge effort in commissioning and

maintenance for each energy applicator combination, as well as the interruption of combined photon and electron treatments to mount and dismount the heavy add-on applicators. In addition, treatment errors due to accidentally using a wrong cutout are present. All these issues negatively affect the workflow in clinical routine. In addition, the fabrication of Cerrobend cutouts requires special safety precautions such as a fume hood because the alloy contains hazardous materials such as lead and cadmium, and during its melting, toxic gases are emitted.

Using MLC-based collimation devices for electron beams offers great advantages for present standard electron treatments. Using the photon MLC has the additional benefit that the MLC is already part of the treatment head of the linear accelerator and overcomes the above-mentioned limitations (Mueller et al., 2018). Owing to the position of the MLC within the linear accelerator head and the scattering of the electrons in air, the SSD should be reduced as much as possible.

On the other hand, the use of a photon MLC for electron beam treatments increases the complexity of the beam defining

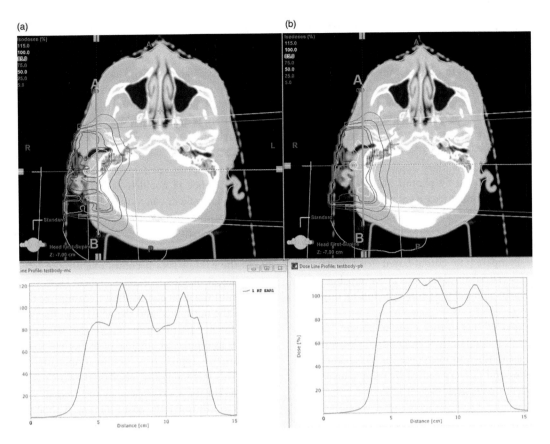

FIGURE 12.6 Calculated dose distributions using (a) MC algorithm and (b) PB algorithm in a commercial TPS using a 2 mm calculation voxel size and no smoothing. It is seen that the PB dose calculation algorithm is not able to accurately predict dose changes caused by the inhomogeneity. The dose profiles start at point A and end at point B as shown in the axial CT images.

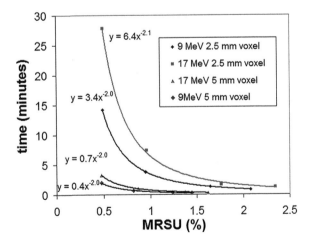

FIGURE 12.7 Timing results for a clinical XiO workstation (Linux OS, 8 CPU, 3 GHz each, 8.29 GB of RAM) illustrating the general behavior.

system and treatment planning. In particular, accurate dose calculation is not trivial. However, using an MC-based beam model and MC dose calculation for predicting the dose distribution of electron beams in radiotherapy with the electron field size shaped with the photon MLC was demonstrated to accurately account for tissue inhomogeneities, particle scattering at the photon MLC, and dose predictions for SSDs not used for commissioning measurements (Henzen et al., 2014c; Mihaljevic et al., 2011; Mueller et al., 2018). It was shown that for different cancer sites standard applicator-based electron treatments can be replaced by photon MLC-based electron treatments, achieving similar treatment plan quality (Mueller et al., 2018). However, for low electron beam energies, laterally located organs at risk (OARs) receive an increased dose contribution especially for extended SSDs because of increased in-air scatter compared to cutout-based treatments. This drawback is less crucial in cases where electron beams are used to boost a subvolume of the primary planning target volume (PTV).

Thus, it is expected that in future use of photon MLCs as an alternative collimation device for standard electron treatments will be established to overcome the disadvantages to clinical workflow and delivery caused by using conventional cutout-based electron applicators.

12.5.2 MERT Using Photon MLC

Most tumor sites including malignancies such as chest wall, small intact breast, paraspinal lesions, or some head and neck cases, e.g., parotid gland, are presently treated using photon beams. However, the shallow dose gradient is a considerable drawback as it leads to augmented dose in OARs and normal tissue, resulting in not optimal treatments. Such superficial tumors MERT is becoming a promising treatment modality, as electron beams have a well-defined therapeutic range and a rapid dose falloff. In MERT, a segmented delivery approach is typically applied in which a set of apertures are superimposed. Thereby, each aperture is associated with an electron beam energy, a shape, and a weight (corresponding to the number of MUs). Thus, MERT is especially of interest when using the photon MLC for shaping the electron field. For this configuration, MC-based dose calculations were investigated by several groups (Henzen et al., 2014a; Henzen et al., 2014b; Klein et al., 2008; Mihaljevic et al., 2011; Salguero et al., 2009; Surucu et al., 2010). However, there have been also contributions to this field with configurations using alternative collimating systems such as a few leaf electron collimator (Al-Yahya et al., 2005; Al-Yahya et al., 2007; Alexander et al., 2010) or a dedicated electron MLC (Engel and Gauer, 2009; Jin et al., 2014; Vatanen et al., 2009).

For radiotherapy with electrons, the electron beam energy is a degree of freedom in plan optimization and more relevant than the intensity modulation of the individual apertures (Joosten et al., 2018). This additional degree of freedom has to be appropriately taken into account during the optimization. While initially a forward planning of MERT was investigated,

it became clear that an inverse planning tool for MERT is needed to gain full benefit from the dosimetric behavior of electrons (Henzen et al., 2014b; Klein et al., 2009; Surucu et al., 2010). One typical approach is a beamlet-based optimization to estimate an optimal fluence also called fluence map optimization (FMO). For this purpose, the energy fluence impinging on the patient is divided into small areas, called beamlets. The dose distribution for each beamlet is then calculated mainly with MC techniques. Using the dose per beamlet and a cost function, an optimized fluence can be determined. As the resulting fluence is not deliverable generally, an MLC leaf sequencer is applied, converting the fluence in a set of deliverable apertures with a certain weight (Salguero et al., 2009). This leaf sequencing task is more complex in cases of electron beams compared to photon beams, as the fluence perturbation due to the MLC leaves becomes much more important and ray-tracing strategies cannot be applied. However, a multiobjective minimization can be used to minimize the difference between the optimized and the deliverable fluence, minimizing the fluence perturbation due to the MLC and to obtain apertures as regular as possible to reduce electron scattering. An alternative approach is the direct aperture optimization (DAO). This method is also based on the dose per beamlet distribution, but the aperture shapes and weights are iteratively changed to minimize the cost function. Although the resulting treatment plan is deliverable with respect to the aperture shapes and weights, the perturbation of the MLC, e.g., due to scattering, is considered in the final MC dose calculation typically followed by a reoptimization of the aperture weights.

Treatment plan comparisons for different sites show advantages for OARs in the high-dose region at the expense of an increased low-dose contribution to the OARs and a more inhomogeneous dose distribution to the target when compared to plans using solely photon beams (Henzen et al., 2014a; Klein et al., 2009; Salguero et al., 2010; Surucu et al., 2010). A potential disadvantage is an increased skin dose for MERT compared with that of photon only treatment techniques. However, more studies are needed to investigate this issue. An improved alternative is to combine electron and photon beams, leading to the MBRT as outlined in the next paragraph.

12.5.3 Mixed Electron and Photon Beam Radiotherapy

The idea in MBRT using electrons and photons is to benefit from the advantageous properties of both particle types such as the distal steep falloff of electron beams as well as the narrow penumbra and deep penetration of photon beams. Combining photon and electron beams results in improved treatment plan quality compared to MERT, IMRT, or VMAT for superficial tumors.

The concept of mixed photon and electron beams is well known in clinical practice. Typically, a single electron beam is applied as a boost using an electron applicator mounted on the gantry. However, this approach does not take advantage of inverse planning optimization, which is of special interest when the photon MLC is used to shape both photon and electron beams, as already discussed in the previous paragraphs. Thus, combining MERT and intensity modulated photon beams is a promising approach, where a simultaneous optimization is desirable in to use the optimizer to exploit the particle type as a degree of freedom during the optimization and finding the optimal mix of electron and photons (Miguez et al., 2017; Mueller et al., 2017; Palma et al., 2012; Renaud et al., 2017).

Studies using MC calculated beamlet-based simultaneous FMO and deliverable DAO were performed (Mueller et al., 2017; Palma et al., 2012; Renaud et al., 2017). As already discussed for MERT, the FMO results in optimized fluence distributions; however, the associated treatment plans are not deliverable without applying a leaf sequencing algorithm, which unfortunately degrades the plan quality. For DAO, the simulated-annealing or the column-generation approach was used for simultaneous optimization in MBRT (Mueller et al., 2017; Renaud et al., 2017). The resulting treatment plans demonstrate that the dose contribution from the electron part is not negligible, i.e., the treatment plans benefit from the dosimetric properties of the electron beams. In addition, these contributions are not intuitive, thus suggesting the limited usability for forward and sequential optimization, in which typically an electron dose distribution is set as a basis on top of which the photon contribution is optimized.

In general, studies demonstrated that MBRT treatment plans are superior to treatment plans using the MERT, IMRT, or VMAT techniques (Mueller et al., 2017; Renaud et al., 2017). More specifically, MBRT treatment plans achieve a target coverage and dose homogeneity which are practically the same as for IMRT or VMAT but often compromised in MERT. In addition, in MBRT, the sparing of OARs is improved throughout compared with that in the other treatment techniques. For example, the dosimetric comparison of treatment plans for breast MBRT resulted in similar target coverage and homogeneity but an improved OAR sparing when compared with photon IMRT (Palma et al., 2012). Another example showed at least the same coverage and homogeneity of the target dose for MBRT treatment plans as for VMAT plans, while mean dose to parallel OARs and dose to 2% of the organ volume, D2%, for serial OARs were reduced on average by 54% and 26%, respectively. In addition, MBRT treatment plans typically result in a lower dose bath when compared with VMAT plans (Mueller et al., 2017).

Overall, MBRT is a promising delivery technique which overcomes limitations and disadvantages from photon or electron only techniques. Owing to MC techniques, accurate dose distributions for MBRT treatment plans can be calculated in this complex scenario. Additional studies should further evaluate and quantify the potential for patients with tumors extending from superficial to larger depths. Those studies should also include investigations about the robustness of the

dose distributions by comparing the planned dose distributions with corresponding measurements.

12.6 Summary

Commercial MC-based TPSs are common in clinics, which has resulted in more accurate dose calculations for electron beams. Although both PB and MC calculations are able to predict dose distribution accurately in homogeneous phantoms, MC provides significant improvements for patient treatment planning. In addition, to commission an electron beam in a commercial TPS is simpler for a MC system as beam data at only one standard SSD are needed for each energy. These can be used for accurate dose and MU calculations at any arbitrary SSD, as demonstrated in the study by Cygler et al. (2004) and Ding et al. (2005). In heterogeneous phantoms, MC dose calculation methods offer superior accuracy compared to the PB algorithm, especially where small 3D heterogeneities or overlying low- and high-density materials such as the air and bone are present. MC results generally show much better agreement with measurements, especially in high-dose gradient regions caused by the perturbation of adjacent 3D inhomogeneities. The accurate dose distributions allow clinicians to use electron beams with more confidence.

Owing to the ability of MC to accurately estimate the dose distributions, new techniques of using the photon MLC to shape the electron fields as well as more advanced techniques such as MERT and MBRT could be used in the future, which in turn may increase the routine usage of electron beams in clinical practice. Another field of application for eMC in the near future might be Flash. Recently, the first patient was treated with this high dose rate technique using electrons (Bourhis et al., 2019).

In the meantime, it is important for the clinical user to become familiar with MC dose calculation techniques and to understand the MC implementation in a particular commercial TPS to select appropriate calculation settings for a given clinical situation. Because commercial implementations use various approximations in source models, it is worth noting that there is no absolute guarantee that the calculation result from the MC-based TPS is always accurate.

References

Al-Yahya K, Schwartz M, Shenouda G, Verhaegen F, Freeman C and Seuntjens J. 2005. Energy modulated electron therapy using a few leaf electron collimator in combination with IMRT and 3D-CRT: Monte Carlo-based planning and dosimetric evaluation. *Med Phys* **32**: 2976–86.

Al-Yahya K, Verhaegen F and Seuntjens J. 2007. Design and dosimetry of a few leaf electron collimator for energy modulated electron therapy. *Med Phys* **34**: 4782–91.

Alexander A, DeBlois F and Seuntjens J. 2010. Toward automatic field selection and planning using Monte

Carlo-based direct aperture optimization in modulated electron radiotherapy. *Phys Med Biol* **55**: 4563–76.

Ali O A, Willemse C A, Shaw W, O'Reilly F H and du Plessis F C. 2011. Monte carlo electron source model validation for an Elekta Precise linac. *Med Phys* **38**: 2366–73.

Allison J, et al. 2006. Geant4 developments and applications. *IEEE Trans Nucl Sci* **53**: 270–8.

Bazalova M and Verhaegen F. 2007. Monte Carlo simulation of a computed tomography x-ray tube. *Phys Med Biol* **52**: 5945–55.

Bazalova M, Zhou H, Keall P J and Graves E E. 2009. Kilovoltage beam Monte Carlo dose calculations in sub-millimeter voxels for small animal radiotherapy. *Med Phys* **36**: 4991–9.

Bourhis J, et al. 2019. Treatment of a first patient with FLASH-radiotherapy. *Radiother Oncol* **139**: 18–22.

Brahme A. 1984. Dosimetric precision requirements in radiation therapy. *Acta Radiol Oncol* **23**: 379–91.

Briesmeister B A. 1997. MCNP-A general Monte Carlo N-particle transport code. Version 4C, LA-13709-M, Los Alamos National Laboratory.

Brualla L, Rodriguez M and Lallena A M. 2017. Monte Carlo systems used for treatment planning and dose verification. *Strahlenther Onkol* **193**: 243–59.

Chetty I J, et al. 2007. Report of the AAPM Task Group No. 105: issues associated with clinical implementation of Monte Carlo-based photon and electron external beam treatment planning *Med Phys* **34**: 4818–53.

Cygler J, Battista J J, Scrimger J W, Mah E and Antolak J. 1987. Electron dose distributions in experimental phantoms: a comparison with 2D pencil beam calculations. *Phys Med Biol* **32**: 1073–86.

Cygler J, Li X A, Ding G X and Lawrence E. 1997. Practical approach to electron beam dosimetry at extended SSD. *Phys Med Biol* **42**: 1505–14.

Cygler J E, Daskalov G M, Chan G H and Ding G X. 2004. Evaluation of the first commercial Monte Carlo dose calculation engine for electron beam treatment planning. *Med Phys* **31**: 142–53.

Ding G X. 1995. *An Investigation of Radiotherapy Electron Beams Using Monte Carlo Techniques*. Ottawa, Canada: Carleton University.

Ding G X, Cygler J E, Yu C W, Kalach N I and Daskalov G. 2005. A comparison of electron beam dose calculation accuracy between treatment planning systems using either a pencil beam or a Monte Carlo algorithm. *Int J Radiat Oncol Biol Phys* **63**: 622–33.

Ding G X, Cygler J E, Zhang G G and Yu M K. 1999. Evaluation of a commercial three-dimensional electron beam treatment planning system. *Med Phys* **26**: 2571–80.

Ding G X, Duggan D M, Coffey C W, Shokrani P and Cygler J E. 2006. First macro Monte Carlo based commercial dose calculation module for electron beam treatment planning--new issues for clinical consideration. *Phys Med Biol* **51**: 2781–99.

Ding G X and Rogers D W O. 1995. Energy spectra, angular spread, & dose distributions of electron beams from various accelerators used in radiotherapy. *National Research Council of Canada Report*.

Edimo P, Clermont C, Kwato M G and Vynckier S. 2009. Evaluation of a commercial VMC++ Monte Carlo based treatment planning system for electron beams using EGSnrc/BEAMnrc simulations and measurements. *Phys Med* **25**: 111–21.

Engel K and Gauer T. 2009. A dose optimization method for electron radiotherapy using randomized aperture beams. *Phys Med Biol* **54**: 5253–70.

Fippel M. 1999. Fast Monte Carlo dose calculation for photon beams based on the VMC electron algorithm. *Med Phys* **26**: 1466–75.

Fippel M, Kawrakow I and Friedrich K. 1997. Electron beam dose calculations with the VMC algorithm and the verification data of the NCI working group. *Phys Med Biol* **42**: 501–20.

Fippel M, Laub W, Huber B and Nusslin F. 1999. Experimental investigation of a fast Monte Carlo photon beam dose calculation algorithm. *Phys Med Biol* **44**: 3039–54.

Fix M K, Frei D, Volken W, Neuenschwander H, Born E J and Manser P. 2010. Monte Carlo dose calculation improvements for low energy electron beams using eMC. *Phys Med Biol* **55**: 4577–88.

Fix M K, et al. 2013. Generalized eMC implementation for Monte Carlo dose calculation of electron beams from different machine types. *Phys Med Biol* **58**: 2841–59.

Fragoso M, Pillai S, Solberg T D and Chetty I J. 2008. Experimental verification and clinical implementation of a commercial Monte Carlo electron beam dose calculation algorithm. *Med Phys* **35**: 1028–38.

Francescon P, Cavedon C, Reccanello S and Cora S. 2000. Photon dose calculation of a three-dimensional treatment planning system compared to the Monte Carlo code BEAM. *Med Phys* **27** 1579–87.

Francescon P, Cora S and Cavedon C. 2008. Total scatter factors of small beams: a multidetector and Monte Carlo study. *Med Phys* **35**: 504–13.

Francescon P, Cora S, Cavedon C and Scalchi P. 2009. Application of a Monte Carlo-based method for total scatter factors of small beams to new solid state micro-detectors. *J Appl Clin Med Phys* **10**(1): 2939.

Gardner J K, Siebers J V and Kawrakow I. 2007. Comparison of two methods to compute the absorbed dose to water for photon beams. *Phys Med Biol* **52**: N439–47.

Henzen D, et al. 2014a. Beamlet based direct aperture optimization for MERT using a photon MLC. *Med Phys* **41**: 121711.

Henzen D, et al. 2014b. Forward treatment planning for modulated electron radiotherapy (MERT) employing Monte Carlo methods. *Med Phys* **41**: 031712.

Henzen D, et al. 2014c. Monte Carlo based beam model using a photon MLC for modulated electron radiotherapy *Med Phys* **41**: 021714.

Hogstrom K R and Almond P R 1983. Comparison of experimental and calculated dose distributions. Electron beam dose planning at the M.D. Anderson Hospital. *Acta Radiol Suppl* **364**: 89–99.

Hogstrom K R, Kurup R G, Shiu A S and Starkschall G. 1989. A two-dimensional pencil-beam algorithm for calculation of arc electron dose distributions. *Phys Med Biol* **34**: 315–41.

Hogstrom K R, Mills M D and Almond P R 1981. Electron beam dose calculations. *Phys Med Biol* **26**: 445–59.

Hogstrom K R, et al. 1984. Dosimetric evaluation of a pencil-beam algorithm for electrons employing a two-dimensional heterogeneity correction. *Int J Radiat Oncol Biol Phys* **10**: 561–9.

Huang J Y, Dunkerley D and Smilowitz J B. 2019. Evaluation of a commercial Monte Carlo dose calculation algorithm for electron treatment planning. *J Appl Clin Med Phys* **20**(6): 184–93.

Iaccarino G, et al. 2011. Monte Carlo simulation of electron beams generated by a 12 MeV dedicated mobile IORT accelerator. *Phys Med Biol* **56**: 4579–96.

ICRU-29. 1978. Dose specifications for reporting external beam therapy with photons and electrons. *ICRU: Report 29*. International Commission on Radiation Units and Measurements, Bethesda, MD.

ICRU-71. 2004. Prescribing, recording, and reporting elecron beam therapy. *ICRU: Report 71*. Oxford University Press, Oxford.

Jarry G, Graham S A, Moseley D J, Jaffray D J, Siewerdsen J H and Verhaegen F. 2006. Characterization of scattered radiation in kV CBCT images using Monte Carlo simulations. *Med Phys* **33**: 4320–9.

Javaid U, Souris K, Dasnoy D, Huang S and Lee J A. 2019. Mitigating inherent noise in Monte Carlo dose distributions using dilated U-Net. *Med Phys* **46**: 5790–8.

Jin L, et al. 2014. Measurement and Monte Carlo simulation for energy- and intensity-modulated electron radiotherapy delivered by a computer-controlled electron multileaf collimator. *J Appl Clin Med Phys* **15**(1): 4506.

Joosten A, et al. 2018. A dosimetric evaluation of different levels of energy and intensity modulation for inversely planned multi-field MERT. *Biomed Phys & Eng Exp* **4**: 045003.

Kawrakow I. 1997. Improved modelling of multiple scattering in the Voxel Monte Carlo model. *Med Phys* **24**: 505–17.

Kawrakow I. 2000. Accurate condensed history Monte Carlo simulation of electron transport. I. EGSnrc, the new EGS4 version. *Med Phys* **27**: 485–98.

Kawrakow I. *Advanced Monte Carlo for Radiation Physics, Particle Transport Simulation and Applications*, (Berlin, Heidelberg, 2001//2001, vol. Series) ed. A Kling, et al. Berlin, Heidelberg: Springer, pp. 229–36.

Kawrakow I, Fippel M and Friedrich K. 1996. 3D electron dose calculation using a Voxel based Monte Carlo algorithm (VMC). *Med Phys* **23**: 445–57.

Kawrakow I and Rogers D W O. 2002. The EGSnrc Code System: Monte Carlo Simulation of Electron and Photon Transport. *NRCC Report Pirs-701*.

Keall P. 2002. Dm rather than Dw should be used in Monte Carlo treatment planning. Against the proposition. *Med Phys* **29**: 923–4.

Keall P J and Hoban P W. 1996. Super-Monte Carlo: a 3-D electron beam dose calculation algorithm. *Med Phys* **23**: 2023–34.

Keall P J, Siebers J V, Libby B and Mohan R. 2003. Determining the incident electron fluence for Monte Carlo-based photon treatment planning using a standard measured data set. *Med Phys* **30**: 574–82.

Khan F M, et al. 1991. Clinical electron-beam dosimetry: report of AAPM Radiation Therapy Committee Task Group No. 25. *Med Phys* **18**: 73–109.

Klein E E, Mamalui-Hunter M and Low D A. 2009. Delivery of modulated electron beams with conventional photon multi-leaf collimators. *Phys Med Biol* **54**: 327–39.

Klein E E, Vicic M, Ma C M, Low D A and Drzymala R E. 2008. Validation of calculations for electrons modulated with conventional photon multileaf collimators. *Phys Med Biol* **53**: 1183–208.

Liu H H. 2002. Dm rather than Dw should be used in Monte Carlo treatment planning. For the proposition. *Med Phys* **29**: 922–3.

Ma C M, Faddegon B A, Rogers D W and Mackie T R. 1997. Accurate characterization of Monte Carlo calculated electron beams for radiotherapy. *Med Phys* **24**: 401–16.

Ma C M, et al. 2000. Monte Carlo verification of IMRT dose distributions from a commercial treatment planning optimization system. *Phys Med Biol* **45**: 2483–95.

Miguez C, et al. 2017. Clinical implementation of combined modulated electron and photon beams with conventional MLC for accelerated partial breast irradiation. *Radiother Oncol* **124**: 124–9.

Mihaljevic J, Soukup M, Dohm O and Alber M. 2011. Monte Carlo simulation of small electron fields collimated by the integrated photon MLC. *Phys Med Biol* **56**: 829–43.

Mueller S, et al. 2017. Simultaneous optimization of photons and electrons for mixed beam radiotherapy. *Phys Med Biol* **62**: 5840–60.

Mueller S, et al. 2018. Electron beam collimation with a photon MLC for standard electron treatments. *Phys Med Biol* **63**: 025017.

Nelson W R, Hirayama H and Rogers D W O. 1985. EGS4 code system. *Report SLAC-265*. Menlo Park, CA, United States.

Neuenschwander H and Born E J. 1992. A macro Monte Carlo method for electron beam dose calculations. *Phys Med Biol* **37**: 107–25.

Neuenschwander H, Mackie T R and Reckwerdt P J. 1995. MMC--a high-performance Monte Carlo code for electron beam treatment planning. *Phys Med Biol* **40**: 543–74.

Ojala J, Kapanen M and Hyodynmaa S. 2016. Full Monte Carlo and measurement-based overall performance assessment of improved clinical implementation of eMC algorithm with emphasis on lower energy range. *Phys Med* 32: 801–11.

Palma B A, et al. 2012. Combined modulated electron and photon beams planned by a Monte-Carlo-based optimization procedure for accelerated partial breast irradiation. *Phys Med Biol* 57: 1191–202.

Papanikolaou N, et al. 2004. Tissue inhomogeneity corrections for megavoltage photon beams. *AAPM Report No 85*; Task Group No. 85; Task Group No. 65.

Pasciuti K, et al. 2011. Tissue heterogeneity in IMRT dose calculation for lung cancer. *Med Dosim* 36 219–27.

Pimpinella M, Mihailescu D, Guerra A S and Laitano R F. 2007. Dosimetric characteristics of electron beams produced by a mobile accelerator for IORT. *Phys Med Biol* 52: 6197–214.

Popple R A, et al. 2006. Comprehensive evaluation of a commercial macro Monte Carlo electron dose calculation implementation using a standard verification data set. *Med Phys* 33: 1540–51.

Renaud M A, Serban M and Seuntjens J. 2017. On mixed electron-photon radiation therapy optimization using the column generation approach. *Med Phys* 44: 4287–98.

Reynaert N, et al. 2007. Monte Carlo treatment planning for photon and electron beams. *Radiation Physics and Chemistry* 76: 643–86.

Richmond N, Allen V, Wyatt J and Codling R. 2019. Evaluation of the RayStation electron Monte Carlo dose calculation algorithm. *Med Dosim*.

Rogers D W. 2006. Fifty years of Monte Carlo simulations for medical physics. *Phys Med Biol* 51: R287–301.

Rogers D W, Faddegon B A, Ding G X, Ma C M, We J and Mackie T R. 1995. BEAM: a Monte Carlo code to simulate radiotherapy treatment units. *Med Phys* 22: 503–24.

Salguero F J, Arrans R, Palma B A and Leal A. 2010. Intensity- and energy-modulated electron radiotherapy by means of an xMLC for head and neck shallow tumors. *Phys Med Biol* 55: 1413–27.

Salguero F J, Palma B, Arrans R, Rosello J and Leal A. 2009. Modulated electron radiotherapy treatment planning using a photon multileaf collimator for post-mastectomized chest walls. *Radiother Oncol* 93: 625–32.

Schreiber E C and Faddegon B A. 2005. Sensitivity of large-field electron beams to variations in a Monte Carlo accelerator model. *Phys Med Biol* 50: 769–78.

Seltzer S M. 1991. Electron-photon Monte Carlo calculations: the ETRAN code. *Int J Radiat Appl Instrum. Part A. Appl Radiat Isot* 42: 917–41.

Sempau J, Acosta E, Baro J, Fernández-Varea J M and Salvat F 1997. An algorithm for Monte Carlo simulation of coupled electron-photon transport. *Nucl Instrum Methods Phys Res Sect B: Beam Interact Mater Atoms* 132: 377–90.

Sheikh-Bagheri D and Rogers D W. 2002a. Monte Carlo calculation of nine megavoltage photon beam spectra using the BEAM code. *Med Phys* 29 391–402.

Sheikh-Bagheri D and Rogers D W. 2002b. Sensitivity of megavoltage photon beam Monte Carlo simulations to electron beam and other parameters. *Med Phys* 29: 379–90.

Shiu A S and Hogstrom K R. 1991. Dose in bone and tissue near bone-tissue interface from electron beam. *Int J Radiat Oncol Biol Phys* 21: 695–702.

Snyder J E, Hyer D E, Flynn R T, Boczkowski A and Wang D. 2019. The commissioning and validation of Monaco treatment planning system on an Elekta VersaHD linear accelerator. *J Appl Clin Med Phys* 20(1): 184–93.

Starkschall G, Shiu A S, Bujnowski S W, Wang L L, Low D A and Hogstrom K R. 1991. Effect of dimensionality of heterogeneity corrections on the implementation of a three-dimensional electron pencil-beam algorithm. *Phys Med Biol* 36: 207–27.

Surucu M, Klein E E, Mamalui-Hunter M, Mansur D B and Low D A. 2010. Planning tools for modulated electron radiotherapy. *Med Phys* 37: 2215–24.

Traneus E, et al. 2001. Application and verfication of a coupled multi-source electron beam source model for Monte Carlo based treatment planning. *Radiother Oncol* 61: S102.

Vatanen T, Traneus E and Lahtinen T. 2009. Comparison of conventional inserts and an add-on electron MLC for chest wall irradiation of left-sided breast cancer. *Acta Oncol* 48: 446–51.

Wieslander E and Knoos T. 2007. A virtual-accelerator-based verification of a Monte Carlo dose calculation algorithm for electron beam treatment planning in clinical situations. *Radiother Oncol* 82: 208–17.

Zhang G G, Rogers D W, Cygler J E and Mackie T R. 1998. Effects of changes in stopping-power ratios with field size on electron beam relative output factors. *Med Phys* 25: 1711–6.

Zhang G G, Rogers D W, Cygler J E and Mackie T R. 1999. Monte Carlo investigation of electron beam output factors versus size of square cutout. *Med Phys.* 26: 743–50.

13

Protons: Clinical Considerations and Applications

Harald Paganetti
*Massachusetts General
Hospital and Harvard
Medical School*

13.1 Introduction

13.1.1 Short Introduction to Proton Therapy Beam Delivery

Proton therapy delivery is typically done either with passively scattered beams or with scanned beams (there are also hybrid methods such as wobbling). While older proton therapy facilities are based on the passive scattering technique, today most proton therapy patients are being treated using magnetic beam scanning. Both techniques can deliver a spread-out Bragg peak (SOBP) with a homogeneous target dose for each field, while the latter technique also allows intensity-modulated proton beam therapy (IMPT) with the potential to delivery highly inhomogeneous doses for a single field. The basic principles are outlined in Figure 13.1 (see Ref. [1] for details).

To deliver a uniform dose to the target for a given treatment field (i.e., beam angle), a SOBP ensures coverage along the thickness of the target volume in beam's eye view. It is created by combining pristine Bragg curves with different beam energies entering the patient. This can be achieved using a fixed beam energy and rotating absorber consisting of steps of various water-equivalent thicknesses (as in passive scattering) or by discretely changing the beam energy before the

beam enters the room (as in beam scanning). In some scanning systems and particularly for lower energies, the beam energy might be adjusted using a range shifter in the treatment head. In order to cover the target volume laterally, the beam needs to be scanned magnetically in x and y directions (beam scanning) or the proton beam needs to be broadened by using scatterers and absorbers consisting of various material combinations in order to achieve a laterally flat dose profile (passive scattering). In passive scattering, further shaping of the field is accomplished by apertures that confine the lateral dimension of the field and range compensators that are sculpted to conform the dose distribution to the distal shape of the target volume. For beam scanning, no patient-specific hardware is needed. However, depending on the width of the delivered pencil beams, an aperture might be used to improve the lateral penumbra of the beam.

13.1.2 Proton Physics

The dose in proton therapy is deposited via ionizations, excitations, multiple Coulomb scattering, and nuclear interactions. The energy loss of protons slowing down in tissue can be calculated using the Bethe-Bloch equation. When tracking

DOI: 10.1201/9781003211846-16

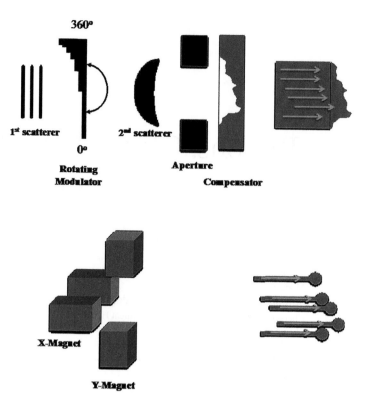

FIGURE 13.1 Illustration of beam delivery in passive scattered proton therapy and beam scanning. Upper, passive scattering: The beam enters from the left and is broadened by a double-scattering system and modulated to an SOBP by a rotating modulator. The field is confined by an aperture. A range compensator modulates the range so that the distal edge of the SOBP follows the distal edge of the target. Lower, beam scanning: Two magnets are used to scan a pencil beam in *x* and *y* directions. The beam energy is ideally modified prior to entering the treatment head (an aperture might still be used to improve the field penumbra).

particles, Monte Carlo simulations typically employ multiple scattering in condensed history class II methods [2]. For instance, Molière theory predicts the scattering angle distribution and the Lewis method [3] allows calculation of moments of lateral displacement, angular deflection, and correlations of these quantities. The specific implementation of the multiple scattering theory can differ slightly from Molière's theory [4–6]. Multiple Coulomb scattering is the main reason behind the broadening of the beam with depth. The Bragg peak is caused by a combination of two effects, i.e. protons slowing down with energy loss due to ionizations being inversely proportional to proton velocity and the reduction of the number of protons with depth because they stop or underwent a nuclear collision.

Depending on the interaction type, the physics is based on models, parameterizations, experimental data, and combinations of these. For nuclear interactions, cross sections as a function of proton energy may not be available for all reaction channels and need to be approximated. For secondary particle emission, interactions between the proton and a nucleus can be modeled as an intranuclear cascade. Once the energy of the particles in a cascade has reached a lower limit, a pre-equilibrium model can be applied with the probability of secondary particle emission. Although nuclear interactions do not significantly influence the shape of the Bragg peak, they do have a significant impact on the depth-dose distribution because they cause a reduction in the proton fluence as a function of depth as well as secondary protons contributing significantly to dose in the entrance region of the Bragg curve [7]. Figure 13.2 shows the significant contribution of secondary protons to the dose in the entrance region of a Bragg curve and the proton fluence reduction due to nuclear interactions as a function of depth. As a rule of thumb, about 1% of primary protons undergo a nuclear interaction per cm beam range. Large angle scattering as well as secondary particles emitted in nuclear interactions cause a 'nuclear halo' around each pencil. The contribution can be significant when adding multiple pencils in beam scanning [8–10]. Consequently, Monte Carlo simulations need to predict the nuclear halo correctly [10,11].

Production cuts for secondary particles can influence the simulated energy loss and thus the simulation results [13]. For proton dose calculation, primary and secondary protons account for about 98% of the dose, depending on the beam energy and including the energy lost via secondary electrons created by ionizations [7]. The maximum range of the delta electrons in water is about 2.5 mm for a 250 MeV proton. The highest energy electrons are preferentially ejected in a forward direction. The electron distribution is much skewed toward lower energies, and the energy of most electrons in a proton beam is much less than 300 keV, which corresponds to

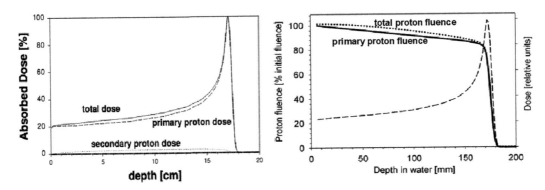

FIGURE 13.2 Depth-dose (Bragg) curve of a 160 MeV proton beam impinging on a water phantom and contribution of primary protons to the total dose (left) and to the total proton fluence (right). (Adapted from Ref. [7,12].)

a range of <1 mm in water. The explicit tracking of secondary electrons can thus be neglected in most cases. For applications other than dose calculation, additional particles might need to be considered, e.g., neutrons.

A Monte Carlo code might allow different physics settings from which the user can choose. Default settings might not be tailored to the energy domain of proton therapy and therefore require adjustment [14–16]. The dependencies of proton therapy-related simulations on different physics settings can be considerable [17–20]. Even with the 'correct' settings, values for the mean excitation energy might have uncertainties in the order of 5%–15%. Such an uncertainty for beam shaping materials can lead to a few mm uncertainty in the predicted beam range in water [21,22]. It is important to consider this uncertainty when simulating energy loss in thick targets such as beam absorbers [13,21,23,24]. Range uncertainties in patients will be discussed in Section 13.3 below. In particular for nuclear interactions, there are significant uncertainties in physics cross sections or models. For example, according to the ICRU [25], angle-integrated emission spectra for neutron and proton interactions are known only within 20%–30%. Despite these uncertainties for some specific reaction channels, dose calculation can typically be done with an accuracy of ~1%–2%.

13.1.3 Codes for Proton Monte Carlo

Various Monte Carlo codes can be used for simulations in proton therapy, e.g., FLUKA [26,27], Geant4 [28,29], MCNPX [30,31], VMCpro [32], Shield-Hit [33], and many others. Typically, a Monte Carlo program is a software executable for which the user has to write an input file depending on the specific problem. There are also other approaches, like in Geant4, where the code is a toolkit providing an assembly of object-oriented libraries with the functionality to simulate different process organized in different functions within a C++ class structure. This requires programming knowledge by the user. Monte Carlo codes differ in the level of control over tracking parameters. The ability to control every parameter (e.g., physics settings, step sizes, and material constants) adds flexibility but also requires a more knowledgeable user.

Most of these codes are multipurpose Monte Carlo codes that were not necessarily designed for radiation therapy applications. In order to tailor these codes for radiation therapy and in order to make them available to non-experts, frameworks (wrappers) have been developed. Examples are GAMOS [34], GATE [35,36], PTsim [37], TOPAS [38,39] (all based on GEANT4), and FICTION [40] (based on FLUKA). These systems extend the functionality of the basic codes and provide user-friendly interfaces or provide code templates. Most recently, TOPAS was designed and supported within the 'Informatics Technologies for Cancer Research' initiative of the National Cancer Institute in the USA. It can model physics and biology experiments for radiation therapy in order to allow non-experts and non-physicists to perform complex Monte Carlo simulations even without the need to engage in programming.

13.2 Simulating the Radiation Field Incident on a Patient or an Experimental Setup

13.2.1 Characterizing Proton Beams at the Treatment Head Entrance

Proton therapy facilities typically consist of a cyclotron or synchrotron, which is connected to one or several treatment rooms by a beamline. The basic beamline elements between the accelerator and the treatment head that might be simulated using Monte Carlo are bending magnets and energy degraders. Monte Carlo beam transport through Carbon and Beryllium degraders has been performed with the goal of improving beam characteristics [13]. However, beam optics calculations are often done numerically [41] despite specialized Monte Carlo codes that simulate magnetic beam steering (see also chapter by Grevillot) [42].

In typical applications for radiation therapy, Monte Carlo simulations either start at the entrance or exit of the treatment head. If the Monte Carlo simulation starts at the treatment head entrance, a reliable parameterization of the beam at this position is needed. The variables are beam energy (E), energy spread (ΔE), beam spot size (σ_x, σ_y), and beam angular distribution ($\sigma_{\theta x}$, $\sigma_{\theta y}$) [43,44]. There are various methods to measure these parameters. The beam spot size can usually be determined with a segmented transmission ionization chamber located at the treatment head entrance. The size of a proton beam is usually in the order of 2–8 mm in σ_x or σ_y, while the angular spread can be in the order of 2–5 mm-mrad for a beam coming from a cyclotron. It can also be parameterized using the emittance of the beam, defined as the product of the size and angular divergence of the beam in a plane perpendicular to the beam direction. The significance of the beam spot size is high for beam scanning systems but is typically less critical for passive scattering. The energy and energy spread can be obtained with sufficient accuracy by measuring range and shape of Bragg peaks in water [45,46] or using an elastic scattering technique [47]. Their values might influence the flatness of an SOBP because of the peak-to-plateau ratio of the individual Bragg peaks that form an SOBP. The energy spread of proton beams entering the treatment head from a cyclotron is typically <1% (DE/E) while a synchrotron may extract beams with an energy spread two orders of magnitude lower.

The parameters describing the beam are typically correlated. Such a correlation, for example, between the particle's position within a beam spot and its angular momentum, needs to be taken into account when modelling proton beam scanning. A full parameterization of the phase space at treatment head entrance might be defined based on the knowledge of the magnetic beam steering system, i.e., from first principles, or by fitting measured data [48]. For passive scattered delivery, any correlation will most likely be blurred by the scattering material in the treatment head [43]. For the same reason, the angular spread at treatment head entrance has little influence on the beam exiting the treatment head in passive scattering.

13.2.2 Monte Carlo Modelling of the Therapy Treatment Head

Proton therapy treatment heads are designed to deliver either scanned beams (most newer facilities) or passively scattered beams (most older facilities, some of which are capable of providing both modes). The distinction is not binary considering methods such as wobbling or the use of range shifters and apertures in scanning systems.

There are numerous reports on Monte Carlo treatment head simulations [14,15,43,45,49–59]. Down to which accuracy one needs to model the treatment head in a Monte Carlo system depends on the purpose of the simulation. If the purpose is calculating dose in the patients, it is sufficient to have only the main beam shaping devices included in the simulation. For beam scanning these are the scanning magnets and potentially a patient-specific aperture. For passive scattering, these include the double- or single-scattering system, modulator wheel, aperture, and compensator.

The accuracy with which the treatment head elements can be modeled depends on the available information, e.g., whether drawings of geometries are provided by the vendor. In addition, it might depend on our knowledge of exact material compositions and material properties [45]. As in photon Monte Carlo, the modelling of machine-specific components in the treatment head can be done using manufacturer's blueprints [43,48,50]. Some of the devices can be modeled as a combination of regular geometrical objects. This holds for example in the case of ionization chambers or scattering foils. Irregular shaped objects can be contoured scatterers, modulator wheels (Figure 13.3), and apertures and compensators. These can be modeled if the Monte Carlo code is capable of reading CAD (computer-aided design) format or if the Monte Carlo allows the definition of objects by importing boundary points in other formats. If these options do not exist, devices need to be constructed by combining regularly shaped objects.

The rotational motion of the modulator can be considered either by adding a large amount of individual Monte Carlo runs based on distinct geometries or by changing the geometry dynamically applying a four-dimensional Monte Carlo technique [60]. Related, in some delivery systems the beam current is modulated at the cyclotron level, which can be incorporated into the Monte Carlo code as well for instance by using look-up tables.

(a)   (b)

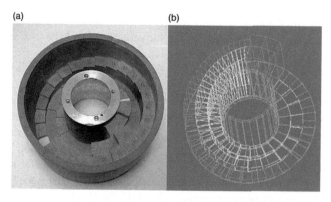

FIGURE 13.3   (a) : Picture of one of the modulator wheels consisting of three tracks used at Massachusetts General Hospital, Boston. (b) : Simulation of a modulator wheel using the Geant4 Monte Carlo code.

In passive scattering, field-specific apertures and compensators are prescribed by the planning system. They are often defined in files that parameterize the geometry for use in a milling machine. How the geometries are imported into the Monte Carlo code depends on the format of the files as well as on the capabilities of the Monte Carlo system. An aperture can be described by a set of points following the inner shape of the aperture opening. If the Monte Carlo code is capable of translating this information into a 3D object representing the aperture, the information can be used directly. The apertures need to be modeled explicitly because of the secondary radiation they produce [61] and the effects of edge scattering [62].

To simulate pencil beam scanning, the magnetic field settings are typically prescribed by a treatment control system based on information by the planning system [63]. These settings are often relative numbers that need to be translated into magnetic strengths in Tesla. The relationship is known from results of commissioning measurements. If the Monte Carlo code is capable of modelling magnetic fields, this information can be incorporated by defining the field strengths as a function of position in the treatment head geometry [43]. This requires a parameterization of the magnetic field lines. One can also use an approximation by defining a constant magnetic field as either on or off in a defined area (i.e., assume perfect dipoles). Attention needs to be paid to the step size while tracking protons through the magnetic field because the field is applied on a step-by-step basis. Large steps lead to considerable errors when simulating the curved path of particles through the field. To simulate an entire scanning pattern, four-dimensional Monte Carlo techniques can be applied [60,63,64]. This allows studying the details of beam scanning delivery parameters [49,64,65].

If dose calculation is not the only purpose of the treatment head model, devices other than beam shaping devices may have to be included in the simulation, e.g., ionization chambers for detector studies or absolute dosimetry [66,67], or the housing of devices to study scattering or shielding effects. To simulate dose in patients, simulating treatment head ionization chambers in detail might not be necessary because they cause only a little scattering and energy loss of the beam. Simulating plain or segmented ionization chambers is done for the purpose of designing ionization chambers or studying beam steering, as well as calculating the absolute dose in machine monitor units [14,52,53,68]. There can be other devices, e.g., detectors, one might want to model for research purposes. For instance, Monte Carlo simulations have been performed to design a prompt gamma detector for quality assurance [69,70] or to optimize image reconstruction for proton computed tomography [71–73]. Monte Carlo simulations have also been used to study aperture scattering [62,74,75] or multi-leaf collimators to replace patient-specific apertures [76]. Quality assurance in radiation therapy is based on well-defined experimental studies that are repeated frequently. Nevertheless, Monte Carlo simulations can assist clinical quality assurance procedures by reducing required experimental studies [77–79]. By simulating dose distributions and varying beam input parameters, tolerance levels for beam parameters can be defined [43]. Further, slight uncertainties or misalignments in the treatment head geometry that might occur over time can be simulated.

13.2.3 Phase-Space Distributions and Beam Models

A phase space characterizes a large number of particles with defined type, energy, direction, and other user-defined identifiers. Phase spaces are typically defined at a specific plane perpendicular to the central axis of a beam, and they can simply be represented by a file containing the information for typically tens or hundreds of millions of particles. They are used to increase computational efficiency if parts of the simulation are generic, allowing a phase space to be reused for subsequent simulations. For example, a phase space can be defined at the treatment head exit.

Phase-space distributions can also be used to characterize the beam at or near the treatment head exit. For instance, each particle in the source can be characterized by its type T, momentum Ω, and position Σ, i.e., $\phi(T, \Omega x, \Omega y, \Omega z, \Sigma x, \Sigma y, \Sigma z)$. The aim is to minimize calculation time due to reusing data in areas that are not varying for each treatment field. In passive scattered proton therapy, phase spaces are less useful for characterizing the beam at treatment head exit. This is because each field typically has a unique setting of the treatment head. The varying parameters are (at least) beam energy, the settings for first scatterer, second scatterer, modulator wheel, utilized wheel solid angle, and aperture and compensator. For beam scanning, even though the scanning pattern depends on the treatment field, the definition of phase spaces to characterize the radiation field exiting the treatment head is feasible as the variations in beam settings are much lower (beam energy and magnet settings). To scan a particular pattern within the patient, the scanning trajectory for a given energy layer can be discretized. These beam spots, corresponding to a magnet setting, can be sampled in a phase-space file. For treatment simulation, a predefined number of protons are then simulated for each beam spot.

In beam scanning, not only Monte Carlo-generated phase-space files but also analytical beam models are feasible. Beam models allow parameterization of a radiation field to be used instead of a phase space. Proton therapy treatment head settings for passive scattering are very complex, and an accurate beam model might be hard to find. However, for pencil beam scanning a field can be characterized by a fluence map of pencil beams (x, y, beam energy, weight, divergence, and angular spread). Thus, a beam source model can be based on parameterized particle sources for a scanned proton beams using fits of measured data, e.g., fluence distributions of pencil beams in air and depth-dose distribution measured in water [46,80–82]. For example, the beam's energy spread as a function of energy can be deduced by fitting the widths of measured pristine Bragg curves. Secondary particles other than protons generated in the treatment head can typically be neglected [46]. One has to be cautious characterizing a beamlet by a simple Gaussian fluence distribution because this leads to an overestimation of the fluence in the center of a beamlet due to a halo caused by large angle protons from scattering or nuclear interactions [9,46,83]. Small corrections when using beam models might thus be needed, such as the consideration of at least a second Gaussian in the formalism [46,84].

The fact that beam models or phase-space distributions can easily be constructed for pencil beam scanning has important implications when using Monte Carlo for clinical dose calculation. Due to the low efficiency of proton therapy treatment heads for passive scattering, the majority of calculation time is spent tracking particles through the treatment head. While this makes Monte Carlo for passive scattered delivery less attractive for routine clinical use, the time frame for beam scanning simulations allows the use of Monte Carlo routinely in the clinic [48]. Note that for some scanned beam deliveries one might want to use an aperture to reduce the beam penumbra. Edge scattering at the aperture can have significant effects on the dose distribution [62] and should thus be included in the beam model.

13.2.4 Uncertainties and Benchmarking

As discussed above (Proton Physics), there are uncertainties in the underlying physics settings of a Monte Carlo code

because various parameters are not necessarily known down to a sufficient level of accuracy. Since direct experimental validation of cross sections is often not feasible in proton therapy institutions where the main users are located, benchmarking is mostly based on Monte Carlo simulations of less fundamental quantities, e.g., dose [85,86]. Benchmarking studies should not be too complicated in terms of the underlying geometry so that discrepancies can be attributed to differences in physics and not to shortcoming in simulating the geometry.

There are various studies comparing experimental deduced depth-dose distributions and beam profiles with results from Monte Carlo codes and for different treatment heads [43,50,87]. In a well-designed Monte Carlo code, measured dose distributions of an SOBP in water should be reproduced with accuracies within typically ~1 mm in range and ~3 mm in modulation width [43]. A benchmark example using an inhomogeneous phantom setup is shown in Figure 13.4 with a comparison between measured (ionization chamber) and Monte Carlo-predicted dose profile.

A more complex validation of a Monte Carlo system can be based on passive scattering treatment head due to the variety of geometries and materials. It has been shown [45] that the SOBP range is most sensitive to changes in the density of materials in the modulator wheel. Materials commonly used are polyethylene (lexan) and lead. A 10% variation in density could result in range changes on the order of 1 mm. Another material used in scatterers of treatment heads is carbon. Here,

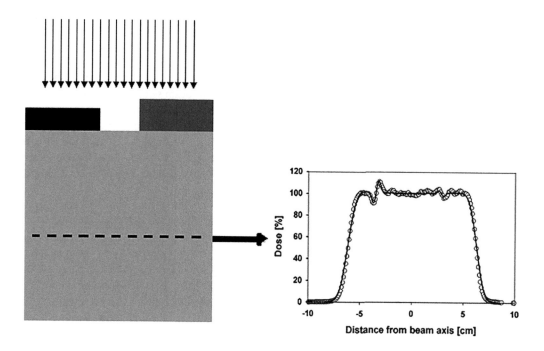

FIGURE 13.4 Comparison of a Monte Carlo-predicted dose distribution (open circles) with measured values using an ionization chamber (solid line). The left side depicts the geometry (black: bone-equivalent material (thickness: 3 cm); dark gray: lung-equivalent material (thickness: 5 cm); light gray: water). The dashed line indicates the plane corresponding to the shown dose distribution (note that the disturbances due to scattering are more pronounced downstream of the bone slab because of scattering and the fact that the range reduction from bone is larger than from lung). An SOBP (full modulation) was used for the beam profile [48].

the nominal density as specified by the manufacturer can vary substantially from the actual density because carbon is available in various specifications leading to larger uncertainties compared to other materials. This can have an impact on the predicted proton beam range as well. Not only the beam range but also the modulation width is affected by uncertainties in the Monte Carlo settings. The flatness of the SOBP might be sensitive to the initial energy spread and spot size at nozzle entrance. These parameters influence the peak-to-plateau ratio of the individual Bragg peaks that form an SOBP. It has been shown that the effect is small for at least certain modulator wheel designs [45]. The impact of the beam spot size depends on the width of each step used on the modulator wheel.

The correct modelling of nuclear interactions is essential for simulating the primary and secondary particle spectrum and for dose calculation, especially if Monte Carlo is used for absolute or relative dosimetry [68]. In general, nuclear interaction components are difficult to study separately in an experiment solely measuring dose but large angle scattering, and thus, the contribution of nuclear interaction products to dose distributions can be measured and compared with Monte Carlo [10,11]. A device to study total inelastic cross sections is the multilayer Faraday cup [18,88]. It measures the longitudinal charge distribution of primary and secondary particles and is capable of separating the nuclear interaction component from the electromagnetic component. Various Monte Carlo physics models for the simulation of electromagnetic and nuclear interactions were validated against the measured charge distribution from the Faraday cup [19,89]. The Faraday cup can only validate total nuclear cross sections, not specific branching ratios. For treatment head simulations and beam characterization, total and differential cross sections for materials of beam shaping devices are required to compensate for fluence loss due to nuclear interactions. For primary standards and reference dosimetry, these cross sections with high accuracy are needed for a limited set of detector materials. Another example is the simulation of nuclear activation of tissues relying on isotope production cross sections for human tissues [90–92].

13.3 Simulating Dose to the Patient

13.3.1 Statistical Accuracy of Dose Modelling

Other than with analytical methods, the precision of Monte Carlo dose calculations depends on the chosen number of histories and thus the calculation time. While efficiency for many applications might not be critical, it becomes very important for the clinical use of Monte Carlo, particularly in the context of treatment optimization. The efficiency of Monte Carlo simulations depends on many factors such as the number of histories or the number of steps each particle takes. For dose calculations, the maximum step size is limited by the size of the CT voxels. This is because the physics settings are typically different for different voxels due to their difference in Hounsfield unit and thus material characteristics.

Typically, the desired statistical uncertainty for dose calculations should be less than 2%, at least for the target volume. The number of histories required to reach this condition differs between proton and photon therapy, as fewer protons are needed to reach the same dose. Photons are indirectly ionizing, and their secondary electrons have a lower linear energy transfer and range. A certain statistical precision in the target volume does not guarantee the same precision in organs at risk because of the lower dose and thus potentially fewer particles involved. On the other hand, the impact of statistical imprecision is less for dose-volume analysis in organs at risk because the dose distribution is less homogeneous and the dose-volume histograms are shallower [93,94].

In order to allow predicting dose distributions based on fewer histories, it has been suggested that Monte Carlo-generated dose distributions can be smoothed [95–99]. This smoothing or de-noising is a technique well known in imaging. These methods need to be applied with caution because regions of low signal are, other than in imaging, not noise but may contain valid information. Some de-noising techniques tend to soften dose falloffs, which could have a negative impact when used in proton therapy. Furthermore, other than in photon therapy, dose is not directly proportional to particle fluence in proton therapy because the linear energy transfer may vary considerably.

In order to decrease the statistical uncertainty, one might also be tempted to interpolate the CT grid to a larger grid. Treatment planning systems often present dose distributions on a more course grid than the one provided by the patient's CT scan. The problem with resampling the CT grid is that averaging of material compositions is not well defined. Thus, to avoid resampling, the Monte Carlo should operate on the actual CT scan, which is typically in the order of 0.5–5 mm. Monte Carlo simulations might require to operate on a non-uniform CT grid as often used clinically [100]. Resampling to improve statistical uncertainty can still be done after the dose calculation, where weighting of doses that contribute to a given voxel on a grid can be done accurately based on volume averaging. The speed of the Monte Carlo simulation certainly depends on the grid size. However, assuming that a step size above 1 mm is typically not warranted, a larger grid size does not translate into a huge gain.

13.3.2 CT Conversion

For photon or electron beams, electron density is being used in analytical dose calculation engines because the dominant energy loss process is interaction with electrons. Protons lose energy by ionizations, multiple Coulomb scattering, and non-elastic nuclear reactions. Each interaction type has a different relationship with the material characteristics obtained from the CT scan [101,102]. Consequently, for analytical proton dose calculations, instead of electron density or mass density, relative stopping power is being used to define water-equivalent tissue properties and a conversion from CT numbers to relative stopping power is being applied [103,104]. The accuracy of dose calculations is affected significantly by the ability to precisely define tissues based on CT scans [104,105]. Not only can the absolute dose vary but also the proton beam range might depend on the accuracy of the CT conversion. For head and neck treatments, it was shown that CT conversion schemes can influence the proton beam range in the order of 1–2 mm [106].

For Monte Carlo dose calculation, a conversion from CT numbers to material compositions and mass densities is done for each tissue. A conversion scheme can be deduced by scanning tissue phantom materials or animal tissues [104,107–109]. Stoichiometric calibrations of Hounsfield numbers with mass density and elemental weights allow CT conversion [110]. A conversion table can be extended to higher Hounsfield units in order to deal with high-Z implant materials in the patient [91]. A robust division of most soft tissues and skeletal tissues can be done, but soft tissues in the CT number range between 0 and 100 can only be poorly distinguished because CT numbers of soft tissues with different elemental compositions are similar. A relationship between a certain CT number and a combination of materials is not unique because a CT number reflects the attenuation coefficient of human tissues to diagnostic X-rays and may be identical for several combinations of elemental compositions, elemental weights, and mass densities [106,109].

In CT conversion schemes, tissues are often grouped into different tissues sharing the same material composition (and ionization potential). The number of groups certainly affects the dose calculation accuracy and is typical between 5 and 30

[91,100,105]. While grouping material compositions is justified, it is typically not sufficient to assign a unique density to each group; i.e., the number of densities is typically the same as the number of gray values (CT numbers) [100].

A comparison of two different CT conversion methods [109,111] is shown in Figure 13.5 to illustrate the magnitude of differences one might expect. The largest difference appears at $H = -119$ for carbon and oxygen because of the defined thresholds at the transition of air and adipose tissue. For hydrogen and calcium, the greatest difference is located at $H = +126$ at the transition from soft tissue to bone. To demonstrate the impact these differences might have on proton dose calculation, Figure 13.6 shows the mass stopping power ratio as a function of CT number for the two methods presented in Figure 13.5.

When implementing a conversion scheme for Monte Carlo dose calculation, a normalization to the specific scanner that is being used for imaging in the department is needed to compare the results from a Monte Carlo dose calculation with those from an analytical algorithm in a treatment planning system. The normalization for the Monte Carlo can be found either by doing a separate stoichiometric calibration or, as an approximation, by simulating relative stopping power values in the Monte Carlo based on an existing CT conversion and then compare the results with the planning system conversion curve. Normalization can then be achieved within the Monte Carlo by slightly adjusting material compositions or, easier, mass densities [48,90,100].

It has been shown that slight discrepancies in mass density assignments play only a minor role in the target region

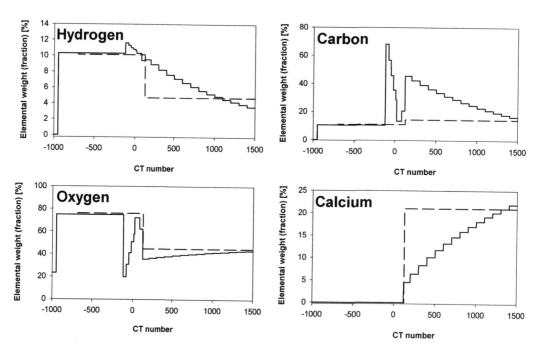

FIGURE 13.5 Elemental weights of four tissue elements (hydrogen, carbon, oxygen, and calcium) as a function of CT number for the CT conversion method of Schneider et al [109] (solid line) and by Rogers et al [111] (dashed line) [105].

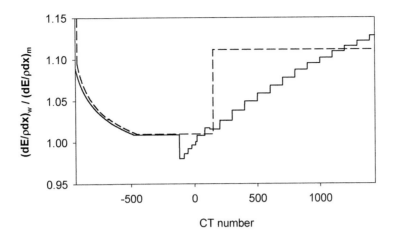

FIGURE 13.6 Water-to-tissue mass stopping power ratio for 100 MeV protons as a function of CT number when applying CT conversion methods by Schneider et al [109] (solid line) and Rogers et al [111] (dashed line).

whereas more significant effects are caused by different assignments in elemental composition [105]. Not only can the absolute dose vary but also the proton beam range might depend on the accuracy of the CT conversion and the CT resolution [106,112,113].

In addition to a proper CT conversion method, mean excitation energies for each tissue are required. These can be interpolated based on the atomic weight of the tissue elements using Bragg's rule. Mean excitation energies for various elements [114] and averaged values for tissues [115,116] are tabulated and are subject to uncertainties on the order of 5–15% for tissues. This is reflected in uncertainties in predicting the proton range [21,117,118].

The tissue and density assignment can be improved when using dual-energy CT as it provides two different attenuation maps to determine relative electron densities and effective atomic numbers thus allowing more accurate material composition maps for Monte Carlo dose calculation [118–124].

13.3.3 Absolute Dose

Some planning systems prescribe dose in terms of Gy per GigaProtons, which makes conversion to Monte Carlo input straightforward. Others may prescribe dose in machine monitor units; i.e., absolute doses are reported as cGy per MU [48,68]. In proton therapy, this is done either by calibration measurements, by using analytical algorithms, or by relying on empirical data. A monitor unit typically corresponds to a fixed amount of charge collected in a transmission ionization chamber incorporated in the treatment head whose reading is related to a dose at a reference point in water [1,125].

Using Monte Carlo, one can actually simulate the ionization chamber output charge when tracking particles through the treatment nozzle. Here, an exact model of the proton beam delivery nozzle, including devices used for dosimetry, is required [68]. The disadvantage of this method is that it typically requires a large number of histories to be simulated,

as the energy deposited in a small reference chambers in air per particle is typically quite small. Nevertheless, Monte Carlo-predicted output factors can be part of an extended quality assurance program because the influence of specific devices in the treatment head on the output factor reading can be studied.

Alternatively, absolute dosimetry for Monte Carlo dose calculation can also be done by simply relating the number of protons at nozzle entrance to the dose in an SOBP in water. With an accurate model of the treatment head, this method is equivalent to a direct output simulation because instead of indirectly 'measuring' the number of protons at a given plane in the treatment head (in an ionization chamber), one 'measures' the number of protons at nozzle entrance.

13.3.4 Dose to Water and Dose to Tissue

Traditionally, dose in radiation therapy is reported as dose-to-water because analytical dose calculation engines, as the ones incorporated in most commercial planning systems, calculate dose by modelling physics relative to water. Furthermore, quality assurance and absolute dose measurements are done in water. Arguments in favor of using dose-to-water instead of dose-to-tissue include the fact that clinical experience is based on dose-to-water, quality assurance and absolute dose measurements are done in water, and the tumor cells in the human body consist mostly of water [126]. Dose constraints in treatment planning are based on our experience with dose-to-water as well.

Monte Carlo dose calculation systems calculate dose-to-tissue because they allow the material properties, as converted from CT numbers, to be modeled using explicit material composition, mass density, and ionization potential. Consequently, in order to allow a proper comparison between Monte Carlo- and pencil-beam-generated dose distributions one has to convert one dose metric to the other. The difference in mean dose roughly scales linear with the average CT number and thus seems to be clinically insignificant for soft tissues but needs to be taken into account for bony anatomy [127]. Dose-to-water can be higher by ~10%–15% compared to dose-to-tissue in bony anatomy. For soft tissues, the differences are only on the order of 2%. Thus, in most cases it is sufficiently accurate (even within ~1%) to do a conversion to dose-to-water retroactively by simply multiplying the dose with energy-independent relative stopping powers [127].

13.3.5 Impact of Nuclear Interaction Products on Patient Dose Distributions

Protons lose energy via electromagnetic and nuclear interactions. The latter can account to well over 10% of the total dose, specifically in the entrance region of the Bragg curve (see Figure 13.2) [7,102,128–130], where secondary protons cause a dose buildup due to predominant forward emission from nuclear interactions. Their impact on fluence reduction of the primary beam is also noticeable (Figure 13.2).

Nuclear interaction cross sections show a maximum at a proton energy of around 20 MeV and decrease sharply if the energy is decreased. As a rule of thumb, the average proton energy in the Bragg peak is about 10% of the initial energy. The contribution of dose due to nuclear interactions becomes thus negligible close to the Bragg peak position of a pristine Bragg curve because of the decreasing proton fluence and a sharply decreasing cross section. For an SOBP, nuclear interactions still play a role in the peak because dose regions proximal to the Bragg peak contribute. This leads to a tilt of an SOBP dose plateau if their contribution is neglected [7]. Figure 13.7 shows the dose contribution from secondary protons compared to the total dose for a treatment field used to treat a nasopharyngeal carcinoma.

Standard analytical dose calculation algorithms in treatment planning systems that are solely based on dose kernels incorporate nuclear interaction contributions intrinsically because they are based on measured or Monte Carlo simulated depth-dose curves. Particularly when used for beam scanning, the nuclear halo has to be considered explicitly in dose kernels (see Section 13.2.3) [9, 131].

Depending on the application, it might not be necessary to track all secondary particles whereas for others it might be sufficient for some types to deposit the energy locally (to ensure proper energy balance). If secondary doses for out of field dose calculations are being studied, one also has to consider neutrons and the protons they cause. Specifically for patient-related dose calculations, solely secondary protons from proton-nuclear interactions need to be tracked whereas the energy deposited via other nuclear secondaries may be deposited locally. If the projected range of secondary particles is smaller than the region of interest (e.g., the size of a voxel in the patient), it might be sufficient to deposit the energy (dose) locally.

13.3.6 Differences between Proton Monte Carlo and Analytical Dose Calculation

While Monte Carlo-based dose calculation is becoming more attractive even for commercial treatment planning systems, the majority of treatment plans still rely on analytical dose calculation algorithms. Monte Carlo simulations have been used to validate analytical algorithms [48,132,133] or for commissioning of treatment planning systems [134–136]. Dosimetric differences between a Monte Carlo and a pencil-beam algorithm can be significant [8,48,133,137–143], particularly in areas of high tissue heterogeneities in the skull [144–146], breast [147,148], or lung [140,149–151]. In fact, the routine use of Monte Carlo dose calculation in proton therapy could result in a reduction of treatment planning margins [117]. The difference between the two dose calculation methods is mainly

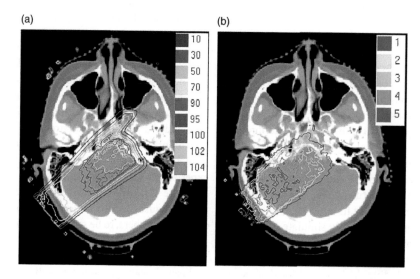

FIGURE 13.7 Impact of nuclear secondaries on the dose distribution in a patient. (a) : Dose distribution for one field (out of three for this treatment plan) for a head and neck case (doses in %). (b): Contribution of secondary protons to the dose distribution in % of the target dose.

caused by the way multiple Coulomb scattering is considered. Analytical algorithms are less sensitive to complex geometries and in-beam density variations, i.e., bone-soft tissue, bone-air, or air-soft tissue interfaces, and often neglect the position of inhomogeneities relative to the Bragg peak depth [137,152–154]. The disequilibrium of scattering events causes protons to be preferentially scattered out of the higher density material into the lower density material. Consequently, deviations from the Monte Carlo dose distribution occur specifically in the penumbra if the beam direction is tangential to an interface connecting high and low CT numbers. Furthermore, small discrepancies in local energy deposition can result in changes in range over the entire beam path [100] causing range degradation [131,155], which is correctly predicted using Monte Carlo. Analytical algorithms predict a much sharper distal falloff of the Bragg peak. Note also that some pencil-beam models do not consider aperture edge scattering [62].

Figure 13.8 demonstrates one of the weaknesses of an analytical algorithm, namely the inaccurate consideration of interfaces parallel to the beam path. This patient was treated for a spinal cord astrocytoma with three coplanar fields. The figure only shows one of the fields. The Monte Carlo dose calculation was based on a CT with $176 \times 147 \times 126$ slices with voxel dimensions of $0.932 \times 0.932 \times 2.5–3.75 \, mm^3$ (slice thickness varied).

Whether Monte Carlo dose calculation is needed or analytical dose distributions are sufficiently accurate depends on the treatment geometry. Very good agreement can be expected for sites such as liver because of the relatively homogeneous geometry. Treatment planners are typically aware of dose calculation uncertainties and take these into consideration when prescribing fields, e.g., by avoiding pointing a beam toward a critical structure and by applying safety margins. Furthermore, one avoids pointing the beam toward a critical structure because of uncertainties in the position and gradient

of the distal dose falloff. Because of such precautions, differences between pencil-beam algorithms and Monte Carlo dose calculations turn out to be small when analyzing dose-volume histograms from carefully designed plans. It appears that the differences are not always clinically significant [8,48].

As demonstrated above, Monte Carlo dose calculations are particularly valuable for studying dosimetric effects in regions with a high gradient in density. Extreme cases are metallic implants in the patient, e.g., dental implants in head and neck cancer patients. Monte Carlo has also been used to study tantalum markers used to stabilize bony anatomy after surgery or as markers for imaging. These can lead to significant dose perturbations typically not predicted accurately by analytical algorithms [156]. Markers implanted in the patient for setup or motion tracking are typically not modeled accurately in pencil-beam algorithms due to their high-Z nature.

13.3.6.1 Differences in the Predicted Range

The range of a clinical proton field is usually defined as R_{90}, i.e., the position of the 90% dose level in the distal falloff region of a spread-out Bragg peak. Differences in overall range between analytical algorithms and Monte Carlo in patients are typically very small because both systems have been commissioned using measurements in water. However, locally, significant differences can occur because analytical algorithms have deficiencies when predicting scattering [117,139]. In clinical practice, range margins of the order of 3.5%+1mm are often assigned to proton fields to cover errors induced by the approximations made in the analytical algorithms [117]. While a generic margin is typically applied in treatment planning, range uncertainties depend on the patient geometry. It has been estimated that margins can be reduced to 2.7%+1.2mm for largely homogeneous patient geometries (such as liver) but should be increased to as high as 6.3%+1.2mm for geometries with lateral inhomogeneities (such as head and neck, lung, and

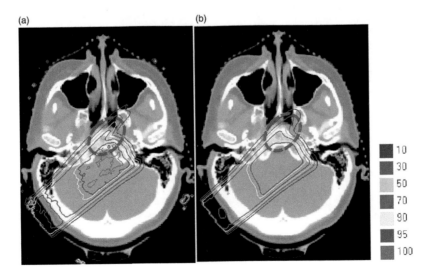

FIGURE 13.8 Axial views of dose distributions calculated using Monte Carlo (a) and a commercial planning system based on a pencil-beam algorithm (b). Doses are in %. The red dashed circles show the dosimetric impact of an interface parallel to the beam path [48].

breast patients). These numbers assume that tumor coverage is to be maintained for each treatment field, which may be conservative depending on the number of fields prescribed [139]. If Monte Carlo techniques were to be used routinely for dose calculations, these margins could be reduced to 2.4%+1.2 mm independent of the complexity of the patients' geometry [117].

Figure 13.9 shows the difference in range as predicted by an analytical dose calculation engine in a commercial treatment planning system versus a Monte Carlo dose calculation engine. Both algorithms were perfectly matched to doses in water as well as the range in water based on experimental commissioning data. The shown head and neck field has an overall range difference at the 50% dose level of less than 0.1%. However, locally, substantial range differences can occur in areas where the beam passed through inhomogeneities. Also shown is an example for the lung where low-density tissue as well as scattering in the chest wall amplifies range differences.

13.3.6.2 Differences in the Predicted Dose

The fact that Monte Carlo algorithms predicted larger scattering components compared to analytical algorithms causes an overall broadening of the dose distribution and thus a decrease in dose in the medium- to high-dose region. Consequently, the dose in the target from Monte Carlo simulations is generally somewhat lower than the one predicted by analytical algorithms. The differences in the mean target dose can be up to 4% for head and neck and lung patients [140,149] or as low as 2% for breast and liver patients [140]. It also depends on the field size as small fields are more sensitive [157–160]. The underestimation of scattering in analytical algorithms depends on the location of tissues with different densities. It can be seen clearly in dose calculations for lung because protons are scattered in the chest wall, which causes the beam to diverge with distance [149]. Analytical algorithms calculate the spread of the beam for a water-equivalent path length and not the physical distance. Figure 13.10 shows an example of this overall dose discrepancy even when doses in water from commissioning data were matched for both dose calculation methods.

13.3.7 Clinical Implementation

Monte Carlo simulations have been used when commissioning planning systems [54,134,135] but also to recalculate plans for quality assurance based on treatment log files [161]. If proton Monte Carlo dose calculation is not part of a commercial treatment planning system, one might implement an in-house Monte Carlo system in the clinic, which can then

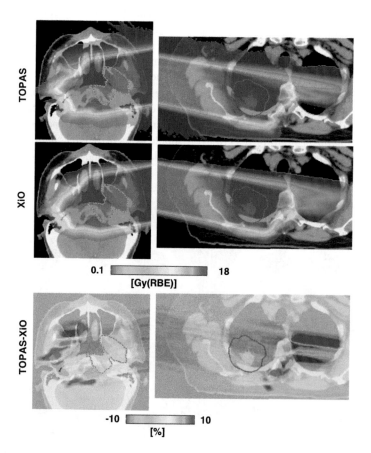

FIGURE 13.9 Range discrepancies in a head and neck treatment field (left column) and a lung treatment field (right column). Top: TOPAS Monte Carlo prediction; Middle: Prediction by an analytical dose calculation method (XiO); Bottom: Difference. (Adapted from Ref. [139].)

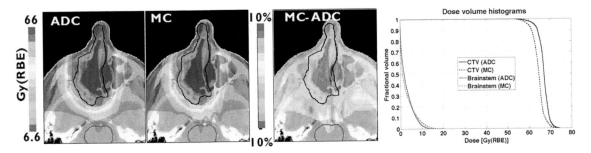

FIGURE 13.10 Head and neck patient with a prescription dose for the CTV of 66Gy(RBE). Shown are the dose distributions from an analytical dose calculation (ADC), a Monte Carlo prediction (MC), and the difference (from left to right). The far right graph shows the dose-volume histograms for the CTV and the selected OARs (Adapted from Ref. [140].)

be used for routine dose calculation in dose verification or for research purposes. This is now being done at the majority of the leading proton centers [81,162–169]. A user interface tailored to the planning system might be needed to facilitate the data flow from the planning system to the Monte Carlo system. The plan information is best transferred to the Monte Carlo code using a DICOM-RTion interface. Regarding the patient's CT, a Monte Carlo code would typically accept a DICOM stream directly [170,171]. Results of the Monte Carlo simulation might be analyzed in the planning system (if this is agreed upon with the vendor) or using a standalone visualization tool.

The information provided to the Monte Carlo dose calculation engine will be translated into specific settings in the Monte Carlo code such as the gantry angle, the patient couch angle, the isocenter position of the CT in the coordinate system of the planning program, the number of voxels and slice dimensions in the CT coordinate system, the size of the air gap between treatment head and patient, and the prescribed dose [48]. This can either be done internally within the Monte Carlo or using a dedicated program that uses the plan information and translates it into an input file readable by the Monte Carlo code.

If the radiation field for proton beam scanning is to be simulated, the planning system will most likely provide a matrix of beam spot energy, beam spot weight, and beam spot position, which can be translated into Monte Carlo settings in a straightforward manner. If passive scattered delivery is simulated, there are two ways of how the radiation field is characterized. First, the planning system prescribes the settings of various treatment head devices and the beam at treatment head entrance directly. These settings are beam energy at nozzle entrance, settings of the first and second scatterer (one scatterer if single scattering is used for small fields), and a combination of scatterer steps in a modulator wheel or a ridge filter. Second, the planning system prescribes range and modulation width only, leaving it to the hardware control system of the facility to translate this information into treatment head settings. In this case, it might be warranted to incorporate the control system algorithms into the Monte Carlo code [43,48]. Planning systems also define aperture and compensator for

passive scattered delivery. These are typically given as files readable by a milling machine.

13.3.8 Monte Carlo-Based Treatment Planning

Monte Carlo treatment planning needs to be incorporated into a framework for treatment plan optimization [8,172,173]. The use of Monte Carlo in an inverse planning framework has stricter requirements in terms of computational efficiency because the dose has to be calculated more than once in the optimization loop. For example, for intensity-modulated proton therapy each potentially desired beamlet would have to be precalculated before evoking the optimization algorithm. Each of these would have to be calculated to high statistical precision because the weight of that particular pencil (i.e., its contribution to the final dose distribution) would not be known a priori. Fast Monte Carlo simulation tools are therefore required (see next section). If the difference between Monte Carlo and analytical algorithm is small, one might base the optimization on the analytical algorithm and use Monte Carlo simulations for fine-tuning only. Alternatively, one might utilize a Monte Carlo dose engine only at a limited number of checkpoints during the optimization process.

The main treatment planning system vendors are currently working toward Monte Carlo-based treatment planning [174–177]. Improvements in both the computational efficiency of Monte Carlo software and computer hardware have made it feasible to use Monte Carlo dose calculation routinely.

13.3.9 Improving Monte Carlo Efficiency

Efficiency in Monte Carlo simulations is particularly important for dose calculation. Improvements have been achieved specifically for tracking particles in a voxelized geometry [178]. Particle tracking in a voxel geometry is computationally inefficient because each particle has to stop when a boundary between two different volumes is crossed in order to react on the change in the underlying physics settings. Algorithms have been developed to address this by re-segmenting the geometry [179,180]. For sure, the efficiency of a Monte Carlo dose calculation depends on the grid size, but assuming that a step size above 1 mm is typically not warranted, a larger grid size typically does not translate into a huge gain.

Some Monte Carlo codes have been specifically designed for fast patient dose calculations using approximations to improve computational efficiency. A dedicated Monte Carlo code, VMCpro, was introduced solely optimized for dose calculation in human tissues for proton therapy [32]. A significant speed improvement compared to established Monte Carlo codes was achieved by introducing approximations in the multiple scattering algorithm and using density scaling functions instead of actual material compositions. Furthermore, nuclear interactions were considered explicitly but in parameterizations relying on distribution sampling. Nevertheless, the agreement with other Monte Carlo codes was found to be excellent. Other codes have been optimized for speed by only tracking primary protons through the treatment head and the patient and treating contributions from nuclear interactions in analytical approximations without the tracking of secondary particles [138]. Depending on the beam energy, primary and secondary protons account for roughly 98% of the dose [7]. This includes the energy lost via secondary electrons created by ionizations. As the range of most electrons in a clinical proton beam is typically less than 1 mm in water, explicit tracking of electrons may not be necessarily required for dose calculations on a typical CT grid.

Kohno et al [181,182] developed a simplified Monte Carlo method which uses the depth-dose distributions in water measured in a broad proton beam to calculate the energy loss in materials based on a water-equivalent model. At each voxel, the proton's residual range is reduced according to the local material property and a corresponding amount of energy is deposited. Multiple Coulomb scattering is modeled by sampling scattering angles from a normal distribution parameterized by theoretical considerations. There is no clear boundary between what one would consider a full-blown Monte Carlo and an analytical algorithm as some fast Monte Carlo codes utilize analytical approaches for certain physics phenomena. Also applied have been hybrid methods using spot decomposition techniques where multiple ray tracings are performed for a single pencil [8,133]. One thus might perform Monte Carlo calculations only on a subset of rays that go through a highly heterogeneous geometry.

Other methods are based on track repeating algorithms in which precalculated proton tracks and their interactions in material are tabulated and used in dose calculations [183–185]. The changes in location, angle, and energy for every transport step and the energy deposition along the track are recorded for all primary and secondary particles and reused in subsequent Monte Carlo calculations [183,186]. Each particle trajectory stored consists of a set of steps, and for each step, the direction, step length, and energy loss. During the simulation, at each step there is a track segment loaded from the database that resembles the scenario for the tracked particle. Because complex physics model calculations are avoided, the method can be more efficient compared to standard Monte Carlo calculations.

In addition to condensed history techniques or tracking limitations, variance reduction techniques aim at improving computational efficiency by giving more emphasis to certain physical quantities of interest. For example, the splitting of primary particles in regions of high interest is often used. By splitting at specific locations, one predominantly considers particles that have a high likelihood of contributing in regions of interest. If primary particles are split, their simulated effects are subsequently treated with a weighting factor <1 (which will also be inherited by subsequent secondary particles). A related technique is Russian roulette, which, instead of splitting, combines several particles into one particle with a combined cumulative statistical weight. Another method is the importance sampling, i.e., a method to oversample certain regions of interest while reducing the weight of events to maintain the statistical balance. Some of these techniques have been implemented for proton therapy applications [187–189].

The biggest improvement in Monte Carlo efficiency has however not been made by these techniques but by improved hardware. A graphics processing unit (GPU) is a dedicated hardware designed for accelerated processing of graphics information. It has a much higher number of processing units sharing a common memory space compared to a conventional central processing unit (CPU). A GPU executes a program in groups of parallel threads. Whether a task can be significantly accelerated using GPU hardware depends on the mathematical formalism. Analytical dose calculations are very well suited because they can be formulated as vector operations which take advantage of multi-threading but Monte Carlo algorithms have been implemented on GPU cards as well. Particularly, track-repeating algorithms can benefit from the multi-threading environment of a GPU [183,190]. In addition to implementing existing algorithms on a GPU, there have been efforts to develop proton Monte Carlo dose calculation algorithms that are tailored to GPU use and are thus very efficient [174,190–199]. While particle tracking with respect to electromagnetic interactions can be done explicitly in a GPU-based code, approximations may be necessary for nuclear interactions because of required data tables that would exceed the memory capability of some GPUs. For instance, the gPMC code [191,194,195] considers proton propagation by a Class II condensed history simulation scheme using the continuous slowing down approximation while nuclear interactions are modeled using an analytical approach [32]. While other secondary particles are neglected, the emitted secondary protons are tracked in the same way as primaries. The code has been compared with CPU-based full Monte Carlo codes for patient dose calculation and found to be in excellent agreement with a gamma index passing rate of ~99% using a 2%/2 mm gamma index criteria. Efficiency improvements when using GPU codes for proton Monte Carlo are significant allowing dose calculation down to 1%

statistical accuracy based on an existing treatment plan to be done in seconds instead of hours. The application of proton Monte Carlo dose calculation on GPUs has been reviewed by Jia et al [200].

13.4 Other Proton Monte Carlo Applications

13.4.1 4D Dose Calculations

Monte Carlo simulations have been used to simulate time-dependent patient geometries for studies of breathing motion in proton therapy of the lung [201–205]. For instance, simulations on the interplay between breathing motion and beam motion in proton beam scanning have shown that the effect is biggest for delivery systems with a small spot size [201,203]. The impact of beam delivery parameters such as scanning speed or the time it takes to switch energies has been studied using Monte Carlo simulations as well [205]. The requirement to simulate dose to various geometries (given, for instance, by 4D CT images) reduces the computational efficiency. GPU-based Monte Carlo has been suggested for such studies [206,207]. The use of Monte Carlo for motion studies is particularly important in proton therapy because small uncertainties in scattering and range can have a large impact in low-density lung tissue [149].

13.4.2 Simulating Proton-Induced Gamma Emission for Range Verification

Protons undergo nuclear interactions in the patient, which can lead to the formation of positron emitters or excited nuclei emitting prompt gamma rays. Thus, PET imaging and prompt gamma imaging both have been suggested to verify the range in proton beams. Monte Carlo simulations play a vital role in this case [208]. However, both PET and prompt gamma-based range verification suffer from substantial

uncertainties in cross section data needed for accurate Monte Carlo simulations [209–215]. The considerable uncertainties are due to lack of nuclear physics experiments in the relevant energy range and for the relevant materials. Experiments at physics laboratories typically focus on thin targets whereas, in a patient, particle energy distributions have to be considered. Most cross-section data are gathered at a few energies only [216–219]. This topic is described in more detail in other chapters in this book.

13.4.3 Simulating LET Distributions for Radiobiological Considerations

Proton doses are prescribed relative to photon doses by applying the concept of relative biological effectiveness (RBE) for proton therapy [220]. The RBE of protons depends on the linear energy transfer (LET) among other parameters such as dose, biological endpoint, and dose rate [221]. In the target area, the dose is typically delivered homogeneous. This however does not guarantee a homogeneous distribution of LET values [222–225]. When considering primary particles only, elevated LET values will appear in the distal end of a proton field. Furthermore, nuclear interaction products may cause elevated LET values even outside of the target volume, i.e., in the entrance region of the Bragg curve [226,227]. One can simulate dose-averaged LET distributions in a patient geometry using Monte Carlo (Figure 13.11 shows an example) to identify potential hot spots of biological effectiveness and use this information for biologically driven optimization [224,225,227,228].

When calculating the dose-averaged LET during a Monte Carlo run, one needs to record each energy loss, dE, of a particle and the length of the particle step that leads to the energy deposition event, dx. Note that when simulating LET using Monte Carlo simulations, the cutoff defined to stop further tracking of the proton can have a significant influence [229–231].

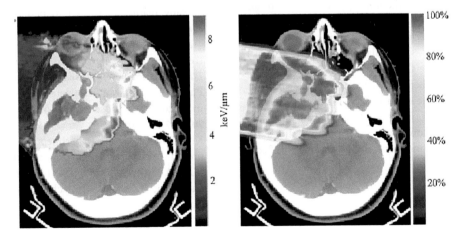

FIGURE 13.11 Distribution of dose-averaged linear energy transfer (left) and dose (right) for a single proton field in a chordoma patient. (Adapted from Ref. [227].)

13.4.4 Track Structure Simulations

In order to understand the biological effect of radiation in more detail, Monte Carlo simulations can even be used to study the interactions of particles with biological structures like the DNA by modelling subcellular geometries [232–249]. The particle track structure describes the pattern of energy deposition events of proton tracks, including the secondary electrons, on a nanometer scale. DNA damage can then be associated with a specific energy imparted per track length or per assumed subcellular volume [250]. Simulations may assume a proportionality between ionization frequency and lesion type in the DNA so that predictions can be made on the likelihood of DNA damage and damage clustering [251], which in turn can be used to make assumptions about repair probability for certain DNA damages. Analyzing the distance of two energy deposition events or clustering of events on a nanometer scale can give insight into lesion complexity [249,252,253].

Track structure can be simulated with the same codes that are used for macroscopic dose simulations. However, many specific codes have been developed particularly to deal with low-energy particle tracks and with the δ-electrons produced by proton tracks [250,254–258]. For instance, PARTRAC includes an accurate representation of the chromatin and of the physical and physiochemical processes associated with the energy deposition by radiation [235,259,260]. The Monte Carlo code PHITS has been modified to allow the simulation of radiolysis [261] to be combined with the MKM model in order to predict biological effects [262]. More importantly, recent developments also suggest the use of Monte Carlo techniques to model detailed radiochemistry more mechanistically [263–265].

Specifically designed for Monte Carlo simulations on subcellular level for radiobiology is the Geant4DNA code, which extends the physics of Geant4 to encompass very low-energy interactions for organic materials [266–268]. Similar to TOPAS as a wrapper around Geant4 (see above), there is now also a wrapper for Geant4DNA (TOPAS-nBio [245, 269]) to simplify Monte Carlo simulations for radiobiology. TOPAS-nBio has been connected to DNA repair models to simulate the entire chain from energy deposition through chemistry and biological repair mechanisms [248]. Simulations on the DNA level pose challenges for Monte Carlo simulations because differential cross sections to either predict radial dose distributions around ion tracks or single electron emission events are subject to considerable uncertainties and typically only available for water [270]. Furthermore, these types of simulations are typically computationally inefficient due to small-scale effects being considered.

References

1. Kooy, H.M., et al., The prediction of output factors for spread-out proton Bragg peak fields in clinical practice. *Phys Med Biol*, 2005. **50**: 5847–56.

2. Kawrakow, I. and A.F. Bielajew, On the condensed history technique for electron transport. *Nucl Instrum Methods Phys Res B*, 1998. **142**(3): 253–80.

3. Lewis, H.W., Multiple scattering in an infinite medium. *Phys Rev*, 1950. **78**: 526–9.

4. Andreo, P., J. Medin, and A.F. Bielajew, Constraints of the multiple-scattering theory of Moliere in Monte Carlo simulations of the transport of charged particles. *Med Phys*, 1993. **20**: 1315–25.

5. Gottschalk, B., et al., Multiple Coulomb scattering of 160 MeV protons. *Nucl Instrum Methods Phys Res*, 1993. **B74**: 467–90.

6. Urban, L., Multiple scattering model in Geant4. CERN report, 2002. **CERN-OPEN-2002-070**.

7. Paganetti, H., Nuclear interactions in proton therapy: Dose and relative biological effect distributions originating from primary and secondary particles. *Phys Med Biol*, 2002. **47**: 747–64.

8. Soukup, M. and M. Alber, Influence of dose engine accuracy on the optimum dose distribution in intensity-modulated proton therapy treatment plans. *Phys Med Biol*, 2007. **52**: 725–40.

9. Sawakuchi, G.O., et al., Monte Carlo investigation of the low-dose envelope from scanned proton pencil beams. *Phys Med Biol*, 2010. **55**(3): 711–21.

10. Lin, L., et al., Experimentally validated pencil beam scanning source model in TOPAS. *Phys Med Biol*, 2014. **59**(22): 6859–73.

11. Hall, D.C., et al., Validation of nuclear models in Geant4 using the dose distribution of a 177 MeV proton pencil beam. *Phys Med Biol*, 2016. **61**(1): N1–10.

12. Paganetti, H., *Proton Therapy Physics*. Taylor & Francis / CRC Press 2012; ISBN-10: 1439836442, 2012.

13. van Goethem, M.J., et al., Geant4 simulations of proton beam transport through a carbon or beryllium degrader and following a beam line. *Phys Med Biol*, 2009. **54**(19): 5831–46.

14. Herault, J., et al., Monte Carlo simulation of a proton-therapy platform devoted to ocular melanoma. *Med Phys*, 2005. **32**(4): 910–9.

15. Stankovskiy, A., et al., Monte Carlo modelling of the treatment line of the Proton Therapy Center in Orsay. *Phys Med Biol*, 2009. **54**(8): 2377–94.

16. Pia, M.G., et al., Physics-related epistemic uncertainties in proton depth dose simulation. *IEEE Trans Nucl Sci*, 2010. **57**(5): 2805–30.

17. Kimstrand, P., et al., Experimental test of Monte Carlo proton transport at grazing incidence in GEANT4, FLUKA and MCNPX. *Phys Med Biol*, 2008. **53**(4): 1115–29.

18. Paganetti, H. and B. Gottschalk, Test of Geant3 and Geant4 nuclear models for 160 MeV protons stopping in CH_2. *Med Phys*, 2003. **30**: 1926–31.

19. Zacharatou Jarlskog, C. and H. Paganetti, Physics settings for using the Geant4 toolkit in proton therapy. *IEEE Trans Nucl Sci*, 2008. **55**: 1018–25.

20. Resch, A.F., et al., Evaluation of electromagnetic and nuclear scattering models in GATE/Geant4 for proton therapy. *Med Phys*, 2019. **46**(5): 2444–56.

21. Andreo, P., On the clinical spatial resolution achievable with protons and heavier charged particle radiotherapy beams. *Phys Med Biol*, 2009. **54**(11): N205–15.

22. Besemer, A., H. Paganetti, and B. Bednarz, The clinical impact of uncertainties in the mean excitation energy of human tissues during proton therapy. *Phys Med Biol*, 2013. **58**(4): 887–902.

23. Moyers, M.F., et al., Calibration of a proton beam energy monitor. *Med Phys*, 2007. **34**(6): 1952–66.

24. Gottschalk, B., On the scattering power of radiotherapy protons. *Med Phys*, 2010. **37**(1): 352–67.

25. ICRU, Nuclear data for neutron and proton radiotherapy and for radiation protection. International Commission on Radiation Units and Measurements, Bethesda, MD, 2000. **Report No. 63**.

26. Battistoni, G., et al., The FLUKA code: An accurate simulation tool for particle therapy. *Front Oncol*, 2016. **6**: 116.

27. Ferrari, A., et al., FLUKA: a multi-particle transport code. CERN Yellow Report CERN 2005-10; INFN/TC 05/11, SLAC-R-773 (Geneva: CERN), 2005.

28. Agostinelli, S., et al., GEANT4- a simulation toolkit. Nuclear Instruments and Methods in Physics Research, 2003. A 506: p. 250–303.

29. Allison, J., et al., Geant4 developments and applications. *IEEE Trans Nucl Sci*, 2006. **53**: 270–8.

30. Pelowitz, D.B.E., MCNPX user's manual, Version 2.5.0. Los Alamos National Laboratory, 2005. **LA-CP-05-0369**.

31. Waters, L., *MCNPX User's Manual*. Los Alamos National Laboratory, Los Alamos, 2002.

32. Fippel, M. and M. Soukup, A Monte Carlo dose calculation algorithm for proton therapy. *Med Phys*, 2004. **31**(8): 2263–73.

33. Dementyev, A.V. and N.M. Sobolevsky, SHIELD-universal Monte Carlo hadron transport code: scope and applications. *Radiat Meas*, 1999. **30**: 553–7.

34. Arce, P., et al., GAMOS: a GEANT4-based easy and exible framework for nuclear medicine applications. *2008 IEEE Nuclear Science Symposium and Medical Imaging conference (2008 NSS/MIC)*, Dresden, Germany, 2008: pp. 3162–3168.

35. Jan, S., et al., GATE V6: a major enhancement of the GATE simulation platform enabling modelling of CT and radiotherapy. *Phys Med Biol*, 2011. **56**(4): 881–901.

36. Jan, S., et al., GATE: a simulation toolkit for PET and SPECT. *Phys Med Biol*, 2004. **49**(19): 4543–61.

37. Akagi, T., et al., The PTSim and TOPAS projects, bringing Geant4 to the particle therapy clinic. *Prog Nucl Sci Technol*, 2011. **2**: 912–7.

38. Perl, J., et al., TOPAS: an innovative proton Monte Carlo platform for research and clinical applications. *Med Phys*, 2012. **39**(11): 6818–37.

39. Polster, L., et al., Extension of TOPAS for the simulation of proton radiation effects considering molecular and cellular endpoints. *Phys Med Biol*, 2015. **60**(13): 5053–70.

40. Bohlen, T.T., et al., A Monte Carlo-based treatment-planning tool for ion beam therapy. *J Radiat Res*, 2013. **54 Suppl 1**: 77–81.

41. Kurihara, D., et al., A 300-MeV proton beam line with energy degrader for medical science. *Jap J Appl Phys*, 1983. **22**: 1599–605.

42. Brown, K.L., et al., Transport, a computer program for designing charged particle beam transport systems. CERN 73-16 (1973) & CERN 80-04, 1980.

43. Paganetti, H., et al., Accurate Monte Carlo simulations for nozzle design, commissioning and quality assurance for a proton radiation therapy facility. *Med Phys*, 2004. **31**(7): 2107–18.

44. Hsi, W.C., et al., Energy spectrum control for modulated proton beams. *Med Phys*, 2009. **36**(6): 2297–308.

45. Bednarz, B., et al., Uncertainties and correction methods when modelling passive scattering proton therapy treatment heads with Monte Carlo. *Phys Med Biol*, 2011. **56**(9): 2837–54.

46. Grassberger, C., A. Lomax, and H. Paganetti, Characterizing a proton beam scanning system for Monte Carlo dose calculation in patients. *Phys Med Biol*, 2015. **60**(2): 633–45.

47. Brooks, F.D., et al., Energy spectra in the NAC proton therapy beam. *Radiat Prot Dosim*, 1997. **70**: 477–80.

48. Paganetti, H., et al., Clinical implementation of full Monte Carlo dose calculation in proton beam therapy. *Phys Med Biol*, 2008. **53**(17): 4825–53.

49. Peterson, S.W., et al., Experimental validation of a Monte Carlo proton therapy nozzle model incorporating magnetically steered protons. *Phys Med Biol*, 2009. **54**(10): 3217–29.

50. Titt, U., et al., Assessment of the accuracy of an MCNPX-based Monte Carlo simulation model for predicting three-dimensional absorbed dose distributions. *Phys Med Biol*, 2008. **53**(16): 4455–70.

51. Fontenot, J.D., W.D. Newhauser, and U. Titt, Design tools for proton therapy nozzles based on the double-scattering foil technique. *Radiat Prot Dosim*, 2005. **116**(1–4 Pt 2): 211–5.

52. Herault, J., et al., Spread-out Bragg peak and monitor units calculation with the Monte Carlo code MCNPX. *Med Phys*, 2007. **34**(2): 680–8.

53. Koch, N., et al., Monte Carlo calculations and measurements of absorbed dose per monitor unit for the treatment of uveal melanoma with proton therapy. *Phys Med Biol*, 2008. **53**(6): 1581–94.

54. Newhauser, W., et al., Monte Carlo simulations of a nozzle for the treatment of ocular tumours with high-energy proton beams. *Phys Med Biol*, 2005. **50**: 5229–49.

55. Polf, J.C. and W.D. Newhauser, Calculations of neutron dose equivalent exposures from range-modulated proton therapy beams. *Phys Med Biol*, 2005. **50**(16): 3859–73.

56. Piersimoni, P., et al., Optimization of a general-purpose, actively scanned proton beamline for ocular treatments: Geant4 simulations. *J Appl Clin Med Phys*, 2015. **16**(2): 5227.

57. Prusator, M., S. Ahmad, and Y. Chen, TOPAS Simulation of the Mevion S250 compact proton therapy unit. *J Appl Clin Med Phys*, 2017. **18**(3): 88–95.

58. Lee, E., J. Meyer, and G. Sandison, Collimator design for spatially-fractionated proton beams for radiobiology research. *Phys Med Biol*, 2016. **61**(14): 5378–89.

59. Liu, H., et al., TOPAS Monte Carlo simulation for double scattering proton therapy and dosimetric evaluation. *Phys Med*, 2019. **62**: 53–62.

60. Paganetti, H., Four-dimensional Monte Carlo simulation of time-dependent geometries. *Phys Med Biol*, 2004. **49**(6): N75–81.

61. Zacharatou Jarlskog, C., et al., Assessment of organ-specific neutron equivalent doses in proton therapy using computational whole-body age-dependent voxel phantoms. *Phys Med Biol*, 2008. **53**(3): 693–717.

62. Titt, U., et al., Monte Carlo investigation of collimator scatter of proton-therapy beams produced using the passive scattering method. *Phys Med Biol*, 2008. **53**(2): 487–504.

63. Paganetti, H., H. Jiang, and A. Trofimov, 4D Monte Carlo simulation of proton beam scanning: modelling of variations in time and space to study the interplay between scanning pattern and time-dependent patient geometry. *Phys Med Biol*, 2005. **50**(5): 983–90.

64. Shin, J., et al., A modular method to handle multiple time-dependent quantities in Monte Carlo simulations. *Phys Med Biol*, 2012. **57**(11): 3295–308.

65. Peterson, S., et al., Variations in proton scanned beam dose delivery due to uncertainties in magnetic beam steering. *Med Phys*, 2009. **36**(8): 3693–702.

66. Baumann, K.S., et al., Monte Carlo calculation of beam quality correction factors in proton beams using TOPAS/GEANT4. *Phys Med Biol*, 2020. **65**(5): 055015.

67. Baumann, K.S., et al., Comparison of penh, fluka, and Geant4/topas for absorbed dose calculations in air cavities representing ionization chambers in high-energy photon and proton beams. *Med Phys*, 2019. **46**(10): 4639–53.

68. Paganetti, H., Monte Carlo calculations for absolute dosimetry to determine machine outputs for proton therapy fields. *Phys Med Biol*, 2006. **51**(11): 2801–12.

69. Kang, B.-H. and J.-W. Kim, Monte Carlo design study of a gamma detector system to locate distal dose falloff in proton therapy. *IEEE Trans Nucl Sci*, 2009. **56**: 46–50.

70. Polf, J.C., et al., Detecting prompt gamma emission during proton therapy: the effects of detector size and distance from the patient. *Phys Med Biol*, 2014. **59**(9): 2325–40.

71. Li, T., et al., Reconstruction for proton computed tomography by tracing proton trajectories: a Monte Carlo study. *Med Phys*, 2006. **33**(3): 699–706.

72. Schulte, R.W., et al., A maximum likelihood proton path formalism for application in proton computed tomography. *Med Phys*, 2008. **35**(11): 4849–56.

73. Chen, X., et al., A novel design of proton computed tomography detected by multiple-layer ionization chamber with strip chambers: a feasibility study with Monte Carlo simulation. *Med Phys*, 2020. **47**(2): 614–25.

74. Kimstrand, P., et al., Parametrization and application of scatter kernels for modelling scanned proton beam collimator scatter dose. *Phys Med Biol*, 2008. **53**(13): 3405–29.

75. van Luijk, P., et al., Collimator scatter and 2D dosimetry in small proton beams. *Phys Med Biol*, 2001. **46**(3): 653–70.

76. Bues, M., et al., Therapeutic step and shoot proton beam spot-scanning with a multi-leaf collimator: a Monte Carlo study. *Radiat Prot Dosim*, 2005. **115**(1–4): 164–9.

77. Guterres Marmitt, G., et al., Platform for automatic patient quality assurance via Monte Carlo simulations in proton therapy. *Phys Med*, 2020. **70**: 49–57.

78. Toscano, S., et al., Impact of machine log-files uncertainties on the quality assurance of proton pencil beam scanning treatment delivery. *Phys Med Biol*, 2019. **64**(9): 095021.

79. Winterhalter, C., et al., Validating a Monte Carlo approach to absolute dose quality assurance for proton pencil beam scanning. *Phys Med Biol*, 2018. **63**(17): 175001.

80. Kimstrand, P., et al., A beam source model for scanned proton beams. *Phys Med Biol*, 2007. **52**(11): 3151–68.

81. Fracchiolla, F., et al., Characterization and validation of a Monte Carlo code for independent dose calculation in proton therapy treatments with pencil beam scanning. *Phys Med Biol*, 2015. **60**(21): 8601–19.

82. Grevillot, L., et al., A Monte Carlo pencil beam scanning model for proton treatment plan simulation using GATE/GEANT4. *Phys Med Biol*, 2011. **56**(16): 5203–19.

83. Sawakuchi, G.O., et al., Experimental characterization of the low-dose envelope of spot scanning proton beams. *Phys Med Biol*, 2010. **55**(12): 3467–78.

84. Hirayama, S., et al., Evaluation of the influence of double and triple Gaussian proton kernel models on accuracy of dose calculations for spot scanning technique. *Med Phys*, 2016. **43**(3): 1437–50.

85. Faddegon, B.A., et al., Experimental depth dose curves of a 67.5 MeV proton beam for benchmarking and validation of Monte Carlo simulation. *Med Phys*, 2015. **42**(7): 4199–210.

86. Testa, M., et al., Experimental validation of the TOPAS Monte Carlo system for passive scattering proton therapy. *Med Phys*, 2013. **40**(12): 121719.

87. Clasie, B., et al., Assessment of out-of-field absorbed dose and equivalent dose in proton fields. *Med Phys*, 2010. **37**(1): 311–21.

88. Gottschalk, B., R. Platais, and H. Paganetti, Nuclear interactions of 160 MeV protons stopping in copper: a

test of Monte Carlo nuclear models. *Med Phys*, 1999. **26**(-12): 2597–601.

89. Rinaldi, I., et al., An integral test of FLUKA nuclear models with 160 MeV proton beams in multi-layer Faraday cups. *Phys Med Biol*, 2011. **56**(13): 4001–11.

90. Parodi, K., et al., Clinical CT-based calculations of dose and positron emitter distributions in proton therapy using the FLUKA Monte Carlo code. *Phys Med Biol*, 2007. **52**(12): 3369–87.

91. Parodi, K., et al., PET/CT imaging for treatment verification after proton therapy: a study with plastic phantoms and metallic implants. *Med Phys*, 2007. **34**(2): 419–35.

92. Parodi, K., et al., Patient study of in vivo verification of beam delivery and range, using positron emission tomography and computed tomography imaging after proton therapy. *Int J Radiat Oncol Biol Phys*, 2007. **68**(3): 920–34.

93. Jiang, S.B., T. Pawlicki, and C.-M. Ma, Removing the effect of statistical uncertainty on dose-volume histograms from Monte Carlo dose calculations. *Phys Med Biol*, 2000. **45**: 2151–62.

94. Keall, P.J., et al., The effect of dose calculation uncertainty on the evaluation of radiotherapy plans. *Med Phys*, 2000. **27**(3): 478–84.

95. Deasy, J.O., Denoising of electron beam Monte Carlo dose distributions using digital filtering techniques. *Phys Med Biol*, 2000. **45**: 1765–79.

96. Deasy, J.O., M. Wickerhauser, and M. Picard, Accelerating Monte Carlo simulations of radiation therapy dose distributions using wavelet threshold denoising. *Med Phys*, 2002. **29**: 2366–73.

97. Fippel, M. and F. Nuesslin, Smoothing Monte Carlo calculated dose distributions by iterative reduction of noise. *Phys Med Biol*, 2003. **48**: 1289–1304.

98. Kawrakow, I., On the de-noising of Monte Carlo calculated dose distributions. *Phys Med Biol*, 2002. **47**: 3087–103.

99. De Smedt, B., et al., Investigation of geometrical and scoring grid resolution for Monte Carlo dose calculations for IMRT. *Phys Med Biol*, 2005. **50**: 4005–19.

100. Jiang, H. and H. Paganetti, Adaptation of GEANT4 to Monte Carlo dose calculations based on CT data. *Med Phys*, 2004. **31**: 2811–8.

101. Matsufuji, N., et al., Relationship between CT number and electron density, scatter angle and nuclear reaction for hadron-therapy treatment planning. *Phys Med Biol*, 1998. **43**: 3261–75.

102. Palmans, H. and F. Verhaegen, Assigning nonelastic nuclear interaction cross sections to Hounsfield units for Monte Carlo treatment planning of proton beams. *Phys Med Biol*, 2005. **50**: 991–1000.

103. Mustafa, A.A.M. and D.F. Jackson, The relation between x-ray CT numbers and charged particle stopping powers and its significance for radiotherapy treatment planning. *Phys Med Biol*, 1983. **28**: 169–76.

104. Schaffner, B. and E. Pedroni, The precision of proton range calculations in proton radiotherapy treatment planning: experimental verification of the relation between CT-HU and proton stopping power. *Phys Med Biol*, 1998. **43**: 1579–92.

105. Jiang, H., J. Seco, and H. Paganetti, Effects of Hounsfield number conversion on CT based proton Monte Carlo dose calculations. *Med Phys*, 2007. **34**(4): 1439–49.

106. Espana, S. and H. Paganetti, The impact of uncertainties in the CT conversion algorithm when predicting proton beam ranges in patients from dose and PET-activity distributions. *Phys Med Biol*, 2010. **55**(24): 7557–71.

107. du Plessis, F.C.P., et al., The indirect use of CT numbers to establish material properties needed for Monte Carlo calculation of dose distributions in patients. *Med Phys*, 1998. **25**: 1195–201.

108. Schneider, U., E. Pedroni, and A. Lomax, The calibration of CT Hounsfield units for radiotherapy treatment planning. *Phys Med Biol*, 1996. **41**: 111–24.

109. Schneider, W., T. Bortfeld, and W. Schlegel, Correlation between CT numbers and tissue parameters needed for Monte Carlo simulations of clinical dose distributions. *Phys Med Biol*, 2000. **45**: 459–78.

110. Goma, C., I.P. Almeida, and F. Verhaegen, Revisiting the single-energy CT calibration for proton therapy treatment planning: a critical look at the stoichiometric method. *Phys Med Biol*, 2018. **63**(23): 235011.

111. Rogers, D.W.O., et al., BEAMnrc user manual. NRCC Report PIRS-0509, 2002.

112. Chvetsov, A.V. and S.L. Paige, The influence of CT image noise on proton range calculation in radiotherapy planning. *Phys Med Biol*, 2010. **55**(6): N141–9.

113. Unkelbach, J., T.C. Chan, and T. Bortfeld, Accounting for range uncertainties in the optimization of intensity modulated proton therapy. *Phys Med Biol*, 2007. **52**(10): 2755–73.

114. ICRU, Stopping powers and ranges for protons and alpha particles. International Commission on Radiation Units and Measurements, Bethesda, MD, 1993. **Report No. 49**.

115. ICRU, Tissue substitutes in radiation dosimetry and measurement. International Commission on Radiation Units and Measurements, Bethesda, MD, 1989. **Report No. 44**.

116. ICRU, Photon, electron, proton and neutron interaction data for body tissues. International Commision on Radiation Units and Measurements, Bethesda, MD, 1992. **Report No. 46**.

117. Paganetti, H., Range uncertainties in proton therapy and the role of Monte Carlo simulations. *Phys Med Biol*, 2012. **57**(11): R99–117.

118. Yang, M., et al., Theoretical variance analysis of single- and dual-energy computed tomography methods for calculating proton stopping power ratios of biological tissues. *Phys Med Biol*, 2010. **55**(5): 1343–62.

119. Hunemohr, N., et al., Tissue decomposition from dual energy CT data for MC based dose calculation in particle therapy. *Med Phys*, 2014. **41**(6): 061714.

120. Lalonde, A. and H. Bouchard, A general method to derive tissue parameters for Monte Carlo dose calculation with multi-energy CT. *Phys Med Biol*, 2016. **61**(22): 8044–69.

121. Hudobivnik, N., et al., Comparison of proton therapy treatment planning for head tumors with a pencil beam algorithm on dual and single energy CT images. *Med Phys*, 2016. **43**(1): 495.

122. Bazalova, M., et al., Dual-energy CT-based material extraction for tissue segmentation in Monte Carlo dose calculations. *Phys Med Biol*, 2008. **53**(9): 2439–56.

123. Almeida, I.P., et al., Monte Carlo proton dose calculations using a radiotherapy specific dual-energy CT scanner for tissue segmentation and range assessment. *Phys Med Biol*, 2018. **63**(11): 115008.

124. Taasti, V.T., et al., Validation of proton stopping power ratio estimation based on dual energy CT using fresh tissue samples. *Phys Med Biol*, 2017. **63**(1): 015012.

125. Kooy, H., et al., Monitor unit calculations for range-modulated spread-out Bragg peak fields. *Phys Med Biol*, 2003. **48**: 2797–808.

126. Liu, H.H. and P. Keall, D_m rather than D_w should be used in Monte Carlo treatment planning. *Med Phys*, 2002. **29**: 922–924.

127. Paganetti, H., Dose to water versus dose to medium in proton beam therapy. *Phys Med Biol*, 2009. **54**: 4399–421.

128. Carlsson, C.A. and G.A. Carlsson, Proton dosimetry with 185 MeV protons. Dose buildup from secondary protons and recoil electrons. *Health Phys*, 1977. **33**: p. 481–4.

129. Laitano, R.F., M. Rosetti, and M. Frisoni, Effects of nuclear interactions on energy and stopping power in proton beam dosimetry. *Nucl Instrum Methods A*, 1996. **376**: 466–76.

130. Medin, J. and P. Andreo, Monte Carlo calculated stopping-power ratios, water/air, for clinical proton dosimetry (50–250 MeV). *Phys Med Biol*, 1997. **42**: 89–105.

131. Sawakuchi, G.O., et al., Density heterogeneities and the influence of multiple Coulomb and nuclear scatterings on the Bragg peak distal edge of proton therapy beams. *Phys Med Biol*, 2008. **53**(17): 4605–19.

132. Sandison, G.A., et al., Extension of a numerical algorithm to proton dose calculations. I. Comparisons with Monte Carlo simulations. *Med Phys*, 1997. **24**: 841–9.

133. Soukup, M., M. Fippel, and M. Alber, A pencil beam algorithm for intensity modulated proton therapy derived from Monte Carlo simulations. *Phys Med Biol*, 2005. **50**: 5089–104.

134. Koch, N. and W. Newhauser, Virtual commissioning of a treatment planning system for proton therapy of ocular cancers. *Radiat Prot Dosim*, 2005. **115**(1–4): 159–63.

135. Newhauser, W., et al., Monte Carlo simulations for configuring and testing an analytical proton dose-calculation algorithm. *Phys Med Biol*, 2007. **52**(-15): 4569–84.

136. Zhu, X.R., et al., Commissioning dose computation models for spot scanning proton beams in water for a commercially available treatment planning system. *Med Phys*, 2013. **40**(4): 041723.

137. Petti, P.L., Evaluation of a pencil-beam dose calculation technique for charged particle radiotherapy. *Int J Radiat Oncol Biol Phys*, 1996. **35**: 1049–57.

138. Tourovsky, A., et al., Monte Carlo dose calculations for spot scanned proton therapy. *Phys Med Biol*, 2005. **50**: 971–81.

139. Schuemann, J., et al., Site-specific range uncertainties caused by dose calculation algorithms for proton therapy. *Phys Med Biol*, 2014. **59**: 4007–31.

140. Schuemann, J., et al., Assessing the clinical impact of approximations in analytical dose calculations for proton therapy. *Int J Radiat Oncol Biol Phys*, 2015. **92**(5): 1157–64.

141. Molinelli, S., et al., Impact of TPS calculation algorithms on dose delivered to the patient in proton therapy treatments. *Phys Med Biol*, 2019. **64**(7): 075016.

142. Huang, S., et al., Validation and application of a fast Monte Carlo algorithm for assessing the clinical impact of approximations in analytical dose calculations for pencil beam scanning proton therapy. *Med Phys*, 2018. **45**(12): 5631–42.

143. Yepes, P., et al., Comparison of Monte Carlo and analytical dose computations for intensity modulated proton therapy. *Phys Med Biol*, 2018. **63**(4): 045003.

144. Sasidharan, B.K., et al., Clinical Monte Carlo versus Pencil Beam Treatment Planning in Nasopharyngeal Patients Receiving IMPT. *Int J Part Ther*, 2019. **5**(4): 32–40.

145. Jia, Y., et al., Proton therapy dose distribution comparison between Monte Carlo and a treatment planning system for pediatric patients with ependymoma. *Med Phys*, 2012. **39**(8): 4742–7.

146. Yamashita, T., et al., Effect of inhomogeneity in a patient's body on the accuracy of the pencil beam algorithm in comparison to Monte Carlo. *Phys Med Biol*, 2012. **57**(22): 7673–88.

147. Tommasino, F., et al., Impact of dose engine algorithm in pencil beam scanning proton therapy for breast cancer. *Phys Med*, 2018. **50**: 7–12.

148. Liang, X., et al., A comprehensive dosimetric study of Monte Carlo and pencil-beam algorithms on intensity-modulated proton therapy for breast cancer. *J Appl Clin Med Phys*, 2019. **20**(1): 128–36.

149. Grassberger, C., et al., Quantification of proton dose calculation accuracy in the lung. *Int J Radiat Oncol Biol Phys*, 2014. **89**(2): 424–30.

150. Saini, J., et al., Advanced Proton Beam Dosimetry Part I: review and performance evaluation of dose calculation algorithms. *Transl Lung Cancer Res*, 2018. **7**(2): 171–9.

151. Maes, D., et al., Advanced proton beam dosimetry part II: Monte Carlo vs. pencil beam-based planning for lung cancer. *Transl Lung Cancer Res*, 2018. **7**(2): 114–21.

152. Pflugfelder, D., et al., Quantifying lateral tissue heterogeneities in hadron therapy. *Med Phys*, 2007. **34**(4): 1506–13.

153. Petti, P.L., Differential-pencil-beam dose calculations for charged particles. *Med Phys*, 1992. **19**: 137–49.

154. Urie, M., M. Goitein, and M. Wagner, Compensating for heterogeneities in proton radiation therapy. *Phys Med Biol*, 1984. **29**(5): 553–66.

155. Urie, M., et al., Degradation of the Bragg peak due to inhomogeneities. *Phys Med Biol*, 1986. **31**: 1–15.

156. Newhauser, W., et al., Monte Carlo simulations of the dosimetric impact of radiopaque fiducial markers for proton radiotherapy of the prostate. *Phys Med Biol*, 2007. **52**(11): 2937–52.

157. Bednarz, B., J. Daartz, and H. Paganetti, Dosimetric accuracy of planning and delivering small proton therapy fields. *Phys Med Biol*, 2010. **55**(24): 7425–38.

158. Geng, C., et al., Limitations of analytical dose calculations for small field proton radiosurgery. *Phys Med Biol*, 2017. **62**(1): 246–57.

159. Bueno, M., et al., An algorithm to assess the need for clinical Monte Carlo dose calculation for small proton therapy fields based on quantification of tissue heterogeneity. *Med Phys*, 2013. **40**(8): 081704.

160. Magro, G., et al., Dosimetric accuracy of a treatment planning system for actively scanned proton beams and small target volumes: Monte Carlo and experimental validation. *Phys Med Biol*, 2015. **60**(17): 6865–80.

161. Winterhalter, C., et al., Log file based Monte Carlo calculations for proton pencil beam scanning therapy. *Phys Med Biol*, 2019. **64**(3): 035014.

162. Verburg, J.M., et al., Automated Monte Carlo simulation of proton therapy treatment plans. *Technol Cancer Res Treat*, 2015. **15**: NP35–46.

163. Kozlowska, W.S., et al., FLUKA particle therapy tool for Monte Carlo independent calculation of scanned proton and carbon ion beam therapy. *Phys Med Biol*, 2019. **64**(-7): 075012.

164. Shin, W.G., et al., Independent dose verification system with Monte Carlo simulations using TOPAS for passive scattering proton therapy at the National Cancer Center in Korea. Phys Med Biol, 2017. **62**(19): 7598–616.

165. Beltran, C., et al., Clinical implementation of a proton dose verification system utilizing a GPU accelerated Monte Carlo engine. *Int J Part Ther*, 2016. **3**(2): 312–9.

166. Lima, T.V., et al., Monte Carlo calculations supporting patient plan verification in proton therapy. *Front Oncol*, 2016. **6**: 62.

167. Bauer, J., et al., Integration and evaluation of automated Monte Carlo simulations in the clinical practice of scanned proton and carbon ion beam therapy. *Phys Med Biol*, 2014. **59**(16): 4635–59.

168. Mairani, A., et al., A Monte Carlo-based treatment planning tool for proton therapy. *Phys Med Biol*, 2013. **58**(8): 2471–90.

169. Grevillot, L., et al., GATE as a GEANT4-based Monte Carlo platform for the evaluation of proton pencil beam scanning treatment plans. *Phys Med Biol*, 2012. **57**(13): 4223–44.

170. Kimura, A., et al., DICOM data handling for Geant4-based medical physics application. *IEEE Nucl Sci Symp Conf Record*, 2004. **4**: 2124–7.

171. Kimura, A., et al., DICOM interface and visualization tool for Geant4-based dose calculation. *IEEE Nucl Sci Symp Conf Record*, 2005. **2**: 981–4.

172. Moravek, Z., et al., Uncertainty reduction in intensity modulated proton therapy by inverse Monte Carlo treatment planning. *Phys Med Biol*, 2009. **54**(15): 4803–19.

173. Li, Y., et al., A new approach to integrate GPU-based Monte Carlo simulation into inverse treatment plan optimization for proton therapy. *Phys Med Biol*, 2017. **62**(1): 289–305.

174. Saini, J., et al., Dosimetric evaluation of a commercial proton spot scanning Monte-Carlo dose algorithm: comparisons against measurements and simulations. *Phys Med Biol*, 2017. **62**(19): 7659–81.

175. Lin, L., et al., A benchmarking method to evaluate the accuracy of a commercial proton monte carlo pencil beam scanning treatment planning system. *J Appl Clin Med Phys*, 2017. **18**(2): 44–9.

176. Chang, C.W., et al., A standardized commissioning framework of Monte Carlo dose calculation algorithms for proton pencil beam scanning treatment planning systems. *Med Phys*, 2020. **47**: 1545–57.

177. Wagenaar, D., et al., Validation of linear energy transfer computed in a Monte Carlo dose engine of a commercial treatment planning system. *Phys Med Biol*, 2020. **65**(2): 025006.

178. Schumann, J., et al., Efficient voxel navigation for proton therapy dose calculation in TOPAS and Geant4. *Phys Med Biol*, 2012. **57**(11): 3281–93.

179. Sarrut, D. and L. Guigues, Region-oriented CT image representation for reducing computing time of Monte Carlo simulations. *Med Phys*, 2008. **35**(4): 1452–63.

180. Hubert-Tremblay, V., et al., Octree indexing of DICOM images for voxel number reduction and improvement of Monte Carlo simulation computing efficiency. *Med Phys*, 2006. **33**(8): 2819–31.

181. Kohno, R., et al., Simplified Monte Carlo dose calculation for therapeutic proton beams. *Jap J Appl Phys*, 2002. **41**: L294–7.

182. Kohno, R., et al., Experimental evaluation of validity of simplified Monte carlo method in proton dose calculations. *Phys Med Biol*, 2003. **48**: 1277–88.

183. Li, J.S., et al., A particle track-repeating algorithm for proton beam dose calculation. *Phys Med Biol*, 2005. **50**(-5): 1001–10.

184. Yepes, P., et al., A track-repeating algorithm for fast Monte Carlo Dose calculations of proton radiotherapy. *Nucl Technol*, 2009. **168**(3): 736–740.

185. Fix, M.K., et al., Macro Monte Carlo for dose calculation of proton beams. *Phys Med Biol*, 2013. **58**(7): 2027–44.

186. Yepes, P.P., et al., Validation of a track repeating algorithm for intensity modulated proton therapy: clinical cases study. *Phys Med Biol*, 2016. **61**(7): 2633–45.

187. Ramos-Mendez, J., et al., Geometrical splitting technique to improve the computational efficiency in Monte Carlo calculations for proton therapy. *Med Phys*, 2013. **40**(4): 041718.

188. Ramos-Mendez, J., et al., Flagged uniform particle splitting for variance reduction in proton and carbon ion track-structure simulations. *Phys Med Biol*, 2017. **62**(15): 5908–25.

189. Mendez, J.R., et al., Improved efficiency in Monte Carlo simulation for passive-scattering proton therapy. *Phys Med Biol*, 2015. **60**(13): 5019–35.

190. Yepes, P.P., D. Mirkovic, and P.J. Taddei, A GPU implementation of a track-repeating algorithm for proton radiotherapy dose calculations. *Phys Med Biol*, 2010. **55**(23): 7107–20.

191. Jia, X., et al., GPU-based fast Monte Carlo dose calculation for proton therapy. *Phys Med Biol*, 2012. **57**: 7783–98.

192. Ma, J., et al., A GPU-accelerated and Monte Carlo-based intensity modulated proton therapy optimization system. *Med Phys*, 2014. **41**(12): 121707.

193. Kohno, R., et al., Clinical implementation of a GPU-based simplified Monte Carlo method for a treatment planning system of proton beam therapy. *Phys Med Biol*, 2011. **56**(22): N287–94.

194. Giantsoudi, D., et al., Validation of a GPU-based Monte Carlo code (gPMC) for proton radiation therapy: clinical cases study. *Phys Med Biol*, 2015. **60**(6): 2257–69.

195. Qin, N., et al., Recent developments and comprehensive evaluations of a GPU-based Monte Carlo package for proton therapy. *Phys Med Biol*, 2016. **61**(20): 7347–62.

196. Wan Chan Tseung, H., J. Ma, and C. Beltran, A fast GPU-based Monte Carlo simulation of proton transport with detailed modelling of nonelastic interactions. *Med Phys*, 2015. **42**(6): 2967–78.

197. Maneval, D., B. Ozell, and P. Despres, pGPUMCD: an efficient GPU-based Monte Carlo code for accurate proton dose calculations. *Phys Med Biol*, 2019. **64**(8): 085018.

198. Mein, S., et al., Fast robust dose calculation on GPU for high-precision (1)H, (4)He, (12)C and (16)O ion therapy: the FRoG platform. *Sci Rep*, 2018. **8**(1): 14829.

199. Schiavi, A., et al., Fred: a GPU-accelerated fast-Monte Carlo code for rapid treatment plan recalculation in ion beam therapy. *Phys Med Biol*, 2017. **62**(18): 7482–504.

200. Jia, X., P. Ziegenhein, and S.B. Jiang, GPU-based high-performance computing for radiation therapy. *Phys Med Biol*, 2014. **59**(4): R151–82.

201. Grassberger, C., et al., Motion interplay as a function of patient parameters and spot size in spot scanning proton therapy for lung cancer. *Int J Radiat Oncol Biol Phys*, 2013. **86**(2): 380–6.

202. Paganetti, H., et al., Monte Carlo simulations with time-dependent geometries to investigate effects of organ motion with high temporal resolution. *Int J Radiat Oncol Biol Phys*, 2004. **60**(3): 942–50.

203. Grassberger, C., et al., Four-dimensional Monte Carlo simulations of lung cancer patients treated with proton beam scanning to assess interplay effects. *Med Phys*, 2012. **39**: 3998.

204. Dowdell, S., C. Grassberger, and H. Paganetti, Four-dimensional Monte Carlo simulations demonstrating how the extent of intensity-modulation impacts motion effects in proton therapy lung treatments. *Med Phys*, 2013. **40**(12): 121713.

205. Dowdell, S., et al., Interplay effects in proton scanning for lung: a 4D Monte Carlo study assessing the impact of tumor and beam delivery parameters. *Phys Med Biol*, 2013. **58**(12): 4137–56.

206. Botas, P., et al., Density overwrites of Internal Tumor Volumes in Intensity Modulated Proton Therapy plans for mobile lung tumors. *Phys Med Biol*, 2017. **63**(3): 035023.

207. Pepin, M.D., et al., A Monte-Carlo-based and GPU-accelerated 4D-dose calculator for a pencil beam scanning proton therapy system. *Med Phys*, 2018. **45**(11): 5293–304.

208. Kraan, A.C., Range verification methods in particle therapy: underlying physics and Monte Carlo modelling. *Front Oncol*, 2015. **5**: 150.

209. Verburg, J.M., H.A. Shih, and J. Seco, Simulation of prompt gamma-ray emission during proton radiotherapy. *Phys Med Biol*, 2012. **57**(17): 5459–72.

210. Dedes, G., et al., Assessment and improvements of Geant4 hadronic models in the context of prompt-gamma hadrontherapy monitoring. *Phys Med Biol*, 2014. **59**(7): 1747–72.

211. Jeyasugiththan, J. and S.W. Peterson, Evaluation of proton inelastic reaction models in Geant4 for prompt gamma production during proton radiotherapy. *Phys Med Biol*, 2015. **60**(19): 7617–35.

212. Pinto, M., et al., Assessment of Geant4 prompt-gamma emission yields in the context of proton therapy monitoring. *Front Oncol*, 2016. **6**: 10.

213. Testa, M., et al., Range verification of passively scattered proton beams based on prompt gamma time patterns. *Phys Med Biol*, 2014. **59**(15): 4181–95.

214. Schumann, A., et al., Simulation and experimental verification of prompt gamma-ray emissions during proton irradiation. *Phys Med Biol*, 2015. **60**(10): 4197–207.

215. Hueso-Gonzalez, F., et al., A full-scale clinical prototype for proton range verification using prompt

gamma-ray spectroscopy. *Phys Med Biol*, 2018. **63**(18): 185019.

216. Espana, S., et al., The reliability of proton-nuclear interaction cross-section data to predict proton-induced PET images in proton therapy. *Phys Med Biol*, 2011. **56**(9): 2687–98.

217. Bauer, J., et al., An experimental approach to improve the Monte Carlo modelling of offline PET/CT-imaging of positron emitters induced by scanned proton beams. *Phys Med Biol*, 2013. **58**(15): 5193–213.

218. Rohling, H., et al., Comparison of PHITS, GEANT4, and HIBRAC simulations of depth-dependent yields of beta(+)-emitting nuclei during therapeutic particle irradiation to measured data. *Phys Med Biol*, 2013. **58**(18): 6355–68.

219. Seravalli, E., et al., Monte Carlo calculations of positron emitter yields in proton radiotherapy. *Phys Med Biol*, 2012. **57**(6): 1659–73.

220. Paganetti, H., et al., Relative biological effectiveness (RBE) values for proton beam therapy. *Int J Radiat Oncol Biol Phys*, 2002. **53**(2): 407–21.

221. Paganetti, H., Relative biological effectiveness (RBE) values for proton beam therapy. Variations as a function of biological endpoint, dose, and linear energy transfer. *Phys Med Biol*, 2014. **59**(22): R419–72.

222. Paganetti, H. and M. Goitein, Radiobiological significance of beamline dependent proton energy distributions in a spread-out Bragg peak. *Med Phys*, 2000. **27**(5): 1119–26.

223. Paganetti, H. and T. Schmitz, The influence of the beam modulation technique on dose and RBE in proton radiation therapy. *Phys Med Biol*, 1996. **41**(9): 1649–63.

224. Unkelbach, J., et al., Reoptimization of intensity modulated proton therapy plans based on linear energy transfer. *Int J Radiat Oncol Biol Phys*, 2016. **96**(5): 1097–106.

225. Giantsoudi, D., et al., LET-Guided Optimization in IMPT: feasibility study and clinical potential. *Int J Radiat Oncol Biol Phys*, 2013. **87**: 216–222.

226. Kempe, J., I. Gudowska, and A. Brahme, Depth absorbed dose and LET distributions of therapeutic 1H, 4He, 7Li, and 12C beams. *Med Phys*, 2007. **34**(1): 183–92.

227. Grassberger, C. and H. Paganetti, Elevated LET components in clinical proton beams. *Phys Med Biol*, 2011. **56**(20): 6677–91.

228. Grassberger, C., et al., Variations in linear energy transfer within clinical proton therapy fields and the potential for biological treatment planning. *Int J Radiat Oncol Biol Phys*, 2011. **80**(5): 1559–66.

229. Wilkens, J.J. and U. Oelfke, Analytical linear energy transfer calculations for proton therapy. *Med Phys*, 2003. **30**: 806–815.

230. Granville, D.A. and G.O. Sawakuchi, Comparison of linear energy transfer scoring techniques in Monte Carlo simulations of proton beams. *Phys Med Biol*, 2015. **60**(14): N283–91.

231. Cortes-Giraldo, M.A. and A. Carabe, A critical study of different Monte Carlo scoring methods of dose average linear-energy-transfer maps calculated in voxelized geometries irradiated with clinical proton beams. *Phys Med Biol*, 2015. **60**(7): 2645–69.

232. Nikjoo, H., et al., Quantitative modelling of DNA damage using Monte Carlo track structure method. *Radiat Environ Biophys*, 1999. **38**: p. 31–8.

233. Nikjoo, H., et al., Computational approach for determining the spectrum of DNA damage induced by ionizing radiation. *Radiat Res*, 2001. **156**: 577–83.

234. Nikjoo, H., et al., Monte Carlo track structure for radiation biology and space applications. *Phys Med*, 2001. **17 Suppl 1**: 38–44.

235. Friedland, W., et al., Simulation of DNA damage after proton irradiation. *Radiat Res*, 2003. **159**: 401–10.

236. Friedland, W., et al., Simulation of DNA damage after proton and low LET irradiation. *Radiat Prot Dosim*, 2002. **99**(1–4): 99–102.

237. Friedland, W., et al., Simulation of light ion induced DNA damage patterns. *Radiat Prot Dosim*, 2006. **122**(1–4): 116–20.

238. Goodhead, D.T., et al., Track structure approaches to the interpretation of radiation effects on DNA. *Radiat Protect Dosim*, 1994. **52**: 217–23.

239. Charlton, D.E., H. Nikjoo, and J.L. Humm, Calculation of initial yields of single- and double-strand breaks in cell nuclei from electrons, protons and alpha particles. *Int J Radiat Biol*, 1989. **56**(1): 1–19.

240. Holley, W.R. and A. Chatterjee, Clusters of DNA damage induced by ionizing radiation: formation of short DNA fragments. I. Theoretical modelling. *Radiat Res*, 1996. **145**: 188–99.

241. Moiseenko, V.V., A.A. Edwards, and N. Nikjoo, Modelling the kinetics of chromosome exchange formation in human cells exposed to ionising radiation. *Radiat Environ Biophys*, 1996. **35**(1): 31–5.

242. Carlson, D.J., et al., Combined use of Monte Carlo DNA damage simulations and deterministic repair models to examine putative mechanisms of cell killing. *Radiat Res*, 2008. **169**(4): 447–59.

243. Semenenko, V.A. and R.D. Stewart, A fast Monte Carlo algorithm to simulate the spectrum of DNA damages formed by ionizing radiation. *Radiat Res*, 2004. **161**(4): 451–7.

244. Semenenko, V.A. and R.D. Stewart, Fast Monte Carlo simulation of DNA damage formed by electrons and light ions. *Phys Med Biol*, 2006. **51**(7): 1693–706.

245. McNamara, A., et al., Validation of the radiobiology toolkit TOPAS-nBio in simple DNA geometries. *Physica Medica*, 2017. **61**(16): 5993–6010.

246. Lazarakis, P., et al., Comparison of nanodosimetric parameters of track structure calculated by the Monte Carlo codes Geant4-DNA and PTra. *Phys Med Biol*, 2012. **57**(5): 1231–50.

247. McNamara, A.L., et al., A comparison of X-ray and proton beam low energy secondary electron track structures using the low energy models of Geant4. *Int J Radiat Biol*, 2012. **88**(1–2): 164–70.

248. Zhu, H., et al., A parameter sensitivity study for simulating DNA damage after proton irradiation using TOPAS-nBio. *Phys Med Biol*, 2020. 65: 085015.

249. Bueno, M., et al., Influence of the geometrical detail in the description of DNA and the scoring method of ionization clustering on nanodosimetric parameters of track structure: a Monte Carlo study using Geant4-DNA. *Phys Med Biol*, 2015. **60**(21): 8583–99.

250. Ottolenghi, A., M. Merzagora, and H.G. Paretzke, DNA complex lesions induced by protons and alpha-particles: track structure characteristics determining linear energy transfer and particle type dependence. *Radiat Environ Biophys*, 1997. **36**: 97–103.

251. Dos Santos, M., et al., Influence of the DNA density on the number of clustered damages created by protons of different energies. *Nucl Instrum Methods Phys Res*, 2013. **B 298**: 47–54.

252. Gonzalez-Munoz, G., et al., Monte Carlo simulation and analysis of proton energy-deposition patterns in the Bragg peak. *Phys Med Biol*, 2008. **53**(11): 2857–75.

253. Backstrom, G., et al., Track structure of protons and other light ions in liquid water: applications of the LIonTrack code at the nanometer scale. *Med Phys*, 2013. **40**(6): 064101.

254. Nikjoo, H., et al., Track structure in radiation biology: theory and applications. *Int J Radiat Biol*, 1998. **73**: 355–64.

255. Nikjoo, H., et al., Track-structure codes in radiation research. *Radiat Meas*, 2006. **41**(9–10): 1052–74.

256. Michalik, V., Particle track structure and its correlation with radiobiological endpoint. *Phys Med Biol*, 1991. **36**: 1001–12.

257. Michalik, V., Model of DNA damage induced by radiations of various qualities. *Int J Radiat Biol*, 1991. **62**: 9–20.

258. Ottolenghi, A., F. Monforti, and M. Merzagora, A Monte Carlo calculation of cell inactivation by light ions. *Int J Radiat Biol*, 1997. **72**: 505–13.

259. Friedland, W., et al., Simulation of DNA fragment distributions after irradiation with photons. *Radiat Environ Biophys*, 1999. **38**: 39–47.

260. Friedland, W., et al., Monte Carlo simulation of the production of short DNA fragments by low-linear energy transfer radiation using higher-order DNA models. *Radiat Res*, 1998. **150**(2): 170–82.

261. Tomita, H., et al., Monte Carlo simulation of physicochemical processes of liquid water radiolysis. The effects of dissolved oxygen and OH scavenger. *Radiat Environ Biophys*, 1997. **36**(2): 105–16.

262. Sato, T., et al., Fluence-to-dose conversion coefficients for neutrons and protons calculated using the PHITS code and ICRP/ICRU adult reference computational phantoms. *Phys Med Biol*, 2009. **54**(7): 1997–2014.

263. Tian, Z., S.B. Jiang, and X. Jia, Accelerated Monte Carlo simulation on the chemical stage in water radiolysis using GPU. *Phys Med Biol*, 2017. **62**(8): 3081–96.

264. Ramos-Mendez, J., et al., Monte Carlo simulation of chemistry following radiolysis with TOPAS-nBio. *Phys Med Biol*, 2018. **63**(10): 105014.

265. Peukert, D., et al., Validation and investigation of reactive species yields of Geant4-DNA chemistry models. *Med Phys*, 2019. **46**(2): 983–98.

266. Incerti, S., et al., The Geant4-DNA project. *Int J Model, Simul, Sci Comput*, 2010. **1**: 157–78.

267. Bernal, M.A., et al., Track structure modelling in liquid water: a review of the Geant4-DNA very low energy extension of the Geant4 Monte Carlo simulation toolkit. *Phys Med*, 2015. **31**(8): 861–74.

268. Incerti, S., et al., Review of Geant4-DNA applications for micro and nanoscale simulations. *Phys Med*, 2016. **32**(-10): 1187–1200.

269. Schuemann, J., et al., A New Standard DNA Damage (SDD) data format. *Radiat Res*, 2019. **191**(1): 76–92.

270. Incerti, S., et al., Simulating radial dose of ion tracks in liquid water simulated with Geant4-DNA: a comparative study. *Nucl Instrum Methods Phys Res*, 2014. **B 333**: 92–8.

Monte Carlo as a QA Tool for Advanced Radiation Therapy

JinSheng Li and
C.-M. Charlie Ma
Fox Chase Cancer Center

14.1 Introduction

Advanced radiation therapy (RT), such as intensity-modulated radiation therapy (IMRT) (Burman et al., 1997, Webb, 1998, Boyer and Yu, 1999, Purdy, 1996) and intensity-modulated rotational therapy including volumetric modulated arc therapy (VMAT) and RapidArc (Yu, 1995, Otto, 2008, Lagerwaard et al., 2008), has been widely accepted as the standard treatment strategy for many treatment sites in the field of radiation oncology because of the ability to provide quality conformal dose distributions. Advanced RT treatment techniques offer better sparing for the surrounding normal tissues than conventional treatment methods and, therefore, lead to less normal tissue complications (Salama et al., 2006, Studer et al., 2008, Zelefsky et al., 2008) and the possibility for dose escalation to the treatment target (Al-Mamgani et al., 2009). To provide a conformal dose distribution, advanced RT treatment techniques require a more complex treatment planning process utilizing computers in the plan optimization and a more complex beam delivery system, which may utilize a multileaf collimator (MLC) with complex leaf motion sequences. For intensity modulation purposes, the radiation field is created with a number of small beamlets with a typical size of 10×10 or 5×5 mm^2 delivered through many small, irregular and asymmetric MLC fields, which obscure the relationship between the accelerator monitor unit (MU) setting and the radiation dose received by the patient. When online images or other localization and tracking systems are used for treatment guidance and intervention, the beam delivery may become even more complex. Furthermore, patient anatomy heterogeneity, organ motion and deformation may add additional uncertainties to the actual dose distribution received by the patient. Overall, potential errors associated with advanced RT include dose calculation inaccuracies, plan transfer errors, beam delivery errors and target localization uncertainties due to patient setup errors and organ motion during the treatment. Considering the serious consequences of these errors, comprehensive quality assurance (QA) should be performed before and/or during the patient treatment.

The ultimate goal of RT QA is to ensure that the patient will receive the planned dose distribution. However, in clinical routine practice, only part of the radiotherapy process is usually verified. Methods for RT QA vary from facility to facility in current practices, which can be categorized as measurement-based, calculation-based, and simulation-based. The measurement-based method includes measuring and verifying the fluence map for all fields using film or a 2-dimensional (2D) detector array (Iori et al., 2007, Pallotta et al., 2007, Grein et al., 2002, Greer and Popescu, 2003) to measure and compare point doses or a 2D dose distribution using an ion-chamber or a 2D detector array in a dummy phantom (Dong et al., 2003, Ma et al., 2003, Dobler et al., 2010). The measurement-based method can verify the treatment plan, dose calculation in the phantom, and plan transfer and delivery. However, it is unable to verify the treatment dose to the patient, especially for a treatment site with heterogeneity in which the absolute dose distribution can be significantly different from that in a homogeneous phantom. In addition, patient setup error, organ motion and deformation, and their effects on the patient's dose distribution cannot be adequately evaluated. Other concerns include the requirement of treatment machine time and the off-hour workload to perform these QA measurements. The calculation-based method can be performed in the patient geometry as specified by the

DOI: 10.1201/9781003211846-17

patient's CT scan or other image data set. This technique can be performed at any time using computers, making it less labor intensive. Most calculation-based QA methods currently used clinically employ simple dose/monitor unit (MU) calculation algorithms. The effects of patient heterogeneity, setup error, and organ motion on the dose distribution are not included in the calculation. They are frequently used as a secondary MU check tool. In addition, plan transfer and beam delivery cannot be verified by this method. Computerized Monte Carlo simulation can be used as a comprehensive method for RT treatment QA. The simulation-based method can perform many tasks that cannot be done with measurement-based methods, such as determining the dose in a patient, and it can provide more information in an accurate manner. Together with measured or recorded beam delivery information, it can verify the plan transfer and treatment delivery as well.

The accuracy of the Monte Carlo method for radiation dose calculation in a complex and heterogeneous geometry has been extensively benchmarked (Andreo, 1991, Rogers, 1990, Rogers, 2006). Compared with those correction-based and even model-based methods used in conventional algorithms, the Monte Carlo method can model the details of the beam delivery system more precisely because each individual particle is simulated separately in the accelerator head and patient geometry (Rogers et al., 1995). Recent developments in computer techniques and efficiency-enhancing techniques have made the Monte Carlo method practical for clinical use. Therefore, the Monte Carlo method can be employed as an accurate and comprehensive QA tool for advanced radiation therapy applications.

The following sections will discuss Monte Carlo dose calculation, techniques and implementations, and the clinical applications of Monte Carlo-based QA for advanced radiotherapy treatments.

14.2 Techniques and Implementation Methods

14.2.1 Overview of Monte Carlo-Based QA

Monte Carlo dose calculation is performed by transporting the primary radiation particles and their descendants through the patient phantom geometry and recording their energy deposition along the path with a random sampling-based computer simulation process. One can perform Monte Carlo-based patient dose calculation for radiation therapy starting with the electrons exiting from the accelerator vacuum window and following them and their descendants through the rest of the accelerator head including the beam collimation and monitoring devices and then the patient geometry. This method is very inefficient, since most of the particles are blocked by the collimators. Realistically, the radiation beam can be represented by a phase-space data set (Rogers et al., 1995) or a source model (Ma et al., 1997) at the plane below the fixed components of the accelerator head. The phase-space data are generated by direct

Monte Carlo simulations of the patient-independent components in the accelerator head, and the source model parameters are derived from the phase-space data (Ma et al., 1997, Ma and Jiang, 1999, Deng et al., 2000, Chetty, 2007, Chetty et al., 2000, Fix et al., 2001c) or measured beam data (Jiang et al., 2001). The phase-space data and the source model can be used for all patients to be treated on this accelerator. The properties of the particles, including type, energy, direction, and location, can be retrieved from the phase-space data file or reconstructed from the source model. Monte Carlo-based RT QA can start from the phase-space plane or the source model and follow all the particles and their descendants through the patient-dependent accelerator components (e.g., jaws, compensators and MLC) and the patient geometry. This is referred to as direct Monte Carlo simulation of the beam delivery. Because many particles will be blocked by the beam collimation system, direct Monte Carlo simulation for advanced radiotherapy QA is not efficient. An alternative method is to use the beam intensity map. Instead of simulating a particle through the collimation system, the particle is assigned a weighting factor based on the probability for it to go through the collimation system. The distribution of the weighting factor is defined by the transmission properties of the compensator or the MLC leaf geometry and the sequence of the leaf movements. The particle's weighting factor will determine the particle's contribution to the dose. The intensity map can be obtained from the 2D fluence measurement or reconstruction from the treatment delivery information as discussed in the following sections.

14.2.2 Treatment Delivery Information

As illustrated in Figure 14.1, in addition to the phase-space file/source model and the patient geometry as specified by a CT or other imaging data set, the treatment delivery information is required by the Monte Carlo-based QA process. First, the radiation beam and its direction relative to the patient geometry need to be established. The phase-space data or source model will be selected based on the actual radiation beam used for the treatment. The beam direction is usually described using the gantry, collimator, and couch angles for a conventional medical linear accelerator. It is sometimes described differently for other treatment machines, such as the nodes for the CyberKnife or the rotation angles for a TomoTherapy machine. This information can be found in the treatment plan, record and verify (R&V) system, and the treatment log file, as illustrated in Figure 14.1. The jaw setting or MLC leaf positions, which define the shape of a radiation field or an MLC field segment, and the MUs used for the radiation field are required by the direct Monte Carlo simulation and the intensity map reconstruction program for the entire treatment. The MU information can be found in the treatment plan, the R&V record and/or the machine treatment log file. The jaw setting and MLC leaf position information can be found in the treatment plan or the machine treatment log file,

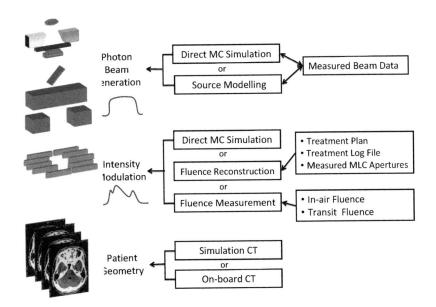

FIGURE 14.1 Treatment beam information used for Monte Carlo-based dose verification.

or measured directly with film, an electronic portal imaging device (EPID) or a 2D detector array.

The calculation-based QA methods can verify the various steps of a radiation treatment procedure. If the treatment field information is taken from the treatment plan, the dosimetry accuracy of the treatment plan can be verified but the plan transfer among different systems involved in the treatment delivery, e.g., the treatment planning system, the R&V system, and the accelerator control system, is not ensured. If the treatment field information is taken from the R&V database and/or the accelerator log file, both the treatment plan and the plan transfer accuracy are verified but the treatment delivery is still not ensured because the actual dose received by the patient also depends on the treatment machine conditions (beam output variation and collimator positional errors) and the patient geometry (setup uncertainty and inter- and intrafractional organ motion). If the treatment field information is derived from the measured portal images of the treatment beam, the treatment delivery can also be verified. Combined with the Monte Carlo dose calculation in the patient geometry with motion and setup errors considered (e.g., if patient's 3D image data are available in real time), the actual dose received by the patient can be determined and compared with that of the original treatment plan. This will verify the entire radiotherapy process for a particular treatment, which will be the most complete patient-specific treatment QA procedure.

14.2.3 Direct Monte Carlo Simulation

Direct Monte Carlo simulation means to simulate the particle transport and interaction in the patient-dependent field-defining components in the treatment delivery system together with patient dose calculation using the Monte Carlo method. The radiation particles and their descendants are transported in the beam collimation system and the patient geometry until they exhaust their energy or escape to the free space. Geometry parameters of the physical components of the treatment delivery system are determined based on the treatment delivery sequence (jaw setting, MLC leaf positions and MUs for individual segments) and the geometry of machine components. During the simulation, the beam collimation geometry can be considered to be stationary at any particular moment (treatment segment). This works for "stationary" beam delivery, e.g., conventional radiotherapy and step-and-shoot IMRT, and for "dynamic" beam delivery, such as dynamic IMRT and IMAT/VMAT/RapidArc. The particles are weighted by the MUs of each treatment segment and, thus, produce "absolute" rather than "relative" dose distributions. Various beam modifiers, such as the block, wedge and MLC, have been incorporated in the patient simulation geometry for IMRT dose verification studies by several research groups (Aaronson et al., 2002, Heath and Seuntjens, 2003, Kim et al., 2001, Keall et al., 2001, Siebers et al., 2002, Leal et al., 2003, Li et al., 2000, Fragoso et al., 2009, Spezi et al., 2001, Fix et al., 2001b, Fix et al., 2001a, Liu et al., 2001). With direct Monte Carlo simulation, the details of the beam collimation device, such as the tongue-and-groove structure and the leaf-end shape of the MLC, can be considered in the patient dose calculation accurately if the treatment delivery system is modeled precisely. The direct Monte Carlo simulation method usually requires enormous computing power because the Monte Carlo method is CPU time consuming. Furthermore, the field shape defined by the beam collimation system varies during the treatment, and most particles are blocked before they can reach the patient geometry. To improve simulation efficiency, a variety of strategies have been applied to the particle transport in the beam collimation system. One method is to terminate the particle transport in the beam collimation system using high cutoff energies for photon and electron particles. Another method is to disregard the secondary Compton scattering photons (Siebers et al., 2002) or the secondary electrons (Tyagi et al., 2007) because they deposit most of their energy locally. Some researchers have also investigated methods to model the radiation transport at the field edge only to account for the effects of partial transmission and photon scattering (Chetty et al., 2000, Fix et al., 2001c, Fix et al., 2000). Significant time saving has been observed using these methods.

14.2.4 Beam Intensity Map for Monte Carlo-Based QA

As mentioned above, instead of performing direct Monte Carlo simulations, the beam intensity maps obtained by direct measurements or analytical reconstruction methods can be used in Monte Carlo patient dose calculation.

14.2.4.1 Beam Intensity Map Measurement

The beam intensity map for each beam angle can be measured using film, EPID, or 2D detector arrays, such as MapCHECK (Sun Nuclear, Melbourne, FL) and MatriXX (IBA, Schwarzenbruck, Germany). Measurements can be performed at various locations, e.g., the beam exit window, the isocenter plane, or any extended distances from the isocenter, pretreatment without a patient in the beam or during the treatment with a patient in the beam. All of the measured intensity maps without a patient in the beam can be used for dose calculation after some corrections are made (e.g., the detector response and back scattering from surrounding materials). The map measured online behind the patient is called the transmission (transit) intensity map, which also includes the effect of beam attenuation and scattering by the patient. This effect must be

corrected or removed from the transit intensity maps before they can be used for dose calculation. Ray-tracing algorithms are frequently used to derive the entrance intensity map in front of the patient from the transit images. Overall, careful calibration of the 2D detector array and proper consideration of the patient scattering and attenuation will improve the accuracy of the patient dose distribution calculated this way. Some investigators have reported the use of the EPID for routine machine QA, pretreatment dose verification, and in vivo dosimetry (Kroonwijk et al., 1998, Pasma et al., 1999b, Pasma et al., 1999a, van Zijtveld et al., 2006, Boellaard et al., 1998).

14.2.4.2 Intensity Map Reconstruction

In addition to the use of direct measurements, the beam intensity map can also be reconstructed based on the treatment delivery information. Reconstructing a 2D intensity map based on the recorded MLC leaf sequence and MU information for advanced radiotherapy dose calculation can save a great deal of simulation time compared with direct Monte Carlo simulations and can be more accurate and versatile compared with online intensity map measurements because the corrections are patient dependent. The beam delivery information including MU and MLC leaf sequences can be derived from various sources as described in Section 14.2.2.

The 2D beam intensity map can be built at the midplane of the MLC and expressed at the isocenter plane for convenience extending over the maximum treatment field dimensions. The pixel size of the beam intensity map may affect the simulation results of the patient dose distribution. It was shown (Li et al., 2010) that the pixel size should be as small as 0.2 mm. Figure 14.2 demonstrates that the MUs for each MLC segment are added to the beam intensity values of all pixels in the open area of a MLC segment (particles B, D and H). Variable fractions of the MUs are added to the beam intensity values of all

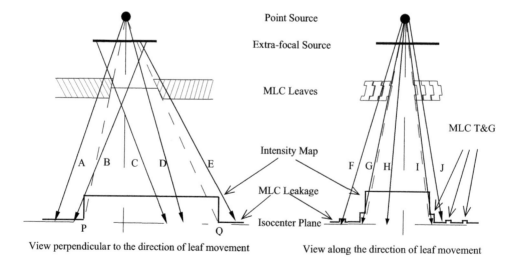

FIGURE 14.2 Diagram showing how the intensity map is built based on the leaf sequence information to include the effects of photon partial transmission through the MLC leaf end and the tongue-and-groove geometry. A to J are photons emitted from the point source and the extra-focal source.

affected pixels, based on the attenuation and scattering effect of the MLC leaves, e.g., using the effective transmission factors of the MLC leaf (particles A, C, E and J) and the MLC leaf tongue-and-groove structure (particles F, G and I). The dimensions of the leaf opening are set according to the treatment beam information plus an offset to consider the partial transmission effect of the leaf end or leaf side. The leaf offset can be calculated based on a picket-fence film measurement as described in the literature (Li et al., 2010). The transmission factor for the areas under the photon jaws can be assumed to be zero. The dimensions of the photon jaw opening are set according to the actual treatment setups for each MLC segment. The leaf sequences are read sequentially, and the beam intensity map is built for every beam direction. It was shown (Li et al., 2010) that it is essential to carefully consider the effects of the details of the beam delivery system, including the source distribution, MLC thickness, MLC transmission and scattering, MLC tongue-and-groove structure and the leaf-end shape. These factors can affect the doses at various field locations differently due to the complexity of the MLC leaf sequence, and, thus, they can result in not only different mean doses but also different dose uncertainties at individual field locations. Their effects are case dependent because the MLC leaf sequence is specific to the treatment plan of a particular patient.

14.2.5 Monte Carlo Dose Calculation

During a Monte Carlo simulation, the spatial source distribution and its effects can be considered by sampling particles from multiple sub-sources, which include a primary point source at the target location, an extra-focal source to account for photon scattering in the primary collimator and flattering filter (Jiang et al., 2001) and an electron contamination source at a location above the jaws to account for the electron component in the beam (Yang et al., 2004). The MLC collimation can be considered either by a direct Monte Carlo simulation of particle transport and interaction in the MLC leaf geometry or by ray tracing through the MLC leaf geometry to determine whether the particles can go through the MLC opening (Li et al., 2010).

When the beam intensity map was initially implemented for Monte Carlo-based dose calculation, the weight of a phase-space particle was altered based on the value of the pixel in the intensity map through which the particle was travelling (Ma et al., 1999, Ma et al., 2000). The gantry/couch/collimator angles were changed automatically after the simulation of each treatment field. A 3-dimensional (3D) voxelized rectilinear phantom converted from the patient/phantom CT scan (Ma et al., 1999, Chetty et al., 2007) has been used for the dose calculation. The density and material for each voxel are converted based on the CT number. The conversion function for any individual CT scanner is unique and is obtained based on the CT calibration with a standard phantom. To avoid the spatial averaging effects (Ai-Dong et al., 2005), a proper voxel

size needs to be used when heterogeneity presents. Since the Monte Carlo method is based on random sampling, the calculated doses always contain statistical uncertainties. The doses calculated using analytical methods do not have statistical uncertainties; however, they usually contain larger systematic errors in heterogeneous phantoms compared to the Monte Carlo method. By simulating a sufficient number of particles, the statistical uncertainty of a Monte Carlo result can be reduced to such a degree that will be considered clinically insignificant. Such a result will be useful and meaningful clinically. Otherwise, large statistical uncertainties may lead to confusing results as shown by some researchers (Jiang et al., 2000, Jeraj and Keall, 2000).

Another issue that is often encountered in the patient dose calculation is whether the dose should be reported as dose-to-water or dose to the local medium. By default, the initial calculated Monte Carlo dose is dose to the local medium. Compared with the results calculated by conventional methods, significant differences are usually seen since conventional methods are reporting dose-to-water by default. If one converts the dose value calculated by Monte Carlo in a bone voxel to dose-to-water using the mass electron stopping power ratio between bone and water, more significant differences may result in the dose value (Siebers et al., 2000, Ma and Li, 2011). It is still debatable as to whether to use dose-to-water or dose-to-medium for advanced radiation therapy since it depends actually on the situation (e.g., the particle type, the voxel size and, for electrons, the treatment depth).

14.3 Clinical Applications

The radiation treatment procedure for a patient includes treatment planning using a treatment planning system (TPS), treatment plan transfer and verification using an R&V system, and treatment delivery on a clinical treatment machine. Accordingly, as shown in Figure 14.3, the Monte Carlo method can be applied in the TPS commissioning and verification, patient-specific plan QA, and online/offline patient treatment QA.

14.3.1 MC for TPS QA

Since computers were employed in radiation therapy treatment planning, many dose calculation algorithms have been developed, which become more accurate but also more sophisticated and computation intensive. Those early correction-based algorithms can compute the patient dose distributions based on the dose data measured in water for a clinical treatment machine with simple corrections for the differences in beam attenuation and scattering between a uniform water phantom irradiated by radiation beams with regular field sizes and the heterogeneous patient geometry irradiated by the actual treatment beams with irregular field shapes. More recent model-based algorithms predict patient dose distributions from primary particle fluence and a dose kernel, which can be derived from direct measurements, analytical calculations or

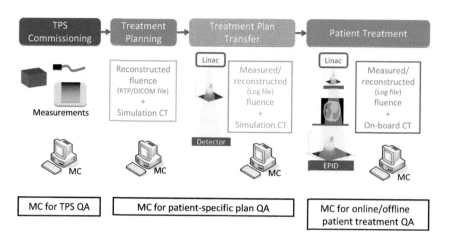

FIGURE 14.3   Clinical applications of Monte Carlo-based QA tool.

Monte Carlo simulations. The accuracy of these algorithms as implemented in different TPS depends critically on the details of the beam fluence modelling for individual treatment delivery systems and the kernel variation to account for the effect of particle attenuation and scattering in the patient's heterogeneous geometry. The same applies to the Monte Carlo dose calculation method, in which the dose calculation accuracy can be affected by the implementation details such as the particle phase-space reconstruction and patient geometry setup. To ensure the accuracy of the dose calculation algorithms of a TPS, a series of tests must be performed during the commissioning process before its clinical use.

Source modelling and beam commissioning are important elements of TPS commissioning, especially for model-based algorithms and Monte Carlo techniques. The source can be modeled as a single point source or multiple sub-sources in which the source size and spatial source distribution can be considered in the dose calculation. One application of Monte Carlo-based QA is to evaluate the accuracy of the source modelling and commissioning by comparing the dose distributions calculated by the TPS and the Monte Carlo code for some representative fields defined by the jaws and/or MLC. Percentage depth-dose (PDD) distributions for different square fields, lateral dose profiles at different depths, and beam output factors for different field sizes are usually used for the evaluation.

The dose calculation algorithms implemented in a TPS can be verified by standard phantom measurements. The QA measurements for a typical patient's treatment plan are usually performed at a reference point with an ionization chamber and a plane with film or a 2D detector array in a phantom. Assorted treatment plans, including 3D conformal RT, IMRT and dynamic rotational therapy for conventional accelerators or other kinds of treatment plans for various specialty treatment machines such as CyberKnife and TomoTherapy, must be tested thoroughly. Though measurement is a good standard for dose verification, it has some major limitations. First, it is still a challenge to measure the 3D dose distribution in a phantom, although some 3D detector arrays and image-based measurement techniques (e.g., gel dosimetry) have been developed. It is also difficult to mimic the real patient geometry using a heterogeneous phantom. In vivo dosimetry is typically performed with detectors on the patient's surface and cavities or inserted invasively in the patient's body. Corrections have to be made to the detector readings to account for the variation of detector response and other factors due to non-calibration conditions. In comparison, the Monte Carlo method can be used to simulate the 3D dose distribution in the patient geometry represented by the patient's CT or in a heterogeneous phantom that can represent the patient geometry accurately. The dose-volume histograms (DVH) for the treatment targets and organs at risk (OAR) can be compared together with the isodose distributions between the TPS and the Monte Carlo method. This comparison can also be used to verify the MUs for the patient's treatment plan as calculated by the TPS, which serves as a part of the independent physics check in the treatment planning QA process. Several research groups have published their results of this application for different TPS (Ma et al., 2000, Li et al., 2001, Weber and Nilsson, 2002, Francescon et al., 2003, Wang et al., 2006, Bush et al., 2008, Sarkar et al., 2008).

The details of the beam delivery system and their effects on the patient dose distributions may not be precisely accounted for in the TPS dose calculation due to simplification in the implementation and uncertainties resulting from the commissioning procedure. The TPS usually employs a global correction factor (a fudge factor) based on the mean ratio of measured point doses for a number of test plans to those predicted by the TPS calculation. This correction factor can only reduce but not eliminate the mean dose difference because this point-dose ratio is case dependent and position dependent as a result of the leaf sequence for a specific treatment plan. The Monte Carlo method is more advantageous in the verification of the absolute dose and for determination of the correction factor since the mean dose ratio can be evaluated based on the dose in a region rather than at one or more points.

14.3.2 MC for Patient-Specific Plan QA

Dose measurement in a phantom is frequently used to verify the dosimetry accuracy of the phantom plan, plan transfer and machine delivery for a specific patient. However, this plan QA process only ensures the dosimetry accuracy of the phantom plan, which is calculated based on the patient's treatment plan and the phantom geometry. This process does not verify the dose to be received by the patient. Monte Carlo dose calculation based on the patient geometry and the treatment delivery information can verify the dose distribution to be received by the patient, and, thus, it can be utilized as a patient-specific QA tool.

The treatment delivery information obtained at the various steps of the radiotherapy process can be used for the Monte Carlo QA calculation. Good agreement between the patient dose distribution predicted by the TPS and that by Monte Carlo has been observed for various treatment sites and different treatment planning systems. However, significant differences in the doses to the target and the critical structures have been observed for some treatment cases, especially in the region where heterogeneity exists. The difference is mainly caused by the systematic error of the simple dose calculation algorithm in the heterogeneous region and inadequate consideration of the effects of the details of the beam delivery system, such as the thickness of the MLC leaf, the shape of the leaf end and tongue-and-groove structure. The faults and inadequate considerations in the implementation and programming of the treatment optimization, dose calculation and leaf sequencing algorithms can, of course, produce dose errors. Many researchers have proposed and demonstrated their methods for Monte Carlo-based patient dose verification for different treatment modalities. For example, Ma et al. (2000), Pawlicki and Ma (2001), Wang et al. (2002), Li et al. (2004), Yang et al. (2005) and Jiang et al. (2006) reconstructed the intensity map using the leaf sequence file or RTP file from the TPS for Linac-based treatments. Luo et al. (Luo et al., 2006) used the treatment log file and R&V record and Lin et al. (2009) used the measured EPID images for intensity map reconstruction for Linac-based IMRT treatments. Katsuta et al. (2017) investigated Elekta log file-based patient-specific QA using Monte Carlo dose calculation for VMAT. Stanhope et al. (2018) validated log file-based Monte Carlo patient-specific QA with ArcCHECK for Linac. For other treatment modalities, such as CyberKnife, TomoTherapy and proton treatment machine, several groups also implemented Monte Carlo-based patient-specific QA in their clinic. Reynaert et al. (2016) developed a Monte Carlo-based treatment plan validation platform for CyberKnife and TomoTherapy treatments. Mackeprang et al. (2017) also developed a Monte Carlo-based independent dose calculation framework to perform patient-specific QA for MLC-based CyberKnife treatment plans. Fracchiolla et al. (2015), Winterhalter et al. (2018) and Huang et al. (2018) implemented a Monte Carlo platform in clinical practice to provide independent dose validation/QA in proton therapy treatments

with pencil beam scanning. All reported results show that the Monte Carlo method is a useful and practical tool for patient dose verification. Together with measurement or recorded beam delivery information, the treatment delivery and plan transfer can also be verified.

For online adaptive radiation therapy QA, a group from Washington University developed a fast Monte Carlo dose calculation system for validating online re-optimized tri-Co60 IMRT adaptive plan based on real-time magnetic resonance imaging (Wang et al, 2017). The Monte Carlo system is based on the dose planning method (DPM) code with further simplification of electron transport and consideration of external magnetic fields. A vendor-provided head model was incorporated into the code. Both GPU acceleration and variance reduction techniques were implemented.

Based on the patient's 4D CT data, dose calculation can be performed for each phase of the breathing cycle. After performing image deformation and registration for all the breathing phases, the dose can be summed up for the entire treatment with a proper weighting factor based on the patient's average breathing cycle and, thus, a 4D dose distribution can be obtained (Keall et al., 2004). The time association between the beam-on time and the breathing phase can be associated with the real-time positioning management (RPM) system or other respiratory tracking signals. Although dosimetry uncertainties still exist due to the uncertainties of the 4D CT images, the correlation between the external surrogate and internal organ/target motion, and the deformation registration of CT images at various phases, 4D dose calculation using the Monte Carlo method is still valuable for the investigation of the dosimetric effects of patient's intrafractional motion and organ deformation.

14.3.3 MC for Online and Offline Treatment QA

Utilizing film, 2D detector arrays, EPID and 3D patient imaging for pretreatment or online treatment measurement, the Monte Carlo method can be used for pretreatment, online and offline patient treatment QA for advanced RT. As described in previous sections, film, 2D detector arrays and EPID can be used to measure the entrance fluence map in absence of the patient or to reconstruct the entrance fluence map based on the transit intensity map measured with the patient in place during the treatment. The entrance fluence map as directly measured by film, 2D detector arrays and EPID must be corrected for the detector response variation and other effects due to attenuation and scattering by the measurement device itself. The entrance fluence map can be derived from the transit intensity maps by ray tracing after correcting the beam attenuation and scattering from the patient, or reconstructed based on the MLC leaf positions and MUs recorded by the R&V system or the treatment log file. Subsequently, online or offline dose calculation using the Monte Carlo method can be performed based on this information. Patient CT images acquired with the in-room CT (cone-beam or CT-on-rails) or other 3D imaging systems (ultrasound and MRI) before and during the treatment can be used for the online dose calculation, and the calculated dose distribution can be used to guide the treatment. Adaptive radiotherapy and online QA is possible if the dose calculation and image processing is sufficiently fast (Wang et al, 2017). The COMPASS system developed by Boggula et al. (2011) reconstructed the 3D dose distribution for treatment verification using the fluence maps measured offline or online with a 2D detector array system.

14.4 Conclusions

Advanced RT techniques have received widespread clinical applications for cancer treatments. A comprehensive and practical QA program is essential to ensure the dosimetry accuracy of treatment planning, plan transfer and treatment delivery for advanced RT. The Monte Carlo method can serve as a useful QA tool in the various steps of the advanced RT process. Techniques and methods to perform Monte Carlo-based QA have been discussed in this chapter. With these methods, a patient's treatment can be verified specifically with confidence. Offline and online dose verification can be performed with online imaging and real-time measurement techniques, which, in turn, can provide useful information for the oncologist to make online or offline treatment adjustments. The Monte Carlo calculated dose distributions incorporated into the patient's setup and organ motion/deformation information during the entire treatment course can be used for treatment assessment and outcome analysis, which will be essential to the design and execution of any retrospective and prospective clinical trials utilizing advanced RT treatment techniques.

References

Aaronson, R. F., Demarco, J. J., Chetty, I. J. & Solberg, T. D. (2002) A Monte Carlo based phase space model for quality assurance of intensity modulated radiotherapy incorporating leaf specific characteristics. *Med Phys*, 29, 2952–8.

Ai-Dong, W., Yi-Can, W., Sheng-Xiang, T. & Jiang-Hui, Z. (2005) Effect of CT image-based voxel size on Monte Carlo dose calculation. *Conf Proc IEEE Eng Med Biol Soc*, 6, 6449–51.

Al-Mamgani, A., Heemsbergen, W. D., Peeters, S. T. H. & Lebesque, J. V. (2009) Role of intensity-modulated radiotherapy in reducing toxicity in dose escalation for localized prostate cancer. *Int J Radiat Oncol Biol Phys*, 73, 685–91.

Andreo, P. (1991) Monte Carlo techniques in medical radiation physics. *Phys Med Biol*, 36, 861–920.

Boellaard, R., Essers, M., Van Herk, M. & Mijnheer, B. J. (1998) New method to obtain the midplane dose using portal in vivo dosimetry. *Int J Radiat Oncol Biol Phys*, 41, 465–74.

Boggula, R., Jahnke, L., Wertz, H., Lohr, F. & Wenz, F. (2011) Patient-specific 3D pretreatment and potential 3D online dose verification of Monte Carlo-calculated IMRT prostate treatment plans. *Int J Radiat Oncol Biol Phys*, 81, 1168–75.

Boyer, A. L. & Yu, C. X. (1999) Intensity-modulated radiation therapy with dynamic multileaf collimators. *Semin Radiat Oncol*, 9, 48–59.

Burman, C., Chui, C. S., Kutcher, G., Leibel, S., Zelefsky, M., Losasso, T., Spirou, S., Wu, Q., Yang, J., Stein, J., Mohan, R., Fuks, Z. & Ling, C. C. (1997) Planning, delivery, and quality assurance of intensity-modulated radiotherapy using dynamic multileaf collimator: a strategy for large-scale implementation for the treatment of carcinoma of the prostate. *Int J Radiat Oncol Biol Phys*, 39, 863–73.

Bush, K., Townson, R. & Zavgorodni, S. (2008) Monte Carlo simulation of RapidArc radiotherapy delivery. *Phys Med Biol*, 53, N359–70.

Chetty, I. (2007) Virtual source modelling in Monte Carlo-based clinical dose calculations: methods and issues associated with their development and use. *Radiother Oncol*, 84, S38–39.

Chetty, I., Demarco, J. J. & Solberg, T. D. (2000) A virtual source model for Monte Carlo modelling of arbitrary intensity distributions. *Med Phys*, 27, 166–72.

Chetty, I. J., Curran, B., Cygler, J. E., Demarco, J. J., Ezzell, G., Faddegon, B. A., Kawrakow, I., Keall, P. J., Liu, H., Ma, C. M., Rogers, D. W., Seuntjens, J., Sheikh-Bagheri, D. & Siebers, J. V. (2007) Report of the AAPM Task Group No. 105: issues associated with clinical implementation of Monte Carlo-based photon and electron external beam treatment planning. *Med Phys*, 34, 4818–53.

Deng, J., Jiang, S. B., Kapur, A., Li, J., Pawlicki, T. & Ma, C. M. (2000) Photon beam characterization and modelling for Monte Carlo treatment planning. *Phys Med Biol*, 45, 411–27.

Dobler, B., Streck, N., Klein, E., Loeschel, R., Haertl, P. & Koelbl, O. (2010) Hybrid plan verification for intensity-modulated radiation therapy (IMRT) using the 2D ionization chamber array I'mRT MatriXX--a feasibility study. *Phys Med Biol*, 55, N39–55.

Dong, L., Antolak, J., Salehpour, M., Forster, K., O'Neill, L., Kendall, R. & Rosen, I. (2003) Patient-specific point dose measurement for IMRT monitor unit verification. *Int J Radiat Oncol Biol Phys*, 56, 867–77.

Fix, M. K., Keller, H., Ruegsegger, P. & Born, E. J. (2000) Simple beam models for Monte Carlo photon beam dose calculations in radiotherapy. *Med Phys*, 27, 2739–47.

Fix, M. K., Manser, P., Born, E. J., Mini, R. & Ruegsegger, P. (2001a) Monte Carlo simulation of a dynamic MLC based on a multiple source model. *Phys Med Biol*, 46, 3241–57.

Fix, M. K., Manser, P., Born, E. J., Vetterli, D., Mini, R. & Ruegsegger, P. (2001b) Monte Carlo simulation of a dynamic MLC: implementation and applications. *Z Med Phys*, 11, 163–70.

Fix, M. K., Stampanoni, M., Manser, P., Born, E. J., Mini, R. & Ruegsegger, P. (2001c) A multiple source model for 6 MV photon beam dose calculations using Monte Carlo. *Phys Med Biol*, 46, 1407–27.

Fracchiolla F., Lorentini S., Widesott L. & Schwarz M (2015) Characterization and validation of a Monte Carlo code for independent dose calculation in proton therapy treatments with pencil beam scanning, *Phys Med Biol.* 60(21): 8601–19.

Fragoso, M., Kawrakow, I., Faddegon, B. A., Solberg, T. D. & Chetty, I. J. (2009) Fast, accurate photon beam accelerator modelling using BEAMnrc: a systematic investigation of efficiency enhancing methods and cross-section data. *Med Phys*, 36, 5451–66.

Francescon, P., Cora, S. & Chiovati, P. (2003) Dose verification of an IMRT treatment planning system with the BEAM EGS4-based Monte Carlo code. *Med Phys*, 30, 144–57.

Greer, P. B. & Popescu, C. C. (2003) Dosimetric properties of an amorphous silicon electronic portal imaging device for verification of dynamic intensity modulated radiation therapy. *Med Phys*, 30, 1618–27.

Grein, E. E., Lee, R. & Luchka, K. (2002) An investigation of a new amorphous silicon electronic portal imaging device for transit dosimetry. *Med Phys*, 29, 2262–8.

Heath, E. & Seuntjens, J. (2003) Development and validation of a BEAMnrc component module for accurate Monte Carlo modelling of the Varian dynamic Millennium multileaf collimator. *Phys Med Biol*, 48, 4045–63.

Huang, S, Kang, M, Souris, K, Ainsley, C, Solberg, TD, McDonough, JE, Simone, CB 2nd & Lin, L (2018) Validation and clinical implementation of an accurate Monte Carlo code for pencil beam scanning proton therapy *J Appl Clin Med Phys.* 19(5): 558–72

Iori, M., Cagni, E., Nahum, A. E. & Borasi, G. (2007) IMAT-SIM: a new method for the clinical dosimetry of intensity-modulated arc therapy (IMAT). *Med Phys*, 34, 2759–73.

Jang, S. Y., Liu, H. H., Wang, X., Vassiliev, O. N., Siebers, J. V., Dong, L. & Mohan, R. (2006) Dosimetric verification for intensity-modulated radiotherapy of thoracic cancers using experimental and Monte Carlo approaches. *Int J Radiat Oncol Biol Phys*, 66, 939–48.

Jeraj, R. & Keall, P. (2000) The effect of statistical uncertainty on inverse treatment planning based on Monte Carlo dose calculation. *Phys Med Biol*, 45, 3601–13.

Jiang, S. B., Boyer, A. L. & Ma, C. M. (2001) Modelling the extrafocal radiation and monitor chamber backscatter for photon beam dose calculation. *Med Phys*, 28, 55–66.

Jiang, S. B., Pawlicki, T. & Ma, C. M. (2000) Removing the effect of statistical uncertainty on dose-volume histograms from Monte Carlo dose calculations. *Phys Med Biol*, 45, 2151–61.

Katsuta, Y, Kadoya, N, Fujita, Y, Shimizu, E, Matsunaga, K, Sawada, K, Matsushita, H, Majima, K & Jingu, K. (2017) Patient-specific quality assurance using Monte Carlo dose calculation and Elekta log files for prostate volumetric-modulated arc therapy. *Technol Cancer Res Treat*. 16(6): 1220–25

Keall, P. J., Siebers, J. V., Arnfield, M., Kim, J. O. & Mohan, R. (2001) Monte Carlo dose calculations for dynamic IMRT treatments. *Phys Med Biol*, 46, 929–41.

Keall, P. J., Siebers, J. V., Joshi, S. & Mohan, R. (2004) Monte Carlo as a four-dimensional radiotherapy treatment-planning tool to account for respiratory motion. *Phys Med Biol*, 49, 3639–48.

Kim, J. O., Siebers, J. V., Keall, P. J., Arnfield, M. R. & Mohan, R. (2001) A Monte Carlo study of radiation transport through multileaf collimators. *Med Phys*, 28, 2497–506.

Kroonwijk, M., Pasma, K. L., Quint, S., Koper, P. C., Visser, A. G. & Heijmen, B. J. (1998) In vivo dosimetry for prostate cancer patients using an electronic portal imaging device (EPID); demonstration of internal organ motion. *Radiother Oncol*, 49, 125–32.

Lagerwaard, F. J., Verbakel, W. F. A. R., van der Hoorn, E., Slotman, B. J. & Senan, S. (2008) Volumetric modulated arc therapy (RapidArc) for rapid, non-invasive stereotactic radiosurgery of multiple brain metastases. *Int J Radiat Oncol Biol Phys*, 72, S530.

Leal, A., Sanchez-Doblado, F., Arrans, R., Capote, R., Carrasco, E., Lagares, J. I., Rosello, J., Perucha, M., Molina, E. & Terron, J. A. (2003) Influence of the MLC leaf width on the dose distribution: a Monte Carlo study. *Radiother Oncol*, 68, S97.

Li, J. S., Lin, T., Chen, L., Price, R. A., Jr. & Ma, C. M. (2010) Uncertainties in IMRT dosimetry. *Med Phys*, 37, 2491–500.

Li, J. S., Pawlicki, T., Deng, J., Jiang, S. B., Mok, E. & Ma, C. M. (2000) Validation of a Monte Carlo dose calculation tool for radiotherapy treatment planning. *Phys Med Biol*, 45, 2969–85.

Li, J. S., Wang, L., Chen, L., Yang, J. & Ma, C. M. (2004) Monte Carlo dose verification for IMRT plan delivered using micro-multileaf collimators. *Med Phys*, 31, 1844.

Li, X. A., Ma, L., Naqvi, S., Shih, R. & Yu, C. (2001) Monte Carlo dose verification for intensity-modulated arc therapy. *Phys Med Biol*, 46, 2269–82.

Lin, M. H., Chao, T. C., Lee, C. C., Tung, C. J., Yeh, C. Y. & Hong, J. H. (2009) Measurement-based Monte Carlo dose calculation system for IMRT pretreatment and on-line transit dose verifications. *Med Phys*, 36, 1167–75.

Liu, H. H., Verhaegen F., Dong, L. (2001) A method to simulate Dynamic Multileaf Collimators using Monte Carlo techniques for Intensity Modulated Radiation Therapy. *Phys Med Biol*, 46, 2283–2298.

Luo, W., Li, J., Price, R. A., Jr., Chen, L., Yang, J., Fan, J., Chen, Z., Mcneeley, S., Xu, X. & Ma, C. M. (2006) Monte Carlo based IMRT dose verification using MLC log files and R/V outputs. *Med Phys*, 33, 2557–64.

Ma, C. M., Faddegon, B. A., Rogers, D. W. O. & Mackie, T. R. (1997) Accurate characterization of Monte Carlo calculated electron beams for radiotherapy. *Med Phys*, 24, 401–16.

Ma, C. M. & Jiang, S. B. (1999) Monte Carlo modelling of electron beams from medical accelerators. *Phys Med Biol*, 44, R157–89.

Ma, C. M., Jiang, S. B., Pawlicki, T., Chen, Y., Li, J. S., Deng, J. & Boyer, A. L. (2003) A quality assurance phantom for IMRT dose verification. *Phys Med Biol*, 48, 561–72.

Ma, C. M. & Li, J. (2011) Dose specification for radiation therapy: dose to water or dose to medium? *Phys Med Biol*, 56, 3073–89.

Ma, C. M., Mok, E., Kapur, A., Pawlicki, T., Findley, D., Brain, S., Forster, K. & Boyer, A. L. (1999) Clinical implementation of a Monte Carlo treatment planning system. *Med Phys*, 26, 2133–43.

Ma, C. M., Pawlicki, T., Jiang, S. B., Li, J. S., Deng, J., Mok, E., Kapur, A., Xing, L., Ma, L. & Boyer, A. L. (2000) Monte Carlo verification of IMRT dose distributions from a commercial treatment planning optimization system. *Phys Med Biol*, 45, 2483–95.

Mackeprang, P. H., Vuong, D., Volken, W., Henzen, D., Schmidhalter, D., Malthaner, M., Mueller, S., Frei, D., Stampanoni, M. F. M., Dal Pra, A., Aebersold, D. M., Fix, M. K. & Manser, P. (2017) Independent Monte-Carlo dose calculation for MLC based CyberKnife radiotherapy, *Phys Med Biol*. 63(1): 015015.

Otto, K. (2008) Volumetric modulated arc therapy: IMRT in a single gantry arc. *Med Phys*, 35, 310–17.

Pallotta, S., Marrazzo, L. & Bucciolini, M. (2007) Design and implementation of a water phantom for IMRT, arc therapy, and tomotherapy dose distribution measurements. *Med Phys*, 34, 3724–31.

Pasma, K. L., Dirkx, M. L., Kroonwijk, M., Visser, A. G. & Heijmen, B. J. (1999a) Dosimetric verification of intensity modulated beams produced with dynamic multileaf collimation using an electronic portal imaging device. *Med Phys*, 26, 2373–8.

Pasma, K. L., Kroonwijk, M., Quint, S., Visser, A. G. & Heijmen, B. J. (1999b) Transit dosimetry with an electronic portal imaging device (EPID) for 115 prostate cancer patients. *Int J Radiat Oncol Biol Phys*, 45, 1297–303.

Pawlicki, T. & Ma, C. M. (2001) Monte Carlo simulation for MLC-based intensity-modulated radiotherapy. *Med Dosim*, 26, 157–68.

Purdy, J. A. (1996) Intensity-modulated radiation therapy. *Int J Radiat Oncol Biol Phys*, 35, 845–6.

Reynaert, N., Demol, B., Charoy, M., Bouchoucha, S., Crop, F., Wagner, A., Lacornerie, T., Dubus, F., Rault, E., Comte, P., Cayez, R., Boydev, C, Pasquier, D., Mirabel, X., Lartigau, E. & Sarrazin, T. (2016) Clinical implementation of a Monte Carlo based treatment plan QA

platform for validation of Cyberknife and Tomotherapy treatments, *Phys Med*. 32(10): 1225–37.

Rogers, D. W. (2006) Fifty years of Monte Carlo simulations for medical physics. *Phys Med Biol*, 51, R287–301.

Rogers, D. W., Faddegon, B. A., Ding, G. X., Ma, C. M., We, J. & Mackie, T. R. (1995) BEAM: a Monte Carlo code to simulate radiotherapy treatment units. *Med Phys*, 22, 503–24.

Rogers, D. W. O. & A. F. Bielajew. (1990) Monte Carlo techniques of electron and photon transport for radiation dosimetry. In Kase, K., Bjarngard, B. and Attix, F. H. (Ed.) *Dosimetry of Ionizing Radiation*. Academic Press, New York.

Salama, J. K., Mundt, A. J., Roeske, J. & Mehta, N. (2006) Preliminary outcome and toxicity report of extended-field, intensity-modulated radiation therapy for gynecologic malignancies. *Int J Radiat Oncol Biol Phys*, 65, 1170–76.

Sarkar, V., Stathakis, S. & Papanikolaou, N. (2008) A Monte Carlo model for independent dose verification in serial tomotherapy. *Technol Cancer Res Treat*, 7, 385–92.

Siebers, J. V., Keall, P. J., Kim, J. O. & Mohan, R. (2002) A method for photon beam Monte Carlo multileaf collimator particle transport. *Phys Med Biol*, 47, 3225–49.

Siebers, J. V., Keall, P. J., Nahum, A. E. & Mohan, R. (2000) Converting absorbed dose to medium to absorbed dose to water for Monte Carlo based photon beam dose calculations. *Phys Med Biol*, 45, 983–95.

Spezi, E., Lewis, D. G. & Smith, C. W. (2001) Monte Carlo simulation and dosimetric verification of radiotherapy beam modifiers. *Phys Med Biol*, 46, 3007–29.

Stanhope, C. W., Drake, D. G., Liang, J., Alber, M., Söhn, M., Habib, C., Willcut, V. & Yan, D. (2018) Evaluation of machine log files/MC-based treatment planning and delivery QA as compared to ArcCHECK QA, *Med Phys*. 45(7): 2864–74.

Studer, G., Graetz, K. W. & Glanzmann, C. (2008) Outcome in recurrent head neck cancer treated with salvage-IMRT. *Radiat Oncol*, 3, 1–7.

Tyagi, N., Moran, J. M., Litzenberg, D. W., Bielajew, A. F., Fraass, B. A. & Chetty, I. J. (2007) Experimental verification of a Monte Carlo-based MLC simulation model for IMRT dose calculation. *Med Phys*, 34, 651–63.

Van Zijtveld, M., Dirkx, M. L., De Boer, H. C. & Heijmen, B. J. (2006) Dosimetric pre-treatment verification of IMRT using an EPID; clinical experience. *Radiother Oncol*, 81, 168–75.

Wang, L., Li, J., Paskalev, K., Hoban, P., Luo, W., Chen, L., Mcneeley, S., Price, R. & Ma, C. (2006) Commissioning and quality assurance of a commercial stereotactic treatment-planning system for extracranial IMRT. *J Appl Clin Med Phys*, 7, 21–34.

Wang, L., Yorke, E. & Chui, C. S. (2002) Monte Carlo evaluation of 6 MV intensity modulated radiotherapy plans for head and neck and lung treatments. *Med Phys*, 29, 2705–17.

Wang, Y., Mazur, T. R., Park, J. C., Yang, D., Mutic, S. & Li, H. H. (2017) Development of a fast Monte Carlo dose calculation system for online adaptive radiation therapy quality assurance, *Phys Med Biol*, 62(12): 4970–90.

Webb, S. (1998) Intensity-modulated radiation therapy: dynamic MLC (DMLC) therapy, multisegment therapy and tomotherapy. An example of QA in DMLC therapy. *Strahlenther Onkol*, 174(Suppl 2): 8–12.

Weber, L. & Nilsson, P. (2002) Verification of dose calculations with a clinical treatment planning system based on a point kernel dose engine. *J Appl Clin Med Phys*, 3, 73–87.

Winterhalter, C., Fura, E., Tian, Y., Aitkenhead, A., Bolsi, A., Dieterle, M., Fredh, A., Meier, G., Oxley, D., Siewert, D., Webwe, D. C., Lomax, A. & Safai, S. (2018) Validating a Monte Carlo approach to absolute dose quality assurance for proton pencil beam scanning, *Phys Med Biol*. 63(17): 175001.

Yang, J., Li, J., Chen, L., Price, R., Mcneeley, S., Qin, L., Wang, L., Xiong, W. & Ma, C. M. (2005) Dosimetric verification of IMRT treatment planning using Monte Carlo simulations for prostate cancer. *Phys Med Biol*, 50, 869–78.

Yang, J., Li, J. S., Qin, L., Xiong, W. & Ma, C. M. (2004) Modelling of electron contamination in clinical photon beams for Monte Carlo dose calculation. *Phys Med Biol*, 49, 2657–73.

Yu, C. X. (1995) Intensity-modulated arc therapy with dynamic multileaf collimation: an alternative to tomotherapy. *Phys Med Biol*, 40, 1435–49.

Zelefsky, M. J., Yamada, Y., Kollmeier, M. A., Shippy, A. M. & Nedelka, M. A. (2008) Long-term outcome following three-dimensional conformal/intensity-modulated external-beam radiotherapy for clinical stage T3 prostate cancer. *Eur Urol*, 53, 1172–79.

15

Monte Carlo Applications in Total Skin Electron Therapy

George X. Ding
*Vanderbilt University
School of Medicine*

Joanna E. Cygler
*The Ottawa Hospital
Cancer Centre*

15.1 Introduction

This chapter describes an application of Monte Carlo simulation technique in total skin electron therapy (TSET). It illustrates the unique capability of the Monte Carlo method in providing full details of patient dose distributions that the experimental methods are not able to obtain.

15.1.1 Total Skin Electron Therapy

TSET has been for decades one of the most effective treatments for cutaneous T-cell lymphoma (mycosis fungoides). TSET is a special radiation treatment procedure as it generally requires equipment modifications to deal with its specific needs [1–9]. The patient population requiring TSET is relatively small, and the technique is usually available only in academic or larger centers.

15.1.2 Requirement of Beam Characteristics for Delivering TSET

TSET aims to irradiate the patient's whole skin while sparing all other organs from any appreciable radiation dose [4]. The field size of the composite electron beam at the patient treatment plane must be approximately 200 cm in height by 80 cm in width to encompass the largest patient with uniformity within this rectangular field [1]. The composite electron beam may require dual fields with nominal SSD range of 300–500 cm to achieve such large size. The uniformity of

dose achieved in the treatment plane in phantom studies may be reproduced during the patient treatment [1]. More details on beam requirements can be found in the AAPM Report 23 [1]. The maximum dose should occur at the surface, while the depth of 80% of the prescribed dose should be >0.4 cm and the dose should fall off to <20% of the prescribed dose at a depth of 2 cm [4]. The most commonly used nominal electron beam energy ranges from 4 to 9 MeV [1,2,4,10].

15.1.3 Irradiation Techniques in TSET

TSET delivery techniques aim to provide a uniform dose distribution to the patient's whole skin with maximum dose at the surface. During the early development of TSET, investigations were carried out to determine the dose distributions obtained for single-field, multi-field, translation, arc, and patient rotational techniques [1]. Phantom studies indicated that patient rotation, using a rotating platform, provided the best dose uniformity over large portions of the body surface, although the eight-field technique has proved to be almost as good. The six-field technique is adequate and simpler to carry out [11]. Currently, two commonly used techniques are six static dual fields and rotational dual fields as shown in Figure 15.1. For rotational dual fields, the patient stands on a platform that rotates, while for 6 static dual fields the patient is positioned in six different orientations: anterior, posterior, lateral (R and L), and two of the angled dual field [1].

DOI: 10.1201/9781003211846-18

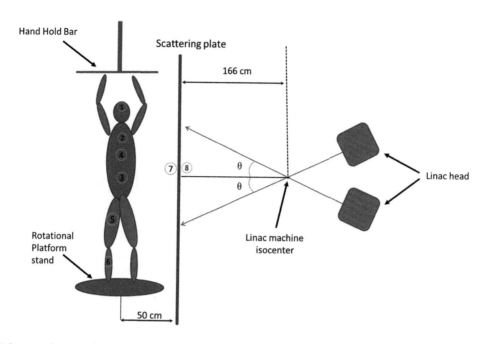

FIGURE 15.1 Schematic drawing of a treatment setup of the rotational total skin electron irradiation (RTSET) technique and placement of in vivo dosimeters where #1–6 are placed in different locations on patient skin (#4 is at the patient's back) and #7 and #8 are placed on each side of the acrylic scatter plate. Angle Θ is ~21°. (Reproduced with permission from Ding et al. [12])

15.1.4 Advantages of Monte Carlo Application in TSET

Historically, the dose distributions for TSET were obtained through experimental studies in phantoms using films, TLDs, ionization chambers, etc. [1]. In vivo doses to patients are typically determined through dosimeters placed on the patient's skin. Due to the complexity of the large field treatment geometry with composite incident beams, the Monte Carlo method has only recently been used to obtain the patient dose distributions [12]. In this chapter, we discuss the application of the EGSnrc Monte Carlo system [13,14], to the generation of incident beams and calculation of dose distributions in TSET therapy.

15.2 Monte Carlo Simulation of Incident Beams Used in TSET

The electron beam field sizes used for TSET are defined by linac jaws without electron applicator attached. In the following example, 6 MeV beams were modeled from Varian Clinac 21EX and Varian TrueBeam. The simulated beams were scored at distance SSD=100 cm and stored in phase-space files which were used as the incident beams for dose calculations. The accuracy of the beam simulation was validated by comparisons between the Monte Carlo calculated and measured dose distributions.

15.2.1 Generation of Incident Beams for Dose Calculation

The simulation of incident beams from a Varian Clinac 21EX machine started with electrons exiting from the vacuum

window of the linac head. The details of the linac head geometry including electron scattering foils and beam defining system were obtained from the manufacturer. For electron beams from a Varian TrueBeam accelerator, the incident beam simulations start from the phase-space files which were provided by the manufacturer. These Varian-provided phase-space files were scored at the plane just above the linac x-y jaw collimators. In all cases, the simulated incident beams containing the position, energy, angle, charge, and weight of particles were stored in phase-space files scored at a plane 100 cm from the source. The stored phase-space files were used as input for the user code DOSXYZnrc [13,14] dose calculations. The default EGSnrc parameter settings are used in all Monte Carlo simulations with AE=ECUT=0.521 MeV, AP=PCUT=0.010 MeV, and no photon interaction forcing, and consistent with previous studies, no Rayleigh scattering was used [15,16].

Figure 15.2 shows measured and calculated percentage depth-dose curves and dose profiles in water for 6 MeV beams from Varian TrueBeam and Clinac accelerates with a field size of 36 cm×36 cm and 40 cm×40 cm at SSD=100 cm, respectively. The good agreement between measured and calculated depth-dose curve underscores the accuracy of the Monte Carlo simulations. In the Monte Carlo simulation of the incident beams, the energy and spot size of the electron beam before hitting the exit vacuum window in the linac head were adjusted to obtain the best agreement between measurements and calculations for the Varian Clinac. For beams from a Varian TrueBeam accelerator, there is no adjustment as it is based on the phase-space files provided by Varian.

It is worthy to note that Monte Carlo results by Ding et al. [12] did not observe a significant increase of the bremsstrahlung

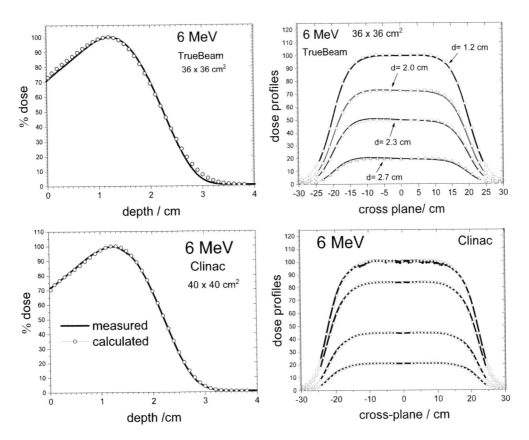

FIGURE 15.2 Comparison of measured and Monte Carlo calculated dose distributions in a water phantom for 6 MeV electron beams from a Varian Clinac and a TrueBeam, respectively. Solid lines and red circles correspond to measured and Monte Carlo calculated, respectively. A diode detector was used in the measurements. (Reproduced with permission from Ding et al. [12])

doses at extended distances which is contrary to that bremsstrahlung dose is up to 5% at an extended distances reported by Das et al. [18]. Although Chen et al. [7] also found that bremsstrahlung dose was only ~ 1% even at SSD > 500 cm for a 6 MeV beam from Varian 21EX-S, they assumed that the discrepancy with the data of Das et al. [18] was caused by different 6 MeV beams generated from a Varian 21 EX-S as opposed to Saturne-I machine reported by Das et al. [18]. In order to resolve the discrepancy, Ding et al. [17] evaluated validity of the reported 5% bremsstrahlung dose extended distances. The Monte Carlo result by Ding et al. [17] confirms the results of Chen et al. and concludes that the reported high bremsstrahlung doses at extended distances by Das et al. [18] are inaccurate. The errors in their reported very high bremsstrahlung doses might be due to a detector for which the readings were dominated by poor signal to noise ratio at extended distances [17]. In addition, the Monte Carlo simulated realistic beams at extended distances were also used by Ding [29] to investigate the stopping-power ratios for accurate dosimetry.

Figure 15.3 presents energy spectra of electrons and bremsstrahlung photons as well as the planar fluence as a function of off-axis position from a Clinac 21EX (a) and a TrueBeam (b), respectively. The phase-space files were generated in air at SSD=100 cm for a 40×40 cm² field. Only electrons and photons inside 40×40 cm² field are accounted in obtaining the

spectra distributions. The bremsstrahlung photons' fluence as a function of off-axis position is consistent with spacial distribution measured by Das et al. [18].

15.2.2 Validation of Simulated Beams in TSET Delivery Geometry

Although the Monte Carlo method is the gold standard for dose calculations, the accuracy of calculated dose distribution depends on many factors. Therefore, the experimental validation of the simulation is essential, especially when the radiation is delivered by combined dual beams to the rotating phantom geometry, as shown in Figure 15.1.

A large plastic plate, served as a beam degrader to reduce electron energy and increase scatter, was placed between the incident beam and the rotating platform. The plastic plate is made of clear acrylic with dimensions of 90 cm×200 cm. The acrylic plate was mounted on a movable wood frame on wheels.

The measurements were performed with nanoDot detectors made by LANDAUER's Optically Stimulated Luminescence (OSL) technology on a water equivalent plastic cylinder (29 cm in diameter, 30 cm in length) with two layers of 5 mm thick water equivalent bolus. The dosimeters were placed on the phantom surface and between the layers of the bolus.

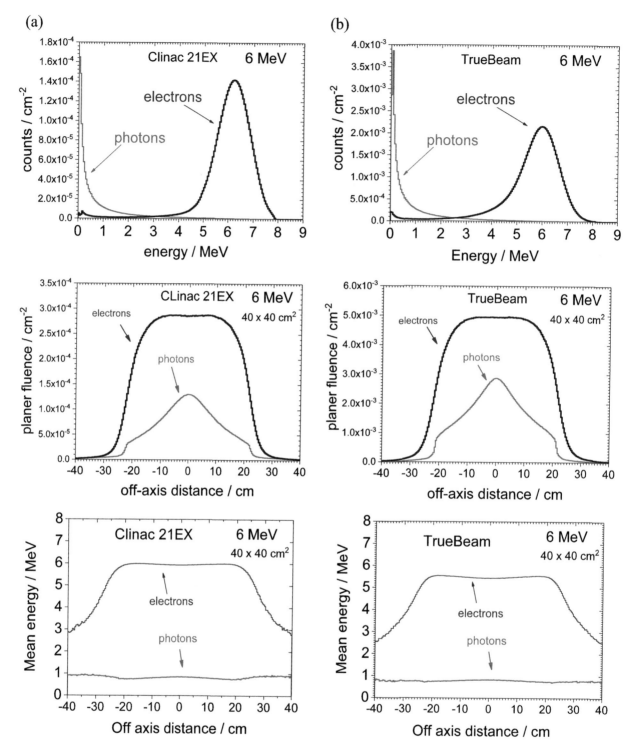

FIGURE 15.3 Energy spectra of electrons and photons, fluence, and mean energy as a function of off x-axis position for a $40 \times 40 \, \text{cm}^2$ field from Clinac 21EX (a) and TrueBeam (b), respectively. The phase-space files were generated in air at SSD = 100 cm.

Figure 15.4 shows the experimental setup of the cylindrical phantom placed on the top of a stool which is placed on the rotating platform which rotates during beam delivery.

Figure 15.5 shows the Monte Carlo calculated dose distributions for the cylinder phantom in color wash in axial plane resulting from the rotational dual field technique with

the scatter plate removed and the dose profiles along the two orthogonal lines AB and CD, respectively.

Figures 15.6 presents comparisons between measured and Monte Carlo calculated dose as a function of depth for the cylindrical phantom surface for the following three scenarios: with an acrylic scatter plate of 9 mm, with a scatter plate of 3 mm,

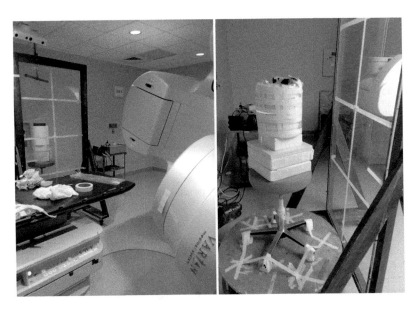

FIGURE 15.4 The rotational dual field TSET technique geometry and the phantom experimental setup with the clear wood-mounted acrylic plate frame on wheels. Reproduced with permission from Ding et al. [12].

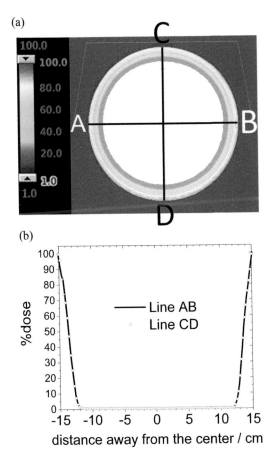

FIGURE 15.5 Monte Carlo calculated dose distributions in color wash (a) in axial plane for the cylinder phantom and dose profiles (b) along the Line AB (solid) and Line CD (empty symbol), respectively. Note that the dose distribution is completely symmetric for circular cylinder geometry.

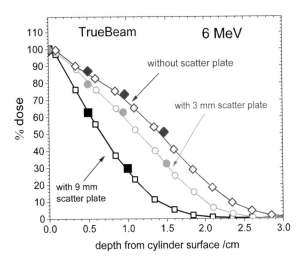

FIGURE 15.6 Comparisons between the Monte Carlo calculated (open symbols) and measured doses (solid symbols) as a function of depth from the surface of the cylindrical phantom for beam delivery with the dual rotational beam technique. The measurements were made with OSLD dosimeters. (Reproduced with permission from Ding et al. [12])

and with the scatter plate removed. The depth-dose coverage dependency on the thickness of the beam degrader can be useful for clinicians in choosing the appropriate thickness of beam scatter (degrader) plate to achieve the desired skin depth-dose coverage. The dose maximum occurs at the surface even in the scenario without the scatter plate. This is the result of incident electrons hitting the cylindrical phantom from multiple directions as the phantom rotates during the radiation delivery. If the phantom is stationary during beam delivery, dose buildup is observed as shown in Figure 15.7, which is consistent with the measured depth-dose curve by Chen et al. [9]

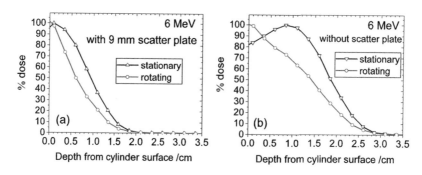

FIGURE 15.7 Comparison between stationary and rotating phantom PDDs for the cylindrical phantom for the dual field technique for with (a) and without (b) a beam scatter plate, respectively.

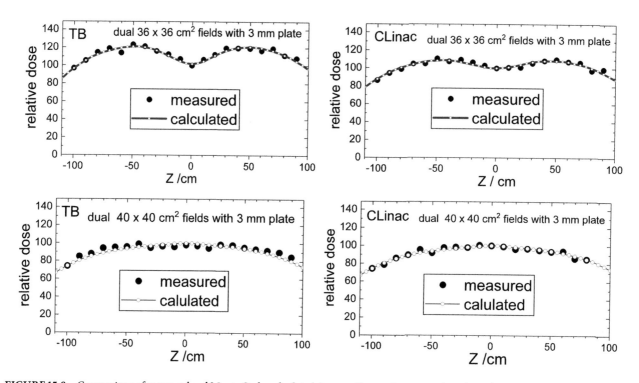

FIGURE 15.8 Comparison of measured and Monte Carlo calculated dose profiles on the scatter plate along the Z-axis direction. The coordinate zero is at the lateral laser level and the scatter plate is 166 cm away from the isocenter in the horizontal direction. Upper and lower graphs are results of dual 36 cm×36 cm field with the gantry angle of 2θ=43° and dual 40 cm×40 cm field with gantry angle of 2θ=42°, respectively. (Reproduced with permission from Ding et al. [12])

To confirm the accuracy of the simulated dual incident beams in the vertical direction, nanoDot dosimeters were placed on the surface of the scatter plate to measure the dose profiles of the dual electron beams. The backscatter [19] from the plate is assumed to be proportional to the surface dose. Therefore, the shape of the profiles is not affected by backscatters.

Figure 15.8 shows the comparison of measured and Monte Carlo calculated dose profiles in the vertical direction on the scatter plate. The dose profiles are very sensitive to the electron beam field size. This is because the dose uniformity in the vertical direction where the dose is measured results from the contribution of dual fields. It is worth noting that for the 40 cm×40 cm field, the dose is lower at the edge of the field

than for the 36 cm×36 cm field when measuring dose profiles in the vertical direction from dual fields. This is because the dual fields contribute to the dose at the center while at the edge the dose is contributed by a single field only. The good agreement between simulated and measured dose validates the simulation accuracy. Our results are consistent with the measurements reported in the literature [20–23].

15.2.3 Use of Simulated Beams for Dose Calculation in TSET

After experimental validation of the Monte Carlo simulations, the simulated beams are used to obtain the dose distributions in phantom and CT image-based patient anatomy.

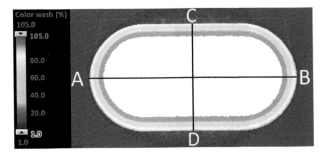

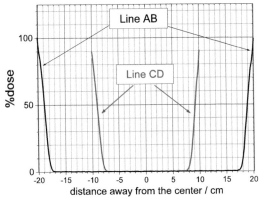

FIGURE 15.9 Upper graph shows the Monte Carlo calculated dose distributions in color wash for an oval cylinder with 40 long and 20 cm short axes in axial planes with beam delivery using the dual rotational beam technique with a 3 mm-thick acrylic scatter plate. Lower graph shows the dose profiles along the Line AB and Line CD, respectively.

15.3 Uniformity of Dose Distributions from Dual Fields

15.3.1 Dose Distributions from Rotational Fields on an Oval Cylindrical Phantom

Since human torso resembles an oval shape, we will use a long oval cylinder phantom to evaluate the dose distribution for a relatively uniform geometry. The water-equivalent oval cylindrical phantom has 40 cm long axis and 20 cm short axis in axial planes and vertical length of 150 cm. This simple geometry is suitable to evaluate the uniformity of dose distributions resulting from the dual beams.

Figure 15.9 shows the Monte Carlo calculated dose distributions in color wash for an oval cylinder with beam delivery using the dual rotational beam technique with a 3 mm-thick acrylic scatter plate. To quantify the dose as a function of depth, the dose profiles are shown along the long axis (Line AB) and short axis (Line CD), respectively. Compared to a circular cylinder geometry where the dose is uniform at the circular surface, significant lower doses (see in Line CD) are observed for the oval cylindrical geometry phantom at surfaces of short axis. This leads to lower skin dose at patient anterior and posterior areas in torso. Although the dose is lower at the surface in the short axis, the depth where dose falls to 50%

of the surface value remains approximately the same. This is because the doses at larger depths from the surface are contributed by near-normal incident beams, while the oblique incident beams cannot reach larger depth from the surface.

15.3.2 Patient Skin Dose Distributions Resulting from Rotational Dual Fields or 6 Static Dual Fields

Figure 15.10a shows the Monte Carlo calculated dose distributions on CT image-based patient anatomy in axial, coronal, and sagittal planes for delivery using the dual rotational beam technique with scatter plate. The displayed isodose lines shown in red, white, and blue correspond to 90%, 75%, and 50%, respectively. As shown in coronal plane, the 90% isodose line is in air between head and arm. This is caused by partial shielding by the raised arms. Due to the irregular body geometry, large dose variations (>20%) at the skin are observed. This result is consistent with in vivo measurements reported in the literature [9,24,25]. Figure 15.10b shows the dose-volume-histograms (DVHs) for skin at three different depth intervals: 1st 5 mm (0–5 mm), 2nd 5 mm (5–10 mm), and 3rd 5 mm (10–15 mm) from the skin surface when using the 3 and 9 mm thick acrylic scatter plates, respectively.

The in vivo patient skin dose verifications are typically performed by using dosimeters placed on the patient skin [9,24–26]. At Vanderbilt University Medical Center, OSLD dosimeters (nanoDot) are placed on the patient's skin for the first treatment fraction at locations shown in Figure 15.1. The measured in vivo skin dose variations at different body locations were consistent with calculation predicted variations of 20% and reported in vivo dose measured variations [9,24–28].

Although the examples presented are for TSET with beams delivered by using rotational dual-field technique, we also calculated dose distributions for the 6 static dual-field technique. Very similar patient skin dose distributions were observed for TSET delivery using 6 static dual fields. The comparable skin dose distributions between using rotational dual-field and 6 static dual-field techniques are expected as reported in the literature although the patient assumes different positions [1,5,7].

15.4 Summary

In this chapter, we demonstrated the unique capability of the Monte Carlo simulation method applied to TSET to obtain the details of the dose distributions that experimental methods are unable to obtain, such as DVH analysis at different depths from the skin surface. The application of Monte Carlo simulations [12,17] to the beam used in TSET has challenged the validity of the commonly accepted perception that bremsstrahlung dose for a 6 MeV electron beam reaches 5% at an extended distance of 500 cm [18]. The study by Ding et al. [17] showed that the that the bremsstrahlung doses of a 6 MeV electron beam are 0.5%

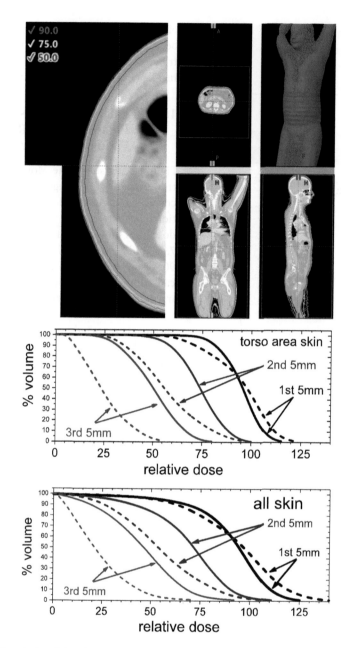

FIGURE 15.10 (a) Monte Carlo calculated skin dose distributions shown in axial, coronal, and sagittal planes. The displayed isodose lines are 90% (red), 75% (white), and 50% (blue). The insert on the left is zoomed in part of the axial plane where the distance between 90% and 50% isodose lines is ~5 mm. Due to the irregular body geometry, large (>20%) dose variations at the skin are observed. The upper right graph shows the 90% isodose surface in red. (b) DVHs showing dose to the 1st 5 mm (0–5 mm), 2nd (5–10 mm), and 3rd 5 mm (10–15 mm) skin depths (middle) for torso area skin (upper) and the entire skin (lower), where solid and dash lines represent the 3 and 9 mm thick acrylic scatter plates, respectively. (Reproduced with permission from Ding et al. [12])

to 1% for SSD from 100 to 700 cm and that the common belief of up to 5% bremsstrahlung dose at large extended distances is incorrect. This finding has impact on the practice in TSET as bremsstrahlung dose contributes dose to the deeper tissues. A incorrect high bremsstrahlung dose estimation could limit its use in TSET. The Monte Carlo results also provide new understanding of the entire patient skin dose distributions for patients treated with TSET. This can be used by clinicians when choosing the appropriate thickness of beam scatter (degrader) plate to achieve the desired skin depth-dose coverage. Based on the results of the Monte Carlo investigation by Ding et al. [12], a 3 instead of 9 mm thick scatter plate and 40×40 cm² field size for optimum skin depth dose coverage should be used. Treatment outcomes for patients cohort treated with the 3 mm degrader at Vanderbilt University Medical Center are the subject of an ongoing prospective study for clinical effectiveness.

The knowledge obtained by Monte Carlo application to TSET can have a significant impact on dosimetry in TSET and improve the patient treatment outcomes.

References

1. C. J. Karzmark, "Total skin electron therapy: technique and dosimetry," AAPM, Report of Task Group 30 Radiationtherapy Committee AAPM, 1987, vol. AAPM Report No. 23.

2. E. B. Podgorsak, C. Pla, M. Pla, P. Y. Lefebvre, and R. Heese, "Physical aspects of a rotational total skin electron irradiation," *Med Phys*, vol. 10, no. 2, pp. 159–68, Mar-Apr 1983, doi: 10.1118/1.595296.

3. C. R. Freeman et al., "Clinical experience with a single field rotational total skin electron irradiation technique for cutaneous T-cell lymphoma," *Radiother Oncol*, vol. 24, no. 3, pp. 155–62, Jul 1992. [Online]. Available: https://www.ncbi.nlm.nih.gov/pubmed/1410569.

4. E. B. Podgorsak, E. B. Podgorsak, Ed. Chapter 15.4. Total Skin Electron Irradiation (TSEI) in Review of Radiation Oncology Physics: A Handbook for Teachers and Students. International Atomic Energy Agency Vienna, Austria, 2003.

5. M. D. Evans, C. Hudon, E. B. Podgorsak, and C. R. Freeman, "Institutional experience with a rotational total skin electron irradiation (RTSEI) technique-A three decade review (1981–2012)," *Rep Pract Oncol Radiother*, vol. 19, no. 2, pp. 120–34, Mar 2014, doi: 10.1016/j.rpor.2013.05.002.

6. T. R. Heumann et al., "Total skin electron therapy for cutaneous T-cell lymphoma using a modern dual-field rotational technique," *Int J Radiat Oncol Biol Phys*, vol. 92, no. 1, pp. 183–91, May 1 2015, doi: 10.1016/j.ijrobp.2014.11.033.

7. M. Chowdhary, A. M. Chhabra, S. Kharod, and G. Marwaha, "Total skin electron beam therapy in the treatment of Mycosis Fungoides: a review of conventional and low-dose regimens," *Clin Lymphoma Myeloma Leuk*, vol. 16, no. 12, pp. 662–671, Dec 2016, doi: 10.1016/j.clml.2016.08.019.

8. E. P. Reynard et al., "Rotational total skin electron irradiation with a linear accelerator," *J Appl Clin Med Phys*, vol. 9, no. 4, p. 2793, Nov 3 2008. [Online]. Available: https://www.ncbi.nlm.nih.gov/pubmed/19020483.

9. Z. Chen, A. G. Agostinelli, L. D. Wilson, and R. Nath, "Matching the dosimetry characteristics of a dual-field Stanford technique to a customized single-field Stanford technique for total skin electron therapy," *Int J Radiat Oncol Biol Phys*, vol. 59, no. 3, pp. 872–85, Jul 1 2004, doi: 10.1016/j.ijrobp.2004.02.046.

10. R. S. Cox, R. J. Heck, P. Fessenden, C. J. Karzmark, and D. C. Rust, "Development of total-skin electron therapy at two energies," *Int J Radiat Oncol Biol Phys*, vol. 18, no. 3, pp. 659–69, Mar 1990, doi: 10.1016/0360-3016(90)90075-u.

11. P. J. Tetenes and P. N. Goodwin, "Comparative study of superficial whole-body radiotherapeutic techniques using a 4-MeV nonangulated electron beam," *Radiology*, vol. 122, no. 1, pp. 219–26, Jan 1977, doi: 10.1148/122.1.219.

12. G. X. Ding, E. C. Osmundson, E. Shinohara, N. B. Newman, M. Price, and A. N. Kirschner, "Monte Carlo study on dose distributions from total skin electron irradiation therapy (TSET)," Physics in medicine and biology, vol. 66, Mar 11 2021, doi: 10.1088/1361-6560/abedd7. https://w3.aapm.org/meetings/2018AM/programInfo/programAbs.php?sid=7817&aid=39361.

13. Kawrakow, E. Mainegra-Hing, D. W. O. Rogers, F. Tessier, and B. R. B. Walters, "The EGSnrc Code System: Monte Carlo simulation of electron and photon transport," NRC Technical Report PIRS-701 v4-2-3-2 (National Research Council Canada, Ottawa, Canada, 2011.

14. D. W. O. Rogers, B. A. Faddegon, G. X. Ding, C. M. Ma, J. We, and T. R. Mackie, "BEAM: a Monte Carlo code to simulate radiotherapy treatment units," Medical physics, vol. 22, no. 5, pp. 503-24, 1995. [Online]. Available: http://www.ncbi.nlm.nih.gov/htbin-post/Entrez/query?db=m&form=6&dopt=r&uid=0007643786

15 B. R. Walters, I. Kawrakow, and D. W. O. Rogers, "DOSXYZnrc Users Manual," National Research Council of Canada, NRCC Report PIRS-794revB, Ottawa, , NRCC Report PIRS-794revB, 2009.

16. G. X. Ding and D. W. O. Rogers, "Energy spectra, angular spread, and dose distributions of electron beams from various accelerators used in radiotherapy," National Research Council of Canada, Report No. PIRS-0439, Ottawa; see also http://www.irs.inms.nrc.ca/inms/irs/papers/PIRS439/pirs439.html, Report No. PIRS-0439, 1995.

17. G. X. Ding, S. K. Dogan, and I. J. Das, "Technical Note: Bremsstrahlung dose in the electron beam at extended distances in total skin electron therapy," *Med Phys*, Accepted, 2021.

18. I. J. Das, J. F. Copeland, and H. S. Bushe, "Spatial distribution of bremsstrahlung in a dual electron beam used in total skin electron treatments: errors due to ionization chamber cable irradiation," *Med Phys*, vol. 21, no. 11, pp. 1733–8, Nov 1994, doi: 10.1118/1.597215.

19. F. Verhaegen, "Interface perturbation effects in high-energy electron beams," *Phys Med Biol*, vol. 48, no. 6, pp. 687–705, Mar 21 2003. [Online]. Available: http://www.ncbi.nlm.nih.gov/entrez/query.fcgi?cmd=Retrieve&db=PubMed&dopt=Citation&list_uids=12699189.

20. E. el-Khatib, S. Hussein, M. Nikolic, N. J. Voss, and C. Parsons, "Variation of electron beam uniformity with beam angulation and scatterer position for total skin irradiation with the Stanford technique," *Int J Radiat Oncol Biol Phys*, vol. 33, no. 2, pp. 469–74, Sep 30 1995, doi: 10.1016/0360-3016(95)00112-C.

21. F. W. Hensley, G. Major, C. Edel, H. Hauswald, and M. Bischof, "Technical and dosimetric aspects of the total skin electron beam technique implemented at Heidelberg University Hospital," *Rep Pract Oncol Radiother*, vol. 19, no. 2, pp. 135–43, Mar 2014, doi: 10.1016/j.rpor.2013.07.002.

22. Q. Bao, B. A. Hrycushko, J. P. Dugas, F. H. Hager, and T. D. Solberg, "A technique for pediatric total skin electron irradiation," *Radiat Oncol*, vol. 7, p. 40, Mar 20 2012, doi: 10.1186/1748-717X-7-40.

23. Y. Xie et al., "Cherenkov imaging for total skin electron therapy (TSET)," *Med Phys*, vol. 47, no. 1, pp. 201–212, Jan 2020, doi: 10.1002/mp.13881.

24. J. A. Antolak, J. H. Cundiff, and C. S. Ha, "Utilization of thermoluminescent dosimetry in total skin electron beam radiotherapy of mycosis fungoides," *Int J Radiat Oncol Biol Phys*, vol. 40, no. 1, pp. 101–8, Jan 1 1998, doi: 10.1016/s0360-3016(97)00585-3.

25. R. D. Weaver, B. J. Gerbi, and K. E. Dusenbery, "Evaluation of dose variation during total skin electron irradiation using thermoluminescent dosimeters," *Int J Radiat Oncol Biol Phys*, vol. 33, no. 2, pp. 475–8, Sep 30 1995, doi: 10.1016/0360-3016(95)00161-Q.

26. G. Guidi, G. Gottardi, P. Ceroni, and T. Costi, "Review of the results of the in vivo dosimetry during total skin electron beam therapy," *Rep Pract Oncol Radother*, vol. 19, no. 2, pp. 144–50, Mar 2014, doi: 10.1016/j.rpor.2013.07.011.

27. C. G. Patel, G. Ding, and A. Kirschner, "Scalp-sparing total skin electron therapy in mycosis fungoides: case report featuring a technique without lead," *Pract Radiat Oncol*, vol. 7, pp. 400–402, Mar 27 2017, doi: 10.1016/j.prro.2017.03.009.

28. A. D. Sherry, G. X. Ding, and A. N. Kirschner, "Diffuse primary cutaneous anaplastic large cell lymphoma treated by rotational total skin electron beam radiotherapy with custom shielding: case report," *J Med Imaging Radiat Sci*, vol. 50, no. 3, pp. 454–459, Sep 2019, doi: 10.1016/j.jmir.2019.05.002.

29. G.X. Ding, "Stopping-power ratios for electron beams used in total skin electron therapy," *Medical Physics*, in Press, 2021.

16

Rowan M. Thomson
*Department of Physics,
Carleton University*

Åsa Carlsson Tedgren
Karolinska University Hospital

Gabriel Fonseca
Maastro Clinic

Guillaume Landry
*Ludwig Maximilian's
University
Maastro Clinic*

Brigitte Reniers
University Hasselt

Mark J. Rivard
*Rhode Island Hospital
Tufts University
School of Medicine*

Jeffrey F. Williamson
*Washington University
VCU Massey Cancer Center*

Frank Verhaegen
Maastro Clinic

Monte Carlo Simulation in Brachytherapy Patients and Applicator Modelling

16.1 Introduction

The use of Monte Carlo (MC) methods in brachytherapy (BT) now goes beyond single-source dosimetry. An important area of current research is the application of MC simulations to perform patient-specific dose calculations. This effort is necessary to overcome the limitations of TG-43-style, source superposition algorithms that neglect interseed attenuation (ISA) and tissue heterogeneities for low-energy seed implants [1–3] and that neglect applicator shielding and partial scattering effects for higher energy BT procedures [4–7]. Several MC dose calculation platforms, generally based on CT images, have been presented in the literature [8–15]. For low-energy sources, the biggest challenge for accurate patient-specific dosimetry is the accurate voxel-by-voxel assignment of photon cross-sections. In this chapter, we cover clinical dose calculation methods for patient-specific treatment planning.

16.2 Departure from TG-43 to MC Dose Calculations for Treatment Planning

16.2.1 Introduction

The TG-43 formalism, based on the work of the Interstitial Collaborative Working Group [16], was a major step forward in BT dose calculation. It replaced semiempirical calculation methods based on apparent activity, equivalent mass of radium, exposure rate constants, and tissue-attenuation coefficients (quantities that are not source model but only radionuclide dependent). By employing dosimetry parameters that depend on the detailed source geometry such as radioactivity distribution, encapsulation, and potential markers for imaging (derived from measured or MC-calculated consensus datasets), the TG-43 formalism improved standardization and accuracy in BT dose calculations. This protocol is still widely cited (over

DOI: 10.1201/9781003211846-19

1,500 citations as of August 2020 [Scopus 24 August 2020, *Times cited = 1,510*], compared to 900 citations for the first edition of this book that went to press in 2013). It is still heavily used in 2020, ensuring consistency, standardization, and comparability of BT dose calculations across institutions worldwide.

Since dose calculations based on the TG-43 formalism rely on the superposition of single-source dose distributions over the dwell or seed positions used for treatment, dose distributions can be obtained with minimal calculation time. This fast and practical method facilitates clinical practices such as transrectal ultrasound (TRUS) image-guided live planning and dose distribution optimization. However, the inherent simplicity of the TG-43 formalism can lead to inaccurate dose distributions when the calculation geometry deviates significantly from the reference water sphere used to derive the underlying dosimetric parameters. Generally, TG-43 dose calculation limitations can be attributed to five phenomena: absorption, attenuation, shielding, scattering, and breakdown of the kerma approximation for absorbed dose. Depending on the source energy and anatomic site, some or all of these phenomena may induce significant dose calculation errors; Table 16.1 lists, for both high- and low-energy BT sources, sites where the formalism leads to significant dose calculation errors [17]. The following sections briefly discuss these aspects.

16.2.2 Absorbed Dose Differences between Water and Human Tissues

At low photon energies, the mass energy–absorption coefficient μ_{en}/ρ varies significantly between tissues due to the importance of the approximately Z^{3-4} dependence of the photoelectric cross section. This means that doses are highly sensitive to tissue elemental composition and also that there are discrepancies in dose to tissue for different media and TG-43 doses, particularly

for lower energy BT sources. For a given photon energy fluence under conditions of charged particle equilibrium (CPE) in a large cavity, the absorbed dose to water and to tissue are related by $D_{tissue}/(\mu_{en}/\rho)_{tissue} = D_{water}/(\mu_{en}/\rho)_{water}$. As seen in Figure 16.1, the ratio $(\mu_{en}/\rho)_{tissue}/(\mu_{en}/\rho)_{water}$ differs from unity for most tissues in the energy range of low-energy sources. At high photon energies, the ratios converge to unity as the incoherent (Compton) cross section becomes more important. This gives rise to significant dependence of the fluence to kerma conversion factor on tissue elemental composition and mass density [18]. A recent review paper summarizes the current knowledge [19].

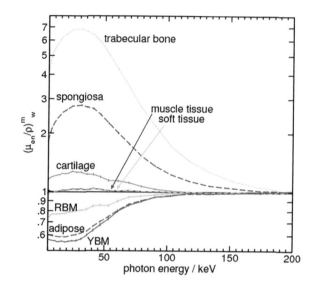

FIGURE 16.1 The EGSnrc [20] user code "g" was used to calculate μ_{en}/ρ for a series of human tissues taken from ICRU Report 46 [21] and ICRP Report 89 [22]. RBM and YBM stand for red and yellow bone marrow from "active and inactive marrow" in Ref. [22]. (Reproduce with permission from Ref. [23].)

TABLE 16.1   List of Anatomic Sites Where the Limitations of the TG-43 Dose Calculation Formalism Lead to Significant Dose Calculation Errors (Indicated by "Y")

Anatomic Site	Source Energy	Absorption	Attenuation	Shielding	Scattering	Breakdown Dose = Kerma Approximation
Prostate	High	N	N	N	N	N
	Low	Y	Y	Y	N	N
Breasts	High	N	N	N	Y	N
	Low	Y	Y	Y	N	N
GYN	High	N	N	Y	N	N
	Low	Y	Y	N	N	N
Skin	High	N	N	Y	Y	N
	Low	Y	N	Y	Y	N
Lungs	High	N	N	N	Y	Y
	Low	Y	Y	N	Y	N
Penis	High	N	N	N	Y	N
	Low	Y	N	N	Y	N
Eyes	High	N	N	Y	Y	Y
	Low	Y	Y	Y	Y	N

Source: From Rivard et al. [17]

16.2.3 Shielding

In multisource implants, photons emitted from a given source can be absorbed by the radiopaque markers (e.g., Au, Ag, and Pb) or the radiopaque components of adjacent sources, causing a lower dose than predicted by the TG-43 formalism. This is generally referred to as ISA in the literature. For high-dose rate (HDR) high-energy sources, ISA is not an issue since a single source steps through the implanted applicators, treating one dwell position at a time. However, applicator materials such as stainless steel can cause deviations. Some applicators also contain high Z, high-density shielding materials such as tungsten, used to protect organs at risk (OARs), e.g., shielded vaginal and intrarectal applicators. As an example of the latter, a recently designed rectal applicator [24], intended to be used with an HDR ^{192}Ir source, is shown in Figure 16.2. Another recent MC study on a highly heterogeneous applicator geometry with rotating Pt shields and a ^{169}Yb source, enabling intensity-modulated BT, has been published [25]. Currently (2020), the only commercially available BT source with internal shielding is from CivaTech Oncology [26–28].

To account accurately for the effects of shielding or ISA, it is usually not acceptable to use voxelized geometries to model the attenuation. MC codes that can model the geometry accurately have to be used, but often these geometries are then embedded in a voxelized phantom, e.g., representing a water phantom or patient. Phase space files can be generated at the outer surface of the source or the outer surface of the applicator, from which particles can be sampled for further transport. Alternatively, the applicator or shielding can be simulated together with the voxel phantom [29]. A special case of modelling applicators is the use of 3D mesh geometries, which use non-Cartesian voxels [30]. This approach is ideally suited for modelling complexly shaped applicators such as the Fletcher–Williamson device or balloon applicators with irregular shapes such as the ones used in breast BT. This approach is now possible in several MC codes, such as MCNP6 and GEANT4. Longer calculation times compared to combinatorial geometries should be expected for transport in these complex voxel shapes.

16.2.4 Scattering Conditions

TG-43 parameters are calculated in water spheres with radii of 15 and 40 cm for low- and high-energy sources, respectively, ensuring that TG-43 calculations represent doses in unbounded water medium over the therapeutically relevant distance range. Situations where full scatter conditions are not met, sources, e.g., close to the skin, will result in deviation of the delivered dose using the TG-43 formalism. Given the longer path length and dominance of incoherent scattering, high-energy photons are more sensitive to geometries with tissue boundaries near the implanted sources. Examples of clinical geometries include the breasts, skin, and lungs [6].

16.2.5 Breakdown of the Kerma Approximation for Absorbed Doses

The vast majority of MC studies deriving TG-43 parameters report collision kerma in water, relying on the equivalence of dose and collision kerma under CPE conditions. For low-energy sources, electronic equilibrium is reached within 0.1 mm of sources and the kerma=dose approximation is

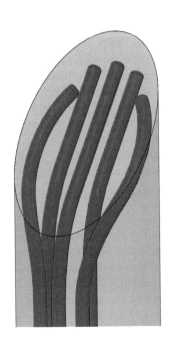
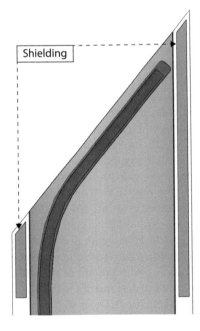

FIGURE 16.2 Example of a recent design of a tungsten-shielded rectal applicator for HDR ^{192}Ir BT with five source channels [24]. (Courtesy of M. Bellezzo (Maastro Clinic. Netherlands).)

accurate [31]. Since secondary electron ranges are larger for high-energy sources, this assumption can lead to dose differences greater than 1% at distances less than 7, 3.5, and 2 mm for ^{60}Co, ^{137}Cs, and ^{192}Ir, respectively [31]. Experimental dosimetry studies rarely report data so close to the source. In addition to the lack of CPE, beta particles emitted from sources can also cause a breakdown of the kerma=dose approximation. Granero et al. investigated the contributions of both betas emitted from ^{192}Ir and the lack of CPE for distances <2.5 mm from the source's center [32]. They found that CPE is reached at 2 mm from the source and that beta contribution to dose is negligible beyond 2.5 mm. Additionally, CPE breakdown may occur at tissue–high-Z interfaces. As also described in the previous chapter, the CPE assumption might break down for some detectors used for experimental BT.

16.2.6 Magnitude of Clinical Impact of Moving from TG-43 to MC Dose Evaluations

Several studies have compared MC dose calculations (considering dose to medium in medium, $D_{m,m}$) to the results of the TG-43 formalism. Meigooni et al. were the first to investigate ISA in 1992 using thermoluminescent dosimeter measurements and estimated that dose reductions of 6% could be expected at the edge of a ^{125}I prostate implant [33]. Chibani et al. and Carrier et al. published results of MC simulations performed in real ^{125}I and [103]Pd prostate implant geometries in 2005 and 2006 respectively, finding D_{90} reductions of 2%–5% due to ISA [3,34]. Chibani et al. also investigated the dosimetric impact of the presence of calcifications in the prostate and found D_{90} reductions of up to 37%. Carrier et al. subsequently performed a retrospective study of 28 prostate cancer patients implanted with ^{125}I using postimplant CT data (Figure 16.3) based upon the International Commission on Radiological Protection (ICRP)-recommended tissue compositions [35]. They found an average D_{90} decrease of 7% due to the combination of ISA and tissue composition heterogeneities [2]. Considering 613 prostate permanent implant BT patients (the largest cohort considered to date), Miksys et al. demonstrated that D_{90} is 6% lower, on average, for patient-specific MC simulations than for TG-43 calculations. Patients with intraprostatic calcifications may exhibit significant underdosing in tissue subvolumes shadowed by calcifications, resulting in D_{90} reductions as large as 25% relative to TG-43 calculations [36]. Miksys *et al.* demonstrated that MC dose evaluations are sensitive to the assumed mass elemental compositions of tissues (e.g., prostate and calcification—both uncertain) and the modelling approach employed (segmentation of prostate tissue and (micro)calcifications), with upward of 10% differences observed in target and OAR dose metrics with different approaches for modelling calcifications. Other studies can be found in [37].

Landry et al. investigated the sensitivity of dose distributions to tissues delineated on post-implant CT images for [125]I

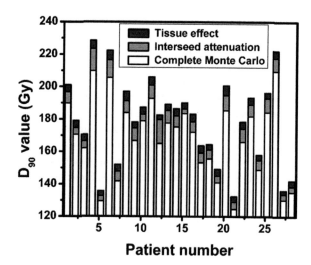

FIGURE 16.3 Impact of interseed attenuation and tissue nonwater equivalence on the clinical parameter D_{90} from post-implant dosimetry of 28 prostate cancer patients treated with[125]I implants. Tissue effect and interseed attenuation represent the D_{90} reductions from the TG-43 estimate from each effect. Complete Monte Carlo represents the D_{90} parameter when accounting for both effects. (Reproduced with permission from Carrier et al. [2].)

prostate and [103]Pd breast implants [38]. Two elemental compositions of prostate tissue were found in the literature, resulting in $D_{w,m,90\%}$ (dose to 90% of the target volume) variations of 3.5% for prostate BT. For the breast implants, Landry *et al.* approximated the heterogeneous anatomy of the breasts by a uniform mixture of adipose and glandular tissues. The proportions of the two tissues varied from 70:30 to 30:70, yielding $D_{w,m,}$ 90% variations of 10% (or about 6% when considering $D_{m,m}$). For a given adipose:glandular proportion, the variability or uncertainty in tissue elemental composition resulted in 10% dose variations. Recent studies of breast BT with low-energy sources suggest segmenting adipose and mammary glands on CT images, as the uniform tissue approximation may fail to reproduce the dose distributions obtained from segmented geometries [39–41]. Only one modern study of the elemental composition of tumors (from various body sites) appears to exist, that of Maughan et al. [42], which demonstrated a large variation in carbon content (8%–32% by weight) and mineral ash (0.9%–3.0% by mass) which results in variations of 20% for $\left(\mu_{en}/\rho\right)_{water}^{tumor}$ at 30 keV. This indicates the need for more studies in this field.

Afsharpour et al. evaluated doses for breast BT with [103]Pd implants and found $D_{90\%}$ reductions ranging from 4% for an all-glandular breast to 35% for an all-adipose breast [43]. This increased sensitivity to the material composition of breast BT compared to prostate implants is caused in part by the lower energy of [103]Pd, but also by the larger deviation of the atomic composition of adipose tissues to that of water compared to the relatively small difference between the atomic composition of prostate and that of water. Miksys et al.

assessed approaches for developing virtual patient models for MC simulation from CT data for [103]Pd breast BT patients. They compared different methods for mitigating metallic artifacts (bright spot and streaking artifacts due to BT seeds) and their interdependence with different tissue assignment schemes. They reported that these modelling choices introduce inherent uncertainties in dose calculations; however, general trends were evident, including that TG-43 overestimates doses in the target ($D_{90\%}$ on average by 10% and by up to 27%) and underestimates doses to the skin, the primary OAR (D_{1cm3} on average by 29% and up to 48%), compared to MC with full-tissue patient models [44].

MC dose evaluations for eye plaque BT (with [103]Pd,[125]I, or [131]Cs seeds) involving realistic nonwater eye tissues and anatomy, as well as the plaque, demonstrated large discrepancies with TG-43 dose evaluations [45,46]. For a 16-mm diameter Collaborative Ocular Melanoma Study plaque, the average dose to the tumor was up to 17% lower for MC (modelling the eye plaque containing sixteen [103]Pd seeds in a nonwater patient) than for TG-43; considerable discrepancies were also seen in normal ocular structures, e.g., lens; average MC dose was up to 34% lower than for TG-43. Administered doses may differ from TG-43 results due to the combined effects of eye plaque radiation collimation and differences in radiation scatter and absorption of eye tissues compared to water.

Melhus and Rivard showed that tissue composition inhomogeneities have negligible effects (<5%) for ^{192}Ir over clinically relevant distances in soft tissues, muscles, and breasts, using point sources in uniform spherical phantoms [47]. This is due to the near-constancy of the electron density and dominance of incoherent scattering at the higher photon energies of [192]Ir. However, more recent studies seem to point to non-negligible spectral shifts with distance to the ^{192}Ir source positions [48]; therefore, dose differences in different tissues may depend on the distance to the source. Poon et al. showed that TG-43 dose distributions for ^{192}Ir breast implants overestimated the skin dose by 5% due to differences from full-scatter conditions with MC simulations [6], confirming earlier findings from Pantelis et al. [49]. Dose differences caused by composition heterogeneities in the liver were also examined [50]. Other MC studies can be found in Refs. 51–54. The rarely considered aspect of transit dose delivered during source motion has also been addressed with MC calculations [55,56].

16.3 MC Dose Calculation Tools

A range of MC dose calculation tools have been developed, mostly for research purposes. Some are described here in some detail while others can be consulted in the literature [12,15,57]. International Working Groups have followed up on TG-186 recommendations by developing approaches to provide data for commissioning MC and other model-based dose calculation algorithms toward clinical implementation [29,58].

16.3.1 Monte Carlo Dose Calculation Tool for Prostate Implants

MCPI (Monte Carlo dose calculation tool for prostate implants) was developed by Chibani and coworkers in 2005 to perform patient-specific prostate implant dose calculations with [103]Pd and [125]I [11]. The code is based on the general-purpose MC software GEPTS [59], using photon-only transport for low-energy (<50 keV) sources. In addition to the track length collision kerma estimator, several variance reduction techniques were developed for MCPI. Photons are transported only once in the source geometry; a phase space file is subsequently used to emit photons from each source surface, while still modelling the source geometry to include ISA. The code simultaneously handles both voxel (rectangular mesh) and seed geometries (nested cylindrical objects, facilitating analytical ray tracing) using a hybrid model where voxel intersecting seeds are flagged, thus avoiding the need to query every seed in the geometry when transporting photons. Finally, MCPI uses ray tracing instead of analog transport of photons; primary and secondary photon trajectories are projected through the entire mesh of voxels, regardless of whether they interacted, thereby enhancing the frequency of energy deposition in voxels located at larger distances, a technique based upon the expected value track length estimator developed by Williamson [60]. Calculation times around 2005 were reported to be about 1 minute for a 2-mm^3 voxel mesh [11].

16.3.2 PTRAN_CT

The correlated sampling technique was implemented in PTRAN_CT [8,61,62]. PTRAN_CT implementation also included the variance reduction techniques of MCPI described above and expected value track length scoring. This code also used a novel generalized correlated sampling scheme whereby photon histories are initially constructed in homogeneous water followed by recalculating the particle weights for each collision. Rather than scoring dose, the difference between the heterogeneous geometry dose and TG-43 dose is scored for each photon history, increasing efficiency by factors of 2–40. Dose calculation times of about 3 seconds for a prostate implant have been reported to achieve 2% statistical uncertainty in a 2-mm^3 voxel mesh. This code can input digital imaging and communications in medicine (DICOM) CT images and currently uses the EGSnrc CTcreate software to assign cross-section files based upon a single-energy CT. An alternative could be the user-interface BrachyGUI, developed at McGill University [63].

16.3.3 BrachyDose

BrachyDose [64] is an EGSnrc [20] user code utilizing the multi-geometry package [65]. BrachyDose can model both photons and electrons with kinetic energies from 1 keV to MeV.

Geometry modelling includes diverse BT sources (low energy, high energy, LDR [low-dose rate], HDR, and miniature X-ray tubes), eye plaque applicators [66], and virtual patient models based on CT data [36,44,67]. Collision kerma is scored with a path length estimator, but dose can also be scored at the cost of longer calculation times. The code has been extensively benchmarked by generating single-source TG-43 parameters for several commercially available ^{125}I, ^{103}Pd, and ^{192}Ir sources [10]. BrachyDose has been used for dose evaluation of eye plaque BT with LDR sources [45,46,66], permanent breast seed implants with ^{103}Pd seeds [44], permanent implant prostate BT [36], and lung mesh BT [67,68]. Calculation times of half a minute a decade ago were reported around 2010 to produce statistical uncertainties less than 2% for a 2-mm^3 voxel mesh [69].

16.3.4 ALGEBRA

ALGEBRA (ALgorithm for heterogeneous dosimetry based on GEANT4 for BRAchytherapy) is based on the general-purpose MC code GEANT4 and the DICOM-RT standard. Initially developed by Carrier and coworkers [2] and referred to as GEANT4/DICOM-RT in the recent review by Rivard et al. [70], the code was later modified by Afsharpour et al. and given its current name [9]. The code can import planning data (seed position, air kerma strength, etc.) from a treatment planning system and structural data (contours). CT images are imported and a semiautomatic segmentation method uses the CT calibration curve to assign densities and organ contours with elemental compositions assigned to each voxel. Seed geometries and voxels do not overlap in ALGEBRA; voxel intersecting seeds are discarded and the missing volume is replaced by water. The layered mass geometry functionality of GEANT4, where the overlap of seeds and voxels is permitted, eliminated the need to perform this step [71]. Dose scoring is performed using the "parallel world" functionality of GEANT4 by creating a scoring mesh, independent of the transport geometry. The resolution of the CT-derived transport geometry can be modified, as well as that of the scoring mesh, and they need not be equal. Secondary electrons are immediately discarded and the collision kerma is scored with a path length estimator. Both kerma in water or medium by transporting photons in medium, $D_{w,m}$, and $D_{m,m}$, can be scored. The use of source phase space files is also possible. Calculation times in ALGEBRA for 2 mm^3 voxels are about 6 and 12 minutes with 2% statistical uncertainty for a breast ^{103}Pd implant and a prostate ^{125}I implant, respectively.

16.3.5 egs_brachy

egs_brachy is a modern EGSnrc application specifically designed for BT applications [14]. It makes use of EGSnrc to model radiation transport [20], and it uses an extended version of egs++ (EGSnrc's C++ class library) to model geometries

and particle source BT [14]. The code can be used for a comprehensive set of BT simulations, including dose calculations, generation of phase space data, and calculation of particle spectra (emitted from a source or within a particular region). The code was initially benchmarked through calculations for different BT sources (^{125}I,^{103}Pd,^{192}Ir), comparing dose distributions and TG-43 parameters with others published and finding excellent agreement [14]. Additionally, eye plaque BT (with ^{125}I) dose distributions have been shown to be in good agreement with the previously published BrachyDose and MCNP results [72].

egs_brachy employs various features to enhance simulation efficiency, including adoption of techniques to efficiently model geometry and radiation transport, using a track-length estimator to calculate collision kerma, phase space sources, and particle recycling (use of one source as a particle generator), as well as particular variance reduction techniques for electronic BT. The use of these techniques results in short calculation times. For example, for prostate and breast permanent implants with 2 mm^3 voxels in the virtual patient model and eye plaque with 1mm^3 voxels spanning the eye, treatment simulations take between 13 and 39 seconds to achieve 2% average statistical uncertainty for doses in the target. The code may be run on multiple cores to achieve even shorter calculation times and/or improved statistical uncertainties.

egs_brachy has been used to update the Carleton Laboratory for Radiotherapy Physics database of TG-43 dosimetry parameters for many source models [73]. Patient-specific dose evaluations based on CT images and treatment data have been carried out for ^{103}Pd permanent breast seed implant [74], as well as the development of a dynamic dose calculation framework applied to permanent implant prostate BT [75]. The code has been released as free, open-source software, distributed with a benchmarked library of source, applicator, and phantom geometries with a graphical user interface to facilitate the integration of clinical data.

16.4 Role of Imaging in in Vivo Cross-Section Assignment for MC Tissue Inhomogeneity Corrections

16.4.1 Computed Tomography

Computed tomography (CT) X-ray image datasets are generally the standard input to MC calculation for BT because intensity approximately tracks the relative linear attenuation coefficient of the underlying tissue. Converting Hounsfield units (HU) to electron or mass density is a relatively straightforward step. For low-energy sources, the issues outlined in the previous sections demand special attention, but for high-energy sources, where tissue composition effects are smaller, CT images provide sufficient information to calculate the absorbed dose within a few percent. Additionally, the location of implanted seeds, catheters, or applicators is easily derived from high-resolution CT images.

Streaking artifacts in CT, especially those caused by the spatial reconstruction of the high-density components of implanted LDR seeds and other foreign metal bodies implanted in the body, can seriously degrade the information required for accurate dose calculations. Metal artifact correction algorithms exist and can alleviate this problem [76]. However, metal artifact reduction cannot be considered to be a solved problem, especially at the level of quantitative accuracy needed to specify tissue composition for low-energy photon-emitting BT sources. Even relatively subtle streaking artifacts can degrade voxel-by-voxel tissue identifications.

16.4.2 Dual-Energy CT

While voxel-based MC simulation may become a clinically feasible treatment planning tool for low-energy BT, a major obstacle is the lack of a robust and accurate imaging tool for noninvasively mapping patient-specific photon cross sections. Our current knowledge of low-energy tissue cross sections is based upon the bulk organ/tissue compositions and densities recommended by International Commission on Radiation Units and Measurements (ICRU) [77] and ICRP [78] which, in turn, are based upon very sparse and old (1930s–1970s) tissue sample measurements [19,79–81]. The primary papers exhibit substantial sample-to-sample variability. For example, Hammerstein's chemical analysis of mastectomy specimens reveals that glandular and adipose tissues exhibit 8%–10% variations in composition by weight of carbon and oxygen, which translate into 8%–15% uncertainties in linear attenuation coefficient at 20 keV [82]. The only modern breast tissue water content study [83] reported large variations in water content: 23%–78% and 18%–94% in adipose and glandular tissues, respectively. Skeletal muscle exhibits surprising variations as well [80] with water content varying from 70% to 78%. The review by Mann-Krzisnik et al. [19] reports $(D_{m,m})_{90}$ differences of 20% for implants in high- vs. low-density breasts and 3.5% for prostate implants (between the two prostate compositions in the literature). The elemental composition assays of tumor specimens reported by Maughan et al. [42] also offer an instructive if cautionary tale. They report variations in mineral ash, oxygen, and carbon content by weight of 0.9%–3.0%, 8.1%–32.1%, and 50.8%–70.8%, respectively. These variations translate into 8% of uncertainties in linear attenuation coefficients in the 20–30 keV range, which implies single seed dose uncertainties of 8% at 1 cm to 50% at 5 cm. In summary, interpatient and intraorgan tissue composition variations introduce uncertainties in MC dose distributions compared to the magnitude of tissue inhomogeneity effects themselves. Hence, clinically useful deployment of MC $D_{m,m}$ calculations is unlikely without the creation of vastly expanded tissue composition datasets or more preferably, accurate CT-based tools for noninvasively imaging patient-specific, low-energy photon cross-section data needed for voxel-based MC without relying on questionable a priori tissue composition assumptions.

As noted above, because of the sensitivity of the photoelectric cross section to atomic composition, single-energy CT imaging is not sufficient. Williamson et al. [84] theoretically demonstrated that energy-dependent cross-section data in the 20–40 keV range can be accurately (1%–2%) extrapolated from CT images acquired at 90 and 140 kVp by means of a two-parameter linear separable model. This suggests that dual-energy CT (DECT) could potentially be used to measure voxel-specific cross-section tables. It is encouraging that image domain material decomposition to transform DECT images into effective atomic number and electron density images have been successfully deployed for proton stopping power ratio (SPR) imaging (see the recent review by Wohlfahrt et al. [85]). These studies suggest that in tightly controlled experimental settings (with matching calibration and patient measurement geometries) and low image noise, 1% accuracy can be achieved. Alternatively, statistical iterative image reconstruction (SIR) [86] for DECT SPR imaging using raw dual-energy projection data exported from commercial scanners is highly promising. The Williamson–O'Sullivan group [87,88] has demonstrated that SPR imaging with subpercentage accuracy independent of phantom size and location can be achieved using a penalized SIR algorithm that uses an accurate model of CT detector signal formation to eliminate beam hardening and photon scatter artifacts.

Unfortunately, low-energy photon cross-section estimation from DECT images is a much more challenging and poorly conditioned problem. Evans et al. [89] was the first to demonstrate experimentally that 30 keV linear attenuation coefficients could be recovered from DECT images with 1% accuracy using image-based material decomposition [84]. However, an error propagation analysis demonstrated that to achieve a cross-section mapping accuracy of 3%, cupping, streaking, and phantom size artifacts must not exceed 0.25%, a specification far beyond the reach of CT vendor–provided preprocessing and reconstruction software. While methods for predicting low-energy cross sections from effective atomic numbers and densities have been proposed [90–94], they rely on vendor-reconstructed images which have residual beam hardening and scatter artifacts far in excess of 0.25%. In contrast, a subsequent study by Evans et al. [95] demonstrated that SIR reconstruction on CT sinograms acquired without beam hardening compensation from a commercial scanner exhibited image intensities that were independent of uniform phantom size and voxel location that met within 0.5%. Yet-to-be-published data (see p. 131 in Zhang [96]) shows that linear attenuation coefficients can be recovered by an integrated DECT-SIR algorithm with an accuracy of 2% down to 30 keV energies, in both head and body size phantoms. However, clinical deployment of these DECT-SIR approaches faces several engineering challenges, including achieving clinically useful reconstruction speeds in 3D spiral imaging, developing CT scanner commissioning procedures that do not require vendor-proprietary knowledge, and mitigating tissue/organ deformation between sequential high- and low-energy scans.

16.4.3 Magnetic Resonance Imaging

With its excellent soft tissue contrast, magnetic resonance imaging (MRI) permits the identification of lesions and OARs not visible on CT and plays an important role in gynecological BT [97]. By differentiating tissues, it permits better calculation of radiation dose distribution at the time of planning. Prostate delineation is also acknowledged to be superior with MRI compared to CT imaging [98]. However, the lack of density information currently limits the use of MRI for MC dose calculations. Work has been performed to employ MRI in external beam radiotherapy (EBRT) treatment planning and for positron emission tomography attenuation correction [99]. The first method, based on anatomy, attempts to deform MR images to a reference MR image and then apply the resulting deformation to a corresponding reference CT image [100]. The voxel-based method attempts to classify MRI voxels into tissue types [101]. Following classification, bulk electron density assignment or direct conversion to HU or electron density is performed. These techniques have not been evaluated in the context of BT dose calculations so far. For prostate EBRT, Lee et al. reported dose calculation errors of less than 2% when segmenting images into the bone and soft tissues compared to using full CT [102]. Given that high-energy photons employed in, e.g., ^{192}Ir BT are relatively insensitive to tissue-type assignment (see Table 16.1), we can expect similar results. For low-energy photons, the added complexity of assigning a correct tissue composition would need to be considered. Geometric distortion of MR images due to magnetic field inhomogeneities, gradient nonlinearity, susceptibility effects, and chemical shifts has also been considered in EBRT [102]. BT is possibly less sensitive to spatial distortions than EBRT, as the distortions are generally small in the field-of-view center where MR compatible applicators or seeds are positioned.

16.4.4 Ultrasound

TRUS images are used to guide needle implantation for HDR and LDR prostate implant BT and to provide information on the location and dimensions of certain organs of interest for permitting TG-43 dose calculations [103]. While US-based tissue identification is an active area of research [104], it does not yet provide patient-specific density and tissue composition information necessary to perform MC simulations. Additionally, the presence of the TRUS probe during planning and treatment deforms the organs of interest. External transducers, permitting noninvasive image acquisition, can avoid this issue but they are currently not used in BT. One approach could be to assign generic densities and compositions to organ contours as explained above for MRI. Using the

TRUS-derived seed positions in an intraoperative setting, ISA could be modeled for LDR prostate implants.

16.5 Dose Specification in Terms of $D_{w,m}$ or $D_{m,m}$

A topic of interest is the dose quantity used to report administered doses. An overview of this issue can be found in the Task Group 186 Report [23]. This issue plays also a role in other radiotherapy applications and is covered in other chapters of this book. For HDR ^{192}Ir BT, dose calculations are still commonly done with the TG-43 formalism, which reports dose to water in water (or simply, dose to water). Advanced dose calculation algorithms may permit reporting $D_{m,m}$ which is the natural way for MC simulations to report dose or $D_{w,m}$ which requires a conversion of the former into the latter. In this notation, the second subscript of $D_{w,m}$ denotes the medium ('m' for the media actually occupying each voxel) in which the photons (and possibly secondary electrons) are transported by MC. The first subscript, e.g., "w," denotes the medium in which dose is scored and reported. These three dose quantities (including $D_{w,w}$ which corresponds approximately to the TG-43 source superposition algorithm) are fundamentally different and may differ numerically. While the numerical differences are small for megavolt radiotherapy (a few percent), they may be much larger in BT, especially for low-energy photon-emitting sources. Discussions on which dose quantity to use, in terms of best predictor for biological damage, are ongoing [105, 106]. The relationship between these dose quantities in low- and high-energy BT can be approximated by various cavity theories, assuming small, intermediate, and large dimensions relative to the secondary electron range.

Carlsson-Tedgren and Carlsson [106] applied Burlin cavity theory to investigate photon energies for which cavity sizes in the range 1 nm–10 mm can be considered small, intermediate, or large. MC methods are ideally suited, even indeed the only accurate way, to derive the factors in the Burlin cavity theory that serves to convert $D_{m,m}$ to $D_{w,m}$ by using average SPRs and average mass energy–absorption coefficient ratios.

$$\frac{D_{w,m}}{D_{m,m}} = d \left(\frac{\overline{S_{coll}}}{\rho} \right)_m^w + (1-d) \left(\frac{\overline{\mu_{en}}}{\rho} \right)_m^w$$

In this, d is a dimensionless weighting factor describing the relative importance of small (first term) versus large (last term) cavity theory in the dose conversion factor. This factor is related to the cavity size and the range of secondary electrons. The terms in parentheses are the spectrum-averaged mass collision SPR and the mass energy–absorption coefficient ratio, water to medium, respectively. Figure 16.4 with data from Ref. 106 shows the d factor versus photon energy for cavities of different mean chord lengths using photon and electron spectra generated with EGSnrc, covering the energy range

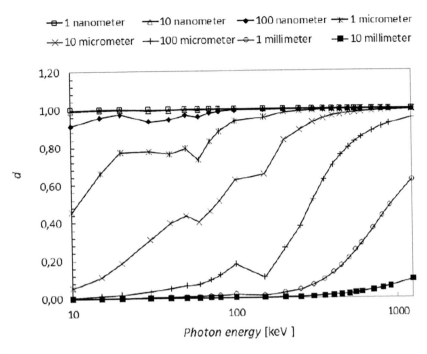

FIGURE 16.4   Burlin cavity weighting factor *d* versus photon energy for various cavity sizes (mean chord lengths). (Data taken from [106] and derived as described therein.)

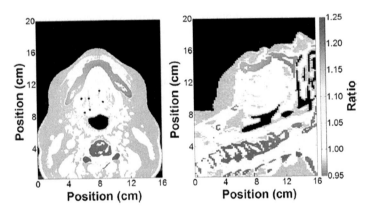

FIGURE 16.5   Ratio $D_{w,m}$(LCT)/$D_{w,m}$(SCT) for HDR ^{192}Ir BT of head & neck cancer. (Reproduced with permission from Ref. [48].)

encountered in BT. Figure 16.4 shows that only water cavities up to 10 nm can be considered small for all photon energies, i.e., the small cavity theory (Bragg–Gray) applies. Cavities of 1–10 mm can be considered large for most BT energies. But most other cavity sizes should be considered intermediate so that the full Burlin theory or MC calculations should be applied. These complexities in dose reporting are not taken into account by current BT treatment planning processes, MC dose calculation algorithms, or radiobiological models.

Fonseca et al. studied the differences between large (LCT) and small cavity theories (SCT) for dose reporting in HDR^{192}Ir BT using the MCNP6 MC code [48]. Figure 16.5 shows the ratio of $D_{w,m}$(LCT)/$D_{w,m}$(SCT). They found that it is mostly the bony and adipose regions where the two cavity theories lead to substantially different results. The ratio between conversion factors from $D_{m,m}$ to $D_{w,m}$ ranges from 1.13 to 1.20

(1.36 for teeth). This clearly shows that in BT the dose conversion from the default MC $D_{m,m}$ needs careful thoughts.

Complications arise when we consider that the radiation response of tumors and healthy tissues correlates with the energy deposited in cell nuclei. The dimensions of mammalian cells (~10 μm) cause them to act as Bragg–Gray or small cavities at ^{192}Ir energies and as intermediate cavities bracketed by large and small cavity theories for the energies of low-energy BT sources [23]. This makes the conversion of $D_{m,m}$ to $D_{w,m}$ using μ_{en} ratios questionable for low-energy sources and means that different conversion methods are required across the BT energy range [107]. Furthermore, research suggests that a water cavity is not an accurate surrogate for representing radiosensitive subcellular targets, including the cell nucleus [105, 108]. However, the DNA molecule is coated in water, and some studies point to a correlation between microdosimetry in the nanometer range and radiobiological effects [109]. The TG-186 recommendation is to require that only $D_{m,m}$ must be reported when performing MC dose calculations for BT, although $D_{w,m}$ and other quantities of interest may optionally be reported [23]. Finally, as discussed in Section 16.4.2, further development of quantitative DECT to improve the accuracy of low-energy photon cross-section imaging is urgently needed.

16.6 Future Use of MC Methods for Brachytherapy Patient and Applicator Modelling

Opportunities for future applications of MC methods for material heterogeneity modelling in BT may be divided into two categories: (1) patient-oriented applications such as for treatment planning and (2) radiobiological evaluations.

16.6.1 Treatment Planning

The first decade of the 21st century has seen the exciting birth of image-based MC treatment planning for BT and of quantitative imaging. Early investigations were focused on MC code variance reduction techniques, patient tissue composition assignment methods and their sensitivity to assumptions, $D_{m,m}$ vs. $D_{w,m}$ dose specification, and retrospective studies attempting to correlate the clinical outcomes with more accurate MC dose assessments.

Future directions for MC-based BT treatment planning will continue with improvements on existing methods, integration of MC computations during the dose optimization phase, use of MC techniques in inverse optimization, parallel processing algorithms, deep learning methods, resolution of imaging artifacts such as from seeds or high-Z applicators, correcting for limited spatial range of image datasets as needed to provide full radiation scatter conditions to simulate the clinical environment, and accounting for changes in BT applicator positioning relative to the relevant tissues in patients. Perhaps, MC methods can even play a role in accounting for organ motion during treatment or between treatments, as has been briefly explored for EBRT. Multi-institutional trials of BT dosimetry comparisons of MC and TG-43 methods are now entirely possible. Radiobiological modelling on the submicron scale for microdosimetry and improved understanding of relative biological effectiveness (RBE) effects, as well as source design for optimal exploitation of RBE effects are also of interest. Given the increased investigation of BT dosimetry and the improved coordination of international research teams, the future is bright for using MC methods in the field of BT.

16.6.2 Radiobiological Evaluations

Knowledge of photon and electron spectra is also required for applying biological damage models. It has long been known that photons below 50 keV induce DNA damage per unit

energy imparted than higher energy photons [110]. This finding, however, is rarely taken into account in clinical BT practice. We discuss as an example an EGSnrc MC model of an electronic BT source operated at 50 kVp. Reniers et al. [111] simulated the initial yield of single- and double-strand DNA breaks (SSB and DSB), which allowed RBE estimation. The MC model was first used to generate X-ray spectra at various distances from the source and to calculate electron spectra from the photon spectra in several media (Figure 16.6). These data were then input into an empirical "MC damage simulation" program derived from track structure theory [112] to estimate the initial yield of SSB and DSB in DNA. From this, the RBE was derived.

The results indicate a substantially increased DSB yield for electronic BT compared to ^{60}Co or ^{192}Ir reference radiations, leading to an RBE for DSB of 1.4–1.5. This elevated RBE was caused mostly by Compton and Auger electrons with energies below 8 keV (Figure 16.6). The RBE estimate for the low-energy X-ray source was found to be very similar to the measured and calculated RBE for the low-energy gamma-ray BT radionuclide ^{125}I [113,114]. These findings should be taken into account if the HDR electronic BT source is intended to replace BT with the commonly used HDR ^{192}Ir radionuclide.

Advanced dose calculation algorithms, such as those employing MC methods, may be coupled with radiobiological models, e.g., to evaluate biologically effective doses (BED) and tumor control probability (TCP). Miksys et al. coupled radiobiological models with patient-specific MC dose calculations for permanent prostate BT [115]. They demonstrated that patients with intraprostatic calcifications have considerably lower predicted BED and TCP when evaluating MC doses in comparison with the TG-43 approach. Future work should assess different treatment sites, consider patients from different institutions, as well as examine normal tissue complication probabilities.

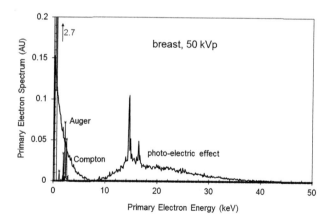

FIGURE 16.6 Initial secondary electron spectrum from an electronic BT source (50 kV), scored at a distance of 0.5 cm from the source's X-ray target in breast tissues. Electrons arising from Compton, photoelectric, and Auger interactions are indicated. (Reproduced with permission from Ref. [111].)

References

1. Burns, G. S., and D. E. Raeside. 1989. The accuracy of single-seed dose superposition for I-125 implants. *Med Phys* 16(4):627–631.

2. Carrier, J. F., M. D'Amours, F. Verhaegen, B. Reniers, A. G. Martin, E. Vigneault, and L. Beaulieu. 2007. Postimplant dosimetry using a Monte Carlo dose calculation engine: A new clinical standard. *Int J Radiat Oncol Biol Phys* 68(4):1190–1198.

3. Chibani, O., J. F. Williamson, and D. Todor. 2005. Dosimetric effects of seed anisotropy and interseed attenuation for 103Pd and 125I prostate implants. *Med Phys* 32(8):2557–2566.

4. Valicenti, R. K., A. S. Kirov, A. S. Meigooni, V. Mishra, R. K. Das, and J. F. Williamson. 1995. Experimental validation of Monte Carlo dose calculations about a high-intensity Ir-192 source for pulsed dose-rate brachytherapy. *Med Phys* 22(6):821–829.

5. Poon, E., B. Reniers, S. Devic, T. Vuong, and F. Verhaegen. 2006. Dosimetric characterization of a novel intracavitary mold applicator for ^{192}Ir high dose rate endorectal brachytherapy treatment. *Med Phys* 33(-12):4515–4526. Research Support, Non-U.S. Gov't.

6. Poon, E., and F. Verhaegen. 2009. Development of a scatter correction technique and its application to HDR ^{192}Ir multicatheter breast brachytherapy. *Med Phys* 36(-8):3703–3713. Research Support, Non-U.S. Gov't.

7. Gifford, K. A., J. L. Horton, Jr., C. E. Pelloski, A. Jhingran, L. E. Court, F. Mourtada, and P. J. Eifel. 2005. A three-dimensional computed tomography-assisted Monte Carlo evaluation of ovoid shielding on the dose to the bladder and rectum in intracavitary radiotherapy for cervical cancer. *Int J Radiat Oncol Biol Phys* 63(2):615–621.

8. Sampson, A., Y. Le, and J. F. Williamson. 2012. Fast patient-specific Monte Carlo brachytherapy dose calculations via the correlated sampling variance reduction technique. *Med Phys* 39(2):1058–1068. Research Support, N.I.H., Extramural.

9. Afsharpour, H., G. Landry, M. D'Amours, S. Enger, B. Reniers, E. Poon, J. F. Carrier, F. Verhaegen, and L. Beaulieu. 2012. ALGEBRA: ALgorithm for the heterogeneous dosimetry based on GEANT4 for BRAchytherapy. *Phys Med Biol* 57(11):3273–3280.

10. Taylor, R. E., G. Yegin, and D. W. Rogers. 2007. Benchmarking brachydose: Voxel based EGSnrc Monte Carlo calculations of TG-43 dosimetry parameters. *Med Phys* 34(2):445–457.

11. Chibani, O., and J. F. Williamson. 2005. MCPI: A subminute Monte Carlo dose calculation engine for prostate implants. *Med Phys* 32(12):3688–3698.

12. Famulari, G., M. A. Renaud, C. M. Poole, M. D. C. Evans, J. Seuntjens, and S. A. Enger. 2018. RapidBrachyMCTPS: A Monte Carlo-based treatment planning system for brachytherapy applications. *Phys Med Biol* 63(17):175007.

13. Mao, X., J. Pineau, R. Keyes, and S. A. Enger. 2020. RapidBrachyDL: Rapid Radiation Dose Calculations in Brachytherapy via Deep Learning. *Int J Radiat Oncol Biol Phys* 108: 802–812

14. Chamberland, M. J., R. E. Taylor, D. W. Rogers, and R. M. Thomson. 2016. egs_brachy: A versatile and fast Monte Carlo code for brachytherapy. *Phys Med Biol* 61(23):8214–8231.

15. Fonseca, G. P., B. Reniers, G. Landry, S. White, M. Bellezzo, P. C. Antunes, C. P. de Sales, E. Welteman, H. Yoriyaz, and F. Verhaegen. 2014. A medical image-based graphical platform -- features, applications and relevance for brachytherapy. *Brachytherapy* 13(6):632–639.

16. Anderson, L. L., R. Nath, and K. A. Weaver. 1990. *Interstitial Brachytherapy: Physical, Biological, and Clinical Considerations*. Interstitial Collaborative Working Group (ICWG), Raven, New York.

17. Rivard, M. J., J. L. Venselaar, and L. Beaulieu. 2009. The evolution of brachytherapy treatment planning. *Med Phys* 36(6):2136–2153. Review.

18. Landry, G., B. Reniers, J. P. Pignol, L. Beaulieu, and F. Verhaegen. 2011. The difference of scoring dose to water or tissues in Monte Carlo dose calculations for low energy brachytherapy photon sources. *Med Phys* 38(-3):1526–1533. Research Support, Non-U.S. Gov't.

19. Mann-Krzisnik, D., F. Verhaegen, and S. A. Enger. 2018. The influence of tissue composition uncertainty on dose distributions in brachytherapy. *Radiother Oncol* 126(3):394–410.

20. Kawrakow, I. 2000. Accurate condensed history Monte Carlo simulation of electron transport. I. EGSnrc, the new EGS4 version. *Med Phys* 27(3):485–498. Research Support, Non-U.S. Gov't.

21. ICRU. 1992. Report 46: Photon, electron, proton and neutron interaction data for body tissues, Bethesda, MD.

22. Valentin, J. 2002. Basic anatomical and physiological data for use in radiological protection: Reference values: ICRP Publication 89. *Annals of the ICRP* 32(3–4):1–277.

23. Beaulieu, L., C. T. A., J. F. Carrier, S. D. Davis, F. Mourtada, M. J. Rivard, R. M. Thomson, F. Verhaegen, W. T.A., and J. Williamson. 2012. Report TG-186 of the AAPM, ESTRO, and ABG on model-based dose calculation techniques in brachytherapy: Status and clinical requirements for implementation beyond the TG-43 formalism. *Med Phys* 39:6208–6236.

24. Bellezzo, M., G. P. Fonseca, R. Voncken, A. S. Verrijssen, C. Van Beveren, E. Roelofs, H. Yoriyaz, B. Reniers, E. J. Van Limbergen, M. Berbee, and F. Verhaegen. 2020. Advanced design, simulation, and dosimetry of a novel rectal applicator for contact brachytherapy with a conventional HDR (192)Ir source. *Brachytherapy* 19(4):544–553.

25. Famulari, G., M. Duclos, and S. A. Enger. 2020. A novel (169) Yb-based dynamic-shield intensity modulated brachytherapy delivery system for prostate cancer. *Med Phys* 47(3):859–868.

26. Aima, M., L. A. DeWerd, M. G. Mitch, C. G. Hammer, and W. S. Culberson. 2018. Dosimetric characterization of a new directional low-dose rate brachytherapy source. *Med Phys*. 45:3848–60.

27. Rivard, M. J. 2017. A directional (103)Pd brachytherapy device: Dosimetric characterization and practical aspects for clinical use. *Brachytherapy* 16(2):421–432.

28. Cohen, G. N., K. Episcopia, S. B. Lim, T. J. LoSasso, M. J. Rivard, A. S. Taggar, N. K. Taunk, A. J. Wu, and A. L. Damato. 2017. Intraoperative implantation of a mesh of directional palladium sources (CivaSheet): Dosimetry verification, clinical commissioning, dose specification, and preliminary experience. *Brachytherapy* 16(6):1257–1264.

29. Ma, Y., J. Vijande, F. Ballester, A. C. Tedgren, D. Granero, A. Haworth, F. Mourtada, G. P. Fonseca, K. Zourari, P. Papagiannis, M. J. Rivard, F. A. Siebert, R. S. Sloboda, R. Smith, M. J. P. Chamberland, R. M. Thomson, F. Verhaegen, and L. Beaulieu. 2017. A generic TG-186 shielded applicator for commissioning model-based dose calculation algorithms for high-dose-rate (192) Ir brachytherapy. *Med Phys* 44(11):5961–5976.

30. Fonseca, G. P., G. Landry, S. White, M. D'Amours, H. Yoriyaz, L. Beaulieu, B. Reniers, and F. Verhaegen. 2014. The use of tetrahedral mesh geometries in Monte Carlo simulation of applicator based brachytherapy dose distributions. *Phys Med Biol* 59(19):5921–5935.

31. Ballester, F., D. Granero, J. Perez-Calatayud, C. S. Melhus, and M. J. Rivard. 2009. Evaluation of high-energy brachytherapy source electronic disequilibrium and dose from emitted electrons. *Med Phys* 36(9):4250–4256. Research Support, Non-U.S. Gov't.

32. Granero, D., J. Vijande, F. Ballester, and M. J. Rivard. 2011. Dosimetry revisited for the HDR^{192}Ir brachytherapy source model mHDR-v2. *Med Phys* 38(1): 487–494.

33. Meigooni, A. S., J. A. Meli, and R. Nath. 1992. Interseed effects on dose for ^{125}I brachytherapy implants. *Med Phys* 19(2):385–390. Research Support, U.S. Gov't, P.H.S.

34. Carrier, J. F., L. Beaulieu, F. Therriault-Proulx, and R. Roy. 2006. Impact of interseed attenuation and tissue composition for permanent prostate implants. *Med Phys* 33(3):595–604.

35. ICRP. 1975. Report of the task group on reference man, ICRP Report 23, Washington D.C.

36. Miksys, N., E. Vigneault, A. G. Martin, L. Beaulieu, and R. M. Thomson. 2017. Large-scale Retrospective Monte Carlo Dosimetric Study for Permanent Implant Prostate Brachytherapy. *Int J Radiat Oncol Biol Phys* 97(3):606–615.

37. Collins Fekete, C. A., M. Plamondon, A. G. Martin, E. Vigneault, F. Verhaegen, and L. Beaulieu. 2015. Calcifications in low-dose rate prostate seed brachytherapy treatment: Post-planning dosimetry and predictive factors. *Radiother Oncol* 114(3):339–344.

38. Landry, G., B. Reniers, L. Murrer, L. Lutgens, E. B. Gurp, J. P. Pignol, B. Keller, L. Beaulieu, and F. Verhaegen. 2010. Sensitivity of low energy brachytherapy Monte Carlo dose calculations to uncertainties in human tissue composition. *Med Phys* 37(10):5188–5198. Research Support, Non-U.S. Gov't.

39. Sutherland, J. G., R. M. Thomson, and D. W. Rogers. 2011. Changes in dose with segmentation of breast tissues in Monte Carlo calculations for low-energy brachytherapy. *Med Phys* 38(8):4858–4865.

40. Afsharpour, H., B. Reniers, G. Landry, J. P. Pignol, B. M. Keller, F. Verhaegen, and L. Beaulieu. 2012. Consequences of dose heterogeneity on the biological efficiency of ^{103}Pd permanent breast seed implants. *Phys Med Biol* 57(3):809–823. Research Support, Non-U.S. Gov't.

41. White, S. A., G. Landry, G. P. Fonseca, R. Holt, T. Rusch, L. Beaulieu, F. Verhaegen, and B. Reniers. 2014. Comparison of TG-43 and TG-186 in breast irradiation using a low energy electronic brachytherapy source. *Med Phys* 41(6):061701.

42. Maughan, R. L., P. J. Chuba, A. T. Porter, E. Ben-Josef, and D. R. Lucas. 1997. The elemental composition of tumors: Kerma data for neutrons. *Med Phys* 24(8):1241–1244.

43. Afsharpour, H., J. P. Pignol, B. Keller, J. F. Carrier, B. Reniers, F. Verhaegen, and L. Beaulieu. 2010. Influence of breast composition and interseed attenuation in dose calculations for post-implant assessment of permanent breast ^{103}Pd seed implant. *Phys Med Biol* 55(16):4547–4561. Research Support, Non-U.S. Gov't.

44. Miksys, N., J. E. Cygler, J. M. Caudrelier, and R. M. Thomson. 2016. Patient-specific Monte Carlo dose calculations for (103)Pd breast brachytherapy. *Phys Med Biol* 61(7):2705–2729.

45. Lesperance, M., M. Inglis-Whalen, and R. M. Thomson. 2014. Model-based dose calculations for COMS eye plaque brachytherapy using an anatomically realistic eye phantom. *Med Phys* 41(2):021717.

46. Lesperance, M., M. Martinov, and R. M. Thomson. 2014. Monte Carlo dosimetry for 103Pd, 125I, and 131Cs ocular brachytherapy with various plaque models using an eye phantom. *Med Phys* 41(3):031706.

47. Melhus, C. S., and M. J. Rivard. 2006. Approaches to calculating AAPM TG-43 brachytherapy dosimetry parameters for ^{137}Cs, ^{125}I, ^{192}Ir, ^{103}Pd, and ^{169}Yb sources. *Med Phys* 33(6):1729–1737. Comparative Study.

48. Fonseca, G. P., A. C. Tedgren, B. Reniers, J. Nilsson, M. Persson, H. Yoriyaz, and F. Verhaegen. 2015. Dose specification for (1)(9)(2)Ir high dose rate brachytherapy in terms of dose-to-water-in-medium and dose-to-medium-in-medium. *Phys Med Biol* 60(11):4565–4579.

49. Pantelis, E., P. Papagiannis, P. Karaiskos, A. Angelopoulos, G. Anagnostopoulos, D. Baltas, N. Zamboglou, and L. Sakelliou. 2005. The effect of finite patient dimensions and tissue inhomogeneities on dosimetry planning of ^{192}Ir

HDR breast brachytherapy: A Monte Carlo dose verification study. *Int J Radiat Oncol Biol Phys* 61(5):1596–1602.

50. Duque, A. S., S. Corradini, F. Kamp, M. Seidensticker, F. Streitparth, C. Kurz, F. Walter, K. Parodi, F. Verhaegen, J. Ricke, C. Belka, G. P. Fonseca, and G. Landry. 2020. The dosimetric impact of replacing the TG-43 algorithm by model based dose calculation for liver brachytherapy. *Radiat Oncol* 15(1):60.

51. Fotina, I., K. Zourari, V. Lahanas, E. Pantelis, and P. Papagiannis. 2018. A comparative assessment of inhomogeneity and finite patient dimension effects in (60)-Co and (192)Ir high-dose-rate brachytherapy. *J Contemp Brachytherapy* 10(1):73–84.

52. White, S. A., G. Landry, F. van Gils, F. Verhaegen, and B. Reniers. 2012. Influence of trace elements in human tissue in low-energy photon brachytherapy dosimetry. *Phys Med Biol* 57(11):3585–3596.

53. Peppa, V., E. Pappas, T. Major, Z. Takacsi-Nagy, E. Pantelis, and P. Papagiannis. 2016. On the impact of improved dosimetric accuracy on head and neck high dose rate brachytherapy. *Radiother Oncol* 120(1):92–97.

54. Abe, K., N. Kadoya, S. Sato, S. Hashimoto, Y. Nakajima, Y. Miyasaka, K. Ito, R. Umezawa, T. Yamamoto, N. Takahashi, K. Takeda, and K. Jingu. 2018. Impact of a commercially available model-based dose calculation algorithm on treatment planning of high-dose-rate brachytherapy in patients with cervical cancer. *J Radiat Res* 59(2):198–206.

55. Fonseca, G. P., R. A. Rubo, R. A. Minamisawa, G. R. dos Santos, P. C. Antunes, and H. Yoriyaz. 2013. Determination of transit dose profile for a (192)Ir HDR source. *Med Phys* 40(5):051717.

56. Fonseca, G. P., G. Landry, B. Reniers, A. Hoffmann, R. A. Rubo, P. C. Antunes, H. Yoriyaz, and F. Verhaegen. 2014. The contribution from transit dose for (192)Ir HDR brachytherapy treatments. *Phys Med Biol* 59(7):1831–1844.

57. Enger, S. A., J. Vijande, and M. J. Rivard. 2020. Model-based dose calculation algorithms for brachytherapy dosimetry. *Semin Radiat Oncol* 30(1):77–86.

58. Ballester, F., A. Carlsson Tedgren, D. Granero, A. Haworth, F. Mourtada, G. P. Fonseca, K. Zourari, P. Papagiannis, M. J. Rivard, F. A. Siebert, R. S. Sloboda, R. L. Smith, R. M. Thomson, F. Verhaegen, J. Vijande, Y. Ma, and L. Beaulieu. 2015. A generic high-dose rate (192)Ir brachytherapy source for evaluation of model-based dose calculations beyond the TG-43 formalism. *Med Phys* 42(6):3048–3061.

59. Chibani, O. 1995. Electron depth-dose distributions in water, Iron and Lead - the GEPTS system. *Nucl Instrum Meth B* 101(4):357–378.

60. Williamson, J. F. 1987. Monte Carlo evaluation of kerma at a point for photon transport problems. *Med Phys* 14(4):567–576.

61. Hedtjarn, H., G. A. Carlsson, and J. F. Williamson. 2002. Accelerated Monte Carlo based dose calculations for brachytherapy planning using correlated sampling. *Phys Med Biol* 47(3):351–376.

62. Dolan, J., Z. Lia, and J. F. Williamson. 2006. Monte Carlo and experimental dosimetry of an ^{125}I brachytherapy seed. *Med Phys* 33(12):4675–4684. Research Support, Non-U.S. Gov't.

63. Poon, E., Y. Le, J. F. Williamson, and F. Verhaegen. 2008. BrachyGUI: An adjunct to an accelerated Monte Carlo photon transport code for patient-specific brachytherapy dose calculations and analysis. *J Phys: Conf Ser* 102(1):012018.

64. Yegin, G., and D. W. O. Rogers. 2004. A fast Monte Carlo code for multi-seed brachytherapy treatments including interseed effects. *Med Phys* 31:1771(abs).

65. Yegin, G. 2003. A new approach to geometry modelling for Monte Carlo particle transport: An application to the EGS code system. *Nucl Instrum Methods Phys Res Sect B: Beam Interact Mater Atoms* 211(3):331–338.

66. Thomson, R. M., R. E. Taylor, and D. W. Rogers. 2008. Monte Carlo dosimetry for ^{125}I and [103]Pd eye plaque brachytherapy. *Med Phys* 35(12):5530–5543. Research Support, Non-U.S. Gov't.

67. Sutherland, J. G., K. M. Furutani, and R. M. Thomson. 2013. Monte Carlo calculated doses to treatment volumes and organs at risk for permanent implant lung brachytherapy. *Phys Med Biol* 58(20):7061–7080.

68. Sutherland, J. G., K. M. Furutani, and R. M. Thomson. 2013. A Monte Carlo investigation of lung brachytherapy treatment planning. *Phys Med Biol* 58(14):4763–4780.

69. Thomson, R., G. Yegin, R. Taylor, J. Sutherland, and D. Rogers. 2010. Fast Monte Carlo dose calculations for brachytherapy with brachydose. 37:3910(abs).

70. Rivard, M. J., L. Beaulieu, and F. Mourtada. 2010. Enhancements to commissioning techniques and quality assurance of brachytherapy treatment planning systems that use model-based dose calculation algorithms. *Med Phys* 37(6):2645–2658.

71. Enger, S. A., G. Landry, M. D'Amours, F. Verhaegen, L. Beaulieu, M. Asai, and J. Perl. 2012. Layered mass geometry: A novel technique to overlay seeds and applicators onto patient geometry in Geant4 brachytherapy simulations. *Phys Med Biol* 57(19):6269–6277.

72. Thomson, R. M., R. E. P. Taylor, M. J. P. Chamberland, and D. W. O. Rogers. 2018. Reply to Comment on 'egs_brachy: A versatile and fast Monte Carlo code for brachytherapy'. *Phys Med Biol* 63(3):038002.

73. Safigholi, H., M. J. P. Chamberland, R. E. P. Taylor, C. H. Allen, M. P. Martinov, D. W. O. Rogers, and R. M. Thomson. 2020. Update of the CLRP TG-43 parameter database for low-energy brachytherapy sources. *Med Phys* 47: 4656–69.

74. Deering, S. G., M. Hilts, D. Morton, D. Batchelar, and R. M. Thomson. 2017. Monte Carlo dose calculations for permanent breast seed implant brachytherapy. *Med Phys.* 44:4384.

75. McCooeye, L., G. Hurd, L. Beaulieu, E. Heath, and R. M. Thomson. 2018. Monte Carlo dose calculations modelling edema resolution in permanent implant prostate brachytherapy. *Radiother Oncol* 129:S82.

76. Xu, C., F. Verhaegen, D. Laurendeau, S. A. Enger, and L. Beaulieu. 2011. An algorithm for efficient metal artifact reductions in permanent seed implants. *Med Phys* 38(-1):47–56. Research Support, Non-U.S. Gov't.

77. ICRU. 1989. *Tissue Substitutes in Radiation Dosimetry and Measurement.* International Commission on Radiation Units and Measurements, Bethesda, MD.

78. ICRP. 2003. *ICRP Publication 89: Basic Anatomical and Physiological Data for Use in Radiological Protection: Reference Values.* International Commission on Radiological Protection, Pergamon Press, Oxford.

79. White, D. R., and H. Q. Woodard. 1988. The effects of adult human tissue composition on the dosimetry of photons and electrons. *Health Phys* 55(4):653–663.

80. Woodard, H. Q., and D. R. White. 1986. The composition of body tissues. *Br J Radiol* 59(708):1209–1218.

81. Hammerstein, G. R., D. W. Miller, D. R. White, M. E. Masterson, H. Q. Woodard, and J. S. Laughlin. 1979. Absorbed radiation dose in mammography. *Radiology* 130(2):485–491.

82. Dance, D. R., C. L. Skinner, and G. A. Carlsson. 1999. Breast dosimetry. *Appl Radiat Isot* 50(1):185–203.

83. Brooksby, B., B. W. Pogue, S. D. Jiang, H. Dehghani, S. Srinivasan, C. Kogel, T. D. Tosteson, J. Weaver, S. P. Poplack, and K. D. Paulsen. 2006. Imaging breast adipose and fibroglandular tissue molecular signatures by using hybrid MRI-guided near-infrared spectral tomography. *Proc Natl Acad Sci USA.* 103(23):8828–8833.

84. Williamson, J. F., S. C. Li, S. Devic, B. R. Whiting, and F. A. Lerma. 2006. On two-parameter models of photon cross sections: Application to dual-energy CT imaging. *Med Phys* 33(11):4115–4129.

85. Wohlfahrt, P., and C. Richter. 2020. Status and innovations in pre-treatment CT imaging for proton therapy. *Br J Radiol* 93(1107): 20190590.

86. Williamson, J. F., B. R. Whiting, J. Benac, R. J. Murphy, G. J. Blaine, J. A. O'Sullivan, D. G. Politte, and D. L. Snyder. 2002. Prospects for quantitative computed tomography imaging in the presence of foreign metal bodies using statistical image reconstruction. *Med Phys* 29(10):2404–2418.

87. Zhang, S. Y., D. Han, D. G. Politte, J. F. Williamson, and J. A. O'Sullivan. 2018. Impact of joint statistical dual-energy CT reconstruction of proton stopping power images: Comparison to image- and sinogram-domain material decomposition approaches. *Med Phys* 45(5):2129–2142.

88. Zhang, S. Y., D. Han, J. F. Williamson, T. Y. Zhao, D. G. Politte, B. R. Whiting, and J. A. O'Sullivan. 2019. Experimental implementation of a joint statistical image reconstruction method for proton stopping power mapping from dual-energy CT data. *Med Phys* 46(1):273–285.

89. Evans, J. D., B. R. Whiting, J. A. O'Sullivan, D. G. Politte, P. H. Klahr, Y. Yu, and J. F. Williamson. 2013. Prospects for in vivo estimation of photon linear attenuation coefficients using postprocessing dual-energy CT imaging on a commercial scanner: Comparison of analytic and polyenergetic statistical reconstruction algorithms. *Med Phys* 40(12):121914.

90. Lalonde, A., and H. Bouchard. 2016. A general method to derive tissue parameters for Monte Carlo dose calculation with multi-energy CT. *Phys Med Biol* 61(22):8044–8069.

91. Landry, G., K. Parodi, J. E. Wildberger, and F. Verhaegen. 2013. Deriving concentrations of oxygen and carbon in human tissues using single- and dual-energy CT for ion therapy applications. *Phys Med Biol* 58(15) 5029.

92. Landry, G., P. V. Granton, B. Reniers, M. C. Ollers, L. Beaulieu, J. E. Wildberger, and F. Verhaegen. 2011. Simulation study on potential accuracy gains from dual energy CT tissue segmentation for low-energy brachytherapy Monte Carlo dose calculations. *Phys Med Biol* 56(19):6257–6278.

93. Landry, G., B. Reniers, P. V. Granton, B. van Rooijen, L. Beaulieu, J. E. Wildberger, and F. Verhaegen. 2011. Extracting atomic numbers and electron densities from a dual source dual energy CT scanner: Experiments and a simulation model. *Radiother Oncol* 100(3):375–379.

94. Landry, G., J. Seco, M. Gaudreault, and F. Verhaegen. 2013. Deriving effective atomic numbers from DECT based on a parameterization of the ratio of high and low linear attenuation coefficients. *Phys Med Biol* 58(19):6851–6866.

95. Evans, J. D., B. R. Whiting, D. G. Politte, J. A. O'Sullivan, P. F. Klahr, and J. F. Williamson. 2013. Experimental implementation of a polyenergetic statistical reconstruction algorithm for a commercial fan-beam CT scanner. *Phys Med* 29(5):500–512.

96. Zhang, S. 2018. Basis vector model method for proton stopping power estimation using dual-energy computed tomography. PhD thesis. Washington University in St. Louis, St. Louis, MO.

97. Potter, R., E. Fidarova, C. Kirisits, and J. Dimopoulos. 2008. Image-guided adaptive brachytherapy for cervix carcinoma. *Clin Oncol (R Coll Radiol)* 20(6):426–432. Review.

98. Viswanathan, A. N., J. Dimopoulos, C. Kirisits, D. Berger, and R. Potter. 2007. Computed tomography versus magnetic resonance imaging-based contouring in cervical cancer brachytherapy: Results of a prospective trial and preliminary guidelines for standardized contours. *Int J Radiat Oncol Biol Phys* 68(2):491–498.

99. Johansson, A., M. Karlsson, and T. Nyholm. 2011. CT substitute derived from MRI sequences with ultrashort echo time. *Med Phys* 38(5):2708–2714. Research Support, Non-U.S. Gov't.

100. Kops, E. R., and H. Herzog. 2007. Alternative methods for attenuation correction for PET images in MR-PET scanners. In *Nuclear Science Symposium Conference Record*, 2007. NSS '07. IEEE. 4327–4330.

101. Keereman, V., Y. Fierens, T. Broux, Y. De Deene, M. Lonneux, and S. Vandenberghe. 2010. MRI-based attenuation correction for PET/MRI using ultrashort echo time sequences. *J Nucl Med* 51(5):812–818. Research Support, Non-U.S. Gov't.

102. Lee, Y. K., M. Bollet, G. Charles-Edwards, M. A. Flower, M. O. Leach, H. McNair, E. Moore, C. Rowbottom, and S. Webb. 2003. Radiotherapy treatment planning of prostate cancer using magnetic resonance imaging alone. *Radiother Oncol* 66(2):203–216.

103. Chew, M. S., J. Xue, C. Houser, V. Misic, J. Cao, T. Cornwell, J. Handler, Y. Yu, and E. Gressen. 2009. Impact of transrectal ultrasound- and computed tomography-based seed localization on postimplant dosimetry in prostate brachytherapy. *Brachytherapy* 8(2):255–264.

104. Zhou, J., P. Zhang, K. S. Osterman, S. A. Woodhouse, P. B. Schiff, E. J. Yoshida, Z. F. Lu, E. R. Pile-Spellman, G. J. Kutcher, and T. Liu. 2009. Implementation and validation of an ultrasonic tissue characterization technique for quantitative assessment of normal-tissue toxicity in radiation therapy. *Med Phys* 36(5):1643–1650.

105. Enger, S. A., A. Ahnesjo, F. Verhaegen, and L. Beaulieu. 2012. Dose to tissue medium or water cavities as surrogate for the dose to cell nuclei at brachytherapy photon energies. *Phys Med Biol* 57(14):4489–4500.

106. Tedgren, A. C., and G. A. Carlsson. 2013. Specification of absorbed dose to water using model-based dose calculation algorithms for treatment planning in brachytherapy. *Phys Med Biol* 58(8):2561–2579.

107. Oliver, P. A. K., and R. M. Thomson. 2017. Cavity theory applications for kilovoltage cellular dosimetry. *Phys Med Biol* 62(11):4440–4459.

108. Thomson, R. M., A. C. Tedgren, and J. F. Williamson. 2013. On the biological basis for competing macroscopic dose descriptors for kilovoltage dosimetry: Cellular dosimetry for brachytherapy and diagnostic radiology. *Phys Med Biol* 58(4):1123–1150.

109. Lindborg, L., M. Hultqvist, A. Carlsson Tedgren, and H. Nikjoo. 2013. Lineal energy and radiation quality in radiation therapy: Model calculations and comparison with experiment. *Phys Med Biol* 58(10):3089–3105.

110. Hill, M. A. 2004. The variation in biological effectiveness of X-rays and gamma rays with energy. *Radiat Prot Dosim* 112(4):471–481. Review.

111. Reniers, B., D. Liu, T. Rusch, and F. Verhaegen. 2008. Calculation of relative biological effectiveness of a low-energy electronic brachytherapy source. *Phys Med Biol* 53(24):7125–7135. Research Support, Non-U.S. Gov't.

112. Semenenko, V. A., and R. D. Stewart. 2004. A fast Monte Carlo algorithm to simulate the spectrum of DNA damages formed by ionizing radiation. *Radiat Res* 161(4):451–457. Research Support, U.S. Gov't, Non-P.H.S.

113. Reniers, B., S. Vynckier, and F. Verhaegen. 2004. Theoretical analysis of microdosimetric spectra and cluster formation for ^{103}Pd and ^{125}I photon emitters. *Phys Med Biol* 49(16):3781–3795. Research Support, Non-U.S. Gov't.

114. Lehnert, S., B. Reniers, and F. Verhaegen. 2005. Relative biologic effectiveness in terms of tumor response of 125I implants compared with 60Co gamma rays. *Int J Radiat Oncol Biol Phys* 63(1):224–229.

115. Miksys, N., M. Haidari, E. Vigneault, A. G. Martin, L. Beaulieu, and R. M. Thomson. 2017. Coupling I-125 permanent implant prostate brachytherapy Monte Carlo dose calculations with radiobiological models. *Med Phys* 44(8):4329–4340.

17

Artificial Intelligence and Monte Carlo Simulation

D. Sarrut and N. Krah
Université de Lyon

Artificial intelligence (AI) and in particular deep neural networks (DNN) have become a major and powerful methodological actor to learn statistical properties and correlation in the past years. The main domain driving this evolution is the computer vision field. However, it can be observed that Monte Carlo (MC) simulation is heavily statistically driven and deals with a lot of data. It thus seems quite natural that DNN could play a role in that field. Indeed, some works are currently emerging.

This chapter will provide a structured overview of the use of AI methods in combination with MC simulation in the context of radiotherapy. We will differentiate between different kinds of applications and how AI methods relate to MC. As virtually all considered works rely on (deep) neural networks (NNs), we will concentrate on those henceforth. We make no attempt to give a comprehensive introduction to deep learning as there is a rich body of literature available. For the sake of clarity, we do, however, introduce the basic concepts behind DNN.

NNs are typically presented as a set of neurons organized in layers. Neurons between layers can be connected with an associated weight. The weights' values need to be determined in what is called the "training." Together, such a network links the input (first layer) and output variables (last layer). Each neuron has an associated activation function that generates the neuron's output, i.e., typically a nonlinear function mapping from an open into a closed real domain (e.g., values bounded between zero and one). The input to a neuron's activation function is the sum of overall outputs of the connected neurons in the upstream layer. The number of layers and neurons, the way they are connected, and the choice of activation functions are referred to as the "network architecture." The term "deep" indicates that the network contains several layers.

Especially when working with image-like data, relevant features are not found in a single-pixel but in an extended region around it. In such a case, suitable convolution operations are mixed in between the neuron layers so the network can capture nonlocal properties of the input data. Networks adopting this strategy are called "convolutional." A common strategy when input and output of a NN are images, i.e., very high-dimensional data, is to first transform the input into a lower-dimensional abstract feature space and then back into an image. Typical architectures employed by several of the works presented in this chapter include the "U-Net" and variational autoencoders (Kingma and Welling 2019, 201). Another class of networks is generative adversarial networks (GAN) which will be discussed more intensively further down.

It is worth drawing some connections to more conventional terminology from applied numerics. A given network architecture corresponds to a well-defined analytical function, namely a concatenation of weighted sums, one for each network layer. The network's ability to adapt even to complex input data is due to the high number of layers and neurons. In this sense, a (deep) NN can also be viewed as a highly flexible high-dimensional nonlinear fitting function, although discontinuities in the data, e.g., in dose distributions, might be a challenge. Nonlinearity arises from the nonlinear activation function. Indeed, if the activation function was linear, the whole network would be nothing more than a huge linear function. The neuron weights correspond to what is usually referred to as fit parameters in a conventional numerical model. Training the network essentially means fitting the model to the data with some appropriate iterative procedures. One iteration is usually referred to as an "epoch" in the AI literature.

DOI: 10.1201/9781003211846-20

A challenge when working with DNN is to select the appropriate network structure for a given problem and adjust the training process. This often involves a series of hyperparameters to be fixed, e.g., the learning rate, penalty weights, and batch size and finding the optimal hyperparameters is far from trivial. It is worth underlining, however, that this problem is not specific to DNN. Also, when fitting a "conventional" complicated nonlinear model to complex data, similar issues arise, e.g., the step length in gradient descent method needs to be fixed, the strength of regularization functions needs to be chosen, and the kind of optimizer often has an impact on convergence. Handcrafted analytical models usually mitigate these difficulties by reducing the number of fit parameters through a knowledge-based choice of the kind of function. This might for example be derived from an understanding of the underlying physical process to be modeled. On the other hand, handcrafted models are task-specific because they are typically able to describe only the kind of phenomenon they are designed for. A DNN is more generic and, as the chapter will show, can adapt to a broader range of problems at once. For example, the same kind of generative network can be used to model different kinds of particle phase spaces or predict photon scatter in various patient geometries.

A practical advantage of applying DNN to a radiotherapy-related problem is that several efficient, well-coded, and well-maintained frameworks exist to build NNs. These include Keras, TensorFlow, PyTorch, and others. Many of these libraries were initially built by large international companies for their own AI applications and then made public to a broader audience. These frameworks have most of the complicated functionality related to training and applying a network built into them and expose it through a computing interface (e.g., via python) which is easy to learn and handle. This alleviates the medical physics researcher's tasks considerably.

We draw attention to yet another way to view NNs; they are mappings from one domain into another. In many of the works, input and output are images in the sense that they represent two-dimensional (2D) or three-dimensional (3D) sets of data organized on a pixel or voxel grid.

A MC simulation in radiotherapy can often also be viewed as a mapping operation, for example from a CT image to a dose distribution. Clearly, there is no explicit analytical expression to describe this mapping as with NNs. Rather, the simulation consists of a concatenation of computational operations, including particle tracking, scoring, and potentially binning, for example into the desired dose grid.

Many of the works presented in this chapter propose NNs trained on MC-generated data. Those networks effectively learn to imitate and replace the MC mapping. On the other hand, we will also present a series of works that aim to replace part of the workflow within a MC simulation, either to speed up computation or facilitate specific tasks. These works are more intimately tied to MC than the previously mentioned ones and the input and/or output data have no immediate

interpretation as images in the larger sense but are rather a list of particle properties (one-dimensional vectors), e.g., representing a phase space and there is.

17.1 AI and Dose Estimation from MC Simulations

Several authors investigated the use of a convolutional neural network (CNN) to estimate dose distribution in various contexts (imaging and internal and external radiation therapy), exploiting MC simulations as input and validation. The aim is to have a fast NN that can replace the lengthy MC simulation. For example, (Lee et al. 2019), proposed deep learning–based methods to estimate the absorbed dose distribution for internal radiation therapy treatments. CNN was trained from PET and CT image patches associated with their corresponding dose distributions computed by GEANT4 application for tomographic emission (GATE) MC simulation and considered as ground truth. The database was composed of ten patients with eight PET/CT time points after intravenous injection of ^{68}Ga-NOTA-RGD, from 1 to 62 minutes postinjection. The network architecture was based on a U-Net structure (Ronneberger et al. 2015), composed of a set of convolutional layers, with the first part performing downsampling operations (contracting path) and the second part upsampling (expansive path). The U-Net was performed patch-based (subparts of the images) rather than image-to-image because of the relatively short range of dose delivery of a source voxel and exploiting both PET and CT as input data to predict the dose. The accuracy was found to be within less than 3% of the reference computation obtained within a few minutes compared to hours with MC. Similarly, Götz et al. 2019 presented a hybrid method based also on a U-Net combined with an empirical mode decomposition technique. It takes as input CT images and corresponding absorbed dose map estimated with medical internal radiation dose protocol (organ S-value) from SPECT images for ^{177}Lu internal radiation therapy treatment. Again, accuracy seems very good, better than the fast dose-volume-kernel method (Bolch et al. 1999) and faster than MC. The computation by CNN of first-order exposure (without scatter) estimate of image-guided X-ray procedures is investigated in Roser et al. 2019. The CNN was trained using smoothed results of MC simulations as output (smoothed to reduce stochastic fluctuations) and ray casting simulations of identical imaging settings and patient models as inputs. In external radiation therapy and brachytherapy, Nguyen et al. 2019 (see also Figure 17.1), Liu et al. 2019, Kalantzis et al. 2011, and Mao et al. 2020 investigated deep learning–based dose prediction models, using structure contours, prescription and delivered doses as training data, for head and neck volumetric modulated arc therapy treatments, and nasopharyngeal cancer helical tomotherapy treatments. Predictions were found to be accurate.

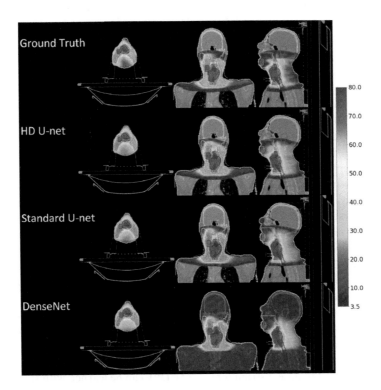

FIGURE 17.1 Dose washes for external beam radiotherapy of one example patient from the test pool. The color bar is shown in units of Gy. The clinical ground truth dose is shown on the top row, followed by the dose predictions of the HD U-Net, standard U-Net, and DenseNet, respectively. Low-dose cutoff for viewing was chosen to be 5% of the highest prescription dose (3.5 Gy). (From Nguyen et al. 2019.)

The works presented in this section are very useful and point into interesting directions to follow. However, a remark seems appropriate here: NNs need a sufficiently large amount of input data to be trained on and MC provides a relatively easy way to achieve this. At the same time, this also implies that NNs can never fully replace MC but will always rely on them for training. The motivation behind many of the presented works is to speed up the computation to clinically reasonable levels, e.g., dose calculations or image reconstructions on the order of minutes rather than hours or days. The principle is that applying a trained network is fast, while the lengthy MC simulations to generate training datasets as well as the training process itself can be run offline and not as part of the treatment planning. Nonetheless, these simulations need to be performed at some point, and this raises the legitimate question of how large and widespread a patient dataset (images, dose distributions, etc.) needs to be able to cover a sufficiently wide range of cases. This calls for care when constructing the training dataset. Furthermore, rigorous validation and thorough investigations on the robustness of new AI-based methods will be mandatory the more they are employed clinically on a larger scale. Another point of interest is that in a pure MC calculation, the statistical uncertainty is known while, with NNs based on MC, this is also implicitly there but invisible. Data augmentation (Shorten and Khoshgoftaar 2019) techniques and transfer learning (Shan et al. 2018; Frégier and Gouray 2019) will likely be of interest in the future.

17.2 AI for Dose Computation Denoising

The methods presented in the previous sections attempt to map from some kind of image data (e.g., patient CT, SPECT images) to a dose distribution. The speedup stems from the fact that applying a network is faster than running a MC simulation. The works in this section are not designed to replace MC but rather add a postprocessing step to reduce noise in the MC output. They thus constitute mappings between dose maps.

MC denoising methods have been studied for a long time and been shown to be able to reduce the dose computation time by smoothing statistical fluctuations (Naqa et al. 2005). The "noise" of the computed dose is related to the variance on the deposited energy in all regions of interest and decreases as the number of simulated particles, N, increases, specifically at a $1/\sqrt{N}$ rate. Hence, a very large number of iterations are required to reach low fluctuation dose estimation, in particular in low-dose regions. On the other hand, being able to lower the number of simulated particles translates into a net gain in the computational speed. Several filtering methods have been employed, such as 3D wavelet-based, advanced mean–median filtering, anisotropic diffusion, and so on. In general, good results were obtained, but the effective acceleration depends significantly on the characteristics of the dose distribution.

More recently, several groups have been studying denoising based on deep learning. The principle is to feed a CNN with pairs of high-noise/low-noise dose distributions obtained from low and high statistics MC simulations with the goal to generate denoised dose maps from noisy ones. Most of the time, CNN architecture is derived from U-Net, but other architectures such as DenseNet or conveying path convolutional encoder-decoder were studied. It has been applied to both photon (Peng et al. 2019b; Fornander 2019; Neph et al. 2019; Kontaxis et al. 2020) and proton doses (Javaid et al. 2019; Madrigal 2018) for various indications, including the brain, head and neck, liver, lungs, and prostate and to dose delocalization due to charged particles interacting with magnetic fields within MRI (MRgRT). Evaluations were performed based on a peak signal-to-noise ratio, gamma index, or dose volume histogram as comparison metrics. Results were generally very encouraging. CNN produced noise-equivalent dose maps with 10–100 times fewer particles than originally needed. Some difficulties still remain; results depend on the size and complexity of the training datasets and it is to be seen how the method can be generalized to other datasets. Furthermore, denoised dose maps must preserve the dose gradient features, and it is not yet fully clear how to guarantee this.

To conclude, we remark that the denoising problem addressed in this section is not limited to dose distributions, but can be found in other areas of radiation therapy as well. Denoising methods investigated for low-dose CT imaging may be a source of inspiration (Wolterink et al. 2017; Yang et al. 2018; Shan et al. 2018; Peng et al. 2019b).

17.3 AI for Imaging Detector and Source Modelling

The works presented so far rely on the output of MC simulations; they do not alter the simulation itself. The works in this section, on the other hand, replace part of the MC simulation in an attempt to accelerate it. More specifically, they model the particle transport through part of the geometrical components implemented in the simulation. In contrast to the previous methods, their input and output cannot be easily viewed as images.

The first case is DNN which model the response of a detector. Instead of explicitly simulating the particle transport in the detector, this is emulated by the network. For example, Sarrut et al. 2018 proposed to use a NN to learn the angular response function (ARF) of a SPECT collimator detector system. The underlying idea is to speed up simulations of SPECT imaging by modelling the collimator detector response function (CDRF) that combines the accumulated effects of all interactions in the imaging head and may be approximated with ARF (Descourt et al. 2010; Ryden et al. 2019; Sarrut et al. 2018; Song et al. 2005). The ARF method replaces the explicit photon tracking in the imaging head with a tabulated model of the CDRF. The tabulated model is derived from a simulation with a source of gamma covering the energy range of the radionuclide of interest and including the complete detector head with a collimator, crystal,

and digitization process. In Sarrut et al. 2018, tabulated data have been replaced by a DNN trained to learn ARF of a collimator detector system. The NN is trained once from a complete simulation including the complete detector head with a collimator, crystal, and digitization process. The simulation records properties of photons crossing a virtual plane in front of the SPECT head, and their energy and direction are used as input to the NN that provides detection probabilities in each energy window. Compared to histogram-based ARF, the artificial neural network method depends less on the statistics of the training data, requires no explicit binning into histograms, provides similar simulation efficiency, and requires less training data. The ARF approach has been shown to be efficient and to provide variance reduction that speeds up the simulation. Speedup compared to analog MC was between 10 and 3,000; ARF methods are more efficient for low-count areas (speedup of 1,000–3,000) than for high-count areas (speedup of 20–300) and more efficient for high-energy radionuclides (such as ^{131}I) that show large collimator penetration. This implementation of ARF by means of NNs is available within the GATE platform. Again, this work is still experimental, and it is not clear if several types of CDRF can be learned with sufficient accuracy.

In a similar direction as the ARF in SPECT imaging, deep learning–based methods have been proposed in PET imaging. Among examples are works that use NNs to estimate the depth of interaction (DOI) and event position within pixelated or continuous monolithic scintillators (Zatcepin et al. 2020; Berg and Cherry 2018; Müller et al. 2019; Oliver et al. 2013; Pedemonte, Pierce, and Van Leemput 2017; T.-Y. Yang 2019). Incorporating the DOI in the image reconstruction improves the quality of the PET images. Training of the networks may be performed on experimentally acquired data of a specific setup or obtained via MC simulation such as in T.-Y. Yang 2019. It is likely that deep learning–based methods will be developed in the future for detector modelling or event selection in other imaging modalities in the context of radiation therapy, such as prompt gamma imaging or ion computed tomography.

While the ARF networks described above use kinematic particle properties which result from MC tracking as input, Sarrut et al. 2019 propose to use a GAN to produce phase space coordinates as output. Specifically, they employ the concept of GAN to learn the phase space distribution obtained from the simulation of a linac. GANs were reported (Goodfellow et al. 2014; Gulrajani et al. 2017) as DNN architectures which allow mimicking a distribution of multidimensional data and have gained large popularity due to their success in realistic image synthesis. GANs learn representations of a training dataset by implicitly modelling high-dimensional distributions of data. Once trained, the resulting model is a NN called "generator G," using random data as input and producing elements that are supposed to belong to the underlying probability distribution of the training data. During the simulation of the linac, the properties (energy, position, and direction) of all particles reaching a plane located at the linac head exit are stored in the phase space file and depend on the detailed properties of the treatment head components, such as their shape and material. Phase space files are typically up to several tens of gigabytes large and inconvenient to use efficiently. Statistical limitations due to particle recycling may also arise when more particles are required than stored in the phase space file. In the proposed method, a GAN is trained using the phase space particles as a learning dataset. At the end of the training process, the resulting network "G" can generate particles that belong to the probability distribution of the phase space while being around 10 MB, instead of a few GB of the initial file. It has been shown that these AI-generated particles can typically be used as source distribution when computing a dose distribution with very good accuracy compared to the real phase space. As a test of feasibility, the authors also applied the GAN method to a brachytherapy treatment where the network learned the source distribution generated by seeds in the prostate region. Simulations performed with the GAN as a phase space generator showed good dosimetric accuracy (Figure 17.2). It is not known, however, if the same GAN architecture can efficiently learn any type of phase space. Moreover, the training process was found to be difficult, involving several hyperparameters to be tuned.

Since the concept of GAN was initially proposed, numerous variants, among them Wasserstein generative adversarial network, have been investigated in the literature with more than 500 papers per month by the end of 2018. Further work is still needed to evaluate which of those variants are of interest for MC simulation and to investigate if the earlier-mentioned difficulty to precisely model sharp features in dose distributions could be overcome. Transfer learning is another concept to be explored: a first network trained for a given phase space could be used as a starting point for the training of another one. Note also that GANs are usually employed for problems with a higher number of dimensions and very different distributions (natural images and speech) than phase space data; experiences are still to be gained to see if they are good tools for this task. Other methods, such as Gaussian mixture models, may also be useful to model phase space files. Finally, the use of GAN to mimic a probability distribution is not necessarily limited to phase space only. There are probably other applications within the MC particle transport process where GAN variants could replace conventional methods.

17.4 CBCT Imaging

Another field intimately linked to radiotherapy where AI methods have been explored and developed is cone beam CT (CBCT) imaging. These works address central shortcomings of CBCT which are poor image quality and artifacts due to scatter. These arise because the imager panel not only captures the attenuated primary photons from the X-ray source but also those originating from coherent and incoherent scatters within the patient. For accurate image reconstruction, the scatter contribution would need to be known and subtracted from the raw projection images. In practice, this is impossible because the imager panel only provides a nondiscriminative cumulative intensity signal. A MC simulation, on the other hand, can specifically tag scattered photons so that perfect scatter-free projections can be obtained via simulation. In fact, some earlier works on CBCT scatter correction rely on MC simulation to estimate the scatter contribution in raw projection (Jarry et al. 2006). However, the direct MC simulation of kV photons is too slow to be integrated into a clinical image reconstruction software, although heavy use of variance reduction techniques might improve this (Mainegra-Hing and Kawrakow 2008).

Recent works propose to use deep convolutional networks which learn from MC-simulated CBCT projections. They generate estimated scatter images (projections) as output based on raw projections as input (Lee et al. 2019; van der Heyden et al. 2020; Lalonde et al. 2020; Maier et al. 2019). The technical details of the networks vary, but all report quite promising results. It is worth mentioning that these methods rely on MC simulations for training where primary photons can be distinguished from scattered ones and could not be trained on experimentally acquired projections which cannot provide explicit scatter images to learn from.

Other authors have reported CBCT scatter correction methods based on deep learning which operates in the image domain. More specifically, they take a CBCT image as input and generate a synthetic CT image as output, i.e., they estimate how a CT image of the patient anatomy described by the CBCT image would have looked like. These synthetic CT images seem to contain much fewer artifacts than the original CBCT images do. Datasets to train the networks consisted in experimentally acquired CT and CBCT images, but in principle, they would also work on MC-generated data.

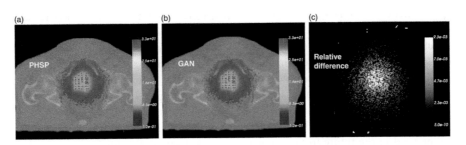

FIGURE 17.2 Slices of CT prostate image with deposited energy overlay (in MeV), computed by phase space (PHSP, a) and GAN-generated (b) particles. The (c) image shows the dose difference relative to the maximum dose. (From Sarrut et al. 2019.)

17.5 Discussion

There is a paradigm shift associated with the use of deep learning methods in medical physics simulation; to some extent, the user needs to abandon the instinct to mathematically master the phenomenon under investigation and instead rely on a sufficiently large amount of data to learn from heuristically. It can be expected that this will also lead to a shift in the skill set within the radiotherapy community, where modelling of radiation physics played a crucial role in the past it will be more and more the ability to implement and handle complex computational tasks and master the underlying mathematics. A word of caution seems to be due; however, as this chapter has shown, many of the works on deep learning in radiotherapy-related applications heavily rely on MC-simulated training data which in turn requires a simulation to be skillfully set up and evaluated in the first place. As the use of AI methods in radiation therapy evolves, physics-driven dataset modelling, i.e., a mix between modelling based on large datasets and understanding of the underlying physics, will become increasingly important.

Continuing this line of argument, it is worth remarking that a MC simulation, be it for dose calculation or an image simulation, always requires a numerical model of the implemented system and its geometry, e.g., the particle phase space and the components of the treatment machine in dose calculation or the radiation source and detector characteristics in CBCT imaging. Any inaccuracy in the model will likely lead to discrepancies between simulated and measured data. A deep learning method trained on MC data will naturally inherit their inaccuracies and can at most be expected to produce data as consistent as the MC itself. We can imagine that generating suitable datasets will become a skill in itself in the future, somewhat similar to commissioning a treatment planning system (TPS) which not only relies on physics but also on suitable experimental data.

Furthermore, we note that there is no intrinsic link between NN and the MC simulation when the latter only serves to generate training data. The network would also work on data acquired otherwise. For example, when mapping between dose distributions, analytical dose calculation methods such as implemented in many commercial TPSs would be perfectly suitable as well from the NN point of view. In some cases, a network requires information for training which can only be obtained via MC simulation. This is true for example for some works on cone-beam CT scatter correction, as we have seen.

Currently, AI is used in the context of MC at a rather high level, either linking input (e.g., a CT image) and output (e.g., a dose distribution) of a simulation or replacing certain intermediate steps of a simulation (e.g., GAN for phase space generation). It is yet to be seen whether AI will also be of interest deeper inside an MC code, i.e., at the level of particle transport. It might be imagined that NNs rather than handcrafted numerical models will be used in the future to fit complex measured data. As a consequence, MC codes might need to implement such trained networks in their physics models,

e.g., to replace analytical probability distribution functions for certain types of interactions.

Whether this will improve the simulation accuracy or the speed is hard to predict at this point.

In this context, it is encouraging to see interest within the high energy physics (HEP) community to explore the use of AI (Radovic et al. 2018; Bourilkov 2019). Applications may be as diverse as deep learning for nuclear interaction modelling (Ciardiello et al. 2020), NNs in condensed matter physics (Carrasquilla and Melko 2017, Shen et al 2018), or the use of GAN (Paganini et al. 2018) for fast simulation of particle showers in electromagnetic calorimeters. Some problems are surprising to those in the context of radiation therapy, such as modelling the response of particle physics detectors at CERN by NNs (Vallecorsa 2018). Exchange among researchers working with MC simulations in radiation therapy on the one hand and HEP on the other would be desirable to share new knowledge.

17.6 Conclusion and Outlook

Those examples are only pioneer's works in that field and further studies are needed to better understand the advantages and drawbacks of those approaches. Even if they are promising, there are still pitfalls, limitations, and unknowns. For example, the training time is still large and heavy computing demanding (Graphics Process Units or GPU cards). The generalization of the learned networks to other datasets than the ones that have been used for the learning stage is still not clear. The final accuracy obtained seems to not always reach yet the one of conventional MC methods. However, this is a very promising field, and it is expected that numerous works will be published in the following years.

References

Berg, Eric, and Simon R. Cherry. 2018. "Using Convolutional Neural Networks to Estimate Time-of-Flight from PET Detector Waveforms." *Physics in Medicine and Biology* 63 (2): 02LT01. https://doi.org/10/ghc3vv.

Bolch, Wesley E., Lionel G. Bouchet, James S. Robertson, Barry W. Wessels, Jeffry A. Siegel, Roger W. Howell, Alev K. Erdi, et al. 1999. "MIRD Pamphlet No. 17: The Dosimetry of Nonuniform Activity Distributions–Radionuclide S Values at the Voxel Level. Medical Internal Radiation Dose Committee." *Journal of Nuclear Medicine* 40 (1): 11S–36S.

Bourilkov, Dimitri. 2019. "Machine and Deep Learning Applications in Particle Physics." *International Journal of Modern Physics A* 34 (35): 1930019. https://doi.org/10/ghd3qg.

Carrasquilla, Juan, and Roger G. Melko. 2017. "Machine Learning Phases of Matter." *Nature Physics* 13 (5): 431–34. https://doi.org/10/gddjbv.

Ciardiello, A., M. Asai, B. Caccia, G. A. P. Cirrone, M. Colonna, A. Dotti, R. Faccini, S. Giagu, A. Messina, P. Napolitani, L. Pandola, D. H. Wright, and C. Mancini-Terracciano. 2020. "Preliminary Results in Using Deep Learning to

Emulate BLOB, a Nuclear Interaction Model." *Physica Medica: European Journal of Medical Physics* 73 (May): 65–72. https://doi.org/10/ghd74s.

Descourt, Patrice, Thomas Carlier, Y Du, X Song, I Buvat, E C Frey, Manuel Bardies, B M W Tsui, and Dimitris Visvikis. 2010. "Implementation of Angular Response Function Modelling in SPECT Simulations with GATE." *Physics in Medicine and Biology* 55 (9): N253–66. https://doi.org/10/bkcct9.

Fornander, Hannes. 2019. "Denoising Monte Carlo Dose Calculations Using a Deep Neural Network." KTH Royal Institute of Technology School of Electrical Engineering and Computer Science.

Frégier, Yaël, and Jean-Baptiste Gouray. 2019. "Mind2Mind : Transfer Learning for GANs." *ArXiv:1906.11613 [Cs, Stat]*, June. http://arxiv.org/abs/1906.11613.

Goodfellow, Ian, Jean Pouget-Abadie, Mehdi Mirza, Bing Xu, David Warde-Farley, Sherjil Ozair, Aaron Courville, and Yoshua Bengio. 2014. "Generative Adversarial Nets." In *Advances in Neural Information Processing Systems*, 2672–2680.

Götz, Theresa Ida, Christian Schmidkonz, Shuqing Chen, Saad Al-Baddai, Torsten Kuwert, and Elmar W. Lang. 2019. "A Deep Learning Approach to Radiation Dose Estimation." *Physics in Medicine and Biology*, December. https://doi.org/10/ggjtbn.

Gulrajani, Ishaan, Faruk Ahmed, Martin Arjovsky, Vincent Dumoulin, and Aaron Courville. 2017. "Improved Training of Wasserstein GANs." In *NIPS 2017*. http://arxiv.org/abs/1704.00028.

van der Heyden, Brent, Martin Uray, Gabriel Paiva Fonseca, Philipp Huber, Defne Us, Ivan Messner, Adam Law, et al. 2020. "A Monte Carlo Based Scatter Removal Method for Non-Isocentric Cone-Beam CT Acquisitions Using a Deep Convolutional Autoencoder." *Physics in Medicine & Biology* 65 (14): 145002. https://doi.org/10.1088/1361-6560/ab8954.

Jarry, Geneviève, Sean A. Graham, Douglas J. Moseley, David J. Jaffray, Jeffrey H. Siewerdsen, and Frank Verhaegen. 2006. "Characterization of Scattered Radiation in KV CBCT Images Using Monte Carlo Simulations." *Medical Physics* 33 (11): 4320–29. https://doi.org/10/fndqjd.

Javaid, Umair, Kevin Souris, Damien Dasnoy, Sheng Huang, and John A. Lee. 2019. "Mitigating Inherent Noise in Monte Carlo Dose Distributions Using Dilated U-Net." *Medical Physics* 46 (12): 5790–98. https://doi.org/10/ghbjz8.

Kalantzis, Georgios, Luis A. Vasquez-Quino, Travis Zalman, Guillem Pratx, and Yu Lei. 2011. "Toward IMRT 2D Dose Modelling Using Artificial Neural Networks: A Feasibility Study." *Medical Physics* 38 (10): 5807–17. https://doi.org/10/dwqrcj.

Kingma, Diederik P., and Max Welling. 2019. "An Introduction to Variational Autoencoders." *Foundations and Trends* in Machine Learning* 12 (4): 307–92. https://doi.org/10/ggfm34.

Kontaxis, C., G. H. Bol, J. J. W. Lagendijk, and B. W. Raaymakers. 2020. "DeepDose: Towards a Fast Dose Calculation Engine for Radiation Therapy Using Deep Learning." *Physics in Medicine and Biology* 65 (7): 075013. https://doi.org/10/ghb8pj.

Lalonde, Arthur, Brian A. Winey, Joost M. Verburg, Harald Paganetti, and Gregory C. Sharp. 2020. "Evaluation of CBCT Scatter Correction Using Deep Convolutional Neural Networks for Head and Neck Adaptive Proton Therapy." *Physics in Medicine & Biology*. doi:10.1088/1361-6560/ab9fcb.

Lee, Min Sun, Donghwi Hwang, Joong Hyun Kim, and Jae Sung Lee. 2019. "Deep-Dose: A Voxel Dose Estimation Method Using Deep Convolutional Neural Network for Personalized Internal Dosimetry." *Scientific Reports* 9 (1): 10308. https://doi.org/10/gf684s.

Liu, Zhiqiang, Jiawei Fan, Minghui Li, Hui Yan, Zhihui Hu, Peng Huang, Yuan Tian, Junjie Miao, and Jianrong Dai. 2019. "A Deep Learning Method for Prediction of Three-Dimensional Dose Distribution of Helical Tomotherapy." *Medical Physics* 46 (5): 1972–83. https://doi.org/10/ggw5v3.

Madrigal, Jorge Ricardo Asensi. 2018. "Deep Learning Approach for Denoising Monte Carlo Dose Distribution in Proton Therapy." Université Catholique de Louvain.

Maier, Joscha, Elias Eulig, Tim Vöth, Michael Knaup, Jan Kuntz, Stefan Sawall, and Marc Kachelrieß. 2019. "Real-Time Scatter Estimation for Medical CT Using the Deep Scatter Estimation: Method and Robustness Analysis with Respect to Different Anatomies, Dose Levels, Tube Voltages, and Data Truncation." *Medical Physics* 46 (1): 238–49. https://doi.org/10/ggc7fd.

Mainegra-Hing, Emesto, and Iwan Kawrakow. 2008. "Fast Monte Carlo Calculation of Scatter Corrections for CBCT Images." *Journal of Physics: Conference Series* 102 (February): 012017. https://doi.org/10/fnkrff.

Mao, Ximeng, Joelle Pineau, Roy Keyes, and Shirin A. Enger. 2020. "RapidBrachyDL: Rapid Radiation Dose Calculations in Brachytherapy Via Deep Learning." *International Journal of Radiation Oncology Biology Physics* 108 (3): 802–12. https://doi.org/10/ghd3pb.

Müller, Florian, David Schug, Patrick Hallen, Jan Grahe, and Volkmar Schulz. 2019. "A Novel DOI Positioning Algorithm for Monolithic Scintillator Crystals in PET Based on Gradient Tree Boosting." *IEEE Transactions on Radiation and Plasma Medical Sciences* 3 (4): 465–74. https://doi.org/10/ghc8kg.

Naqa, I El, I Kawrakow, M Fippel, J V Siebers, P E Lindsay, M V Wickerhauser, M Vicic, K Zakarian, N Kauffmann, and J O Deasy. 2005. "A Comparison of Monte Carlo Dose Calculation Denoising Techniques." *Physics in Medicine and Biology* 50 (5): 909–22. https://doi.org/10/bn8p4j.

Neph, Ryan, Yangsibo Huang, Youming Yang, and Ke Sheng. 2019. "DeepMCDose: A Deep Learning Method for Efficient Monte Carlo Beamlet Dose Calculation by Predictive Denoising in MR-Guided Radiotherapy." In *Artificial Intelligence in Radiation Therapy*, edited by Dan Nguyen, Lei Xing, and Steve Jiang, 11850:137–45.

Cham: Springer International Publishing. doi:10.100 7/978-3-030-32486-5_17.

Nguyen, Dan, Xun Jia, David Sher, Mu-Han Lin, Zohaib Iqbal, Hui Liu, and Steve Jiang. 2019. "3D Radiotherapy Dose Prediction on Head and Neck Cancer Patients with a Hierarchically Densely Connected U-Net Deep Learning Architecture." *Physics in Medicine and Biology* 64 (6): 065020. https://doi.org/10/ggw5v2.

Olivier, A., C. Baillet, S. Truant, A. Béron, A.-C. Deshorghes, F.-R. Pruvot, and D. Huglo. 2013. "Étude de la variabilité de la clairance hépatique de la mébrofénine-99mTc en scintigraphie hépatobiliaire." *Médecine Nucléaire* 37 (9): 379–86. https://doi.org/10/ggjs9h.

Paganini, Michela, Luke de Oliveira, and Benjamin Nachman. 2018. "CaloGAN: Simulating 3D High Energy Particle Showers in Multi-Layer Electromagnetic Calorimeters with Generative Adversarial Networks." *Physical Review D* 97 (1): 014021. https://doi.org/10/ggjtj9.

Pedemonte, Stefano, Larry Pierce, and Koen Van Leemput. 2017. "A Machine Learning Method for Fast and Accurate Characterization of Depth-of-Interaction Gamma Cameras." *Physics in Medicine & Biology* 62 (21): 8376–8401. https://doi.org/10/ghc9pv.

Peng, Zhao, Hongming Shan, Tianyu Liu, Xi Pei, Ge Wang, and X. George Xu. 2019a. "MCDNet – A Denoising Convolutional Neural Network to Accelerate Monte Carlo Radiation Transport Simulations: A Proof of Principle With Patient Dose From X-Ray CT Imaging." *IEEE Access* 7: 76680–89. https://doi.org/10/ghd3p4.

Peng, Zhao, Hongming Shan, Tianyu Liu, Xi Pei, Jieping Zhou, Ge Wang, and X. George Xu. 2019b. "Deep Learning for Accelerating Monte Carlo Radiation Transport Simulation in Intensity-Modulated Radiation Therapy." *ArXiv:1910.07735 [Physics]*, October. http://arxiv.org/abs/1910.07735.

Radovic, Alexander, Mike Williams, David Rousseau, Michael Kagan, Daniele Bonacorsi, Alexander Himmel, Adam Aurisano, Kazuhiro Terao, and Taritree Wongjirad. 2018. "Machine Learning at the Energy and Intensity Frontiers of Particle Physics." *Nature* 560 (7716): 41–48. https://doi.org/10/gdwskn.

Ronneberger, Olaf, Philipp Fischer, and Thomas Brox. 2015. "U-Net: Convolutional Networks for Biomedical Image Segmentation." *ArXiv:1505.04597 [Cs]*, May. http://arxiv.org/abs/1505.04597.

Roser, Philipp, Xia Zhong, Annette Birkhold, Norbert Strobel, Markus Kowarschik, Rebecca Fahrig, and Andreas Maier. 2019. "Physics-Driven Learning of x-Ray Skin Dose Distribution in Interventional Procedures." *Medical Physics* 46 (10): 4654–65. https://doi.org/10/ggw5v7.

Ryden, Tobias, Ida Marin, Martijn van Essen, Johanna Svensson, and Peter Bernhardt. 2019. "Deep Learning Generation of Intermediate Projections and Monte Carlo Based Reconstruction Improves 177Lu SPECT Images

Reconstructed with Sparse Acquired Projections." *Journal of Nuclear Medicine* 60 (supplement 1): 44.

Sarrut, David, Nils Krah, Jean-Noël Badel, and Jean Michel Létang. 2018. "Learning SPECT Detector Angular Response Function with Neural Network for Accelerating Monte-Carlo Simulations." *Physics in Medicine & Biology* 63 (20): 205013. https://doi.org/10/ggjtkb.

Sarrut, David, Nils Krah, and Jean-Michel Letang. 2019. "Generative Adversarial Networks (GAN) for Compact Beam Source Modelling in Monte Carlo Simulations." *Physics in Medicine and Biology*, August. https://doi.org/10/gf82xv.

Shan, Hongming, Yi Zhang, Qingsong Yang, Uwe Kruger, Mannudeep K. Kalra, Ling Sun, Wenxiang Cong, and Ge Wang. 2018. "3D Convolutional Encoder-Decoder Network for Low-Dose CT via Transfer Learning from a 2D Trained Network." *IEEE Transactions on Medical Imaging* 37 (6): 1522–34. https://doi.org/10/ghbjz7.

Shen, Huitao, Junwei Liu, and Liang Fu. 2018. "Self-Learning Monte Carlo with Deep Neural Networks." *Physical Review B* 97 (20): 205140. https://doi.org/10.1103/PhysRevB.97.205140.

Shorten, Connor, and Taghi M. Khoshgoftaar. 2019. "A Survey on Image Data Augmentation for Deep Learning." *Journal of Big Data* 6 (1): 60. https://doi.org/10/ggb3hw.

Song, X, W P Segars, Y Du, B M W Tsui, and E C Frey. 2005. "Fast Modelling of the Collimator–Detector Response in Monte Carlo Simulation of SPECT Imaging Using the Angular Response Function." *Physics in Medicine and Biology* 50 (8): 1791–1804. https://doi.org/10/d6rjs6.

Vallecorsa, Susan. 2018. "Generative Models for Fast Simulation." *Journal of Physics: Conference Series* 1085 (September): 022005. https://doi.org/10/ghdz3w.

Wolterink, Jelmer M., Tim Leiner, Max A. Viergever, and Ivana Isgum. 2017. "Generative Adversarial Networks for Noise Reduction in Low-Dose CT." *IEEE Transactions on Medical Imaging* 36 (12): 2536–45. https://doi.org/10/ggzxvr.

Yang, Qingsong, Pingkun Yan, Yanbo Zhang, Hengyong Yu, Yongyi Shi, Xuanqin Mou, Mannudeep K. Kalra, Yi Zhang, Ling Sun, and Ge Wang. 2018. "Low-Dose CT Image Denoising Using a Generative Adversarial Network with Wasserstein Distance and Perceptual Loss." *IEEE Transactions on Medical Imaging* 37 (6): 1348–57. https://doi.org/10/gfkz9g.

Yang, Ting-Yi. 2019. "Machine Learning for High Resolution 3D Positioning of Gamma-Interactions in Monolithic PET Detectors." Master Thesis, Ghent University.

Zatcepin, Artem, Marco Pizzichemi, Andrea Polesel, Marco Paganoni, Etiennette Auffray, Sibylle I. Ziegler, and Negar Omidvari. 2020. "Improving Depth-of-Interaction Resolution in Pixellated PET Detectors Using Neural Networks." *Physics in Medicine & Biology* 65 (17): 175017. https://doi.org/10/ghc3sg.

Index

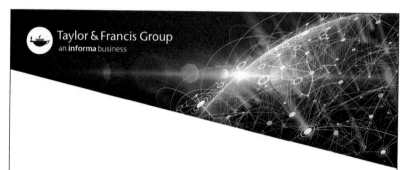